Radiationless Transitions

CONTRIBUTORS

V. E. Bondybey

L. E. Brus

H. G. Drickamer

Karl F. Freed

Robin M. Hochstrasser

W. E. Howard

Edward K. C. Lee

S. H. Lin

Gary L. Loper

D. J. Mitchell

Michael D. Morse

Frank A. Novak

Rudolph P. H. Rettschnick

William Rhodes

Stuart A. Rice

E. W. Schlag

G. B. Schuster

R. Bruce Weisman

Radiationless Transitions

EDITED BY

SHENG HSIEN LIN
Department of Chemistry
Arizona State University
Tempe, Arizona

1980

ACADEMIC PRESS
A Subsidiary of Harcourt Brace Jovanovich, Publishers

New York London Toronto Sydney San Francisco

ACADEMIC PRESS, INC.
111 Fifth Avenue, New York, New York 10003

United Kingdom Edition published by
ACADEMIC PRESS, INC. (LONDON) LTD.
24/28 Oval Road, London NW1 7DX

Library of Congress Cataloging in Publication Data
Main entry under title:

Radiationless transitions.

Includes bibliographies and index.
1. Molecular theory. 2. Radiationless transitions.
3. Excited state chemistry. I. Lin, Sheng Hsien.
QD461.R22 541.3'5 79–26781
ISBN 0–12–450650–X

PRINTED IN THE UNITED STATES OF AMERICA

80 81 82 83 9 8 7 6 5 4 3 2 1

Contents

3 ON ROTATIONAL EFFECTS IN RADIATIONLESS PROCESSES IN POLYATOMIC MOLECULES

Frank A. Novak, Stuart A. Rice, Michael D. Morse, and Karl F. Freed

4 VIBRATIONAL RELAXATION OF ISOLATED MOLECULES

Rudolf P. H. Rettschnick

5 DYNAMIC ASPECTS OF MOLECULAR EXCITATION BY LIGHT

William Rhodes

6 SPECTROSCOPIC AND TIME RESOLVED STUDIES OF SMALL MOLECULE RELAXATION IN THE CONDENSED PHASE

L. E. Brus and V. E. Bondybey

List of Contributors

Numbers in parentheses indicate the pages on which the authors' contributions begin.

V. E. Bondybey (259), Bell Laboratories, Murray Hill, New Jersey 07974

L. E. Brus (259), Bell Laboratories, Murray Hill, New Jersey 07974

H. G. Drickamer (289), School of Chemical Sciences and Materials Research Laboratory, University of Illinois, Urbana, Illinois 61801

Karl F. Freed (135), The Department of Chemistry and The James Franck Institute, The University of Chicago, Chicago, Illinois 60637

Robin M. Hochstrasser (317), Department of Chemistry and Laboratory for Research on the Structure of Matter, University of Pennsylvania, Philadelphia, Pennsylvania 19104

*W. E. Howard** (81), Institut für Physikalische Chemie, Technische Universität München, 8046 Garching, Germany

Edward K. C. Lee (1), Department of Chemistry, University of California, Irvine, California 92717

S. H. Lin (363), Department of Chemistry, Arizona State University, Tempe, Arizona 85281

Gary L. Loper (1), Chemistry and Physics Laboratory, The Aerospace Corporation, Los Angeles, California 90009

D. J. Mitchell (289), School of Chemical Sciences and Materials Research Laboratory, University of Illinois, Urbana, Illinois 61801

Michael D. Morse (135), The Department of Chemistry and The James Franck Institute, The University of Chicago, Chicago, Illinois 60637

Frank A. Novak (135), The Department of Chemistry and The James Franck Institute, The University of Chicago, Chicago, Illinois 60637

Rudolph P. H. Rettschnick (185), Laboratory for Physical Chemistry, University of Amsterdam, Amsterdam, The Netherlands

William Rhodes (219), Department of Chemistry and Institute of Molecular Biophysics, Florida State University, Tallahassee, Florida 32306

Stuart A. Rice (135), The Department of Chemistry and The James Franck Institute, The University of Chicago, Chicago, Illinois 60637

*Present address: Department of Chemistry, University of California, Irvine, California 92717.

E. W. Schlag (81), Institut für Physikalische Chemie, Technische Universität München, 8046 Garching, Germany

G. B. Schuster (289), School of Chemical Sciences and Materials Research Laboratory, University of Illinois, Urbana, Illinois 61801

*R. Bruce Weisman** (317), Department of Chemistry and Laboratory for Research on the Structure of Matter, University of Pennsylvania, Philadelphia, Pennsylvania 19104

*Present address: Department of Chemistry, Rice University, Houston, Texas 77001.

Preface

In the past two decades, there has been very rapid progress in both the experimental and theoretical aspects of radiationless transitions in photochemistry, photophysics, and photobiology. In view of the fact that the theoretical techniques and concepts used in electronic relaxation (or radiationless transition) can be applied to a large number of related processes, the term "radiationless transition" will be used to refer to any nonradiative process that involves a transition between two states (or positions). As a result, it covers a wide diversity of subjects. The purpose of this publication is thus to promote a critical discussion by the leading research workers of several topics (including both theory and experiment) that are of interest now or in the immediate future, each from the author's personal point of view.

The topics covered in this volume are theory and experiment of photophysical processes of single vibronic levels and/or single rovibronic levels (Chapters 1–3), vibrational relaxation of isolated molecules (Chapter 4), the dynamics of molecular excitation by light (Chapter 5), photophysical processes of small molecules in condensed phase (Chapter 6), high pressure effects on molecular luminescence (Chapter 7), and ultrafast processes (Chapters 8 and 9). The editor wishes to thank the other authors for their important contributions. It is hoped that the collection of topics in this volume will prove to be useful, valuable, and stimulating not only to active researchers but also to other scientists in the fields of physics, chemistry, and biology.

Radiationless Transitions

1

Experimental Measurement of Electronic Relaxation of Isolated Small Polyatomic Molecules from Selected States

Edward K. C. Lee

Department of Chemistry
University of California
Irvine, California

Gary L. Loper

Chemistry and Physics Laboratory
The Aerospace Corporation
Los Angeles, California

ISBN 0-12-450650-X

I. INTRODUCTION

Radiationless transitions have been the subject of many recent studies by molecular physicists, chemists, and photobiologists. A large number of papers and reviews has been written over the past 15 years. The most recent review article on the subject of nonradiative electronic relaxation under collision-free conditions, written by Avouris *et al.* (1977), appeared during the preparation of this article. It provides an excellent general overview of recent theoretical and experimental studies on nonradiative transitions. We shall therefore attempt to provide a review with more detailed analysis of recent experimental work on small polyatomic molecules or other studies relevant to the electronic relaxation of these small molecules. We feel strongly that the study of small polyatomic molecules is presently at the infant stage of its development. Experimental measurements obtainable from selective excitation of single rotational and vibronic levels (SRVL) in particular show greater promise of providing significant insight into radiationless processes and photodecomposition (Lee, 1977) than those of single vibronic levels (SVL). Rather than attempt an exhaustive review of the subject, we shall limit ourselves to the review of a number of representative, model cases. We shall focus on the significant findings of very recent experimental studies and make some forecasts about future prospects. A recent review by Robinson (1974) provides a fascinating, historical account of development in the study of molecular radiationless transitions, and we recommend it highly to new investigators. In view of the space available and the authors' interest in small polyatomic molecules, we consider the limited scope of this chapter only practical. Despite our effort to be accurate and objective, there will undoubtedly be some errors and oversight for which we take full responsibility.

II. THEORETICAL BACKGROUND

A. Classification of Radiationless Transitions

A well-resolved, high-resolution electronic absorption spectrum of a polyatomic molecule contains a wealth of information about vibrational and rotational energy levels of the lower and upper electronic states and the allowedness of various radiative transitions between these electronic states. In addition, the spectrum contains less accessible information about the radiationless processes that occur in the excited molecule (Rice, 1929; Herzberg, 1966). The most obvious case of radiationless processes of this type is predissociation. Continuous spectra characterize direct photodissociation within a time scale on the order of 10^{-12}–10^{-13} sec after the initial preparation of an unstable, excited state. Diffuse spectra characterize fast molecular predissociation. The level width associated with the predissociative lifetime is comparable to or greater than the Doppler width; for example, a 1×10^{-10} sec lifetime corresponds to a Heisenberg uncertainty width of 0.05 cm^{-1}, which is comparable to the Doppler width of a molecule like NO_2 at room temperature. Much slower molecular predissociation processes can be more readily detected by a sharp fall-off in the emission line intensities from the diffuse levels or by Doppler-free laser spectroscopy. Of course, it is more common to determine the rate through photoluminescence decay time measurement.

A classification of radiationless transitions helpful for a study of the single rovibronic levels (SRVL) is Herzberg's classification of molecular predissociation into three types (Herzberg, 1950, 1966); case I is predissociation by electronic transition, case II by vibration, and case III by rotation. Electronic predissociation involves a crossing to another electronic configuration (usually of lower energy) with a dissociation continuum whose convergence limit is at a lower energy. Vibrational and rotational predissociations, however, occur on the same electronic manifold. If the initially prepared vibrational level evolves via vibrational energy redistribution to find a dissociation continuum at a lower energy, vibrational predissociation takes place. Unimolecular decomposition rate theories deal precisely with this problem (Marcus, 1952, 1965; Forst, 1973). If the vibrational level excited is slightly below the lowest dissociation limit and if rotational excitation energy becomes available to overcome the dissociation limit plus the rotational barrier, rotational predissociation takes place. This classification was given originally to explain the observation of line broadening in absorption spectra and the weakening of emission line intensities as a manifestation of the coupling of the optically excited discrete levels to a dissociation continuum corresponding to a certain set of photodecomposition products. A further use of this classification is nicely elaborated by Gelbart in a recent review of photodissociation dynamics of polyatomic molecules (Gelbart, 1977).

Selection rules for molecular predissociation are relatively simple when electronic, vibrational, and rotational motions are separable (Herzberg, 1966, pp. 458–463). Selection rules for radiationless transitions are even simpler if no dissociation occurs. The consideration of both angular momentum conservation for a set of well-defined angular momentum quantum numbers and conservation of the electronic, vibrational, and rotational symmetries gives rise to the selection rules for radiationless processes. The total angular momentum **J**, the parities (+, −) and the symmetry with respect to the exchange of identical nuclei (*s*, *a*) are strictly obeyed. Several other selection rules are relaxed, however, with the introduction of the interaction between electronic and nuclear motions, and *heterogeneous* predissociation, involving a change in the electronic symmetry, becomes possible. For the interaction of rotation and electronic motion, the symmetry species of the two electronic states (Γ_i^e and Γ_j^e) differ by the species of a rotation (Γ^r),

$$\Gamma_i^e \times \Gamma^r = \Gamma_j^e. \tag{1}$$

For the interaction of vibration and electronic motion, the symmetry species of the two vibronic states ($\Gamma_i^{ev} = \Gamma_i^e \times \Gamma_i^v$ and $\Gamma_j^{ev} = \Gamma_j^e \times \Gamma_j^v$) remain the same

$$\Gamma_i^{ev} = \Gamma_j^{ev}, \tag{2}$$

where Γ_i^v and Γ_j^v are the vibrational species. An electronic spin–orbit interaction relaxes the conservation rule for the electronic spin angular momentum **S**,

$$\Delta S = 0. \tag{3}$$

A concise summary of the selection rules for radiationless transitions which are strictly obeyed as well as approximately obeyed is given in Table 1. The selection rules for the asymmetric rotor molecules are similar to those for symmetric tops. The components of the electronic orbital angular momentum **L**, the electronic spin angular momentum **S**, and the resultant angular momentum **L** + **S** about the internuclear axis are Λ, Σ, and Ω, respectively. P and K are the components of the total angular momentum, including spin **J** and the molecular rotational angular momentum **N** about the symmetric top axis, respectively.

TABLE 1

Selection Rules for Radiationless Transitions in an Isolated Polyatomic Molecule[a]

Rule	Condition	Linear	Symmetric top
1.	strict	$\Delta J = 0$	$\Delta J = 0$[b]
2.	strict	$+ \not\leftrightarrow -$	$+ \not\leftrightarrow -$
3.	strict	$s \not\leftrightarrow a$	
4.	el–vib–rot separability	$\Gamma_i^e = \Gamma_j^e$	$\Gamma_i^e = \Gamma_j^e$
		$\Gamma_i^v = \Gamma_j^v$	$\Gamma_i^v = \Gamma_j^v$
		$\Gamma_i^r = \Gamma_j^r$	$\Gamma_i^r = \Gamma_j^r$
5.	approximate	$\Delta S = 0$	$\Delta S = 0$
6.	approximate	$\Delta\Lambda = 0$	$\Delta K = 0$[c]
7.	vib–el interaction	$\Gamma_i^{ev} = \Gamma_j^{ev}$	$\Gamma_i^{ev} = \Gamma_j^{ev}$
8.	rot–el interaction	$\Delta\Lambda = \pm 1$	$\Delta K = \pm 1$
9.	Hund's case (a)	$\Delta\Sigma = 0$[d]	
10.	Hund's case (b)	$\Delta N = 0$[e]	see text
11.	Hund's case (c)	$\Delta\Omega = 0$[d]	
12.	$\Delta S \neq 0$	$\Delta\Omega = 0$[d]	see text

[a] Adapted from Herzberg, 1966.
[b] $\Delta P = 0$ also for the component of **J** about the top axis, P.
[c] $\Delta N = 0$ also in large molecules where K is the component of the molecular rotation **N** about the top axis.
[d] $\Delta N = 0$ also (redundant).
[e] $\Delta K = 0$ also (redundant).

B. Rotational Selection Rules

A detailed theoretical study of the *rotational* selection rules for radiationless transitions between two singlet states, a singlet and a triplet state, and two triplet states has been recently given by Howard and Schlag (1978). We shall summarize their elegant results below. The selection rules, $\Delta J = 0$ and $\Delta P = 0$, are strictly obeyed in bound states of isolated molecules for both internal conversion (IC) where $\Delta S = 0$ and intersystem crossing (ISC) where $\Delta S \neq 0$. The states are characterized by a set of quantum numbers, $|\gamma SJNK\rangle$, where S, J, N, and K are good quantum numbers for a large molecule and γ represents all other quantum numbers applicable to a Hund's case (b) angular momentum coupling scheme. This case is represented by the operator equation,

$$\mathbf{N} = -\mathbf{S} + \mathbf{J}. \tag{4}$$

From this, one obtains for a singlet state ($S = 0$),

$$\mathbf{N} = \mathbf{J}. \tag{5}$$

For the case of the singlet–singlet IC, rotational selection rules,

$$\Delta N = 0 \qquad \text{and} \qquad \Delta K = 0, \tag{6}$$

are obtained from the $\Delta J = 0$, $\Delta P = 0$ selection rules. The matrix elements for the singlet–singlet IC process reduce to

$$\langle \Psi_S^j | \tilde{T}_N | \Psi_S^i \rangle = \langle \gamma_j | \tilde{T}_N | \gamma_i \rangle \delta_{N_i, N_j} \delta_{K_i, K_j}, \tag{7}$$

where Ψ_S^i and Ψ_S^j are the singlet wavefunctions for the initial and the final states, respectively, $\tilde{T}_N$ is the nuclear kinetic energy operator, and $\delta_{N_i, N_j}\delta_{K_i, K_j}$ represents delta functions requiring conservation of the quantum numbers N and K.

For the case of the singlet–triplet ISC, the rotational selection rules

$$\Delta N = 0, \pm 1 \qquad \text{and} \qquad \Delta K = 0, \pm 1 \tag{8}$$

are obtained after expanding the case (b) wavefunction in terms of the case (a) wavefunction for the triplet state. The matrix elements for the singlet–triplet ISC process reduce to

$$\langle \Psi_T | \tilde{H}_{so} | \Psi_S \rangle = \sum_{\alpha = 0, \pm 1} (-1)^{N_S + K_S} (2N_T + 1)^{1/2} \times \begin{pmatrix} N_S & N_T & 1 \\ K_S & -K_T & \alpha \end{pmatrix} \langle \gamma_T | \tilde{l}_-^\alpha | \gamma_S \rangle, \tag{9}$$

where Ψ_S and Ψ_T are the singlet and the triplet wavefunctions, respectively, $\alpha = K_T - K_S$, $\tilde{l}_-^\alpha = \sqrt{\tfrac{1}{2}}(\lambda_1 \tilde{l}_1^\alpha - \lambda_2 \tilde{l}_2^\alpha)$ is an antisymmetric combination of the single electron orbital operator, and the quantity in the parentheses is a Wigner 3-J symbol. It is interesting to note that the above matrix element shows an explicit dependence on N and K.

For the case of the triplet–triplet IC process, two matrix elements contribute,

$$\langle \Psi_{\rm T}^{j} | \tilde{\rm T}_{\rm N} | \Psi_{\rm T}^{i} \rangle = \langle \gamma_j | \tilde{\rm T}_{\rm N} | \gamma_i \rangle \delta_{N_i, N_j} \delta_{K_i, K_j}, \tag{10}$$

and

$$\langle \Psi_{\rm T}^{j} | \tilde{H}_{\rm so} | \Psi_{\rm T}^{i} \rangle = \sum_{\alpha} [6(2N_i + 1)(2N_j + 1)]^{1/2} (-1)^{N_i + N_j + S + J + K_j}$$

$$\times \begin{pmatrix} N_i & N_j & 1 \\ K_i & -K_j & \alpha \end{pmatrix} \begin{pmatrix} 1 & 1 & 1 \\ N_i & N_j & J \end{pmatrix} \langle \gamma_j | \tilde{l}_{+}^{\alpha} | \gamma_j \rangle. \tag{11}$$

As in the case of the singlet–singlet IC, Eq. (10) gives the rotational selection rules,

$$\Delta N = 0 \quad \text{and} \quad \Delta K = 0. \tag{12}$$

Equation (11) gives a set of the rotational selection rules, as for the singlet–triplet ISC,

$$\Delta N = 0, \pm 1 \quad \text{and} \quad K = 0, \pm 1. \tag{13}$$

Again as in the case of the singlet–triplet ISC, the matrix elements with the spin–orbit operator show explicit dependence on N and K for the triplet–triplet IC process.

This study shows that the observed structure in the rotationally selected IC rate constants between two singlet states cannot be due to the rotational matrix elements but rather a random energy matching of rotational levels. It also shows that the spin sublevel is conserved always in the triplet–triplet IC process induced by $\tilde{\rm T}_{\rm N}$ but is so only for $\Delta N = 0$ in the singlet–triplet ISC process induced by $\tilde{H}_{so}$. It is clear that theoretical descriptions of *rotational* selection in radiationless processes are very interesting and a comparison to *vibrational* selection should be made.

C. Vibronic Effects

One of the important reasons for measuring the single vibronic level (SVL) decay rates is to determine the energy dependence of the radiationless decay rates associated with the selective optical excitation of specific vibrational modes. For the case of the large molecule limit, the golden rule rates based on a many-phonon theory formalism were obtained (Lin, 1966; Lin and Bersohn, 1968; Englman and Jortner, 1970), showing explicitly the role of the *promoting* normal modes i which are effective in inducing the electronic relaxation and the role of the *accepting* normal modes j which are effective sinks for the electronic energy. The radiationless transition rate from an initial state (bv′) to a final state (av″) of a molecule in the statistical limit is shown below (Lin and Bersohn, 1968),

$$k_{\rm nr}({\rm bv}' \to {\rm av}'') = \frac{2\pi}{\hbar} \sum_{i} |R_i({\rm ab})|^2 \left| \left\langle X_{{\rm av}_i''} \left| \frac{\partial}{\partial Q_i} \right| X_{{\rm bv}_i'} \right\rangle \right|^2$$

$$\times \prod_{j} |\langle X_{{\rm av}_j''} | X_{{\rm bv}_j'} \rangle|^2 \delta(E_{{\rm av}''} - E_{{\rm bv}'}), \tag{14}$$

where $R_i(\mathrm{ab}) = -\hbar^2\langle\Phi_\mathrm{a}|\partial/\partial Q_i|\Phi_\mathrm{b}\rangle$ is the perturbation matrix element responsible for the vibronically induced electronic relaxation between the electronic states Φ_a and Φ_b, the X's are the vibrational wavefunctions of given vibrational modes, $\delta(E_{\mathrm{av}''} - E_{\mathrm{bv}'})$ is a delta function necessary for energy conservation, and the Q_i are the nuclear coordinates.

The subsequent theoretical developments are fully reviewed elsewhere (Freed, 1976a; Avouris *et al.* 1977), and it suffices here to show the radiationless transition rate from an upper electronic state s, with the optical excitation of the promoting mode m_1, to the isoenergetic lower electronic state l, with vibrational modes n, $k_{\mathrm{nr}}(sm_1 \to l)$, given by Avouris *et al.* (1977),

$$k_{\mathrm{nr}}(\mathrm{sm}_1 \to l) = \frac{2\pi\beta sl^2}{\hbar^2\omega_1'} \sum_{n_1} |\langle m_1|n_1\rangle|^2 \sum_{n_3} |\langle 0|n_3\rangle|^2 \, |\langle 0|n_2\rangle|^2_{n_2=\gamma}. \tag{15}$$

Here the 0's in the last two matrix elements represent the values of the quantum numbers m_3 and m_2, respectively. The subscripts 1, 2, and 3 on the vibrational states m and n refer to the promoting mode, the best accepting mode, and the remaining degrees of freedom capable of taking up the electronic energy, respectively. The fundamental frequencies of the mode i on the electronic states s and l are designated ω_i^s and ω_i^l, respectively. The electronic matrix element β_{sl} is equivalent to $R_i(\mathrm{ab})/\hbar^2$ in Eq. (14). The integral quantum number γ is defined by $\Delta E(n_1, n_3)/\hbar\omega_2^l$, where

$$\Delta E(n_1, n_3) = \Delta E_{sl} + m_1\hbar\omega_1^s - n_1\hbar\omega_1^l - n_3\hbar\omega_3^l \tag{16}$$

is the energy available to mode 2 in state l. This simple expression then includes the factors due to the role of the promoting mode, the role of the accepting mode, and consequently the factor attributable to the energy gap law. For example, the ratio of the SVL rates, $k_{sl}(sm_1 \to l)/k_{sl}(s0 \to l)$ can be readily computed from the energy gap ΔE_{sl}, the vibrational frequencies, equilibrium positions, and anharmonicities of each of the accepting modes in each electronic state. A procedure for evaluating $|\langle 0|n_2\rangle|^2$ for the nonintegral values of $n_2 = \gamma$ has been developed by Kühn *et al.* (1978).

Certain normal modes of the optically prepared state are not accessible for direct optical excitation because of small Franck–Condon factors or because they are excluded by the optical selection rules, i.e., they are forbidden transitions. Some of these "dark" vibronic levels may be prepared through Coriolis interaction or Fermi resonance with the optically accessible vibronic levels. Fermi resonance can occur for degenerate levels of the same vibrational species as indicated in Table 1 (rule 4). Coriolis interaction can occur if the product of a vibrational species and a rotational species has the same symmetry as other vibrational species,

$$\Gamma_i^{\mathrm{v}} \times \Gamma_i^{\mathrm{r}} = \Gamma_j^{\mathrm{v}}, \tag{17}$$

provided that the energy is conserved (Wilson *et al.*, 1955; Herzberg, 1945, p. 375). For example, the rotational selection rule for a symmetric top molecule is

$$\Delta K = 0, \pm 1. \tag{18}$$

The vibration–rotation interaction results from the coupling of the vibrational angular momentum with the rotational angular momentum.

D. Preparation of the Excited States

The decay behavior of an excited molecule should be discussed with a careful consideration of the preparation method and conditions employed, since they are interdependent. There are many subtleties, pointed out in several articles (see Avouris *et al.*, 1977). For convenience, the analysis by Bixon and Jortner (1969) will be briefly presented with the energy level scheme for a large molecule, as shown in Fig. 1. Consider the singlet excited state ϕ_s which decays to the nearly degenerate triplet state $\{\phi_l\}$, both of which are zeroth order Born–Oppenheimer (BO) states. The transition moment integral connecting ϕ_s and ϕ_0 (the singlet ground state) is $\mu_{0s} \equiv \langle \phi_s | \tilde{\mu} | \phi_0 \rangle$ and that connecting ϕ_l and ϕ_0, $\langle \phi_l | \tilde{\mu} | \phi_0 \rangle = 0$. Since the molecular eigenstate ψ_n, of the form

$$|\psi_n\rangle = C_n^s|\phi_s\rangle + \sum_l C_n^l|\phi_l\rangle, \tag{19}$$

is an eigenfunction of the full molecular Hamiltonian,

$$H_{\text{mol}}|\psi_n\rangle = E_n|\psi_n\rangle, \tag{20}$$

the oscillator strength carried by ψ_n is proportional to the weight of singlet character in ψ_n, $|C_n^s|^2$. Hence, the optically prepared state can be described in terms of ψ_n.

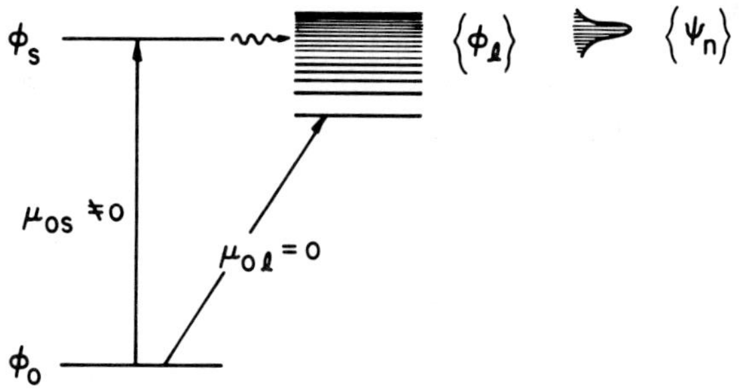

Fig. 1. The molecular energy level model useful in discussing the electronic relaxation of polyatomic molecules (adapted from Jortner *et al.*, 1969). ϕ_0 is the zeroth-order Born–Oppenheimer (BO) molecular state of the ground electronic state vibronic level. ϕ_s is the zeroth-order BO molecular state of the optically accessible, excited vibronic level with a nonzero dipole transition moment $\tilde{\mu}_{0s}$ connecting ϕ_0 and ϕ_s. $\{\phi_l\}$ is the zeroth-order BO molecular state corresponding to a dense manifold of excited vibronic levels whose electronic origin is below that of ϕ_s and with the dipole transition moment connecting ϕ_0 and ϕ_l being zero, $\tilde{\mu}_{0l} = 0$. $\{\psi_n\}$ correspond to the molecular eigenstates, $\psi_n = C_n^s\phi_s + \Sigma_l C_n^l \phi_l$.

For a very short, exciting light pulse, the time-dependent molecular excited state prepared at time $t = 0$ is

$$|\psi(t=0)\rangle \simeq \frac{\mu_{0s}}{ih} \sum_n C_n^s |\psi_n\rangle, \tag{21}$$

and it corresponds to a coherent superposition of the molecular eigenstates $\{\psi_n\}$. Then, the time evolution of ψ is given by

$$|\psi(t)\rangle \simeq \frac{\mu_{0s}}{ih} \sum_n C_n^s \exp\left(\frac{-iE_n t}{\hbar}\right) |\psi_n\rangle. \tag{22}$$

The probability of finding the excited molecule in its initial state ϕ_s at time t, $P_s(t)$, is

$$P_s(t) \equiv |\langle \phi_s | \psi(t) \rangle|^2 = \left| \sum_n (C_n^s)^2 \exp\left(-\frac{iE_n t}{\hbar}\right) \right|^2. \tag{23}$$

When the approximation $E_n = E_s + n\varepsilon$ for $n = 0, \pm 1, \pm 2, \ldots$ is applied, the probability becomes

$$P_s(t) \simeq \left| \sum_n \left\{ \frac{v^2}{n^2\varepsilon^2 + \Delta^2} \exp\left(-\frac{in\varepsilon t}{\hbar}\right) \right\} \right|^2, \tag{24}$$

where, ε is the average energy level spacing, $v = \langle \phi_s | H_{\mathrm{mol}} | \phi_l \rangle$, and $\Delta^2 \equiv (\pi v^2/\varepsilon)^2 + v^2$. When the conditions of $v \gg \varepsilon$ and $t \ll \hbar/\varepsilon$ are satisfied, the probability shows an exponential decay,

$$P_s(t) = \exp(-k_{\mathrm{nr}} t), \tag{25}$$

and the golden rule rate

$$k_{\mathrm{nr}} = 2\pi v^2/\hbar\varepsilon \tag{26}$$

is obtained. This exponential decay behavior is no longer observed beyond the time limit corresponding to the recurrence time, $\tau_{\mathrm{rec}} = \hbar/\varepsilon$. Within the context of the Bixon–Jortner analysis, the large molecule limit is set by the conditions that

$$\tau_{\mathrm{nr}} = \varepsilon\hbar/2\pi v^2 = k_{\mathrm{nr}}^{-1} \ll \tau_{\mathrm{rec}} = \hbar/\varepsilon, \tag{27}$$

and

$$\tau_{\mathrm{max}} \ll \tau_{\mathrm{rec}}, \tag{28}$$

where τ_{max} is the maximum time that the excited molecule is isolated. Therefore, an irreversible radiationless relaxation will occur for a polyatomic molecule with a continuum of levels in the final state, i.e., $\varepsilon \ll v$. Nonexponential decay behavior is obtained if (a) the interactions mixing ϕ_s with the ϕ_l levels or (b) the spacings between ϕ_l levels are not uniform (Gelbart *et al.*, 1975).

E. Isolated Molecules

The upper limit of gas pressure which is expected to assure the integrity of an isolated molecule in an excited state has been often questioned. It is noteworthy that an estimate of the van der Waals interaction energy at a 50 Å distance (i.e., for a collision cross section of 8000 $Å^2$) is 4×10^{-6} cm^{-1} between the electronically excited $C_6H_6(^1B_{2u})$ and He, and 1.7×10^{-4} cm^{-1} between $C_6H_6(^1B_{2u})$ and its ground state $C_6H_6(^1A_{1g})$. In comparison, for a radiative lifetime of $\tau_r = 630$ nsec, the uncertainty broadening of the level width in $C_6H_6(^1B_{2u})$ is 1×10^{-5} cm^{-1} (Robinson, 1967). The observed fluorescence decay time of $C_6H_6(^1B_{2u})$ is ~100 nsec at low pressures, and the half pressure for a collision diameter of 50 Å can be calculated as 1×10^{-2} torr. Therefore, it is necessary to have the benzene pressure at 1×10^{-4} torr (3×10^{12} molecules cm^{-3}) or lower, if greater than 99% of the excited $C_6H_6(^1B_{2u})$ molecules are to be kept collision free before they decay. Robinson estimates the density of vibrational states at 8570 cm^{-1} above the zeroth level of $C_6H_6(^3B_{1u})$ to be 1.8×10^5 states cm^{-1}, or the average level spacing to be $\sim 5 \times 10^{-6}$ cm^{-1}, comparable to the magnitude of the intermolecular interaction at a 50 Å distance. If the rotational levels are very close and the density of the vibrational levels are very high, both rotational and vibrational relaxation can occur at a collision diameter significantly greater than the hard sphere gas kinetic value. Therefore, the condition for irreversibility can be brought about by destroying the integrity of the initially prepared state. Further vibrational and rotational relaxation eventually will put the triplet molecule below the lowest level of the excited singlet state in question. The intermolecular attraction between polar molecules would be greater than that between nonpolar molecules, and hence the collision-free pressure condition is even more demanding. It is obvious that one can satisfy isolated molecule conditions in a molecular beam. Suppose that a flux of 1×10^{14} molecules cm^{-2} sec^{-1} is obtained with a linear velocity of 1×10^5 cm sec^{-1}. If a uniform particle density is assumed throughout in such a beam, the average distance to the closest neighbors is $\sim 1 \times 10^5$ Å.

The selection rules for the collision induced rotational transitions in small molecules are given in a recent review (Oka, 1973). In addition to the simple case of the rotational and vibrational relaxation of the SRVLs, there is the possibility of collision-induced radiationless transitions. A recent theory of Freed (1976b) has successfully dealt with the collision-induced intersystem crossing in glyoxal. In any case, the collisional destruction of the rotationally selected, glyoxal S_1 electronic state by the ground electronic state of glyoxal (S_0) has a large cross section of about 240 $Å^2$ (Rordorf *et al.*, 1978).

III. EXPERIMENTAL STUDIES ON SMALL TRIATOMIC AND TETRA-ATOMIC MOLECULES

In this section, we shall review the recent experimental studies concerning electronic relaxation in triatomic and tetra-atomic molecules NO_2, SO_2, H_2CO, NH_3, C_2HCl, and C_2HBr. There has been an enormous amount of work done on the first two molecules because of their relevance to atmospheric photochemistry and theoretical interest in their anomalously long lifetime behavior—the Douglas effect (Douglas, 1966; Bixon and Jortner, 1969). There have been many photophysical and photochemical studies on formaldehyde because of the interest in single vibronic level (SVL) photochemistry and photochemical laser isotope separation.

A. NO_2

It is not possible to provide here a detailed account of all of the recent significant contributions made by numerous workers in determining the spectroscopic and photophysical properties of the low-lying electronically excited states of NO_2. Therefore, we shall present the highlights of those studies which provide experimental information about electronic relaxation processes following selective excitation. A relatively complete list of literature references is given in a recent paper describing NO_2 fluorescence lifetime measurements near 6000 Å (Donnelly and Kaufman, 1977a), and a comprehensive treatise has recently become available (Monts *et al.*, 1978).

1. Electronic Spectrum

The electronic absorption spectrum of NO_2, extending from 3200 to 10,000 Å, consists of a large number of discrete but irregular vibrational and rotational lines. In 1965, Douglas and Huber were able to complete the rotational analysis of a long progression of bands in the 3700–4600 Å region and assign them to the $^2B_1 \leftarrow {}^2A_1$ transition. However, they failed to obtain the vibrational assignments even with isotopic substitution data (Douglas and Huber, 1965). The band head positions of this $^2B_1 \leftarrow {}^2A_1$ transition in NO_2 are shown in Table 2. In addition to recognizing the difficulty in interpreting the strongly irregular vibration–rotation structure in NO_2, Douglas (1966) measured the fluorescence lifetime (τ_F) of NO_2 and found it to be anomalously longer than the theoretical radiative lifetime based on integrated absorption intensity. He accounted for both of these observations by proposing that an intramolecular vibronic coupling mechanism was operative in NO_2 which redistributed the intensity of the zeroth-order BO state ϕ_s among a set of quasidegenerate, zeroth-order levels $\{\phi_l\}$ carrying no oscillator strength. Bixon and Jortner (1969) have provided a further theoretical treatment for the anomalously long lifetime behavior, or Douglas effect, and predicted the decay mode of NO_2 to be a sum of slowly

TABLE 2

Band Head Positions of the $^2B_1 \leftarrow {}^2A_1$ *Transition in* NO_2[a]

$^{14}NO_2$		$^{15}NO_2$	
λ (Å, air)	ν_0 (cm^{-1}, vac)	λ (Å, air)	ν_0 (cm^{-1}, vac)
3778.5 (diffuse)	26,455	3806.8 (diffuse)	26,259
3909.80 (diffuse)	25,566.4	3938.3 (diffuse)	25,381
		4080.2	24,502
		4231.80	23,620.8
4366.86	22,890.5	4395.82	22,739.7
4544.7	21,994.7	4573.44	21,856.3

[a] Analyzed by Douglas and Huber (1965).

varying exponentials, as for their sparse, intermediate case. Subsequently, the rotational analysis of the ν_2 progression in the resonance fluorescence spectrum of NO_2 excited by Ar^+ laser lines (4880 and 5145 Å) led to the assignment of the symmetry of another excited electronic state as $^2B_2$ (Abe *et al.*, 1971, 1974). This provided evidence for the existence of the $^2B_2 \leftarrow {}^2A_1$ transition overlapping the $^2B_1 \leftarrow {}^2A_1$ transition. The relative strengths of these transitions are of considerable interest. Figure 2 shows their relative strengths as assigned by

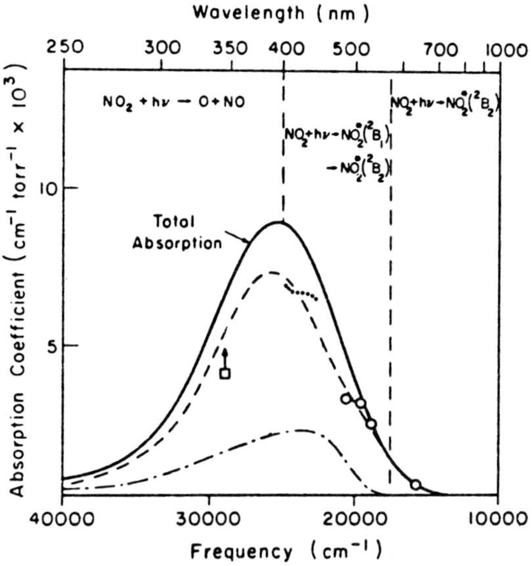

Fig. 2. The proposed partitioning of total absorption intensity between two excited states, (-·-) $^2B_1$ and (---) $^2B_2$ (from Donnelly and Kaufman, 1977a, Fig. 8). The following symbols represent $^2B_2$ absorption: ● (Atherton *et al.*, 1964); ○ (Bird and Marsden, 1974); □ Busch and Wilson, 1972), the arrow signifying the lower limit.

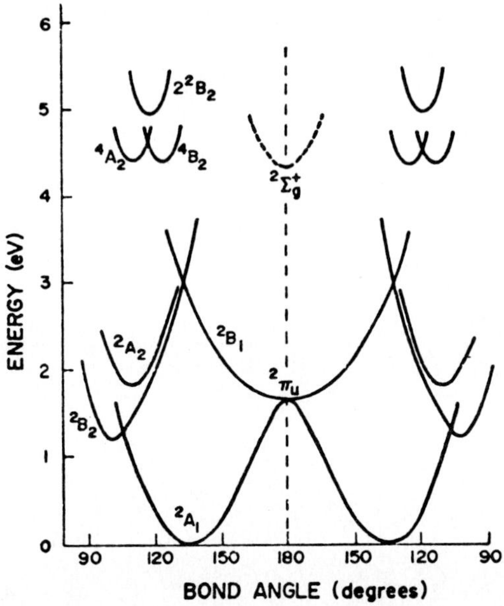

Fig. 3. Electronic energy as a function of bond angle for the low-lying electronic states of NO_2 (from Gillespie *et al.* 1975, Fig. 3).

Donnelly and Kaufman (1977) from previous absorption measurements (Atherton *et al.*, 1964; Busch and Wilson, 1972; Bird and Marsden, 1974). The absorption manifold of the $^2B_1 \leftarrow {}^2A_1$ transition has been estimated with its predicted origin at 678 nm from the data of Douglas and Huber (Hardwick and Brand, 1973). The most recent study of Merer and Hallin (1978) shows that the electronic origin of the $\tilde{A}\,{}^2B_2 \leftarrow \tilde{X}\,{}^2A_1$ transition lies near 10,250 Å (with possible uncertainty of one quantum of $v_2' = 730\ \mathrm{cm}^{-1}$), in agreement with the multiconfiguration SCF calculation of Gillespie and Khan (1976). Earlier, the electronic origin of the $^2B_2 \leftarrow {}^2A_1$ transition had been located at 836.5 nm (Brand *et al.*, 1973a). At this point, it is useful to examine the diagram in Fig. 3, based on the theoretical results of Gillespie *et al.* (1975). This diagram shows how the calculated electronic energies of several low-lying electronic states of NO_2 vary as a function of bond angle.

2. Fluorescence Lifetimes

Recent measurements of the NO_2 fluorescence lifetimes obtained with laser excitation are summarized in Table 3 from several literature references. The collisional relaxation effects become definitely important at $\sim$1 mtorr. In addition, the diffusion of the long-lived excited molecules away from the field of view causes apparent shortening of the lifetime at low pressures when a small viewing region of fluorescence is used in a static experiment. In order to

TABLE 3

Lifetime Data for Laser Excitation of NO_2

Reference[a]	λ_{ex}(Å)	Probable assignment[b]	Pressure (mtorr)	Lifetime (μsec) τ_1	τ_S	τ_L	$R \equiv I_S^0/I_L^0$ [c] 1 mtorr	5 mtorr
S	4365	$^2B_1$	—	—	—	58 ± 25		
SS	4416 (He–Cd)	$^2B_1(^416)$	0.6–2.3	—	36 ± 5	—		
SY	4515 ~ 4605	$^2B_1/^2B_2$	≥0.04	0.5 ~ 3.7	—	62–75		
H	4548 (Δλ = 0.003)	$^2B_1(^011;\,^013)$	≥0.1	—	33 ± 4	—		
SL	4880 (Ar^+)	$^2B_2(^524)$	≥1.5	3.39 ± 0.36	—	—		
PSD	4880 (Ar^+)	$^2B_2(^524)$	<0.1	3.0 ± 1.0	28 ± 1	75 ± 1		
PSD	5145 (Ar^+)	$^2B_2(^414)$	<0.1	—	30 ± 2	105 ± 1.5		
PSD	5145 (Ar^+)	$^2B_2(^615)$	<0.1	—	28 ± 2	100 ± 2		
SSWZ	5935.7 ± 0.1	$^2B_2(1, 5, 0)$	—	—	30 ± 5	115 ± 10		
DK	5782.1 ± 0.5		≥0.01		—	190 ± 13	~0	~0
DK	5850 ± 10		≥0.01		22 ± 5	200 ± 16	0.6	1.4
DK	5934.5 ± 0.5	$^2B_2(1, 5, 0)$	≥0.01		45^{+11}_{-7}	260 ± 35	0.4	1.0
DK	5934.9 ± 0.2	$^2B_2(1, 5, 0Q)$	≥0.01		71 ± 3	217 ± 9	2.2	1.5
DK	5935.5 ± 0.5	$^2B_2(1, 5, 0)$	≥0.01		33 ± 6	226 ± 33	0.5	0.9
DK	5939.5 ± 0.5				—	239 + 47	~0	~0
DK	5936.7 ± 0.2	$^2B_2(1, 5, 0P)$	≥0.01		75 ± 1	~200	1.9	1.2
DK	6032.3 ± 0.5				—	217 ± 13	~0	~0
DK	6123.2 ± 0.5	$^2B_2(2, 0, 1)$	≥0.01		$57 \pm ^{9}_{6}$	179 ± 13	1.3	2.4
DK	6129.7 ± 0.5	$^2B_2(2, 0, 1)$	≥0.01		54^{+20}_{-11}	188 ± 17	0.5	1.1

[a] S (Sidebottom *et al.*, 1972); SS (Schwartz and Senum, 1975); SY (Sackett and Yardley, 1971, 1972); H (Haas *et al.*, 1975); SL (Solarz and Levy, 1974a); PSD (Paech, Schmiedle, and Demtroeder, 1975); SSWZ (Stevens, Swagle, Wallace, and Zare, 1973); DK (Donnelly and Kaufman, 1977a).

[b] For example, (^{524}) is $K' = 5$; $N' = 24$ and (1, 5, 0) is $v_1' = 1$; $v_2' = 5$; $v_3' = 0$. *P* and *Q* are rotational sub-branches. In view of the recent study by Merer and Hallin (1978), the vibrational assignments of the $\tilde{B}\,^2B_2 \leftarrow \tilde{X}\,^2A_1$ transition should be revised accordingly.

[c] See Eq. (29).

avoid these complications, either a combination of a low static pressure with a large viewing region or an appropriate molecular beam fluorescence monitor configuration has been used. In the most recent studies, for example, a 72 liter cell (Donnelly and Kaufman, 1977a), a cooled nozzle beam (Smalley *et al.*, 1974, 1975), and an effusive beam (Paech *et al.*, 1975) were employed successfully. Sackett and Yardley (1971, 1972) had made the first attempt to observe the presence of two decay components of $^{14}NO_2$ in the 4545 Å excitation region where Douglas and Huber identified the $^2B_1 \leftarrow {}^2A_1$ transition (see Table 2). They reported a weak but fast decaying component with a lifetime in the range of 0.5–3.7 μsec similar to those calculated from the integrated absorption coefficient, in addition to a strong component with a lifetime in the range of 62–75 μsec. Since there appear to be three ranges of lifetime values in the literature, we shall classify them as τ_1 for the fast component at $\sim$3 μsec, as found by Sackett and Yardley (1971), τ_S for the short-lived component at $\sim$30 μsec, and τ_L for the long-lived component over 100 μsec, as found by Stevens *et al.* (1973) and Donnelly and Kaufman (1977a).

The lifetime measurements are available for the rovibronic levels of the $^2B_1$ state; 36 $\pm$ 5 μsec for the $^4 16$ ($K' = 4$; $N' = 16$) rovibronic level of $^{14}NO_2$ with 4416 Å line of a He–Cd laser (Schwartz and Senum, 1975) and 33 $\pm$ 4 μsec at 4548 Å with a 30 GHz bandwidth for the $^0 11$ ($K' = 0$, $N' = 11$) and $^0 13$ ($K' = 0$; $N' = 13$) rovibronic levels of $^{14}NO_2$ assigned by Douglas and Huber (Haas *et al.*, 1975). Although Sackett and Yardley did experiments near 4548 Å, the 0.8 Å bandwidth of the laser used was too wide for SRVL excitation. They obtained a value of $\tau_L = 70$ nsec. The above measurements appear then to establish a lifetime of 33–36 μsec for a SRVL of $^2B_1$. It should be noted that Sidebottom *et al.* (1972a) measured the decay times of photoluminescence excited at 4365 (close to the 4.366.85 Å band analyzed by Douglas and Huber; see Table 2), 5324, and 6943 Å. At $\lambda_{ex} = 4365$ and 5324 Å, exponential decays were observed when the emission was monitored at short wavelengths whereas nonexponential decays were observed when the emission was monitored at long wavelengths. Furthermore, lifetimes were pressure dependent. However, at $\lambda_{ex} = 6943$ Å only an exponential decay with little pressure dependence was observed. They attribute this to the possibility that the electronic origin of the $^2B_1 \leftarrow {}^2A_1$ transition is close to 6943 Å. They explained the nonexponential behavior at shorter wavelength excitation as being due to vibrational relaxation in the NO_2 $^2B_1$ electronic state.

A fast decay time of 3.39 μsec was obtained for the excitation of one spin component of the $^5 24$ ($K' = 5$; $N' = 24$) level of $^2B_2$ using a single mode Ar^+ laser line at 4880 Å in a microwave optical double resonance experiment (Solarz and Levy, 1974a). This value has been confirmed in a molecular beam experiment with a 3.0 μsec lifetime from the deconvolution of three exponential decays, $\tau_1 = 3.0 \pm 1.0$ μsec, $\tau_S = 28 \pm 1$ μsec, and $\tau_L = 75 \pm 1$ μsec (Paech *et al.*, 1975). Two other SRVLs of $^2B_2$ were excited by the Ar^+ laser at the 5145 Å line. The $^4 14$ level gave $\tau_S = 30$ μsec and $\tau_L = 105$ μsec; and the $^6 15$ level gave

$\tau_S = 28$ and $\tau_L = 100$ μsec. A similar set of lifetimes $\tau_S = 30$ and $\tau_L = 115$ μsec, were also obtained by the dye laser excitation of the 1, 5, 0 (v'_1; v'_2; v'_3) level of $^2B_2$ at 5935.7 $\pm$ 0.1 Å (Stevens *et al.*, 1973).

Donnelly and Kaufman found the above values of τ_L are short compared to their values, and they attribute this difference to the effect of the cell size reflected in the parameter R defined as I_S^0/I_L^0 in Table 3 (Donnelly and Kaufman, 1977a). I_S^0 and I_L^0 are the preexponential factors in the time-dependent bi-exponential intensity expression

$$I(t) = I_S^0 \exp(-t/\tau_S) + I_L^0 \exp(-t/\tau_L). \tag{29}$$

It is interesting to note that (1) at 5782, 5939. 5, and 6023.2 Å, corresponding to the region of weak absorption peaks, there is little of the short-lived component, (2) at all other wavelengths, corresponding to the region of strong absorption peaks, there are both long- and short-lived components, and (3) at 5934.9 and 5936.7 Å there are coincidences of the peak values of the parameter R, the fluorescence excitation intensity, and the lower values of τ_S when the laser wavelengths are scanned with a bandwidth of 0.5 cm^{-1} (FWHM), as shown in Fig. 4. In the two-state model, it is reasonable to leave out the $^2A_2$ state, because the absence of strong v_3 progressions in the fluorescence emission spectra excited at 593, 612, and 647 nm is indicative of the absence of the electric dipole forbidden,

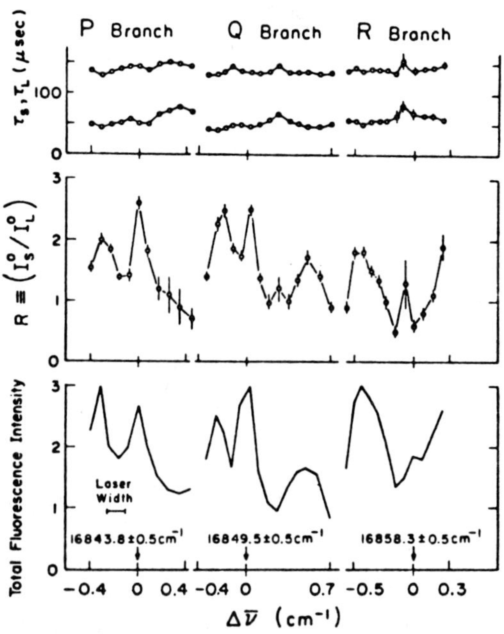

Fig. 4. The correlation of τ_S, τ_L, R, and total fluorescence intensity (at 0.5 mtorr) versus excitation energy of the laser for portions of P, Q, and R branches of the 5936 Å band for NO_2 fluorescence (from Donnelly and Kaufman, 1977a, Fig. 3).

vibronically allowed $^2A_2 \rightarrow {}^2A_1$ transition induced by the $\nu_3(b_2)$ vibration (Brand *et al.*, 1973a; Abe et *al.*, 1971, 1974; Stevens *et al.*, 1973, Stevens and Zare, 1975).

Zare and his co-workers have postulated that the biexponential decay behavior can be explained by assigning τ_L to be $^2B_1$ state and τ_S to the $^2B_2$ state at 5936 Å (Stevens *et al.*, 1973). However, Donnelly and Kaufman find this two-state model unacceptable for explaining their observation that the emission intensity associated with τ_L is greater than that with τ_S, and hence the $^2B_1 \leftarrow {}^2A_1$ transition would have to be predominant over the $^2B_2 \leftarrow {}^2A_1$ transition in the 578–612 nm region. To the contrary, a much stronger intensity is observed for the latter transition in the nozzle beam study at low rotational temperatures (Smalley *et al.*, 1975). Therefore, an alternative, one-state model involving only the $^2B_2$ state is postulated (Donnelly and Kaufman, 1977a). Many discrete vibronic levels with nonuniform energy intervals (ε) and with different lifetimes result, as discussed in Section IID (Gelbart *et al.* 1975), due to the non-uniform energy intervals ε and the variation of the magnitude of the coupling between the $^2B_2$ state and the high vibrational levels of the ground electronic state $^2A_1$. Hence, a nonexponential decay behavior is expected, with the shortest-lived component corresponding to those states with the largest $^2B_2$ character. It should be noted, if the coupling between the $^2B_2$ and $^2A_1$ electronic state is principally of a vibronic interaction, the promoting mode must be the ν_3 (b_2) vibration from Eq. (2). Likewise, vibronic coupling between the $^2B_2$ and $^2B_1$ and that between the $^2B_1$ and $^2A_1$ states require vibrational modes of a_2 and b_1 symmetry, respectively, but there is no such vibrational species in NO_2. Therefore, the radiationless relaxation of the $^2B_1$ state must involve another coupling mechanism, involving presumably *spin* and/or *rotation*, except for consideration of the fact that the $^2B_1$ and $^2A_1$ states are the Renner–Teller components of the linear $^2\Pi_u$ state (see Fig. 3). A further complication arises from the wideband laser excitation populating a large number of rovibronic states imbedded in the dense absorption lines. In any case, the conditions of $v > \varepsilon$ for the sparse intermediate case in the Bixon–Jortner formalism are satisfied for NO_2, where Γ is the radiative linewidth, although neither v nor ε are uniform throughout.

Donnelly and Kaufman have calculated the radiative lifetime (τ) of 1.5 μsec for the $^2B_2$ state at 5936 Å, a factor of 5 greater than the earlier calculated values of 0.3 μsec. They used $\nu_e = 15{,}000\ \text{cm}^{-1}$ in the following expression,

$$\tau_r^{-1} \simeq 5.4 \times 10^{-11} \nu_e^3 \int \frac{k_\nu\, d\nu}{\nu}, \tag{30}$$

where τ is in μsec, ν_e is the frequency for the emission intensity maximum (cm^{-1}), ν is the absorption frequency (cm^{-1}), and k_ν is the absorption coefficient at ν ($\text{cm}^{-1}\ \text{torr}^{-1}$). They estimate an upper limit lifetime of $\sim$100 μsec for the Douglas effect using (a) the above calculated value of τ_r, (b) the factor due to the variation in ν [as appearing in Eq. (24)], and (c) a level density ratio reflecting the dilution of the emitting state. This 100 μsec lifetime value comes within the range of the observed lifetimes. Therefore, the anomalous lifetime behavior in NO_2 is satisfactorily explained with a reasonably quantitative agreement between theory and experiment.

3. Fluorescence Quenching

The collisional quenching of NO_2 fluorescence was much studied in the late 1960s (Donnelly and Kaufman, 1977b). It was noted in the first study of the Ar^+ laser excited fluorescence emission spectra of NO_2 that the intensity ratio of the continuum background to the discrete vibrational structure (superimposed on the continuum) increased with increasing pressure (Sakurai and Broida, 1969). This was the expected behavior of a stepwise vibrational deactivation. Further studies of this ratio by Levy and co-workers showed that the discrete structure was quenched in a magnetic field while the continuum was unaffected (Solarz *et al.*, 1973; Butler *et al.*, 1975; Butler and Levy, 1977). On the basis of these studies, the mechanism of stepwise collisional deactivation was questioned (Butler *et al.*, 1975), but a recent study of the continuum/discrete ratio in the NO_2 fluorescence emission spectra excited with a Nd–YAG laser at 532 nm (Donnelly and Kaufman, 1977b) confirms a stepwise vibrational deactivation model. Donnelly and Kaufman found the electronic quenching rate constants of NO_2^* by NO_2 to be 6.0×10^{-10} and 1.0×10^{-10} cm^3 $molecule^{-1}$ sec^{-1} and those of NO_2^* by He are 3.3×10^{-10} and 0.35×10^{-10} cm^3 $molecule^{-1}$ sec^{-1} for the discrete and continuum emissions, respectively. The quenching of the discrete emission has a high collision efficiency, close to unity. If a quenching cross section of 80 $Å^2$ for the NO_2^*/NO_2 collision pair and a 200 μsec lifetime of NO_2^* are assumed, a value of 5×10^{-4} torr is obtained for the half-quenching pressure. Therefore, it is possible that an appreciable amount of rotational relaxation could occur even at a 0.1 mtorr pressure if the rotational relaxation cross section is much larger than 80 $Å^2$. Further studies with narrow band excitation to select an SRVL at truly collision-free conditions and low rotational temperatures using the technique developed by Levy and co-workers (Smalley *et al.*, 1974, 1975) could lead to a better understanding of the nonexponential behavior and whether the one- or two-state mechanism is most correct. Furthermore, such a study of the relatively unperturbed $^2B_1$ levels could provide detailed information regarding the role rotation plays in electronic relaxation of isolated NO_2 molecules.

4. Fluorescence Quantum Yields

The fluorescence decay times and emission yields from NO_2 were measured as a function of pressure and excitation wavelength λ_{ex} using a 15–30 Å spectral bandwidth of a continuum arc lamp by Kaufman and co-workers (see Keyser *et al.*, 1971) and Schwartz and Johnston (1969). They found that the emission disappeared near 4000 Å due to the predissociation of NO_2 to $O(^3P)$ and $NO(^2\Pi)$.

Douglas and Huber (1965) determined the predissociation limit to be 3979 Å from the rotational diffuseness of the absorption lines of the 3910 Å band of the $^2B_1 \leftarrow {}^2A_1$ transition (see Table 2). In order to observe the effect of rotational excitation, the relative values of the fluorescence quantum yields around 4000 Å were measured between 1 and 10 mtorr using excitation bandwidths of 0.1 and 1.0 Å (Lee and Uselman, 1972). Also, the fluorescence lifetimes were measured at an interval of every 10 Å between 3980 and 4200 Å with a spectral resolution of ~1 Å (Uselman and Lee, 1976b). The wavelength (λ_{ex}) dependent fluorescence yields indicated that nearly all of the rotational excitation energy of NO_2 at room temperature was released to the bond-breaking vibration. The λ_{ex} dependent fluorescence decay times extrapolated to zero pressure were found to be about constant at $\sim 7 \times 10^{-5}$ sec for a fluorescing species, whereas the observed values of Φ_F were certainly λ_{ex} dependent. The apparent lack of expected correlation between the observed fluorescence quantum yield (Φ_F) and the fluorescence lifetime (τ_F),

$$\tau_F = \Phi_F \tau_r, \tag{31}$$

is illustrated in Fig. 5, where τ_r is the radiative lifetime which is expected to be constant here. This apparent discrepancy was rationalized with the explanation that near the predissociation limit there are two types of molecules: (1) a *fluorescing* type with $\tau_F \simeq 7 \times 10^{-5}$ sec and below the predissociation limit, and (2) a *predissociating* type with $\tau \ll 10^{-7}$ sec and above the predissociation limit; and that the observed value of Φ_F is an *average* determined by the fraction of

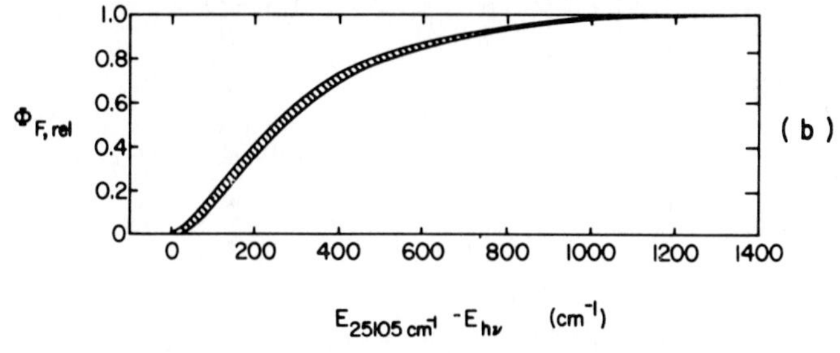

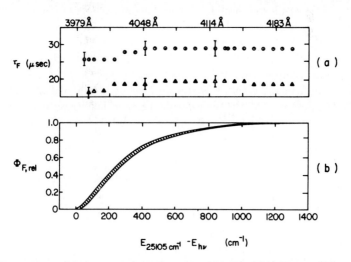

Fig. 5. (a) Observed fluorescence decay time (τ_F) at (⊙) 16 and (△) 26 mtorr NO_2 versus the photon energy deficiency for predissociation ($E' = E_{pd} - E_{h\nu}$). (b) Relative fluorescence quantum yields ($\Phi_{F,rel}$) versus E' for $P = 0.5$–100 mtorr (from Uselman and Lee, 1976b, Fig. 3).

photoexcited molecules which must utilize the available rotational excitation energy at room temperature in order to overcome the predissociation threshold. In Fig. 5, this fraction and $\Phi_{F,rel}$ were observed to have an identical photoexcitation energy dependence. This illustrates the fact that the relationship in Eq. (31) is valid only for a well-behaved SRVL species.

It should be emphasized here that to date no measurement of the absolute fluorescence quantum yield has been reported at collision-free pressures, and this important quantity validating the relationship in Eq. (31) must be measured under usual conditions of interest.

5. $^2A_2$ State

It was discussed earlier that evidence for the $^2A_2 \leftarrow {}^2A_1$ transition at 593, 612, and 647 nm is lacking. A recent study of the resonance fluorescence emission excited at 4416 Å with a He–Cd laser indicates that the upper level has a vibronic symmetry of $^2B_1$ with an odd value of v_3' indicative of the $^2A_2$ electronic state (Senum and Schwartz, 1977). It remains to be seen what role this state might play in the radiationless processes of NO_2.

6. $2\,^2B_2$ State

So far we have been concerned about the low-lying electronic states reached in the visible region. The second $^2B_2$ state shown in Fig. 3 can be excited in the 2400 Å region by the $2\,^2B_2 \leftarrow \tilde{X}\,^2A_1$ transition (Huber, 1966; Coon *et al.*, 1970; Hallin and Merer, 1976). Huber found the onset of rotational line broadening with the 2459 Å band, and more recently Hallin and Merer (1976) found the line broadening even with the 0–0 band at 2491 Å below the dissociation limit for the formation of $O(^1D)$ atom by process C shown in Table 4.

In order to determine if the predissociation is associated with the $O(^1D)$ formation and whether or not rotational excitation energy is utilized in this dissociative process the emission yields and the quantum yields for the formation of $O(^1D)$ atoms, $\Phi[O(^1D)]$, were determined for six optically accessible

TABLE 4
Energetics of NO_2 *Dissociation*[a]

Process	Dissociation product	Energy (cm^{-1})	Onset of diffuseness (cm^{-1})
A	$O(^3P_2) + NO(^2\Pi_{1/2})$	25,105 ± 10	25,125 ± 7
B	$N(^4S) + O_2(^3\Sigma_g)$	36,320 ± 60	
C	$O(^1D) + NO(^2\Pi_{1/2})$	40,973 ± 10	> 40,142 < 40,653

[a] After Douglas and Huber (1965).

vibronic levels (Uselman and Lee, 1976a). It was found that the photochemical threshold for $O(^1D)$ formation was reached at $\sim 40844\ cm^{-1}$, in agreement with the thermodynamic threshold value of $40973\ cm^{-1}$ in Table 4, and the value of $\Phi[O(^1D)]$ reached a plateau value of 0.5 ± 0.1, leaving $\Phi[O(^3P)] = 0.5 \pm 0.1$ also. Furthermore, no photoluminescence could be detected, setting an upper limit of the lifetime of $\sim 10^{11}$ sec, which is consistent with a value of 42 ± 5 psec lifetime estimated from line width measurements (Hallin and Merer, 1976). It is clear that the diffuseness observed for the 0–0 band below the above threshold is unrelated to process C, and that it must be due to rapid radiationless transitions to one or more of the lower-lying electronic states. The assistance of the rotational excitation energy to enhance the photodissociation to give $O(^1D) + NO(^2\Pi)$ by supplementing the photon energy is less apparent with the $2\,^2B_2$ state than in the case of the photodissociation to give $O(^3P) + NO(^2\Pi)$ below 3979 Å (Lee and Uselman, 1972; Uselman and Lee, 1976a).

A vibrational mode specificity effect on the branching ratio between processes C and A in the photodecomposition was not found within the experimental error of this work. It cannot be ruled out, however, that vibrational and rotational effects appear in the electronic relaxation rate of the 2^2B_2 state to the lower-lying electronic states. As far as the photodissociation process is concerned, the asymmetric stretching mode (ν_3') is most important. Therefore, rapid intramolecular energy transfer of vibrational excitation from the symmetric stretch mode (ν_1') or the bending mode (ν_2') to the ν_3' mode could lead to a prompt photodissociation with a lack of mode specificity. This type of process can be facilitated through a Q_1', Q_3' coupling or Q_2', Q_3' coupling. The most recent spectroscopic study of Hallin and Merer (1976) presents some evidence of significant coupling of Q_1' and Q_3' and does not favor a double minimum potential function in Q_3' as suggested by Coon *et al.* (1970). Instead, they suggest anharmonicity effects resulting from a low dissociation energy as the cause for the unusual vibrational structure.

B. SO_2

There are a large number of recent papers on SO_2. Hence we shall limit ourselves mainly to a presentation of experimental and related theoretical studies on the electronic relaxation of the low-lying excited electronic states of SO_2 that have been prepared by selective excitation. To date, there is neither spectroscopic emission nor lifetime data on rotationally selected levels of SO_2, and therefore the selectivity concerns electronic and vibrational aspects only. In the most recent studies, Calvert and co-workers (Su *et al.*, 1977, 1978) obtained a radiative lifetime (τ_r) of 8.1 ± 2.5 msec for the SO_2 ($^3B_1$) state and a fluorescence decay time of $\sim 320\ \mu$sec for the long-lived singlet component of SO_2. Therefore, the experimental difficulties associated with a study of isolated SO_2 molecules is just as severe as for NO_2.

1. Electronic Spectrum

The electronic absorption spectrum of SO_2 between 2000–4000 Å consists of a few overlapping transitions, with varying oscillator strengths and varying degrees of discreteness in the vibration-rotation structure. An electronic energy level diagram and a low-resolution absorption spectrum are shown in Figs. 6 and 7, respectively. A great deal of detailed, high-resolution electronic spectroscopic data useful for the assignment of the molecular electronic states of SO_2 has been obtained recently by Brand and co-workers and by Merer and co-workers. A weak absorption feature in the 3900–3400 Å region is attributed to the spin-forbidden $\tilde{a}\,^3B_1 \leftarrow \tilde{X}\,^1A_1$ transition (see Herzberg, 1966; Brand *et al.*, 1971; Hallin *et al.*, 1976). The rotational analyses of several absorption bands of this $\tilde{a} \leftarrow \tilde{X}$ system have shown that although the first few vibrational bands, 0–0, 1_0^1, 2_0^1, and $1_0^1 2_0^1$, appear quite regular, the higher bands show anharmonicities and gross irregularities in intensities and band contours (Brand *et al.*, 1971 and, 1973c; Hallin *et al.*, 1976). A strong vibronic interaction with a nearby $^3A_2$ state as well as the possibility of a b-axis rotational–electronic Coriolis coupling with a $^3B_2$ state has been suggested for this perturbation (Hochstrasser and Marchetti, 1970; Brand *et al.*, 1973c).

An absorption feature of medium intensity in the 3400–2600 Å region is attributed to two overlapping, forbidden transitions, the $\tilde{B}\,^1B_1 \leftarrow \tilde{X}\,^1A_1$ transition (Herzberg, 1966; Brand and Nanes, 1973) and the $\tilde{A}\,^1A_2 \leftarrow \tilde{X}\,^1A_1$ transition (Dixon and Halle, 1973; Hamada and Merer, 1974, 1975). Much of the complexity of the vibrational and rotational structure is due to the strong vibronic coupling through a $\nu_3(b_2)$ mode between the $\pi\pi^*\tilde{A}\,^1A_2$ electronic state and the $n\pi^*\ \tilde{B}^1B_1$ electronic state. These two electronic states are Renner–Teller components of the $^1\Delta_g$ linear molecule, and again the anomalously long lifetime

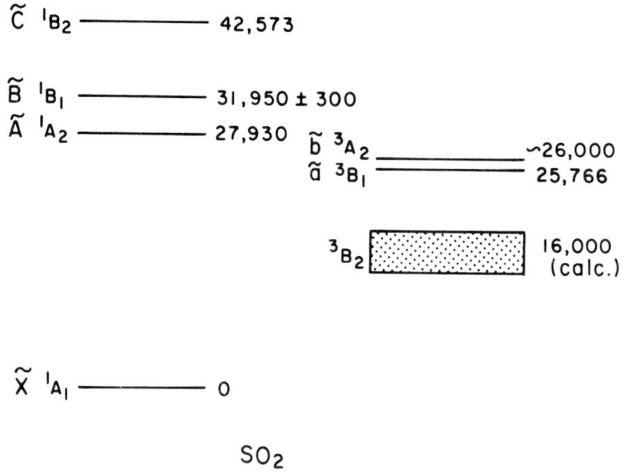

Fig. 6. Energy level diagram for low-lying electronic states of SO_2.

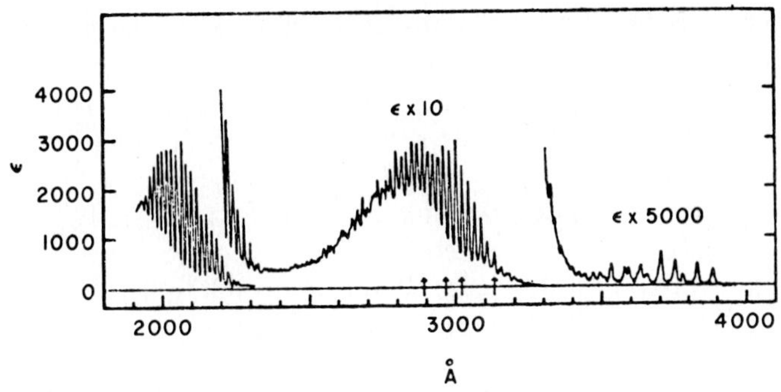

Fig. 7. Electronic absorption spectrum of SO_2 gas (from Strickler and Howell, 1968, Fig. 1).

behavior (Douglas, 1966) has been attributed to the strong perturbation of the $\tilde{B}\,^1B_1$ state by the dense manifold of high vibrational states of the $\tilde{X}\,^1A_1$ ground state (Jungen and Merer, 1976). Furthermore, the $\tilde{A}\,^1A_2 \leftarrow \tilde{X}\,^1A_1$ transition shows strong magnetic rotation (Kusch and Loomis, 1939) and Zeeman effects (Brand *et al.*, 1976a) due to perturbation by the triplet states ($^3B_1$, $^3A_2$, and $^3B_2$).

A strong absorption feature starting at 2350 Å is attributed to the electric dipole allowed $\tilde{C}\,^1B_2 \leftarrow \tilde{X}\,^1A_1$ $[2b_1(\pi^*) \leftarrow 1a_2(\pi)]$ transition (Brand *et al.*, 1972, 1973b, 1976a). The rotational analysis of the absorption and the resonance fluorescence emission showed that this transition was of a parallel, A-type band of an asymmetric, near-prolate rotor (1976a).

2. Phosphorescence Lifetimes and Quantum Yields

Collision-induced population and quenching of triplet SO_2 molecules have been studied by measuring phosphorescence emission spectra as a function of pressure (Mettee, 1968; Strickler and Howell, 1968). Because of the weak absorption intensity of the $T_m \leftarrow S_0$ transition, the early measurements of the phosphorescence emission spectra and decay times involved excitation of $S_n \leftarrow S_0$ transitions and observation of phosphorescence following collision-induced $S_n \rightsquigarrow T_m$ transitions. Otsuka and Calvert (1971) have determined the zero-pressure extrapolated phosphorescence lifetime of 0.88 ± 0.15 msec from the flash photolysis (2400–3200 Å) data collected over a temperature range of 21–100°C and a pressure range of 60–366 mtorr. However, they found that below 60 mtorr the Stern–Volmer plot ($1/\tau$ versus P_{SO_2}) of the SO_2 phosphorescence lifetime (τ_p) showed an upward curvature, indicative of a significant diffusional loss of the excited species. Calvert and co-workers (Sidebottom *et al.*, 1971) have also determined the phosphorescence emission quantum yield $(\Phi_p) = 0.12 \pm 0.09$ at 25°C by directly exciting the forbidden transition, $\tilde{a}\,^3B_1(0, 1, 0) \leftarrow \tilde{X}\,^1A_1(0, 0, 0)$, using 3828.8 Å light of a Raman-shifted, doubled ruby laser (Sidebottom *et al.*, 1971). Because of the long triplet lifetime, all of the above data correspond to thermally equilibrated triplet SO_2 molecules. Subsequent measurements at 289, 297, and 313 nm have given a zero-pressure extrapolated lifetime of 2.7 ± 0.4 msec. These data were taken over a pressure range of 1–500 mtorr at 25°C using a cell with a spherical diameter of 35 cm (Briggs *et al.*, 1975). This cell was much larger than the one used previously by Otsuka and Calvert (1971). Calvert and co-workers (Su *et al.*, 1977) have very recently obtained a collision-free lifetime of 8.1 ± 2.5 msec using the data taken upon excitation of the (0, 3, 0) ← (0, 0, 0) transition of $SO_2(^3B_1)$ at 3631 Å over a pressure range between 1 mtorr and approximately 20 torr. These data were obtained at 26°C using a 22 liter cell. They also made a new estimate of the zero-pressure phosphorescence quantum yield of 0.95 ± 0.29 based upon their earlier data. Since this vibronic level was not rotationally selected, it is not clear what is the extent of the perturbation present.

A theoretical value of the radiative lifetime (τ) can be obtained from the integrated absorption coefficient as follows (Strickler *et al.*, 1976)

$$\frac{1}{\tau_{\text{calc}}} = A_{u\to 1} = 2.880 \times 10^{-9} n^2 \langle \tilde{\nu}_e^{-3} \rangle_{\text{Av}}^{-1} \frac{g_1}{g_u} \frac{\langle \mathbf{M}^2 \rangle_u}{\langle \mathbf{M}^2 \rangle_1} \int \frac{\varepsilon(\tilde{\nu})}{\tilde{\nu}} \, d\tilde{\nu}, \tag{32}$$

where n is the index of refraction for the medium, $A_{u\to l}$ is Einstein coefficient A for the radiative transition from the upper level (u) with the degeneracy of g_u to the lower level (l) with the degeneracy of g_l, ε is the decadic molar extinction coefficient obtained at the absorption frequency $\tilde{\nu}$ measured in wavenumbers, $\langle \tilde{\nu}_e^{-3} \rangle_{\text{Av}}$ is the average of the quantity $(\tilde{\nu}_e)^{-3}$, $\tilde{\nu}_e$ being the emission frequency in wavenumbers, and $\langle \mathbf{M}^2 \rangle_u / \langle \mathbf{M}^2 \rangle_l$ is the ratio of the averages of the quantities

$$\langle \phi_{u0} | |\mathbf{M}_{ul}(Q)|^2 | \phi_{u0} \rangle \qquad \text{and} \qquad \langle \phi_{l0} | |\mathbf{M}_{ul}(Q)|^2 | \phi_{l0} \rangle.$$

M_{ul} is the electronic transition moment for nuclei fixed in the configuration Q. The quantities ϕ_{l0} and ϕ_{u0} refer to the wavefunctions for the zeroth vibrational levels of the lower and the upper electronic states, respectively. The use of this equation gives a value of $\tau_{\text{calc}} = 12.7$ msec (Strickler *et al.*, 1976), while the use of the more approximate, Strickler–Berg relationship gives a value of $\tau_{\text{calc}} =$ 16.9 msec. (Sidebottom *et al.*, 1971). Therefore, the above observed collision-free lifetime of 8.1 ± 2.5 msec for $SO_2(^3B_1)$ (Su *et al.*, 1977a) approaches the theoretical radiative lifetime value. This is consistent with the fact that the $\tilde{a}\,^3B_1(0, 3, 0)$ level at $E_{\text{vib}} \simeq 1770$ cm^{-1} is not high on the vibrational manifold of the $^3B_1$ state and suggests there is no strong perturbation from higher-lying electronic states.

3. Fluorescence Lifetimes

The early studies of Mettee (1968, 1969) and Strickler and Howell (1968) have shown the effectiveness of the collision-induced S ⇝ T intersystem crossing in SO_2. Calvert and co-workers have studied the fluorescence decay at 2660 ± 100, 2662 ± 1, and 2860 ± 160 Å and found that the decay is nonexponential (Sidebottom *et al.*, 1972a). Since the fluorescence cell used in this study was later found to be too small, we shall not discuss the results. However, the observation of the nonexponential decay was confirmed later. The explanation of the nonexponential decay behavior was provided by an extensive study of many vibronic levels of singlet SO_2 between 2600 and 3250 Å (Brus and McDonald, 1973, 1974; Calvert, 1973). It is proposed that the short-lived component with a ~50 μsec lifetime (τ_S) is due to the $\tilde{A}\,{}^1A_2$ state and the long-lived component with a lifetime as long as ~600 μsec (τ_L) is due to the $\tilde{B}\,{}^1B_1$ state correlated to the linear ${}^1\Delta_g$ state. The zero-pressure extrapolated lifetimes (τ_S^0 and τ_L^0) as well as the intensity ratio of the short-lived to the long-lived component, $R \equiv I_S/I_L$, were obtained using an expression like Eq. (29) for a biexponential decay. The data of Brus and McDonald are listed in Table 5 for a comparison with some of the recent data of Calvert and co-workers (Su *et al.*, 1978).

The lifetime results obtained by the two separate groups are in reasonable agreement with each other wherever comparison is possible. However, Su *et al.* (1978) find that the ratio of $R = I_S/I_L$ varies, e.g., from 3.4 to 1.3 for the pressure range of 0.1–8.2 mtorr at 3107 Å, contrary to the observation of Brus and McDonald (1973, 1974) that the ratio is almost independent of pressure. Therefore, Su *et al.* suggest on the basis of their pressure dependence study at 3107, 3211, and 3225 Å, mechanisms which require the collision induced conversion of $\tilde{A} \rightarrow \tilde{B}$ or $\tilde{A} \rightleftarrows \tilde{B}$, rather than Brus and McDonald's mechanism of noninterconversion. But this new suggestion points to a new inconsistency with the tentative assignment of the 0–0 band of the $\tilde{B}\,{}^1B_1 \leftarrow \tilde{X}\,{}^1A_1$ transition between 3100 and 3160 Å (Hamada and Merer, 1975; Shaw *et al.*, 1976a), since the pressure dependence of the ratio of $R = I_S/I_L$ is observed even when the levels below the previously assigned electronic origin of the $\tilde{B}\,{}^1B_1$ state are excited at 3211 and 3225 Å. In order to remove this apparent inconsistency, Calvert and co-workers suggest the electronic origin of the $\tilde{B}\,{}^1B_1 \leftarrow \tilde{X}\,{}^1A_2$ transition is at some wavelength longer than 3273 Å (Su *et al.*, 1978).

A calculated radiative lifetime of ~0.3 μsec is obtained from the oscillator strength ($f \sim 0.006$) of the 3400–2600 Å transition, and this value is considerably shorter than the observed fluorescence lifetimes of $\tau_S \approx 20 \sim 60$ μsec and $\tau_L \approx 100$–600 μsec (see Table 5). A Douglas effect of this magnitude was observed for the case of NO_2. Brus and McDonald (1974) suggest that a zeroth-order BO ${}^1B_1$ state couples with ~100$\tilde{X}\,{}^1A_1$ levels through a strong Renner–Teller interaction. Hamada and Merer (1975) suggest that the perturbation of the $\pi\pi^*\tilde{A}\,{}^1A_2$ levels are probably caused by (1) the lower-lying ${}^3B_2$ state (calculated by Hillier and Saunders, 1971) and the $\tilde{a}\,{}^3B_1$ state and (2) Renner–Teller interaction by the ground $\tilde{X}\,{}^1A_1$ state, transferred from the $n\pi^*\tilde{B}\,{}^1B_1$ state by vibronic coupling, $A_2^e \times b_2^v = B_1^{ev}$. A b-axis Coriolis coupling with the

TABLE 5

Zero-Pressure Extrapolated Fluorescence Lifetime Data for Laser Excitation of SO_2

Reference[a]	λ_{ex} (Å)	Probable assignment[b]	Pressure (mtorr)	Lifetime (μsec) τ_S^0	τ_L^0	$R \equiv I_S/I_L$ at 1 mtorr
BM	3225.9	$^1A_2$ (0, 8, 1)	≥0.8	25	240	19 ± 2.0
BM	3220.8	$^1A_3$ (2, 3, 1)	≥0.8	22	253	4.2 ± 0.3
BM	3198.0	$^1A_2$ (3, 1, 1)	≥0.8	60	532	2.3 ± 0.2
BM	3140.5	—	≥0.8	42	617	1.9 ± 0.2
BM	3043.9	E band	≥0.8	49	308	1.8 ± 0.2
BM	3003.9	G band	≥0.8	42	309	1.4 ± 0.15
BM	2980.8	H band	≥0.8	60	212	0.85 ± 0.1
BM	2961.6	J band	≥0.8	52	184	1.0 ± 0.15
BM	2889.6	O band	≥0.8	58	148	0.4 ± 0.08
BM	2856.7	Q band	≥0.8	59	140	0.4 ± 0.08
BM	2804.5	—	≥0.8	—	114	0.3 ± 0.1
BM	2715.1	—	≥0.8	—	102	0.3 ± 0.1
BM	2617.4	—	≥0.8	—	79	0.25 ± 0.1
S	3273	$^1A_2$ (1, 4, 1)	>0.05	18 ± 6	180 ± 50	
S	3251	$^1A_2$ (2, 2, 1)	>0.05	27 ± 16	110 ± 30	
S	3235	—	>0.05	25 ± 22	150 ± 50	
S	3225	$^1A_2$ (0, 8, 1)	>0.05	19 ± 8	140 ± 30	
S	3211	$^1A_2$ (1, 6, 1)	>0.05	17 ± 8	250 ± 40	
S	3196	$^1A_2$ (3, 1, 1)	>0.05	41 ± 28	200 ± 30	
S	3143	—	>0.05	24 ± 7	220 ± 30	
S	3129	—	>0.05	32 ± 22	220 ± 30	
S	3107	—	0.1–8.2	40 ± 10	320 ± 20	3.4–1.3[c]
S	3065		0.1–8.2	40 ± 38	300 ± 40	

[a] BM (Brus and McDonald, 1974), S (Su *et al.*, 1978).

[b] Vibrational assignment of the upper electronic state, (v_1', v_2', v_3'), as $\tilde{A}\,^1A_2$ (Hamada and Merer, 1975). Alphabetical labeling of the bands are due to J. H. Clements (1935).

[c] R was determined over the pressure range of 0.1–8.2 mtorr.

ground state with the interaction parameter, the b-axis orbital angular momentum matrix element $\zeta_e^{(b)} = \langle \tilde{A}\,^1A_2 | L_b | \tilde{X}\,^1A_1 \rangle$, was also considered possible and was not entirely ruled out.

4. Fluorescence and Phosphorescence Quenching

It is interesting to note that three groups of the above investigators found the collisional self-quenching rates of the two fluorescence decay components are different by an order of magnitude (Mettee, 1969; Brus and McDonald, 1974; Su *et al.*, 1978). The short-lived components are quenched with an efficiency several times the gas kinetic value, and the long-lived components are quenched with near gas kinetic efficiency. The phosphorescence from the direct excitation of $\tilde{a}\,^3B_1$ at 3828.8 Å was quenched by various gases (Ar, N_2, O_2, CO, CO_2, and CH_4) with an activation energy of 2.5 ~ 3.2 kcal/mole (Wampler *et al.*,

1973). It was suggested that this energy pumps the molecule to the levels where it can internally convert to a slightly higher triplet state, i.e., $\tilde{b}\,^3A_2$. The varying efficiency of the fluorescence and phosphorescence quenching by various collision partners and their mechanisms are very interesting, but space will not allow us to discuss this here (see Su *et al.*, 1978).

5. Fluorescence Quantum Yield

The early measurements of the SO_2 fluorescence quantum yield clearly suffer as a consequence of the small cells and the high pressures used, but the zero-pressure extrapolated fluorescence yields were not too far from unity at 2650 Å; 0.55 ± 0.43 (Rao and Calvert, 1970) and 0.41 ± 0.24 (Sidebottom *et al.*, 1972a). The most recent estimates given by Calvert (1973) range from a value of 0.50 to 2.03 with appreciable error limits for several excitation wavelengths, and it has been suggested that these values are quite close to unity at zero pressure. Therefore, it was concluded that radiationless decay of the singlet SO_2 to the ground state or to the triplet states under isolated molecule conditions is unimportant.

6. $\tilde{C}\,^1B_2$ State

Fluorescence lifetimes τ_F and emission quantum yields Φ_F have been measured for ten vibronic levels of the $\tilde{C}\,^1B_2$ state with an excitation bandwidth of 0.81 Å, and from these data radiative and nonradiative rates (k_r and k_{nr}, respectively) have been deduced using the following expressions (Hui and Rice, 1972):

$$k_r = \Phi_F/\tau_F \tag{33a}$$

and

$$k_{nr} = (1 - \Phi_F)/\tau_F. \tag{33b}$$

Their results are tabulated in Table 6. A sudden drop in τ_F and Φ_F around $\lambda_{ex} = 2200$ Å is attributed to the photodissociation to $SO(^3\Sigma^-)$ and $O(^3P)$ (Okabe, 1971).

The most remarkable aspect of the above results is that the several low vibronic levels of the $\tilde{C}\,^1B_2$ state are quite weakly coupled to the isoenergetic, very high vibrational levels of other low-lying electronic states. As a result, nonradiative processes below the dissociation limit are quite unimportant. It should also be recognized that the observed lifetime of 30–40 nsec is fairly close to the theoretical radiative lifetime of 12 nsec calculated with the oscillator strength of $f \sim 0.07$ for the absorption between 2350 and 1650 Å (Brand *et al.*, 1976b). Hui and Rice (1972) suggest that the randomization of the excitation

TABLE 6

Fluorescence Data for Ten Vibronic Levels of $SO_2(\tilde{C}\,^1B_2)$[a]

λ_{ex} (Å)	Assignment ($v'_1 v'_2 v'_3$)[b]	τ_F (nsec)	Φ_F[c]	k_r (10^6 sec^{-1})[d]	k_{nr} (10^6 sec^{-1})[d]
2297.5	(1, 0, 0)	31.8	0.91	28	2.94
2277.1	(1, 1, 0)	32.8	1.0	30	~0
2260.9	(1, 0, 2)	35.3	0.98	28	0.57
2258	(1, 2, 0)	33.7	0.96	29	1.3
2243	α_2 (0, 1)	38.7	0.92	24	2.1
2224.1	α_2 (0, 2)	41.3	0.74	18	6.2
2206.3[d]	α_2 (0, 3)	45.4	0.66	15	7.4
2187.8[d]	α_2 (0, 4)	28.1	0.15	5.4	30
2169.3	α_2 (0, 5)	9.6	0.071	7.4	96
2152.4	α_2 (0, 6)	7.8	0.024	3.1	125

[a] Obtained by Hui and Rice (1972).

[b] Vibrational assignments of Brand *et al.* (1976b), except for the α_2 bands of Duchesne and Rosen (1947).

[c] Values normalized to the value at 2277.1 Å.

[d] The dissociation limit [($SO_2 \rightarrow SO(^3\Sigma^-) + O(^3P)$)] between α_2 (0, 3) and α_2 (0, 4) occurs at ~2900 cm^{-1} above the electronic origin of the $\tilde{C}\,^1B_2$ state, 42,578 cm^{-1}.

energy in the $\tilde{C}$ state is very slow, as in the case of chloroacetylene and bromoacetylene (Evans and Rice, 1972; Evans *et al.*, 1973). They also suggest that the intramolecular transfer of energy from the symmetric stretching mode (v'_1) to the asymmetric stretching mode (v'_3), presumably important in the decomposition, is probably slow. However, this may not be so, since the most recent spectroscopic study shows an appreciable Q'_1, Q'_3 coupling and a large inversion splitting for the levels containing $2v'_3$ due to the double minimum potential for Q'_3 (Brand *et al.*, 1976b).

C. Formaldehyde

Formaldehyde is the simplest stable molecule in the group of carbonyl compounds for which extensive photochemical studies have been made (see Calvert and Pitts, 1966). The ground and excited electronic states of formaldehyde are near prolate tops. One large rotational constant (~9 cm^{-1}) and six vibrational modes with high frequencies contribute partly to the observed, well-resolved rotational and vibrational structures, respectively, in its electronic spectrum. Formaldehyde photodecomposes via two channels to give (1) radical products and (2) molecular products.

$$H_2CO + h\nu \xrightarrow{\text{I}} H + HCO, \qquad (34)$$

$$H_2CO + h\nu \xrightarrow{\text{II}} H_2 + CO. \qquad (35)$$

These molecular properties and kinetic behavior make possible ideal photophysical and photochemical studies with single rovibronic level (SRVL) excitation (Lee, 1977). Many recent studies deal with the single vibronic level (SVL) effects on radiative and nonradiative transitions, selective photoproduct formation, and energy partitioning processes (Yeung and Moore, 1973; Luntz and Maxon, 1974; Miller and Lee, 1975; Marling, 1977; Houston and Moore, 1976).

1. Electronic Spectrum

An excellent review on the ultraviolet spectroscopy of formaldehyde and its excited states is by Moule and Walsh (1975). We shall briefly discuss the two low-lying $n\pi^*$ excited electronic states shown in Fig. 8 and the two electronic transitions connecting each excited state with the ground electronic state. The $\tilde{a}\,^3A_2 \leftarrow \tilde{X}\,^1A_1$ transition has an onset near 3950 Å with an oscillator strength of $f \sim 1.2 \times 10^{-6}$ (DiGiorgio and Robinson, 1959). This extremely weak, spin-forbidden and symmetry-forbidden transition extends beyond 3550 Å, the onset of the weak, symmetry-forbidden $\tilde{A}\,^1A_2 \leftarrow \tilde{X}\,^1A_1$ transition with an oscillator strength of $f \sim 2.4 \times 10^{-4}$. Extended vibrational progressions of the $\tilde{A} \leftarrow \tilde{X}$ system for H_2CO are shown in Fig. 9. A few vibronic bands of the $\tilde{a}\,^3A_2 \leftarrow \tilde{X}\,^1A_1$ transition have been analyzed to show that the triplet excited state is nonplanar as is the $\tilde{A}\,^1A_2$ state (Robinson and DiGiorgio, 1958). Rotational analyses and Stark effect studies have provided numerous molecular constants, and it is known that the transition moment directed along the top axis is ~90% of the total (Birss *et al.*, 1973 and references therein). Although a weak phosphorescence emission was observed in a lifetime study (Luntz and Maxon, 1974), the emission was too weak to be analyzed spectrally.

A nonplanar equilibrium geometry of the $\tilde{A}\,^1A_2$ state was verified by analyses of high resolution fluorescence emission spectra (Brand, 1956;

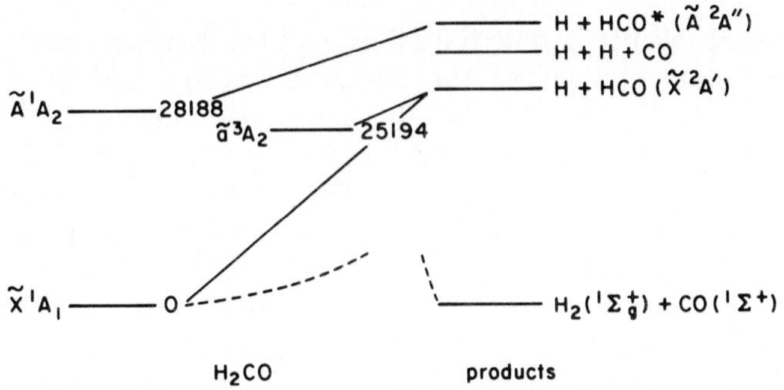

Fig. 8. Energy level diagram for low-lying electronic states and dissociation product states of H_2CO. Product correlations are from Hayes and Morokuma (1972) and Jaffe and Morokuma (1976).

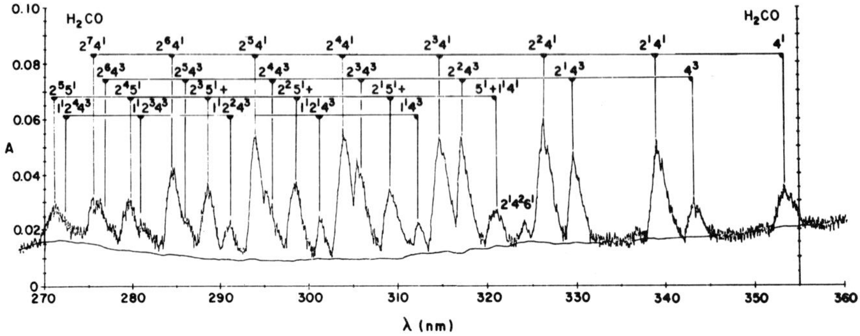

Fig. 9. Electronic absorption spectrum of H_2CO gas (adapted from Miller and Lee, 1978): $P = 7.65$ torr; $l = 5.0$ cm; 23°C. All of the assigned transitions originate from the zeroth vibrational levels. In the 2^n progressions built upon the $5^1_0 + 1^1_0 4^1_0$ transition, $1^1 2^n 4^1$ was omitted after $2^n 5^1$ due to lack of space.

Robinson, 1956). Recently many vibrational levels of H_2CO, HDCO, and D_2CO have been studied in detail by Innes and co-workers (Job *et al.*, 1969; Sethuraman *et al.*, 1970). In the $\tilde{A}\,^1A_2$–$\tilde{X}\,^1A_1$ system, the weak 0–0 band is due to a magnetic dipole transition (Callomon and Innes, 1963), but the more prominent bands to vibronically allowed electric dipole transitions (Brand, 1956). Perturbations due to rotation–vibration Coriolis coupling and Fermi resonances occur with some low-lying vibronic levels, and they have been studied recently by high-resolution absorption spectroscopy (Parkin *et al.*, 1962; Sethuraman *et al.*, 1970). Perturbations due to singlet–triplet mixing in the rovibronic levels of $2^2 4^1$ have been investigated by analyses of magnetic rotation spectra (Birss *et al.*, 1978) and Zeeman spectra (Brand and Stevens, 1973). The rotational line broadening due to predissociation becomes appreciable below 2800 Å, and the line widths in the diffuse region have been measured recently (Baronavski *et al.* 1976).

The assignments of vibronic transitions and the designations of SVLs will be given by the Brand notation. The numerical superscript on the right of the ith numbered vibration specifies the vibrational quantum number of the upper electronic state (n_i') and the numerical subscript on the right specifies the n_i'' of the lower electronic state. The absence of the transition designation for a given vibrational mode means a 0–0 transition for the mode. For example, $2^3_0 4^1_0$ is the transition ($n_2' = 3; n_4' = 1$) ← ($n_2'' = 0; n_4'' = 0$) and $2^3 4^1$ is the vibrational state designation of the upper state. Fundamental vibrational frequencies in the ground (S_0), first singlet (S_1), and first triplet (T_1) states are summarized in Table 7. The $\pi^* \leftarrow n$ nature of the upper electronic state causes large frequency decreases observed for v_2', v_3', v_4', and v_6'. It is clearly seen in Fig. 9 that the $v_4'(b_1)$ vibration induces the $\tilde{A}\,^1A_2 \leftarrow \tilde{X}\,^1A_1$ transition and that the *long* 2^n_0 progressions built upon 4^1_0 and 4^3_0 are most prominent. The broad Franck–Condon envelope, typical of a $\pi^* \leftarrow n$ transition in carbonyl compounds, is due to the lengthening of the C–O bond in the $n\pi^*$ state. Somewhat less prominent 2^n_0 progressions are

TABLE 7

Fundamental Vibrational Frequencies of ν_i (cm^{-1}) *in* H_2CO

ν_i	Mode	Symmetry	$\tilde{X}\,^1A_1$[a]	$\tilde{A}\,^1A_2$[a]	$\tilde{a}\,^3A_2$
ν_1	Symmetric C—H stretching	a_1	2766.4	2847	2871[c]
ν_2	C=O stretching	a_1	1746.1	1173	1281[d]
ν_3	CH_2 scissoring	a_1	1500.6	1290[e]	—
ν_4	⊥ wagging	b_1	1167.3	(124.6)[b]	(30)[c f]
ν_5	antisymmetric C—H stretching	b_2	2843.4	2968	—
ν_6	∥ wagging	b_2	1251.2	904	—

[a] From Job *et al.* (1969).

[b] For the first vibrational level ($v_4' = 1$ or 0^- in a double minimum potential function. $\nu_4' = 1$, 2, 3, and 4 are located at 124.6, 542.4, 948.4, and 1426.7 cm^{-1}, respectively.

[c] From Robinson and DiGiorgio (1958).

[d] From Brand and Stevens (1973e).

[e] This value of ν_3 from Hardwick and Till (1979) replaces the earlier value of 887 cm^{-1}.

[f] For the first vibrational level ($\nu_4' = 1$ or 0^-) in a double minimum potential. $\nu_4' = 1$, 2, 3, and 4 are located at 30, 537, 779, and 1171 cm^{-1}, respectively.

built upon 5^1_0, $1^1_0 4^1_0$, and $1^1_0 4^3_0$, where $2^n_0 5^1_0$ is induced by the $\nu_5'(b_2)$ vibration and is quasi degenerate with $1^1_0 2^n_0 4^1_0$ (Sethuraman *et al.*, 1970).

2. Triplet States ($\tilde{a}\,^3A_2$)

Luntz and Maxon (1974) obtained a laser-induced phosphorescence excitation spectra of D_2CO (~1 torr at −78°C) consisting of six peaks at 4^2_0 (3879 Å), 4^4_0 (3814 Å), 2^1_0 (3762 Å), $3^1_0 4^2_0$ (3723 Å), $2^1_0 4^2_0$ (3700 Å), and $2^1_0 4^4_0$ (3641 Å), which had been assigned previously in absorption (Robinson and DiGiorgio, 1958). No such emission from H_2CO was observed under identical circumstances. An exponential decay time of phosphorescence (τ_p) of 16.9 ± 0.5 μsec for the vibrationally relaxed $D_2CO(\tilde{a}\,^3A_2)$ was observed, and a conservative upper limit for the lifetime of 0.4 μsec for all vibronic levels of $H_2CO(\tilde{a}\,^3A_2)$ was estimated. Assuming the value of the radiative lifetime (τ_{calc}) for D_2CO ($^3A_2$) to be the same as for H_2CO ($\tilde{a}\,^3A_2$) estimated from the integrated absorption, ~2 msec (DiGiorgio and Robinson, 1959), one can estimate the values of the phosphorescence quantum yield (Φ_p) from Eq. (31) to be ~0.01 for $D_2CO(\tilde{a}\,^3A_2)$ and $<2 \times 10^{-4}$ for $H_2CO(\tilde{a}\,^3A_2)$. Luntz and Maxon also obtained a large electronic quenching cross section for $D_2CO(\tilde{a}^3A_2)$ by NO of 22 ± 5 Å^2 as compared to a self-quenching value of <0.01 Å^2. The near gas kinetic efficiency of luminescence quenching by the paramagnetic species NO was considered consistent with the triplet character of the emitter. In view of the low estimated values of Φ_p, it is obvious that the observed value of τ_p is very close to the nonradiative lifetime τ_{nr} [see Eq. (33)].

According to theory, the ground state S_0, and not the triplet state T_1, correlates to the ground states of the molecular products, $H_2(\tilde{X}\,^1\Sigma_g^+)$ and

$CO(\tilde{X}\,^1\Sigma^+)$ (Jaffe and Morokuma, 1976). Both S_0 and T_1 correlate to the ground states of the radical products, $H(^2S)$ and $HCO(^2A')$ (Hayes and Morokuma, 1972), but the low vibronic levels of T_1 cannot energetically dissociate (see Fig. 8). Triplet benzene ($^3B_{1u}$) sensitized decomposition of H_2CO gave an appreciable quantum yield of H_2 and CO, ~ 0.3, with and without radical scavengers such as O_2, (Miller and Lee, 1974). Therefore, this observation was interpreted to mean that the molecular products arise from the unimolecular decomposition of the vibrationally hot $H_2CO(S_0^\dagger)$ produced from the $T_1 \rightsquigarrow S_0$ intersystem crossing (ISC),

$$H_2CO(T_1) \xrightarrow{k'_{ISC}} H_2CO(S_0^\dagger). \tag{36}$$

It appears therefore that the above $T_1 \rightsquigarrow S_0$ ISC process is largely responsible for the nonradiative electronic relaxation of $H_2CO(T_1)$.

Luntz and Maxon (1974) measured the rate of nonradiative decay of $H_2CO(T_1)$, i.e., $\geq 2 \times 10^6\ sec^{-1}$, to be just as efficient as its corresponding S_1 state, i.e., $k_{nr} = 10^6$–$10^7\ sec^{-1}$ (Yeung and Moore, 1973). Therefore, they invoked an explanation based on the golden rule approximation [see Eq. (14)] to account for the comparable rates of $T_1 \rightsquigarrow S_0$ and $S_1 \rightsquigarrow S_0$ processes. They suggested that a substantially larger Franck–Condon factor exists for ISC than for the $S_1 \rightsquigarrow S_0$ internal conversion (IC),

$$H_2CO(S_1) \xrightarrow{k_{IC}} H_2CO(S_0^\dagger), \tag{37}$$

due to a more bent geometry for T_1 than S_1. An approximately 10^3 times smaller electronic coupling matrix element was expected for ISC than for IC, but it was thought to be nearly compensated for by the favorable Franck–Condon factor above, implicating v_4'' as an efficient accepting mode [see Eqs. (14) and (15)]. Therefore, they regarded the comparable values of $k'_{ISC} \approx \tau_P^{-1}$ and $k_{IC} \leq \tau_F^{-1}$ for H_2CO to be quite reasonable theoretically.

3. Fluorescence from SVLs of Singlet State ($\tilde{A}\,^1A_2$)

Yeung and Moore (1973) first studied fluorescence decay from many SVLs of H_2CO, HDCO, and D_2CO at relatively low pressures ($p \geq 50$ mtorr) using a pulsed, tunable dye laser (1 Å bandwidth, 7 nsec pulsewidth). They observed some nonexponentiality in the fluorescence decay, depending on the pressure used. They attributed the long decay component of the nonexponential decay to population of other long-lived vibronic levels via collisional vibrational relaxation processes. They attributed the short decay component to unimolecular decay of the initially prepared state, and the short lifetime τ_A was obtained from the early part of the decay. Zero-pressure SVL lifetimes τ_A^0 were obtained by extrapolation of the short lifetimes in a Stern–Volmer plot ($1/\tau_A$ versus P). The most extensive measurements were made with 15 SVLs of D_2CO; the range of lifetimes, 4.6 μsec for 4^1 to 53 nsec for $2^3 4^3$, show a hundredfold increase in rate for a 4000 cm^{-1} increase in vibrational energy (E_{vib}). The lifetimes of nine SVLs of HDCO were measured, and they decreased from 290 nsec for 4^1 to 8 nsec for $2^3 4^3$. Only three SVLs of H_2CO were measured, since the instrumental limit was close to ~ 10 nsec. These values were 282 nsec for 4^1, 29 nsec for $2^1 4^3$ and 20 nsec for $2^2 4^1$. Since fluorescence quantum yields (Φ_F) were not measured, the rates of radiative and nonradiative processes could not be unambiguously obtained, particularly for D_2CO with a sizable value of Φ_F, using Eqs. (33a) and (33b). The effect of deuteration on lifetime was marked, but H_2CO, HDCO, and D_2CO showed rather similar variations in lifetimes upon excitation of particular combinations of normal modes. Yeung and Moore (1973 and 1974) concluded that the only mechanism of radiationless decay compatible with the energy level structure of formaldehyde and the observed lifetimes was $S_1 \rightsquigarrow S_0$ internal conversion to high vibrational levels of S_0 which are broadened by unimolecular dissociation.

Subsequent measurements of Φ_F for numerous SVLs of H_2CO over the 0.1–5.0 torr range were made by fluorescence excitation spectroscopy (FEX) using a 1.5 Å excitation bandwidth (Miller and Lee, 1975). In the same study, lifetime measurements of these SVLs were made with a single-photon time-correlated fluorescence decay apparatus using pulses from a D_2 flash lamp with 10–30 Å bandwidth and 2 nsec decay. These measurements were then extended to HDCO and D_2CO (Miller and Lee, 1976, 1978). It was found that Φ_F for the higher SVLs became smaller as the vibrational energy (E_{vib}) increased. However, values of Φ_F for the quasi-degenerate levels (5^1; $1^1 4^1$), ($2^1 5^1$; $1^1 2^1 4^1$), and ($2^2 5^1$; $1^1 2^2 4^1$) in H_2CO were unusually high compared to other nearby levels. Furthermore, definitely longer values of τ_F were obtained for these levels compared to other nearby levels. Linear Stern–Volmer plots ($1/\tau$ versus P) were obtained in the pressure range of 0.1–2.0 torr in all cases, although some plots were nearly flat for the low lying levels of H_2CO. From these plots, the zero-pressure extrapolated values of lifetime (τ_F) were obtained. For H_2CO, the values of τ_F were 82 nsec for 4^1, 10.8 nsec for $2^1 4^3$, and 9.8 nsec for $2^2 4^1$, indicating these values were two to three times shorter than those of Yeung and Moore (1973). Similar discrepancies were found for HDCO lifetimes, and fairly close agreement was found for D_2CO lifetimes below $E_{vib} \simeq 3000$ cm^{-1}. In general, τ_F values of Miller and Lee (1975, 1976, 1978) were shorter by a factor of 2–3 than the corresponding τ_A^0 values of Yeung and Moore (1973) and Baronavski (1975) for τ_F shorter than ~400 nsec. A detailed comparison is given elsewhere (Miller and Lee, 1978). It should also be noted that steep downward curvatures were observed in Stern–Volmer plots ($1/\tau_F$ versus P) at formaldehyde pressures below 10–100 mtorr (Luntz, 1975; Baronavski, 1975). The explanation for the τ_F discrepancy is just emerging from the recent studies of Weisshaar *et al.* (1978).

Weisshaar *et al.* (1978) were able to examine the fluorescence decay of the 4^0 and 4^1 levels of H_2CO and D_2CO with an improved laser fluorescence apparatus over a wide range of observation time and at pressures as low as 0.04 mtorr.

The dye laser pulses ($\sim$5 nsec long and $\sim$0.2 Å FWHM spectral width) were used to excite a few rR sub-band heads of 4^0. The third harmonic of a Nd : YAG laser at 3547 Å was used to excite the 4_0^1 transition of H_2CO. The Stern–Volmer plot for D_2CO was found to be linear from 0.1–4 torr with a significant negative curvature below 0.1 torr. The zero-pressure lifetimes of 7.8 ± 0.7 μsec for 4^0 and 6.0 ± 0.4 μsec for 4^1 in D_2CO were obtained. These values are comparable to the theoretically estimated radiative lifetimes and they are somewhat longer than the values obtained earlier for the 4^1 level, 4.6 μsec by Yeung and Moore (1973) and 4.5 μsec by Miller and Lee (1976). In the case of H_2CO, the fluorescence decay plots were found to be biexponential below 0.2 torr, at pressures for which vibrational energy relaxation is negligible. The Stern–Volmer plots for the long-lived decay component ($1/\tau_L$ versus P) in the excitation of the rR sub-bands of 4^0 showed dramatic negative curvature below 1 torr, and they were found to become linear below 20 mtorr with slopes on the order of 10 times the gas kinetic rate. A brief summary of the observations of Weisshaar *et al.* is shown in Table 8. It is clear that the Stern–Volmer slopes and lifetimes depend significantly on the rR subband excited within 4_1^0, and hence on the K' rotational quantum number. Furthermore, their lifetime values for the 4^1 level of H_2CO indicate that $\tau_S = 69$ nsec is comparable to the value of $\tau_F = 82$ nsec reported by Miller and Lee (1975) and $\tau_L = 435$ nsec is comparable to the value of $\tau_A^0 = 282$ nsec reported by Yeung and Moore (1973). It is likely that a poor signal-to-noise (S/N) ratio in the long delay time range inherent in the single-photon time-correlation counting measurement (Ware, 1971) may have prevented Miller and Lee (1975) from observing the long-lived component, whereas a noisy signal in the short decay time range after the laser pulse may have prevented Yeung and Moore (1973) from observing the short-lived component. According to the recent molecular beam study of Luntz (1978), H_2CO in the $K' = 1$ sublevel of 4^0 has a lifetime of 3.6 ± 0.3 μsec, approaching the 3.2 μsec value of the radiative lifetime observed by Miller and Lee (1975). First of all the result of this latest study raises a serious question about the reliability of analyzing what may be multiexponential decay as a biexponential.

TABLE 8
Stern–Volmer Slopes and Zero-Pressure Lifetime τ Data for $H_2CO(\tilde{A})$ [a]

ν_{vac} (cm^{-1})	Vibronic band	rR sub-band[b]	Short-decay slope	Short-decay τ_S (nsec)	Long-decay slope	Long-decay τ_L (nsec)
28,185	4_0^1	Not known	3.4 ± 0.9	69 ± 7	4.5 ± 0.5	435 ± 15
27,037	4_1^0	$K' = 1, J' = 3$–10	1.9 ± 1.2	83 ± 10	$2.2 + 0.6$	550 ± 90
27,113	4_1^0	$K' = 5, J' = 5$–11	6.5 ± 1.9	130 ± 24	3.5 ± 0.7	490 ± 90
27,156	4_1^0	$K' = 7, J' = 7$–11			2.2 ± 0.7	235 ± 22

[a] From Weisshaar *et al.* (1978). Slopes are measured in microseconds per Torr.
[b] r designates $\Delta K = +1$, and R designates $\Delta J = +1$ where K'' is for the lower state and K' is for the upper state. The Stern–Volmer slopes are in units of 10^{-9} cm^3 molecule^{-1} sec^{-1}.

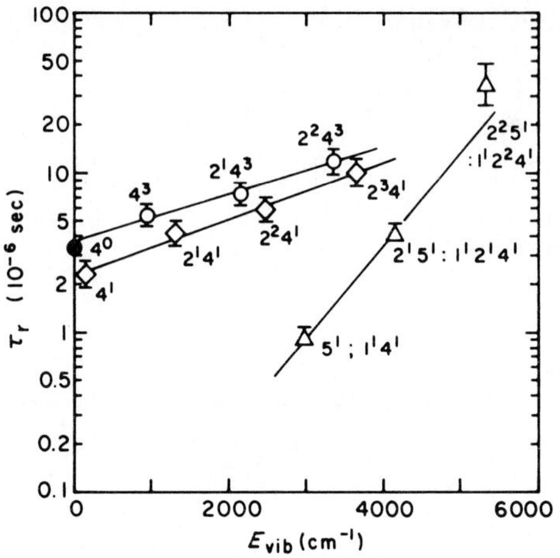

Fig. 10. Variation of radiative lifetimes (τ_r) versus vibrational energy (E_{vib}) in $H_2CO(A\ ^1A_2)$ (from Miller and Lee, 1978, Fig. 8). Diamonds are for the ν_2' progression of 4^1, circles for that of 4^3, and triangles for that of (5^1; 1^14^1).

For the large quenching cross sections at very low pressures and the dramatic curvature of the Stern–Volmer plots observed for H_2CO, Weisshaar *et al.* (1978) offered a mechanistic explanation based on Gelbart and Freed's model of collision-induced electronic quenching with large rotational relaxation cross sections (Gelbart and Freed, 1973; Freed, 1976b, c). They suggest that the large quenching rates and curved Stern–Volmer plots are primarily due to the rotational relaxation of the initially excited levels to slower decaying rotational levels, resulting in rapid decrease of both the apparent lifetime and the fluorescence quantum yield with increased pressure.

The energy dependence of τ_r in H_2CO is shown in Fig. 10. In general, τ_r increases with increasing energy, and the accidentally quasidegenerate pairs of levels in the ν_2' progression of (5^1; 1^14^1) in H_2CO have distinctly higher oscillator strengths. A detailed discussion of this striking behavior is given elsewhere (Miller and Lee, 1978) in regards to the SVL theory of radiative transitions (Fleming *et al.*, 1973; Lin, 1976). The increase in τ_r with energy has been observed in a number of molecular systems, SO_2 (Hui and Rice, 1972), some fluorinated benzenes (Loper and Lee, 1972) and p-$C_6H_4F_2$ (Volk and Lee, 1977). The energy dependence of τ_{nr} in H_2CO, HDCO, and D_2CO is shown in Fig. 11. In general, τ_{nr} decreases with increasing energy, and again the accidentally degenerate pairs of levels in the ν_2' progression of (5^1; 1^14^1) in H_2CO have distinctly slower nonradiative decay rates. The SVL structure diminishes in HDCO and D_2CO. The deuterium isotope effect is 1–3 orders of magnitude for τ_{nr} between $E_{vib} = 0$–$4000\ cm^{-1}$. A recent study of vibrational relaxation of the 4^1 level of H_2CO ($\tilde{A}$) by laser-induced resonance fluorescence emission spectroscopy shows that the half-pressure ($P_{1/2}$) for the $4^1 \rightarrow 4^0$ conversion by $H_2CO(\tilde{X})$ at 23°C is 0.56 torr and the cross section is ~ 150 Å^2

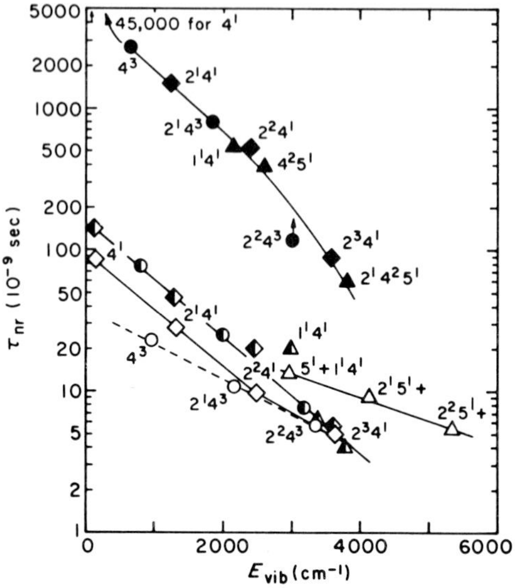

Fig. 11. Variation of nonradiative lifetimes (τ_{nr}) versus vibrational energy (E_{vib}) in $H_2CO(\tilde{A}\ ^2A_2)$ (open symbols) compared with HDCO (half-filled symbols) and D_2CO (filled symbols) (from Miller and Lee, 1978, Fig. 11). Legends are similar to those for Fig. 10.

(Shibuya and Lee, 1978). Therefore, Miller and Lee's values of τ_r and τ_{nr} are certainly values of the original SVLs pumped but presumably rotationally equilibrated by virtue of using a wide band optical excitation over a significant portion of the rotational envelope and using pressures where efficient, collision-induced rotational relaxation can take place. Clearly, an SVL lifetime represents some complicated average of the SRVL lifetimes, which is sensitive to the nature of the light source used, the ambient ground state distribution, and the pressure.

As far as the role of the promoting mode is concerned, Yeung and Moore (1974) and Lin (1976) have shown that the ν_4 mode is most important. We shall not further discuss the role of the promoting mode due to space limitation.

4. Fluorescence from SRVLs of Singlet State ($\tilde{A}\ ^1A_2$)

In the recent study of Tang *et al.* (1977), a large number of rotational levels of the 2^24^1 and 2^34^1 SVLs of H_2CO ($\tilde{A}\ ^2A_2$) were populated by pumping near the tails of their rotational envelopes using a narrowly tuned dye laser (~ 0.2 cm^{-1}). The 2^24^1 band was chosen because it had shown singlet–triplet perturbation (Kusch and Loomis, 1939; Brand and Stevens, 1973; Birss *et al.*, 1978). The 2^34^1 band was chosen because it had shown rotation–vibration Coriolis perturbation (Parkin *et al.*, 1962; Sethuraman *et al.*, 1970).

Portions of the absorption and fluorescence excitation spectra taken of the 2^34^1 level of H_2CO with $\tau_F \simeq 4$–5 nsec are shown in Fig. 12. The intensity ratios

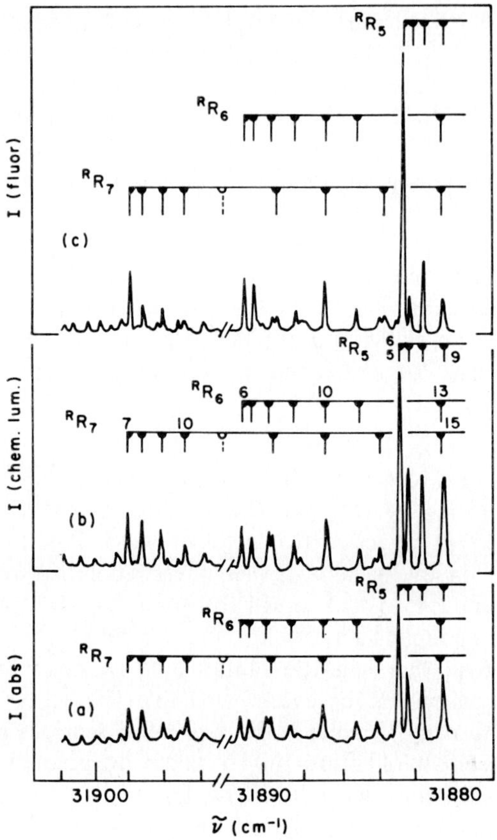

Fig. 12. Portions of H_2CO spectra of the $2_0^3 4_0^1$ transition: (a) absorption spectrum with ordinate in absorbance unit; (b) red HNO* chemiluminescence excitation spectrum taken with the emission at 760 nm (1.0 torr H_2CO/10 torr NO); (c) blue H_2CO* fluorescence excitation spectrum taken with the emission at 380–500 mm (0.20 torr H_2CO). The rotational sub-branch assignment as a near prolate top is ${}^{\Delta K}\Delta J_{K''}$ with J'' shown on top of the line. The line for $J'' = 11$ of ${}^R R_7$ is not shown. (taken from Tang *et al.* 1977, Fig. 1).

of the rotational levels in the progression show apparent irregularity due to the varying degree of localized interaction between the perturbed, optically accessible $2^3 4^1$ level and perturbing, optically inaccessible vibrational level or levels nearby. The rotation–vibration Coriolis coupling requires as in Eqs. (1) and (2),

$$\Gamma_i^v \times \Gamma^r = \Gamma_j^v. \tag{38}$$

Sethuraman *et al.* (1970) suggest that b-axis rotation–vibration Coriolis coupling occurs between $2^3 4^1(b_1^v)$ and the lower-lying, nearly degenerate $2^1 3^2 4^2(a_1^v)$ or $2^1 4^2 6^2(a_1^v)$ levels with a selection rule $\Delta K = \pm 1$. The lack of regularity in the $2^3 4^1$ progression requires that a pure $2^1 3^2 4^2$ or $2^1 4^2 6^2$ level has an intrinsically different value of Φ_F than a pure $2^3 4^1$ level, owing either to a different value of τ_r and/or of τ_{nr}. Tang *et al.* (1977) suggested that the low fluorescence quantum yield of the perturbed $2^3 4^1$ level is simply associated with the greater nonradiative character of the pure $2^1 3^2 4^2$ or $2^1 4^2 6^2$ level which may decay by $S_1 \rightsquigarrow S_0$ more rapidly than the pure $2^3 4^1$ level. More study is needed to obtain a detailed understanding of this interesting observation about the intramolecular, collisionless vibrational energy redistribution by time evolution of an optically prepared vibrational state.

A similar study of the $2^2 4^1$ level of $H_2CO(S_1)$ with $\tau_F \simeq 9.8$ nsec showed that despite the presence of an extensive S–T perturbation no changes in the fluorescence quantum yields of a number of perturbed levels were observed. Therefore, Tang *et al.* suggested that $S_1 \rightsquigarrow T_1$ intersystem crossing (ISC),

$$H_2CO(S_1) \xrightarrow{k_{ISC}} H_2CO(T_1^\dagger), \qquad (39)$$

must have a negligible role in the photochemical mechanism for the $2^2 4^1$ level. Brand and co-workers (Stevens and Brand, 1973; Brand and Stevens, 1973; Liu and Brand, 1974) suggested that most of the S–T perturbations in $2^2 4^1$ can be explained on the basis of (1) an intersystem spin–rotation coupling mechanism between $2^2 4^1$ of $\tilde{A}\,^1A_2$ and $1^1 2^2 4^1$ of $\tilde{a}\,^3A_2$, except for some by (2) a vibronic spin–orbit coupling between $2^2 4^1$ of $\tilde{A}\,^1A_2$ and $1^1 2^2$ of $\tilde{a}\,^3A_2$. Birss *et al.* (1978) attempted to fit their magnetic rotation data by using the selection rules in Table 9. They suggested that more than two triplet levels are needed to explain all of the observed magnetic rotation lines and that Coriolis and Fermi interaction on the triplet manifold also should be taken into account. It is disappointing that no simple conclusion can be drawn, but it is not too surprising in view of the high vibrational energy on the triplet manifold, $\sim 5450\ cm^{-1}$.

TABLE 9

Selection Rules for Spin–Orbit–Orbital Rotation Interaction[a] and Vibronic Spin–Orbit Interaction[b] in Polyatomic Molecules[c]

Case	Selection rule $\Delta J = 0$: ΔN	ΔK	Orbital symmetry restriction
1	$0, \pm 1$	$0, \pm 2$	$\Gamma_S^e = \Gamma_T^e;\ \Gamma_S^{ev} = \Gamma_T^{ev}$
2	$0, \pm 1$	0	$\Gamma_S^{ev} \times \Gamma_T^{ev} \supset R_z$
	$0, \pm 1$	± 1	$\Gamma_S^{ev} \times \Gamma_T^{ev} \supset R_x$ or $\Gamma_S^{ev} \times \Gamma_T^{ev} \supset R_y$

[a] $\langle \Gamma_S^{ev} NJK | H_{so-or} | \Gamma_T^{ev} N'JK' \rangle$.
[b] $\langle \Gamma_S^{ev} NJK | H_{so-ev} | \Gamma_T^{ev} N'JK' \rangle$.
[c] Taken from Stevens and Brand (1973).

In the diffuse region below 2800 Å, Baronavski *et al.* (1976) have estimated lifetimes by rotational line width measurements in the electronic absorption bands. The lifetime values of 10^{-11}–10^{-12} sec were obtained for the range of $E_{\text{vib}} = 7000$–$9000\ \text{cm}^{-1}$, and the shortest range of lifetimes were obtained with HDCO and the longest with D_2CO.

Two vibronic levels of $H_2CO(\tilde{A})$, 5^1 and 1^14^1, are accidentally nearly degenerate with an energy separation of $2.7\ \text{cm}^{-1}$, at $\sim 2970\ \text{cm}^{-1}$ above the $\tilde{A}\,^1A_2$ electronic origin. The rotational levels of these vibronic levels blend into each other, and the weak vibration–rotation coupling between them has been studied by Sethuraman *et al.* (1970). Tang and Lee (1976a) have recently studied laser-induced fluorescence emission spectra from selected rovibronic levels of 5^1 and 1^14^1. They were able to excite some levels which show nearly pure resonance fluorescence emission spectra characteristic of the initially prepared pure rovibronic level. For example, the 3215.1 Å excitation of the $5_0^1\,{}^{\text{P}}Q_3$ $[(J' = 7, K' = 2) \leftarrow (J'' = 7, K'' = 3)]$ band gave an emission spectrum with a set of progressions with $5_1^1 + 2_n^0 4_m^0$ where $n = 0, 1, 2, \ldots$ and $m = 1, 3, \ldots$, and the 3207.0 Å excitation of the $1_0^1 4_0^1\,{}^{\text{r}}Q_3$ $[(J' = 13, K' = 4) \leftarrow (J'' = 13, K'' = 3)]$ gave an emission spectrum with a set of progressions with $1_1^1 + 2_n^0 4_m^1$ where $n = 0, 1, 2, \ldots$, and $m = 0, 2, 4, \ldots$. The SVL lifetime of the $(5^1;\ 1^14^1)$ level measured by Miller and Lee (1975) is ~ 13 nsec. The nearly pure resonance fluorescence emission spectrum from 5^1 or 1^14^1 obtained at 0.1 torr gradually degenerates to the thermalized equilibrium spectrum consisting of both the 5^1 and the 1^14^1 emission lines upon addition of 0.5–100 torr of N_2. One of the most interesting aspects of this observation is that some rovibronic levels of the two nearly degenerate vibrational levels (with the zeroth-level energy difference of only $2.7\ \text{cm}^{-1}$) do not efficiently redistribute their vibrational energy within the time scale of their electronic relaxation lifetime (~ 13 nsec), despite the fact that there is evidence of a localized weak vibration–rotation Coriolis coupling (Sethuraman *et al.*, 1970) and the fact that these levels have a moderate amount of vibrational energy, $E_{\text{vib}} \simeq 2970\ \text{cm}^{-1}$. Therefore, a long-range, low-energy collision can have a significant effect on the radiative and radiationless transitions by mixing these isolated rovibronic levels. Some excellent examples in the cases of benzene $(^1B_{2u})$ and glyoxal (^1A_u) studied by Parmenter and co-workers (Parmenter, 1972a, b; Parmenter and Rordorf, 1978; Parmenter and Tang, 1978; Rordorf *et al.*, 1978) illustrate the point.

5. Photodecomposition Processes

Photolyses of formaldehyde carried out at wavelength longer than 3000 Å and above 1 torr pressure, particularly those in early studies (McQuigg and Calvert, 1969, and references therein), were clearly dominated by collisional processes. Before discussing the excitation wavelength dependence of the branching ratio of the radical-to-molecular processes, $k_{\text{I}}/k_{\text{II}}$, shown by Eqs. (34) and (35), it is instructive to review briefly the recent study of time dependence and product energy partitioning for the molecular decomposition process [Eq. (35)] by Houston and Moore (1976). They obtained the time evolution and energy distribution of the CO photochemical product after laser photolysis pulses. The rate of appearance, relative yields, and the vibrational state distributions of the photolytic CO were measured using either absorption of a probe, cw CO laser tuned to a specific transition or through the detection of infrared emission from the photolytic CO molecules. A nitrogen laser output at 3371 Å was used to pump formaldehyde lines near the $2_0^1 4_0^1$ band, a doubled ruby laser output at 3472 Å was used to pump the 4_0^3 band directly, and tunable doubled dye laser output was used to pump formaldehyde bands in the 294–317 nm region.

The relevant results can be summarized as follows: (1) In the limit of low formaldehyde pressure, the photolytic CO appears at a rate more than 100 times slower than the decay rate of S_1 formaldehyde, indicating the presence of a long-lived intermediate state between S_1 and the molecular products. (2) Collision-induced CO production proceeds after the 3371 Å photolysis pulse at an appearance rate of 4.7×10^{-11} cm^3 $molecule^{-1}$ sec^{-1}, indicating a high collisional efficiency, $\sim\frac{1}{4}$ of gas kinetic. (3) The photolysis yield decreases with the addition of Ar, but increases dramatically upon the addition of NO or O_2 for 3055 Å photolysis, indicating interruption of the kinetics of CO formation. (4) Although the amount of energy appearing in the vibrational modes of CO increases with increasing photoexcitation energy, the CO vibrational energy accounts for only 0.7–4.5% of the total energy available to the products at the photolytic wavelengths.

The two most striking suggestions concerning the above are (1) that there is a long-lived intermediate corresponding to a vibrationally excited $S_0^\dagger$ with a geometry such as $H\ddot{C}OH$, and (2) that a collision is required to form CO and H_2. In regard to the first suggestion, there have been two recent theoretical calculations on the structure and energetics of the photoisomer, $H\ddot{C}OH$, hydroxymethylene. (Altmann *et al.*, 1977; Lucchese and Schaefer, 1978.) Furthermore, a recent study of low temperature matrix photolysis of formaldehyde near 3300 Å indicates that formaldehyde is photoisomerized to hydroxymethylene before undergoing further bimolecular reactions (Sodeau and Lee, 1978). The formaldehyde/Ar matrix photolysis gives glycoaldehyde ($CHO{-}CH_2OH$), CH_3OH + CO, and no product with increasing dilution, and the formaldehyde/CO matrix photolysis gives $HOCH{=}C{=}O$. However, no direct observation of $H\ddot{C}OH$ has been made to date, and further experimental evidence is necessary to prove the overall mechanistic importance of this intriguing intermediate in the gas phase photolyses. Still, the nature of the collision-induced formation of CO and H_2 found by Houston and Moore (1976) is a mystery.

If a long-lived intermediate like $H\ddot{C}OH$ has an important photochemical role in the gas phase, and furthermore if a collision-induced decomposition is required, the dynamics of the photodecomposition processes need further theoretical study beyond what Morokuma *et al.* (Hayes and Morokuma,

1972; Jaffe and Morokuma, 1976) and Heller *et al.* (1978) have done. The conventional models used in the calculation of nonradiative rates (Yeung and Moore, 1974; Heller *et al.*, 1978) may not be adequate in view of the photoisomerization process (PI),

$$H_2CO(S_1) \xrightarrow{k_{PI}} H\ddot{C}OH. \tag{40}$$

It is becoming increasingly clear that only a very limited amount of information about primary photoprocesses can be obtained from conventional measurements of final product quantum yields in formaldehyde photolyses. Nonetheless, these measurements are necessary and useful to applications in atmospheric chemistry (Calvert *et al.*, 1972; Nicolet, 1975) and laser isotope separation (Moore, 1973; Marling, 1977). For example, the photochemical threshold and quantum yield for the radical process (I) has been recently redetermined. Several groups of research workers (Lewis *et al.*, 1976; Becker *et al.*, 1977; Marling, 1977; Clark *et al.*, 978; Horowitz and Calvert, 1978; Moortgat *et al.*, 1978) have established the radial decomposition threshold to be at 3300–3250 Å, corresponding to 85 ± 2 kcal/mole, as obtained earlier by Walsh and Benson (1966). The quantum yield for the radical decomposition Φ_I has also been measured recently as a function of excitation wavelength by several groups of research workers (Marling, 1977; Lewis and Lee, 1978; Clark *et al.*, 1978; Horowitz and Calvert, 1978; Moortgat *et al.*, 1978), and the quantum yield gradually levels off at a value of ~0.8 near 3000 Å. There is minor structure in the wavelength dependent values of Φ_I, but no definite explanation can be offered about its significance.

Tang *et al.* (1977) have tried to measure relative values of Φ_I for a number of rovibronic levels of the $2^3 4^1$ level of $H_2CO(\tilde{A})$ by photoexcited chemiluminescence excitation spectroscopy, using the following reactions:

$$H + NO + M \longrightarrow HNO^*(\tilde{A}\,^1A'') + M, \tag{41}$$

$$H + NO + M \longrightarrow HNO(\tilde{X}\,^1A') + M, \tag{42}$$

$$HNO^*(\tilde{A}\,^1A'') \longrightarrow HNO(\tilde{X}\,^1A') + h\nu(\sim 760\text{ nm}). \tag{43}$$

The results shown in Fig. 12 indicate that the values of I(chem. lum.)/I(abs) are independent of the rovibronic levels pumped. These results could mean that the values of Φ_I are the same from all rovibronic levels pumped, if there was no collision induced rotational relaxation before photodissociation. Under the experimental conditions used, significant rotational relaxation could have taken place if its cross section were as high as 240 Å^2, which was observed for glyoxal (S_1) by Rordorf *et al.* (1978) or if, as previously discussed, the pressure effect on τ_F observed by Weisshaar *et al.* (1978) applies here. Therefore, the above results cannot be unambiguously interpreted at the present time.

Direct time-resolved observation of HCO formed in the radical decomposition channel has been made with intracavity dye laser spectroscopy by Atkinson *et al.* (1973) and Clark *et al.* (1978). In the latter study, rate constants for the reactions of HCO with NO and O_2 have been determined.

The important questions that remain to be answered concerning the photochemistry of formaldehyde are as follows:

(1) What is the precise nature of the collision-induced electronic relaxation at low pressure?

(2) What are the precise roles of rotation in the collision-free decay as well as in the collision-induced decay?

(3) How important is the role of intramolecular vibration–rotation coupling in electronic relaxation?

(4) How important is hydroxymethylene as an intermediate in the gas phase?

(5) What fraction of the primary, radical fragments emerge from the T_1 surface at high excitation energy?

(6) How do the collisional processes affect the production formation?

D. Chloroacetylene and Bromoacetylene

Rice (1975) and Gelbart (1977) provide a fairly comprehensive summary of the dynamics of photophysical and photodissociative processes of cloroacetylene and bromoacetylene studied by Evans and Rice (1972) and Evans *et al.* (1973). For brevity we shall provide a brief summary of the most important conclusions from this work.

1. Electronic Spectra

The electronic spectra of H—C≡C—Cl and H—C≡C—Br have been studied recently by Thomson and Warsop (1969, 1970) and subsequently by Evans *et al.* (1973). The longest-wavelength region of absorption of chloroacetylene consists of a long series of diffuse bands commencing at ~2500 Å and reaching a Franck–Condon maximum near 1950 Å. This transition was analyzed by analogy to the well known 2500 Å system of acetylene which Ingold and King (1953) studied earlier. This transition in acetylene takes place from the linear ground state to a *trans*-planar (bent) $\pi\pi^*$ upper state. Although the vibrational assignments by Evans *et al.* (1973) differ to some extent from those of Thomson and Warsop (1969), we shall use the assignments by Evans *et al.* for this review, for the sake of consistency, since they have studied the radiative and nonradiative processes of selected vibrational levels of both haloacetylenes. The longest wavelength region of absorption of bromoacetylene consists of a series of weak diffuse bands superimposed on a continuum commencing near 3300 Å. Evans *et al.* (1973) tentatively assigned its electronic origin to lie at 29,760 cm^{-1} and its major vibrational progression to be the result of excitation of the ν_5' mode. The lack of well resolved vibrational and rotational structure in both of these haloacetylenes certainly introduces some ambiguity to the assignments of the SVLs for which electronic relaxation was studied, but the fluorescence yield and decay time variations observed merit further discussion.

2. Fluorescence

Evans *et al.* (1972, 1973) measured fluorescence quantum yields Φ_F and decay times τ_F of 17 low-lying vibronic levels of H—C≡C—Cl and nine vibronic levels in H—C≡C—Br at 0.7 and 1.0 torr, respectively, using a single-photon time-correlation fluorescence lifetime apparatus with a 2 Å bandpass. Since the maximum values of Φ_F observed were not large, 0.37 for H—C≡C—Cl and 0.095 for H—C≡C—Br, the nonradiative decay rate contributes much more than the radiative decay rate to the magnitude of the observed lifetime. From the observed values of τ_F and Φ_F, the values of τ_r and τ_{nr} were obtained using Eqs. (32) and (33). Therefore, the experimental uncertainty associated with the measurement of Φ_F would appear in the radiative lifetime τ_r value in direct proportion but to a lesser extent in the nonradiative lifetime τ_{nr}. The values of τ_{nr} as a function of excess energy in the S_1 state are shown in Fig. 13. Note that the CCCl bending mode (ν_5') forms a long progression because the equilibrium geometry changes from linear in the ground state to bent in the upper electronic state. One quantum excitation of the HCC bending mode (ν_4'), 4_0^1, appears as combination bands with the 5_0^n progression. Only one of the nine vibronic bands studied in H—C≡C—Br was assigned.

The most significant observation for H—C≡C—Cl is that τ_{nr} increases gradually from 11.4 nsec at $\Delta E = 639\ \text{cm}^{-1}$ to 60.5 nsec at $\Delta E = 3173\ \text{cm}^{-1}$, rather than decreases as usual with excess energy. In the case of H—C≡C—Br, τ_{nr} is nearly constant at ~ 16 nsecs from $\Delta E \simeq 7700\ \text{cm}^{-1}$ to $\Delta E \simeq 10{,}700\ \text{cm}^{-1}$. The relative magnitude of τ_{nr} for H—C≡C—Cl and H—C≡C—Br led Evans *et al.* (1973) to conclude that the nonradiative decay route involved internal conversion from the excited to the ground state singlet manifolds, followed by case II predissociation, instead of singlet–triplet intersystem crossing. The conclusions by Evans *et al.*, drawn from the above data, can be summarized as follows. The bond dissociation energy for the dissociation,

$$\text{HCCX} \longrightarrow \text{HCC}\cdot + \text{X}\cdot, \tag{44}$$

is not known for X = Cl or Br, but they adapted a value of 33,000 cm^{-1} for the C—Cl bond and 28,000 cm^{-1} for the C—Br bond, considerably below the origins of the vibronic levels excited. To break the C—X bond, the energy must be concentrated in the C—X stretching mode (ν_3), although the CCCl bending mode (ν_5) is pumped. In the ground electronic state, $\nu_3'' = 751\ \text{cm}^{-1}$ is much smaller than $\nu_1'' = 3353\ \text{cm}^{-1}$ (C—H stretch) and $\nu_2'' = 2153\ \text{cm}^{-1}$ (C≡C stretch). For this reason, and because of the smaller change in frequency between the excited and ground states for the ν_3 mode than for other modes, ν_3 was concluded to be

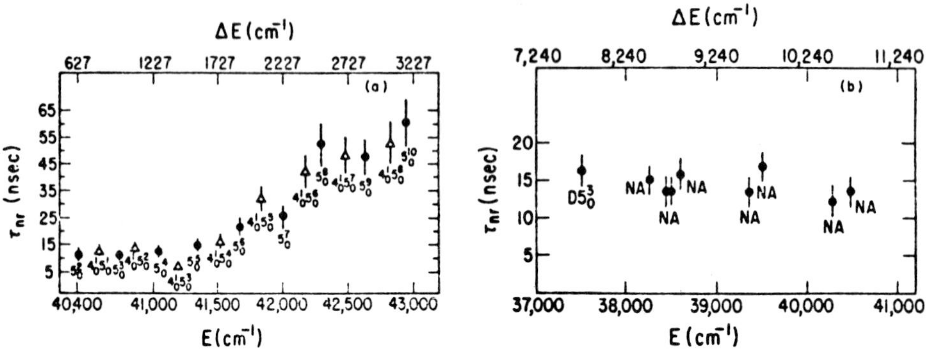

Fig. 13. Variation of nonradiative lifetimes (τ_{nr}) versus excess energy (Δ E) for (a) HC≡CCl (Δ, $4_0^1 5_0^n$; ●, 5_0^n) for ΔE, 0–0 chosen at 39,773 cm^{-1}, and (b) HC≡CBr, for ΔE, 0–0 chosen at 29,760 cm^{-1}; NA = not assigned (from Rice 1975, Fig. 41).

the poorest accepting mode in the internal conversion. Two possible bottlenecks were suggested to make photon emission competitive with decomposition: (1) The rate of internal conversion may be small, and hence rate limiting, or (2) ν_3 may not play a direct role in the internal conversion. Evans *et al.* favor explanation (2) and suggested further that coupling of the vibrational modes must be inefficient and thus not lead to randomization of the excitation energy, i.e., RRKM type behavior. They postulated that the observed increase of τ_{nr} with increasing energy of excitation in H—C≡C—Cl reflects constraints on chemical decomposition imposed by various conservation conditions. By conservation of energy, linear and angular momentum, H—C≡C—Cl is likely to decompose nearly along the line defined by the molecular axis of the linear ground state. Since the probability of finding this molecule instantaneously in a quasilinear configuration decreases with pumping of the higher ν_5 levels, the rate of fragmentation was suggested to decrease.

The energy dependence of τ_{nr} for H—C≡C—Br was suggested to have a different basis. Of the states investigated, those of H—C≡C—Br had much more vibrational excitation than do those of H—C≡C—Cl. The extra energy was suggested to be mostly in ν_4 (619 cm^{-1}). Since ν_4 is quasi degenerate with ν_3 (618 cm^{-1}), they suggested that slow energy transfer from ν_4 to ν_3 by weak Coriolis and two quantum interactions was responsible for the near constancy of τ_{nr} as well as small values of τ_{nr}. It should be noted that, contrary to the above conclusions of Evans *et al.* for H—C≡C—Cl, recent classical trajectory studies of H—C≡C—Cl dissociation, however, claim that energy distribution occurs rapidly, within 10^{-12} sec, and the initially prepared, nonrandom energy distribution is quickly destroyed (Hase and Feng, 1976; Sloan and Hase, 1977).

The dynamics of the photodissociation process discussed in the above studies were presented in the absence of experimental measurement of the quantum yields of radical formation, or a study of the energy distribution of radical fragments formed, as a function of excitation energy. Furthermore, a study of resonance fluorescence emission spectra from the single vibronic levels discussed above is warranted. Such experimental studies should be done, since Evans *et al.* (1972, 1973) regard H—C≡C—Cl as an excellent example of an intermediate case molecule in radiationless processes. This is because this system undergoes a change of geometry upon excitation, it has an intermediate density of states ($\simeq 10^4$–$10^5/\mathrm{cm}^{-1}$) at the energy of the prepared electronic state, and it has a photodissociation channel. The geometry change that occurs between the S_1 and the S_0 states provides an opportunity to study the role of angular momentum conservation in radiationless processes.

E. Ammonia

The predissociation of NH_3 at 2168 Å has been often discussed as a good example of case I electronic predissociation (see Herzberg, 1966; Gelbart, 1977). There have been some recent studies of ammonia photolysis (Koda and Back, 1977; Back and Koda, 1977) because of the current interest in laser induced chemistry and its potential application in laser isotope separation (Moore, 1973; Letokhov, 1977).

1. Electronic Spectrum

The first electronic absorption band of ammonia extends from ~2170 to ~1700 Å (Herzberg, 1966; Calvert and Pitts, 1966). Although this $\tilde{A}\,{}^1A_2'' \leftarrow \tilde{X}\,{}^1A_1$ transition shows well resolved vibronic band structure with a long progression of the out-of-plane bending mode ν_2' of the upper electronic state, resolved rotational fine structure can be seen only from $v_2' = 0$ and 1 levels of $ND_3(\tilde{A})$. A complete vibrational analysis of this system in NH_3 was carried out by Walsh and Warsop (1961). A detailed rotational analysis of the corresponding band in ND_3 was carried out by Douglas (1963) to confirm that the upper electronic state ($\tilde{A}$) is planar and of ${}^1A_2''$ symmetry. Douglas (1963) estimated the rotational line widths of the 0000 and the 0100 levels (v_1', v_2', v_3', v_4') of $ND_3(\tilde{A})$ to be 2.5 and 0.8 cm^{-1}, respectively, corresponding to excited state lifetimes of 2 and 7 psec. Presumably, higher vibrational levels have even shorter lifetimes due to strong predissociation.

2. Emission

A preliminary study of fluorescence emission from ND_3 excited by 2139 Å (Zn I line) and 2144 Å (Cd II lines) was reported by Koda *et al.* (1974). More detailed study (Hackett *et al.*, 1976) has shown that this emission can be explained as an example of resonance scattering midway between Raman and fluorescence scattering in character, in accord with recent theories of resonance scattering. Hackett *et al.* assigned a number of rotationally resolved resonance scattering lines for the vibronic transitions originating from the two ND_3 SVLs populated by the 0000 ← 00^+00 and 0100 ← 01^-00 band of the $\tilde{A} \leftarrow \tilde{X}$ system. Furthermore, they obtained a value of $(1.5 \pm 1) \times 10^{-5}$ for the emission quantum yield at 2139 Å for ND_3, compared to the value of $(8.3 \pm 0.5) \times 10^{-5}$ obtained by Lipsky and Gregory (as quoted by Hackett *et al.*, 1976). They caution that the largest source of error lies in the self-absorption of the first emission bands of ND_3. They argue that their value of emission quantum yield is consistent with the nonradiative rate constant of 5×10^{11} sec^{-1} based on the linewidth measurement and the theoretically estimated radiative rate constant of 1×10^7 sec^{-1} based on the absorption and emission spectra.

3. Photodecomposition Processes

The most interesting information regarding the mechanism of ammonia predissociation is provided in the photodecomposition study of NH_3, NH_2D, NHD_2, and ND_3 (Koda and Back, 1977; Back and Koda, 1977). NH_3 may decompose through two distinct product channels, a radical process requiring 108 kcal/mole and a molecular elimination requiring 94 kcal/mole:

$$NH_3 + h\nu \longrightarrow H + NH_2, \qquad (45)$$

$$NH_3 + h\nu \longrightarrow H_2 + NH. \qquad (46)$$

Koda and Back (1977) found that radical decomposition was the main process. For the ND_3 photolysis, the quantum yield of the molecular dissociation process has been found to be small, in a range from ≤ 0.003 at 2139 Å to ≤ 0.009 at 1849 Å (Back and Koda, 1977). The dissociation quantum yield for all four isotopic molecules was presumed to be unity for 2144, 2139, 2062, and 1849 Å photolyses. Analysis of the H_2 and HD products permitted evaluation of intramolecular competition for the radical decomposition process giving either H or D in the photodissociation of NH_2D and NHD_2. At the shortest wavelengths dissociation giving H was favored by a factor of 2 to 3, while this isotope effect was much larger at 2144 and 2139 Å.

When Back and Koda (1977) studied the photolyses of NH_3 and ND_3 at 2139, 2062, and 1850 Å in the presence of propane and ethylene, they found that more translationally hot H and D atoms were formed at longer photolysis wavelengths. In order to explain this result, they proposed a second, radical dissociation channel of greater endoergicity, producing electronically excited $NH_2(\tilde{A}\,^2A_1)$ or $ND_2(A\,^2A_1)$,

$$NH_3 + h\nu \longrightarrow H(^2S) + NH_2(\tilde{A}\,^2A_1). \qquad (47)$$

For case I electronic predissociation, previously, $NH_3(\tilde{A}\,^1A_2'')$ had been correlated to the product states of $NH_2(\tilde{X}\,^2B_1) + H(^2S)$ and the ground state of $NH_3(\tilde{X}\,^1A_1)$ to $NH_2(\tilde{A}\,^2A_1) + H(^2S)$; assuming photodissociation to occur in the plane of the molecule (Herzberg, 1966, p. 465; Douglas, 1963). Back and Koda (1977), however, proposed a nonplanar dissociation for $ND_3(\tilde{A}\,^1A_2'')$, and

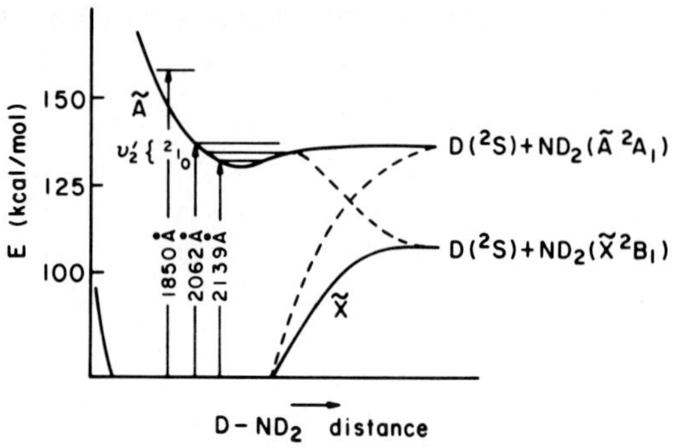

Fig. 14. Nonplanar dissociation model for the predissociation of the Ã $^1A_2''$ state of ND_3, proposed by Back and Koda (1977).

their model is shown in Fig. 14. At low excitation energy, the excited molecule cannot energetically dissociate to give the electronically excited $ND_2(\tilde{A}\,^2A_1)$ and hence is forced to curve cross and produce the ground state of $ND_2(\tilde{X}\,^2B_1)$ with a high degree of translational excitation of the fragments. At higher excitation energy, the excited molecule fragments by case II predissociation, making less energy available for translational excitation. A time evolution study of unimolecular processes in ammonia by photofragment spectroscopic techniques recently developed by Wilson and co-workers (Riley *et al.*, 1974, and references therein), Bersohn and co-workers (Kawasaki *et al.*, 1975, and references therein), and Simons and co-workers (Simons and Tasker, 1973, and references therein) should be able to reveal some interesting dynamical details.

4. Two-Step IR–UV Photodissociation

In order to study isotope-selective photodissociation, $^{14}NH_3$ and $^{15}NH_3$ were employed (Ambartzumian *et al.*, 1974, and references therein). They pumped NH_3 to isotopically selected rotational levels of the $v_2'' = 1$ level from the $v_2'' = 0$ level of the ground electronic state using tuned 9–10 μm CO_2 laser lines, followed by selective pumping to vibrational levels of the upper electronic state from the $v_2'' = 1$ level using a filtered UV pulse. The gaseous NH_3 filter takes out the UV frequencies which would otherwise excite the NH_3 molecules in the $v'' = 0$ level. A complex sequence of radical reactions gives molecular N_2 as a final product, and the isotopic enrichment coefficient $K(^{15}N/^{14}N)$ was found to vary between 2.5 and 6, depending upon experimental conditions.

The multiphoton dissociation of NH_3 by pulsed, high-power IR radiation from a CO_2 laser produces the ground state of $NH_2(\tilde{X}\,^2B_1)$ fragments. Welge and co-workers (Campbell *et al.*, 1976) have studied the laser-induced fluorescence for detecting this fragment under essentially collision free conditions. The dissociation is understood to occur from the high vibrational levels of the ground electronic state.

IV. EXPERIMENTAL STUDIES ON SIX-ATOM AND LARGER MOLECULES

In this section we review experimental studies of electronic relaxation of the simple *trans*-dicarbonyl systems glyoxal [$(CHO)_2$], methylglyoxal [CH_3COCHO], and biacetyl [$(CH_3CO)_2$], as well as the simplest conjugated acetylenic-carbonyl system propynal [$C_2H \cdot CHO$]. The three α-dicarbonyls represent a homologous series of molecules that exhibit similar electronic spectral properties but possess 12, 21, and 30 vibrational degrees of freedom, respectively. The latter difference gives rise to marked differences in the density of vibrational states in the triplet (T_1) manifold isoenergetic to the initially prepared vibrational level(s) of the S_1 state. For example, at an excess T_1 vibrational energy of 3000 cm^{-1} (the S_1–T_1 energy gap for these molecules is less than 3000 cm^{-1}) the calculated density of vibrational states for glyoxal, methylglyoxal, and biacetyl are 2.0, 180, and 8810 states/cm^{-1}, respectively (Coveleskie and Yardley, 1975b). As a result, much comparative study has been conducted on these molecules to determine the effect of triplet state vibrational density on the dynamics of their $S_1 \rightsquigarrow T_1$ ISC processes, both under collision-free conditions and in the presence of collisional perturbation. The α-dicarbonyls and propynal are particularly well suited for experimental studies of this type since they exhibit both S_1 and T_1 vapor phase emission in separate spectral regions. Thus both the initial (S_1) and final (T_1) states involved in the ISC process can be directly monitored through their emission.

The six-atom molecules glyoxal and propynal exhibit well-resolved SVL structure in their first excited singlet electronic states. They have thus been useful molecular systems for investigating the relaxation dynamics of SVL of electronically excited states. Experimental studies on glyoxal and propynal are reviewed in Sections A and B, respectively. Studies on the 12-atom molecule biacetyl are reviewed in Section C. Studies on the nine-atom molecule methylglyoxal, which is structurally intermediate to glyoxal and biacetyl, are reviewed in Section D.

A. Glyoxal

1. Electronic Spectrum

The longest-wavelength region of electronic absorption for glyoxal, $(CHO)_2$, in the gas phase, consists of a series of weak but sharp bands extending from about 5400 to 3700 Å. The absorption in this region was shown by Brand (1954) to actually correspond to two band systems, an extremely weak $\tilde{a}\,{}^3A_u \leftarrow \tilde{X}\,{}^1A_g(\pi^* \leftarrow n)$ system with a 5208 Å (19,196 cm^{-1}) 0–0 band, and a relatively much stronger $\tilde{A}\,{}^1A_u \leftarrow \tilde{X}\,{}^1A_g(\pi^* \leftarrow n)$ system with an absorption onset near 4800 Å and a 4549 Å (21,973 cm^{-1}) 0–0 band. Glyoxal has been shown to exist in nearly the same planar *trans*-configuration (C_{2h}) in both its ground $\tilde{X}\,{}^1A_g$

TABLE 10

Vibrational Frequencies of Glyoxal in Various Electronic States (cm^{-1})[a]

Species	Vibration	Ground state ($\tilde{X}\,^1A_g$)	Excited state Singlet ($\tilde{A}\,^1A_u$)	Excited state Triplet ($\tilde{a}\,^3A_u$)
a_g	ν_1 C—H stretching	2843	2809	—
	ν_2 C=O stretching	1745	1391	1459
	ν_3 C—H rocking	1338	—	1415
	ν_4 C—C stretching	1065	955	963
	ν_5 C—C=O bending	551	509	503
a_u	ν_6 C—H wagging	801	—	—
	ν_7 torsional	127	233	234
b_g	ν_8 C—H wagging	1048	735	—
b_u	ν_9 C—H stretching	2835	—	—
	ν_{10} C=O stretching	1732	—	—
	ν_{11} C—H rocking	1312	—	—
	ν_{12} C—C—O bending	339	380	392

[a] Holzer and Ramsay (1970) and references therein.

and excited $\tilde{A}\,^1A_u$ electronic states (Herzberg, 1966). The $\tilde{a}\,^3A_u \leftarrow \tilde{X}\,^1A_g$ and $\tilde{A}\,^1A_u \leftarrow \tilde{X}\,^1A_g$ glyoxal band systems have been extensively studied in both absorption and emission and are well characterized spectroscopically. Available data on the frequencies of the 12 fundamental modes of vibration of glyoxal in its $\tilde{X}$, $\tilde{A}$, and $\tilde{a}$ electronic states are listed in Table 10.

The glyoxal $\tilde{A}\,^1A_u$–$\tilde{X}\,^1A_g$ transition is relatively weak with an oscillator strength of $f \sim 4 \times 10^{-5}$ (McMurry, 1941), even though it is symmetry allowed. For glyoxal with C_{2h} symmetry, B_u–A_g transitions are also symmetry allowed. Strong vibronic interactions are thus observed in glyoxal for the C–H wagging mode ν_8 since this mode has b_g symmetry and $A_u^e \times b_g^v = B_u^{ev}$. As a result the 8_1^0 band in emission and the 8_0^1 band in absorption are very strong transitions (Herzberg, 1966).

2. Fluorescence Lifetimes

Yardley *et al.* (1971) reported the first observations of time-resolved electronic fluorescence from the $\tilde{A}\,^1A_u$ state of glyoxal. They used pulsed, tunable, dye laser output to directly measure the lifetime of the $\tilde{A}\,^1A_u$ state at low pressures with excitation centered at the 4549 Å 0–0 band. Under these conditions, a collision-free lifetime of 2.16 ± 0.05 μsec was obtained. In addition, no dramatic variations of the glyoxal collision-free lifetime were observed at low resolution when the excitation was varied over the wavelength range 4510–4585 Å (22,173–21,812 cm^{-1}). Besides the 0–0 band, the glyoxal $8_1^1 7_2^2$, 4_1^1, and 7_1^1 $\tilde{A}$–$\tilde{X}$ bands lie in this wavelength range.

The study of the fluorescence lifetimes for glyoxal over much of its $\tilde{A}-\tilde{X}$ absorption band were determined by van der Werf *et al.* (1975) at essentially collision-free pressures with an excitation bandwidth of about 3 Å. They were able to distinguish three general excitation regions, each of which gave rise to different decay behavior.

At low excitation energy (up to 4200 Å), the only emission observed had a lifetime on the microsecond time scale. In the excitation region from 4145 up to 3925 Å they observed, in addition to the long-lived component, an emission component with a lifetime of a few nanoseconds. At high excitation energy, above 3925 Å, the only emission observed had a lifetime of a few nanoseconds. The emission spectra of both decay components were observed to be identical.

These results were interpreted by van der Werf *et al.* (1975) as indicating that, because of an increase in the density of states in the $\tilde{a}\ ^3A_u$ manifold, the extent of vibronic singlet-triplet mixing in excited glyoxal becomes increasingly important as a function of excess vibrational energy. At low excitation energy, the amount of vibronic coupling was believed to be negligible. Under these conditions, the molecule was thought to exhibit behavior in the resonant (pure small molecule) limit, such that the decay rate was determined only by zeroth-order singlet relaxation (probably internal conversion to $\tilde{X}$ or radiative decay only). At intermediate excitation energies, the decay was thought to be determined by both zeroth-order singlet and zeroth-order triplet relaxation. The short fluorescence component was concluded to be due to constructive interference between the molecular eigenstates. At high excitation energy, the singlet–triplet mixing was thought to be very strong and the molecular eigenstates were assumed to be almost pure triplet states. Since the decay of the zeroth-order triplet states was believed to become very large at these energies (probably due to intersystem crossing to $\tilde{X}$ and/or photochemistry), the relaxation was classified as being of almost statistical limit behavior.

Because previous studies had provided information on the photophysical behavior of only a limited number of individual vibronic levels in the glyoxal $\tilde{A}\ ^1A_u$ state, Lineberger and co-workers carried out extensive measurements of the collision-free lifetimes and the glyoxal collisional self-quenching rate constants for 26 single vibronic levels (SVL) of glyoxal-h_2 ($\tilde{A}\ ^1A_u$) (Beyer *et al.*, 1975). The excitation wavelengths used covered the range 4552–4145 Å. Recently they have obtained similar data for 8 and 16 SVLs of glyoxal-hd ($\tilde{A}\ ^1A''$) and glyoxal-d_2 ($\tilde{A}\ ^1A_u$), respectively (Zittel and Lineberger, 1977). The excitation wavelength ranges used for glyoxal-hd and glyoxal-d_2 were 4547–4278 and 4543–4183 Å, respectively. In each of these studies, an N_2 laser pumped tunable dye laser was used as a narrow-band (0.5 Å) excitation source of the glyoxal vapors. SVL emission was resolved by a spectrometer.

The collision-free lifetimes, τ_F^0, and the collisional self-quenching rate constants for these molecules were typically determined over a pressure range of 2–25 mtorr. In these studies, it was found that an emission bandpass large enough to observe the entire vibronic band was desirable to minimize the effect on the observed fluorescence lifetimes of rapid rotational redistribution. This observation was thought to provide evidence of a very rapid collisional rotational relaxation ($\sigma > 100$ Å^2) in the $\tilde{A}\,^1A_u$ state of glyoxal.

The collision-free decay rates, $k_F^0 = (\tau_F^0)^{-1}$, for the fundamental vibrational levels of the three isotopic species of glyoxal are plotted in Fig. 15 as a function of their excess vibrational energy E_{vib}. Fluorescence quantum yield data were unavailable for quantitative separation of the radiative k_r and nonradiative k_{nr} contributions to the decay rates k_F^0. The dependence of k_F^0 on isotopic substitution and on E_{vib}, however, led Zittel and Lineberger (1975) to some conclusions about the relative contributions of k_r and k_{nr} to k_F^0 for the different isotope species. They concluded that the collision-free loss rates of glyoxal-h_2 and glyoxal-hd are primarily nonradiative, while that for glyoxal-d_2 is primarily radiative. These conclusions were based in part on the following considerations. (1) The estimated radiative decay rates based on the integrated absorption coefficients for glyoxal-h_2, -hd, and -d_2 were determined to be nearly identical,

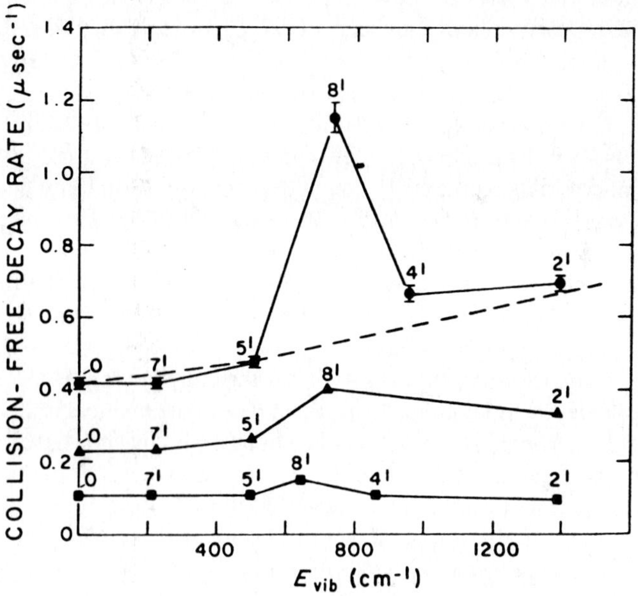

Fig. 15. Comparison of collision-free decay rates for SVLs $\tilde{A}\,^1A_u$ glyoxal-h_2 (●), $\tilde{A}\,^1A''$ glyoxal-hd (▲), and $\tilde{A}\,^1A_u$ glyoxal-d_2 (■). The relative density of $\tilde{X}\,^1A_g$ ($\tilde{X}\,^1A'$) vibrational levels (---) was calculated using Haarhoff's method (Haarhoff, 1963). The dependence of $\tilde{X}\,^1A_g$ ($\tilde{X}\,^1A'$) relative density was nearly identical for all three isotopes over this energy range; thus, the dashed line represents the relative density for all isotopes. Calculated $\tilde{A}\,^1A$ ($\tilde{A}\,^1A'$) densities at $E_{vib} = 0$ were $4.0 \times 10^5/\text{cm}^{-1}$, $7.6 \times 10^5/\text{cm}^{-1}$, and $15.0 \times 10^5/\text{cm}^{-1}$ for glyoxal-h_2, -hd, and -d_2, respectively (from Zittel and Lineberger, 1977, Fig. 7).

but yet the collision-free decay rates for the SVL of the $\tilde{A}$ states of glyoxal-h_2, and -hd are, respectively, nearly fourfold and twofold greater than for the corresponding levels of glyoxal-d_2. The collision-free decay rates for all the SVL of the glyoxal-d_2 $\tilde{A}$ state measured (with the exception of 8^1 and 8^1X^b) are nearly identical ($\sim 0.10\ \mu sec^{-1}$), and are close to the estimated radiative decay rate of 0.05–0.10 μsec^{-1} obtained previously by Anderson *et al.* (1973) for glyoxal-h_2. (2) The collision-free decay rates for SVLs which do not have ν_8 excited show an increase with increasing E_{vib} for glyoxal-h_2 and glyoxal-hd but not for glyoxal-d_2.

An enhancement was observed in the k_F^0 for the 8^1 and 8^1X^b SVL of each isotopic species. As illustrated in Fig. 15, for the 8^1 level, this effect is much greater for glyoxal-h_2 than for glyoxal-d_2. Since the decay rate in glyoxal-h_2 from the vibrationless level was assumed to be primarily nonradiative, the increases in the k_F^0 for its 8^1 and 8^1X^b SVL were proposed also to be essentially nonradiative in nature. In the case of glyoxal-d_2, the increase was assumed to be essentially radiative in nature.

Based upon the assumption that the radiative decay rates of the $\tilde{A}$ state vibrationless level for all three isotopic species are equal, and in the range 0.05–0.10 μsec^{-1}, Zittel and Lineberger (1977) estimated the limits for the nonradiative contributions to k_F^0 for the vibrationless level of the three isotopes: $k_{nr}(h_2) \simeq 0.76\text{–}0.88 k_F^0(h_2)$, $k_{nr}(hd) \simeq 0.56\text{–}0.78\ k_F^0(hd)$, and $k_{nr}(d_2) \leq 0.50 k_F^0(d_2)$. The ratios of the nonradiative decay rates for the vibrationless levels were also estimated as $k_{nr}(h_2)/k_{nr}(hd) = 1.8\text{–}2.9$, and $k_{nr}(h_2)/k_{nr}(d_2) \geq 6.0$.

Zittel and Lineberger (1977) concluded that the dominant nonradiative process taking place from the vibrationless level of the $\tilde{A}$ state for each of the different isotopic species of glyoxal was internal conversion (IC) to the ground state. This conclusion was based on two considerations. First, they argued that the large-isotope effect observed for the glyoxal $\tilde{A}$ state vibrationless level, $k_{nr}(h_2)/k_{nr}(d_2) \geq 6.0$, would seem characteristic of IC processes, involving large energy gaps. Second, they showed that the collision-free k_F^0 for glyoxal-h_2, which was believed to be largely nonradiative in nature, increased slowly with increasing E_{vib}, roughly paralleling the increase in the estimated density of $\tilde{X}\,^1A_g$ vibrational levels, as shown in Fig. 15. The level densities of the glyoxal-d_2 and glyoxal-hd ground states were suggested to have the same energy level dependence as the level densities in the ground state of glyoxal-h_2. The smaller dependence of collision-free k_F^0 on vibrational energy for the a_g symmetric vibrational levels (5^1, 4^1, and 2^1) in glyoxal-hd and glyoxal-d_2 were attributed to the greater relative radiative contribution to the decay rates from the $\tilde{A}$ states of these molecules. The radiative decay rates from these molecules were thought not to be strongly dependent on the vibrational state of these a_g symmetric vibrations.

Even though the nonradiative decay rates of glyoxal-h_2 showed a general increase with increasing density of ^1A_g vibrational states, considerable dependence was observed on the specific vibrational level excited. The k_{nr} for the 8^1 and 8^1X^b-combination levels were observed to be anomalously large, while the 7^b levels exhibited little dependence on excess vibrational energy. Zittel and Lineberger noted that these observations were consistent with the theoretical model of Heller and Freed (1972), Heller *et al.* (1972) and Prais *et al.* (1974) based on the possible role of the optically excited mode acting as an accepting mode in a nonradiative process. According to this model, the Franck–Condon factors for the optically excited vibration can have a strong influence on k_{nr}. When the frequency of such an optically excited vibrational mode in the initial electronic state increases in going to the final electronic state, k_{nr} is predicted to increase with increasing E_{vib}, since this mode should become a progressively better accepting mode. Conversely, when the frequency of such an optically excited mode decreases from the initial to final electronic state, k_{nr} is predicted to decrease or remain constant with increasing vibrational energy. Since $\nu'_8(^1A_u)/\nu''_8(^1A_g) = 0.69$ and $\nu'_7(^1A_u)/\nu''_7(^1A_g) = 1.84$, the anomalous increase observed in k_{nr} for the ν_8 levels and the near constancy observed for the ν_7 levels is believed to be due at least partially to the efficiency of those modes in acting as accepting modes in the glyoxal-h_2 IC process.

Tramer and co-workers (Cossart-Magos *et al.*, 1978) have recently analyzed emission spectra from the 0^0, 8^1, and 2^1 SVLs of the glyoxal-h_2 and -d_2 $\tilde{A}\,^1A_u$ states and thermally equilibrated $\tilde{a}\,^3A_u$ states. They concluded, unlike Zittel and Lineberger (1977), that a large portion of the aforementioned enhancement of the collision-free decay rates of glyoxal-h_2 $\tilde{A}\,^1A_u$ 8^1 and 8^1X^b SVLs result from enhanced radiative contributions. This was attributed to a Herzberg–Teller mechanism in which the glyoxal-h_2 $v_8(b_g)$ vibrational mode facilitates $\tilde{A}\,^1A_u \leftarrow \tilde{X}^1A_g$ intensity borrowing from the higher lying, strongly allowed $^1B_u \leftarrow {}^1A_g$ ($\pi^* \leftarrow \pi$) transition(s). Tramer and co-workers also looked for but found no evidence of singlet or triplet emission from *cis*-glyoxal. They thus concluded that the efficiency of *trans–cis* isomerization is probably low enough to neglect this mechnism as a significant nonradiative deactivation pathway for glyoxal ^1A_u and ^3A_u states.

3. Collision-Induced Intersystem Crossing Following Excitation at 4358 Å

Two early studies of Parmenter and co-workers (Anderson *et al.*, 1971, 1973) have provided a basis for much of the understanding of the photophysics of glyoxal in the vapor phase. They utilized 4358 Å radiation from a low-pressure Hg arc to excite both the 4^1_0 and $8^1_0 7^2_2$ transitions in glyoxal to produce vibronic levels lying 955 and 1201 cm^{-1}, respectively, above the $\tilde{A}\,^1A_u$ zero-point level. At very low pressures of pure glyoxal ($\sim$0.01 torr), only $\tilde{A}\,^1A_u$ emission could be identified. Both $\tilde{A}\,^1A_u$ and $\tilde{a}\,^3A_u$ emissions were observed at glyoxal pressures above about 0.1 torr, and both were increasingly self-quenched with increasing glyoxal pressure. The self-quenching of the ^3A_u emission was suggested as possibly resulting from complex formation between triplet and ground state glyoxal molecules followed by photochemical product formation. With increasing pressures of the added gases cyclohexane, argon, and helium, on the other hand, glyoxal singlet emission was observed to be selectively quenched while triplet emission was enhanced slightly.

The effect of increasing pressures of foreign gases on the glyoxal ^1A_u and ^3A_u emissions was interpreted by Parmenter and co-workers as reflecting a collision induced $\tilde{A}\,^1A_u \rightsquigarrow \tilde{a}\,^3A_u$ intersystem crossing process, and (if present) collision insensitive $\tilde{A}\,^1A_u \rightsquigarrow \tilde{X}\,^1A_g$ internal conversion and $\tilde{a}\,^3A_u \rightsquigarrow \tilde{X}\,^1A_g$ intersystem crossing processes. These conclusions were suggested as being consistent with predictions of the theory of nonradiative transitions in the resonant and statistical limits, respectively.

In the resonant limit, an intramolecular nonradiative transition between two electronic states does not occur in the absence of collisional perturbation. This situation exists for a nonradiative transition between a vibronic level of an initial electronic state and a set of isoenergetic vibronic levels of sparse density in the final electronic state. The resonant limit is usually observed for small molecules and/or in systems with a small energy gap between the initial vibronic level and the zero-point level of the final electronic state.

In the statistical limit, the nonradiative transition between two electronic states takes place intramolecularly at a rate which is virtually insensitive to the pressure of added foreign gas. The statistical limit is observed when the density of the set of vibronic levels of the final electronic state isoenergetic to the initial vibronic level is large. This situation is usually observed for large molecules or for systems in which the aforementioned energy gap is large.

The sensitivity of the $\tilde{A}\,^1A_u \rightsquigarrow \tilde{a}\,^3A_u$ intersystem crossing process and (if they occur) the insensitivity of the $\tilde{A}\,^1A_u \rightsquigarrow \tilde{X}\,^1A_g$ and $\tilde{a}\,^3A_u \rightsquigarrow \tilde{X}\,^1A_g$ nonradiative processes to increasing collisional perturbations are thus in accord with the densities of vibrational levels in the final states estimated by Parmenter and co-workers for each of these processes. They estimated the density of vibronic levels in the $\tilde{a}^3A_u$ state lying approximately 4000 cm^{-1} above its zero-point level (and isoenergetic to the $\tilde{A}^1A_u4^1$ and 8^17^2 levels) to be less than 10 states/cm^{-1}. By contrast, the densities of vibrational levels in the $\tilde{X}^1A_g$ state terminating the possible nonradiative transitions $\tilde{A}^1A_u \rightsquigarrow \tilde{X}^1A_g$ and $\tilde{a}^3A_u \rightsquigarrow \tilde{X}^1A_g$ were estimated to be $\sim 2 \times 10^7$ and $\sim 3 \times 10^6$ levels/cm^{-1}, respectively, reflecting energy gaps of at least 21,973 and 19,196 cm^{-1} between the zero-point level of the $\tilde{X}\,^1A_g$ state and the initial vibronic levels for each of these processes.

The observation that the glyoxal $\tilde{A}^1A_u \rightsquigarrow \tilde{a}^3A_u$ intersystem crossing process exhibits behavior in the resonant limit of nonradiative transitions was found to be in striking contrast to the statistical limit behavior that had been observed previously for the corresponding $S_1 \rightsquigarrow T_1$ intersystem crossing process in biacetyl (Parmenter and Poland, 1969; Anderson and Parmenter,

1970). As will be discussed in Section C, the $S_1 \rightsquigarrow T_1$ nonradiative process in biacetyl is the dominant relaxation channel of isolated S_1 molecules, and its rate is unaffected by collisional perturbation. The replacement of the glyoxal hydrogen atoms by the methyl groups of biacetyl, $(CH_3CO)_2$, increases the number of vibrational degrees of freedom from 12 to 30. Parmenter and co-workers suggested that this effect apparently transforms the $S_1 \rightsquigarrow T_1$ nonradiative process from the resonant limit in glyoxal to the statistical limit in biacetyl.

4. Apparent Vibrational Relaxation of SVL

Parmenter and co-workers reported that vibrational relaxation in the glyoxal $\tilde{A}\,^1A_u$ state was so efficient that about a third of the intensity of even 3 mtorr fluorescence came from the $\tilde{A}\,^1A_u$ zero-point level upon excitation to the 4^1 and 8^17^2 levels. This was noteworthy, since the $\tilde{A}\,^1A_u$ radiative lifetime of about 10–20 μsecs calculated from integrated absorption spectra was shorter than the hard sphere mean collision interval of about 40 μsec at 3 mtorr. Fluorescence emission spectra with dominant contributions from the initially pumped 8^17^2 levels were observed only at pressures below about 10 mtorr. (Although the 4^1 level is excited by 4358 Å radiation, emission from it could not be identified.) It was further found that collisions leading to this vibrational relaxation had a strong selectivity for zero-point population. Fluorescence emission from intermediate vibrational levels could not be observed. Parmenter and co-workers suggested that the very efficient relaxation of vibrationally excited glyoxal $\tilde{A}\,^1A_u$ molecules and the selectivity for their zero-point population upon collision with ground state glyoxal molecules resulted from nearly resonant V–V energy transfer processes, or from a Förster dipole–dipole electronic energy transfer mechanism (Yardley, private communication to Parmenter). The electronic energy transfer mechanism in this case was assumed to be similar to one which has been used to account for the anomalously high efficiency observed for relaxation of the $v' = 1$ vibrational level in the $\tilde{A}\,^2\Sigma^+$ state of NO through collisions with $\tilde{X}^2\Pi$ ground state NO molecules (Melton and Klemperer, 1971; Gordon and Chiu, 1971).

The phenomenon of very efficient vibrational relaxation observed by Parmenter and co-workers for the 4^1 and 8^17^2 vibrational levels of the glyoxal $\tilde{A}\,^1A_u$ state at low pressures was judged by Frad and Tramer (1973) to be also general for its 8^1, 2^1, and 8^14^1 levels with E_{vib} of 735, 1391, and 1690 cm^{-1}, respectively. To further investigate the nature of the efficient vibrational relaxation within the glyoxal $\tilde{A}^1A_u$ state at low pressure, Frad and Tramer (1973) studied the fluorescence spectra of glyoxal-h_2 and glyoxal-d_2 isotopic mixtures upon selective SVL excitation of either species. By an analysis of the relative emission intensities from each species resulting from this selective excitation, they postulated that the Förster electronic energy transfer mechanism was more important than V–V energy transfer in determining the rate of vibrational relaxation within the glyoxal $\tilde{A}\,^1A_u$ state at low pressures.

Time-resolved single vibronic level (TRSVL) spectroscopic studies should provide further information on glyoxal $\tilde{A}\,^1A_u$ state vibrational relaxation processes. Such studies have been initiated by Photos and Atkinson (1975). Rettschnick, Ten Brink, and Langelaar (1978) have conducted TRSVL studies on the 8^1 SVL of the glyoxal $\tilde{A}\,^1A_u$ state. The aim of these studies was to determine the relaxation effect of a single collision between a glyoxal excited state and ground state molecule. Rettschnick and co-workers concluded that the nearly isoenergetic 6^1 level was produced with close to gas kinetic efficiency. However, the rates of other nearly resonant processes such as production of 5^17^1 levels were concluded to be less than 10% as efficient as the $8^1 \rightarrow 6^1$ conversion. Other dominant routes of 8^1 level vibrational relaxation were the processes: $8^1 \rightarrow 0$ ($k_{8-0} = 0.7 \times k_{8-6}$), $8^1 \rightarrow 8^17^1$ ($k_{8-8,7} = 0.7 \times k_{8-6}$), and $8^1 \rightarrow 7^1$ ($k_{8-7} = 0.3 \times k_{8-6}$). The addition of foreign gases was found to have little influence on the branching ratios of the different relaxation routes. This was taken by Rettschnick *et al.* to indicate that vibrational–translational energy transfer is a leading vibrational relaxation mechanism for the glyoxal $\tilde{A}\,^1A_u$ 8^1 level.

5. Collisional Self-Quenching of SVL

The previously mentioned self-quenching rate constants k_q measured by Lineberger and co-workers for 26, 16, and 8 SVLs of the glyoxal-h_2, glyoxal-d_2, and glyoxal-hd $\tilde{A}\,^1A_u$ ($\tilde{A}\,^1A''$) states ranged over excessive vibrational energies of approximately 2200, 2100, and 1400 cm^{-1}, respectively. A strikingly similar k_q dependence on E_{vib} was observed for each of the isotopic species. The rate constants for glyoxal-h_2 and glyoxal-d_2 were found to increase from slightly less than gas kinetic (assuming a 4 Å molecular diameter) up to greater than six times gas kinetic at ~2100 cm^{-1} of excess vibrational energy. This is illustrated for glyoxal-d_2 in Fig. 16. The self-quenching rate constants were thought to primarily reflect the effects of collisionally induced ISC to $\tilde{a}\,^3A_u$($\tilde{a}\,^3A''$) and vibrational relaxation within the $\tilde{A}\,^1A_u$($\tilde{A}\,^1A''$) manifold, since IC to the ground state was concluded to be in the statistical limit and unaffected by collisional perturbation.

As observed in Fig. 16, the glyoxal-d_2 $\tilde{A}\,^1A_u$ rate constants show a large difference between 0 and 7^1, a small difference between 7^1 and 7^2, and nearly equal values for 7^2 and 7^3. These rate constants were observed to be within experimental error of the rate constants for the corresponding levels in glyoxal-h_2 and glyoxal-hd. The large increase in k_q for the 7^b torsional levels over k_q for the vibrationless level was believed to be due primarily to vibrational relaxation. Zittel and Lineberger (1977) similarly concluded that gas kinetic vibrational relaxation efficiencies were likely for most vibrational levels. However, by noting that the relaxation rate constants of the 7^b levels (where $b \geq 1$) were nearly equal,

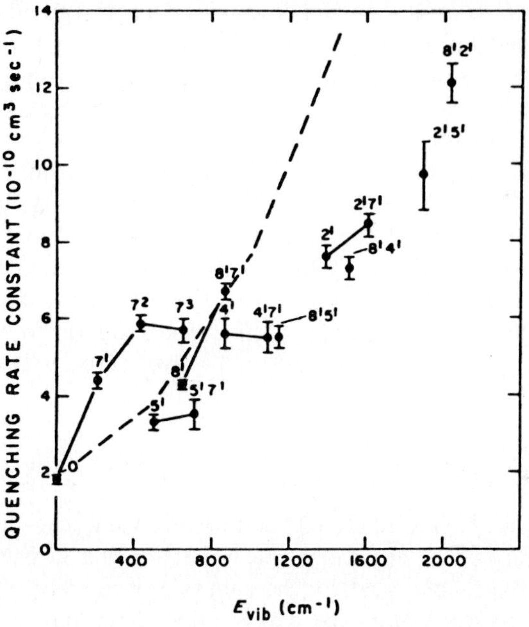

Fig. 16. Collisional quenching rate constants for SVL of Ã ^1A_u glyoxal-d_2. The relative density of ã ^3A_u vibrational levels (---) was calculated using the method of Haarhoff (1963) with vibrational frequencies of Dong and Ramsey (1973) and reasonable estimates for unobserved vibrational frequencies. The ã ^3A_u density at $E_{vib} = 0$, is calculated to be 6/cm^{-1} for glyoxal-d_2 (from Zittel and Lineberger, 1977, Fig. 6).

while those for levels involving the ν_5 and ν_8 fundamentals were much lower than those for roughly isoenergetic ν_7 torsional levels, they concluded that vibrational relaxation probably did not account for the steep energy dependence or large values of k_q observed at large E_{vib}.

The rapid increase in k_q with vibrational energy for high vibrational levels was interpreted by Zittel and Lineberger as apparently being due to a corresponding increase in the collision induced ISC rate to the ã^3 A_u(ã $^3A''$) state. This apparent increase in the collision induced ISC rate was thought to be in accord with the mixed state model for collision-induced ISC of Freed (1976a, b), and an increasing density of vibrational states in the triplet manifold with increasing E_{vib}. The relative density of glyoxal-d_2 ã ^3A_u vibrational levels, ρ, calculated by Zittel and Lineberger using the method of Haarhoff (1963), are also plotted in Fig. 16 as a function of excess vibrational energy in Ã ^1A_u. The observation that the k_q increase less rapidly than ρ, was suggested to perhaps indicate that not all vibrational levels couple intramolecularly to the excited Ã^1A_u levels. Zittel and Lineberger reported that, for highly vibrationally excited glyoxal Ã ^1A_u

state molecules, little emission could be observed from levels other than initially excited levels even at pressures where collisional relaxation is much faster than collision-free decay. As an example, very little emission was observed from levels other than the 8^12^1 level of glyoxal-d_2 ^1A_u state when it was excited at 11 mtorr of pressure, even though collisional relaxation accounts for 80% of the decay at this pressure. The observation that little emission occurs from levels other than the initially prepared 8^12^1 level is in contrast with the aforementioned observations of Parmenter and co-workers, Frad and Tramer, and Rettschnick and co-workers for initially prepared levels with lower excess vibrational energy. These workers reported that a substantial fraction of the emission intensity occurred from the $\tilde{A}\,^1A_u$ zero-point level upon excitation to the 4^1 and 8^17^2 levels, the 8^1, 2^1, and 8^14^1 levels, and the 8^1 level, respectively, at even lower glyoxal pressures.

6. Foreign Gas Quenching of Zero-Point Level $\tilde{A}\,^1A_u$ and $\tilde{a}\,^3A_u$ States

Yardley *et al.* (1971) were the first to measure the quenching efficiencies of glyoxal $\tilde{A}\,^1A_u$ zero-point emission using various collision partners including glyoxal itself. Typical conditions for these measurements were 15 mtorr glyoxal and from 0 to 2 torr of foreign gas. The Stern–Volmer hard sphere quenching efficiencies determined for various quenchers in this postulated collision-induced $\tilde{A}\,^1A_u \rightsquigarrow \tilde{a}\,^3A_u$ process were: He (0.04), D_2 (0.09), Ar (0.10), O_2 (0.14), glyoxal (0.25), and CH_3F (0.32). The cross section for quenching by cyclohexane under the same conditions was found to be $\sim$0.10 hard sphere (Anderson *et al.*, 1973).

Beyer and Lineberger (1975) carried out an extensive study of the collisional quenching of low-lying levels of the $\tilde{A}\,^1A_u$ state of glyoxal using 24 collision partners. The hard sphere quenching cross sections for these collision partners were observed to vary between the values obtained by Yardley *et al.* for He (0.04) and CH_3F (0.32), with the polar collision partners exhibiting the larger cross sections. Studies of $\tilde{a}\,^3A_u$ state production for five of these gases indicated that the principal collisional loss channel from the $\tilde{A}\,^1A_u$ state was to the $\tilde{a}\,^3A_u$ state. It was shown that the quenching by nonpolar molecules was well correlated with the basic molecular parameters for three interaction models. The quenching by polar molecules was also shown to have fair correlation with that model of these three (model of Thayer and Yardley, 1972) which took into account the effect of the permanent dipole moment of the quencher. However, since the model of Thayer and Yardley was postulated for a spin-forbidden process, it was concluded that such correlations have predictive value, but that they could not be taken to verify the quenching model. It was found that the efficiency of glyoxal $\tilde{A}\,^1A_u$ state quenching by glyoxal $\tilde{X}\,^1A_g$ ground state molecules was substantially larger than would be predicted by any of these correlation models. Beyer and Lineberger (1975) suggested that the additional glyoxal–glyoxal interaction that was responsible for this enhanced quenching efficiency was probably due to hydrogen bonding forces. They concluded that the $\tilde{A}\,^1A_u$–$\tilde{X}\,^1A_g$ transition in glyoxal was too weak to enable a Förster dipole–dipole electronic energy transfer mechanism to account for the observed self-quenching efficiency.

The most efficient collision partners (polar quenchers) were observed to have a quenching probability, given by (number of hard-sphere collisions to quench)$^{-1}$, in the range 0.26–0.29 per collision and were concluded to be at essentially a constant limit of efficiency. The dominant intermolecular interaction between glyoxal and these efficient polar quenchers was thought to result either from the strong electric field associated with the quencher's permanent dipole moment at short range or from a hydrogen bonding interaction. Beyer and Lineberger concluded that the quenching limit near 3.7 collisions for deactivation of the glyoxal $\tilde{A}\,^1A_u$ state, implies that about one-fourth of the rovibronic $\tilde{A}\,^1A_u$ levels populated near the zero-point level have significant $\tilde{a}\,^3A_u$ character from spin–orbit or other mixing of the states. The collisional perturbation was concluded to serve to relax the $\tilde{a}\,^3A_u$ state vibrationally and irreversibly and to thus result in quenching of the $\tilde{A}\,^1A_u$ state.

Yardley (1972a, b) carried out two gas phase quenching studies of the glyoxal $\tilde{a}\,^3A_u$ state. In both studies, quenching rate constants were determined by a time-resolved phosphorescence technique using tunable dye laser excitation.

In the first study, Yardley (1972a) determined the collision-free lifetimes for glyoxal-h_2 $\tilde{a}\,^3A_u$ and glyoxal-d_2 $\tilde{a}\,^3A_u$ to be 3.29 and 6.1 msec, respectively. The quenchers used in this study were glyoxal, O_2, NO, and di-*tert*-butylnitroxide $[(t\text{-}C_4H_9)_2NO]$. The self-quenching cross sections for glyoxal-h_2 and glyoxal-d_2 were 8×10^{-2} and 2×10^{-2} Å^2, respectively. Both NO and $(t\text{-}C_4H_9)_2NO$ were observed to be extremely efficient quenchers with cross sections $\sim\frac{1}{6}$ gas kinetic, while O_2 was found to be about 400 times less efficient. These results were concluded to suggest that quenching of glyoxal $\tilde{a}\,^3A_u$ by NO and $(t\text{-}C_4H_9)_2NO$ apparently involves a chemical reaction.

In the second study, Yardley (1972b) used the olefin quenchers ethylene, butene-1, and isobutylene. For these quenchers, significant deviations from Stern–Volmer behavior were observed. These results were suggested to be consistent with the existence of a collisionally accessible intermediate state or complex, with lifetimes possibly much longer than ~100 nsec, from which both collision-induced and unimolecular photochemical decay could take place. The data could not unambiguously distinguish between a C–T complex and a biradical one.

7. Magnetic Field Effects on Radiationless Processes from the $\tilde{A}\,^1A_u$ State

Dong and Kroll (unpublished results cited in Dong and Ramsey, 1973) first examined the effect of an external magnetic field on the fluorescence of glyoxal. They concluded that the fluorescence of glyoxal was partially quenched in the presence of a magnetic field at all pressures. Later Matsuzaki and

Nagakura (1976) used the 4776 Å Ar^+ laser line to excite the $\tilde{A}\,^1A_u(0^0) \leftarrow \tilde{X}\,^1A_g(8^1)$ glyoxal transition and found an ~20% decrease in fluorescence intensity of 5 torr of glyoxal in a magnetic field of ~0.8 kG and above. The absorption intensity of this transition was found to be unaffected by the magnetic field. These observations were taken to indicate that the nonradiative decay of the $\tilde{A}\,^1A_u$ zero-point level was enhanced by the magnetic field. Stern–Volmer plots of the observed decay rates of this level versus glyoxal pressures were reported to give extrapolated collision-free lifetimes of 2.17 ± 0.04 and 1.89 ± 0.04 μsec at magnetic field strengths of 0 and 5 kG, respectively. [The former value was in good agreement with that of 2.16 ± 0.05 μsec obtained by Yardley *et al.* (1971). Both of these zero-field values are lower, however, than that of 2.41 ± 0.06 μsec obtained by Beyer *et al.* (1975).] The quenching rate constants at 0 and 5 kG, unlike the collision-free lifetimes, were reported to remain constant within the limits of experimental error. These observations were taken as additional support that the observed magnetic quenching of fluorescence was due to the enhancement of an intramolecular nonradiative process. Time-resolved emission spectra of 0.8 torr of glyoxal in the presence of a magnetic field at 50 μsec delay were observed to consist of two different emissions. One was assigned to emission from the $\tilde{a}\,^3A_u(0)$ state. The collision-free decay time of the other emission was determined to be 67 μsec and could not be assigned to previously observed emission from glyoxal. Matsuzaki and Nagakura attributed this new emission as being due to phosphorescence from vibrationally excited $\tilde{a}\,^3A_u$ glyoxal [$\tilde{a}\,^3A_u(n)$]. These findings were explained as being due to an enhancement of the intramolecular $\tilde{A}\,^1A_u(0) \rightsquigarrow \tilde{a}\,^3A_u(n)$ ISC process by the magnetic field. The possible mechanism of this magnetic field effect was not discussed.

Recently Schlag and co-workers (Küttner *et al.*, 1977a,b) carried out additional studies of the magnetic field effect on the ISC process in gaseous glyoxal. These studies monitored glyoxal fluorescence quenching with different collision partners as a function of magnetic field strength. Fluorescence and phosphorescence excitation spectra and time-resolved emission spectra were observed and quenching rate constants were determined by measuring fluorescence lifetimes as a function of quencher pressure in the presence and absence of a magnetic field. Fluorescence intensity and lifetime were scanned versus magnetic field at various pressures. In contrast to Matsuzaki and Nagakura (1976), who reported a change in the glyoxal-h_2 collision-free lifetime at a magnetic field strength of 5 kG, Schlag and co-workers found that the collision-free fluorescence lifetimes of glyoxal-h_2 and glyoxal-d_2 were constant over the magnetic field strength ranges employed of 0–2 kG and 0–3.8 kG, respectively. The magnetic field effect was only observed in the presence of colliders. This effect was found to increase gradually with pressure until at 200 mtorr a ~20% reduction in fluorescence lifetime and intensity was obtained when a magnetic field strength of about 1 kG was switched on. The magnetic effect was observed to saturate at a field strength of ~ 1 kG. Schlag and co-workers concluded that ISC was enhanced by the presence of the magnetic field and that collisions did not effect ISC but merely transferred energy, in either a singlet or triplet manifold.

8.   Fluorescence from $\tilde{A}\,^1A_u$ SRVL

In two very recent studies Parmenter and co-workers (Parmenter and Rordorf, 1978; Rordorf et al., 1978) used the 4545 Å line of an Ar^+ laser to selectively excite only a few rotational levels in the zero-point level of the $\tilde{A}\,^1A_u$ state of glyoxal. Parmenter and Rordorf found that the levels excited at low gas pressure could be identified with certainty by the rotational analysis of four bands in fluorescence. The analysis showed that fluorescence from isolated single rotational levels in the $\tilde{A}\,^1A_u$ state could be resolved easily in each of these bands. No dependence of the collision-free emission yield on $K'J'$ could be detected in emission from five $K'J'$ levels. The levels ranged in rotational energy from 89 to 482 cm^{-1} with an equally wide variation in quantum numbers. Approximate rotational constants were obtained for $\tilde{X}\,^1A_g$ glyoxal with $v_2'' = 1$, and an improved value of $v_2'' = 1741.2 \pm 0.5\ cm^{-1}$ was determined.

Rordorf et al. studied the details of rotational energy transfer from a few selected $K'J'$ levels in the zero-point vibrational level of $\tilde{A}\,^1A_u$ glyoxal vapor. An approximately 240 $Å^2$ (or 4.5 times gas kinetic) cross section for destruction of an initial $K'J'$ level by rotational relaxation upon collision with $\tilde{X}\,^1A_g$ glyoxal was observed. This large rotational relaxation cross section is consistent with the rapid rotational relaxation observed by Beyer et al. (1975). Much of the rotational transfer within the $\tilde{A}\,^1A_u$ state was observed to occur with large $\Delta K'$ and $\Delta J'$. No strong propensities for $\Delta K' = 0, \pm 1, \pm 2,$ or ± 3 with small $\Delta J'$ changes were observed for collisions with $\tilde{X}\,^1A_g$ glyoxal.

B.   Propynal

Propynal, $C_2H \cdot CHO$, is similar to glyoxal in its molecular size and the complexity of its electronic structure. Because of these similarities, as in the case of glyoxal, time-resolved emission studies of propynal in its SVLs have been carried out by Yardley and co-workers (Thayer and Yardley, 1972, 1974; Thayer et al., 1975). Recently the dynamics of photodissociation from SVLs of propynal have been also studied (Kumar and Huber, 1976; Huber and Kumar, 1977). As in the case of formaldehyde and glyoxal, molecular elimination of CO takes place as in the following equations:

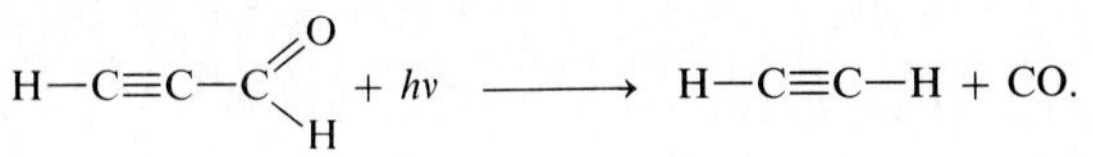

$$H-C \equiv C-C\Big\langle^O_H + h\nu \longrightarrow H-C \equiv C-H + CO. \qquad (48)$$

Due to lack of space as detailed a review as for glyoxal will not be given.

1. Electronic Spectrum

The first electronic absorption band of propynal in the gas phase is structured but weak, commencing at 3820 Å and extending beyond 3000 Å (Howe and Goldstein, 1958). This transition is analogous to the $\pi^* \leftarrow n$ transition of formaldehyde. Vibrational analysis of this transition is nearly complete with 11 of the 12 fundamental modes assigned, and rotational analysis of this band indicates that it is of $\tilde{A}\,^1A'' \leftarrow \tilde{X}\,^1A'$ type with the 0–0 band at 3822 Å (26,163 cm^{-1}) and that both the ground and excited states are planar (Brand *et al.*, 1963). The electronic structure of the lowest triplet state has been studied recently through phosphorescence emission as well as absorption in the $\tilde{a}\,^3A'' \rightarrow \tilde{X}\,^1A'$ system with the 0–0 band at 4144Å (24,128.5 cm^{-1}) (Lin and Moule, 1971a, b, and references therein). Lin and Moule (1971a) obtained the oscillator strengths of the S–S and T–S systems to be 4.8×10^{-4} and 9.6×10^{-5}, respectively. Vibrational frequencies of the ground and excited electronic states of C_2HCHO are listed in Table 11. In the singlet absorption bands, v_4', v_5', v_6', v_9', and v_{12}' are fairly active. Brand *et al.* (1963) has suggested that the singlet π^*–n excitation is not localized in the CHO group on the basis of the frequency changes in v_{11}. Fermi and Coriolis perturbations have been found in some bands. The dipole moment of the $\tilde{A}\,^1A''$ state is 0.7 ± 0.2 Debye as compared to the value of 2.39 Debye in the ground state (Freeman *et al.*, 1966).

TABLE 11

Vibrational Frequencies of Propynal in Various Electronic States (cm^{-1})[a]

		Ground state	Excited state	
Species	Vibration	($\tilde{X}\,^1A'$)[a]	($\tilde{A}\,^1A''$)[b]	($\tilde{a}\,^3A''$)[a]
a′	v^1 CH^2 stretching[c]	3326	—	—
	v_2 CH^1 stretching[c]	2858.2	2952.5	—
	v_3 C≡C stretching	2110.4	1945.8	—
	v_4 C=O stretching	1698.2	1304.0	1323.3
	v_5 C—H rocking	1383.3	1119.5	1118.4
	v_6 C—C stretching	934.9	951.6	987.6
	v_7 CCH bend	662.8	650	664.9
	v_8 CCO bend	614.7	506.9	518.0
	v_9 CCC bend	203.3	189.4	181.9
a″	v_{10} CH^1 wagging[c]	981.2	462.1	384.7
	v_{11} CCH bend	692.7	389.7	422.0
	v_{12} CCC bend	260.6	345.9	340.0

[a] Taken from Lin and Moule (1971a).

[b] Taken from Brand *et al.* (1963).

[c] CH^1 refers to the aldehyde group and CH^2 refers to the acetylenic group.

2. Collision-Free Radiationless Decay Rates

Thayer and Yardley (1972) measured the emission decay time of the zeroth vibrational level of the $\tilde{A}\,^1A''$ state of HC_2CHO and have shown that the lifetime of the excited glyoxal was shortened by collision-induced quenching. A more detailed experimental study by Thayer and Yardley (1974) was carried out in the pressure range of 0.05–1.0 torr in order to determine the rates of several collision-free and collision-induced internal conversion and intersystem crossing processes They measured, for the zeroth vibrational level, a zero-pressure fluorescence quantum yield Φ_F^0 of ~ 0.17 and a collision-free decay rate of $(1.04 \pm 0.05) \times 10^6\ \text{sec}^{-1}$. The radiative decay rate constant obtained was thus $0.08 \times 10^6\ \text{sec}^{-1} \leq k_0 = 0.17 \times 10^6 \leq 0.38 \times 10^6\ \text{sec}^{-1}$, in excellent agreement with the calculated value of $(0.12 \pm 0.03) \times 10^6\ \text{sec}^{-1}$ based on the absorption coefficient. A zero-pressure phosphorescence quantum yield Φ_P^0 of ~ 0.16 was obtained, although a much smaller pressure dependence was observed for the Stern–Volmer plot of $(\Phi_P)^{-1}$ versus pressure than for that of $(\Phi_F)^{-1}$. They deduced collision-free values of k_{IC} for $S_1 \rightsquigarrow S_0$ and k_{ISC} for $S_1 \rightsquigarrow T_1$; $0\ \text{sec}^{-1} \leq k_{IC} = 0.35 \times 10^6 \leq 0.83 \times 10^6\ \text{sec}^{-1}$ and $0.12 \times 10^6\ \text{sec}^{-1} \leq k_{ISC} = 0.69 \times 10^6 \leq 0.99 \times 10^6\ \text{sec}^{-1}$, respectively.

Thayer *et al.* (1975) extended the above study further to include rate measurements for ten higher-lying SVLs excited with dye laser output of 2.0 Å bandwidth (FWHM) in the 3830–3770 Å region and 10 Å bandwidth in the 3700–3639 Å region. They found that the singlet and triplet emission intensities decayed exponentially in time, independent of the excitation wavelength in the pressure range of 0.015–1.0 torr. Furthermore, a linear Stern–Volmer plot of τ_F versus pressure was observed. These observations were taken to support the hypothesis that energy flow out of the fluorescing state is apparently irreversible and must correspond to radiationless transitions into a number of other states. However, they suggest that this does not preclude the possibility of reversible behavior at still lower gas pressures. They have compared the experimental rates with the calculated values of k_{IC} and k_{ISC} based on the theoretical model of Heller *et al.* (1972b). The energy dependence of these rates are compared in Figs. 17 and 18. The agreement for the $S_1 \rightsquigarrow T_1$ process is fair except for the 5^1 level, as they point out.

3. Collision-Induced Processes

Thayer and Yardley (1974) deduced the collision-*induced* rate constants as follows: $(0.0022\ \mu\text{sec}^{-1}\ \text{mtorr}^{-1} \leq k_3 = 0.0094\ \mu\text{sec}^{-1}\ \text{mtorr} \leq 0.015\ \mu\text{sec}^{-1}\ \text{mtorr}^{-1}$ for $S_1 \rightsquigarrow T_1$; $0\ \mu\text{sec}^{-1}\ \text{mtorr}^{-1} \leq k_4 = 0.0047\ \mu\text{sec}^{-1}\ \text{mtorr} \leq 0.013\ \mu\text{sec}^{-1}\ \text{mtorr}^{-1}$ for $S_1 \rightsquigarrow S_0$; $k_7 = (0.135 \pm 0.009) \times 10^{-2}\ \text{msec}^{-1}\ \text{mtorr}^{-1}$ for $T_1 \rightsquigarrow S_0$. They concluded that the first two values correspond to cross sections of 60 and 30 Å^2, respectively. This was thought to reflect the presence

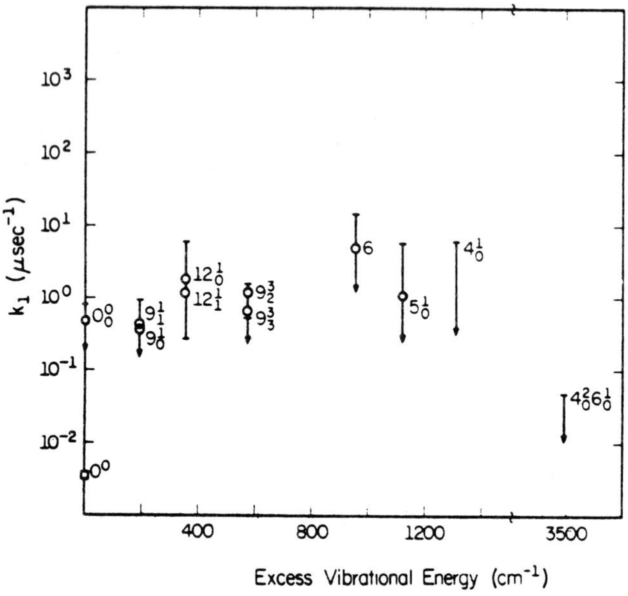

Fig. 17. Internal conversion rate (probably $^1A'' \rightsquigarrow {}^1A'$) as a function of excess vibrational energy for HCCCHO. The notation k_i^m indicates excitation from a state with i quanta in vibrational mode k to a state with m quanta in mode k: ○, experimental results (arrows indicate that error limits include zero as a possibility); □, calculated from the theory (from Thayer *et al.*, 1975, Fig. 3).

of a quasicontinuum of states in the triplet vibronic manifold to which nonradiative decay may take place. Thayer and Yardley (1974) developed a theoretical model to explain the dominance of collision-induced $S_1 \rightsquigarrow T_1$ over collision-induced $S_1 \rightsquigarrow S_0$. A slightly modified dipole–dipole interaction model (Thayer and Yardley, 1972) was shown to be adequate in explaining the quenching of the S_1 state.

4. Photodecomposition

A photodecomposition study, complementary to the photophysical study of Yardley and co-workers (Thayer *et al.*, 1975), has been recently carried out for several SVLs at high vibrational energies, $E_{vib} = 0 \sim 5905$ cm^{-1}, by Kumar and Huber (1976) and Huber and Kumar (1977). They observed for 0.6 torr propynal that (1) Φ_F drops sharply above $\sim 30{,}000$ cm^{-1} of excitation; (2) Φ_P begins dropping gradually around 31,000 cm^{-1}; and (3) the quantum yield of CO formation increases abruptly at both 29,116 cm^{-1} (aldehyde stretch, 2_0^1 excitation) and 31,010 cm^{-1} ($4_0^2 5_1^2$ excitation). They interpreted these observations to mean that with increasing E_{vib} a nonradiative channel, which below the onset of photodissociation, must be $S_1 \rightsquigarrow T_1$ and/or $S_1 \rightsquigarrow S_0$,

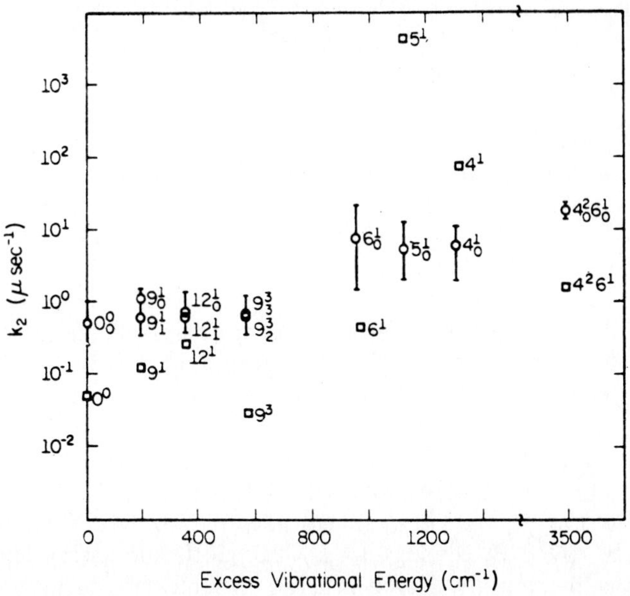

Fig. 18. Intersystem crossing rate ($^1A'' \rightsquigarrow {}^3A''$) as a function of excess vibrational energy for HCCCHO. Notation is the same as for Fig. 17. Taken from Thayer *et al.* (1975), Fig. 4.

is competing successfully with fluorescence emission. They concluded that ISC is the dominant deactivation path below $E_{\mathrm{vib}} = 2950\ \mathrm{cm}^{-1}$ corresponding to the 2^1 level and that k_{ISC} increases with E_{vib}, since the phosphorescence emission intensity (in the photoluminescence excitation spectra) remains constant in the excitation energy region where the fluorescence emission intensity strongly decreases. Above this energy an additional nonradiative channel, IC, was suggested to be important. This energy dependent behavior of the S_1 photochemistry of the SVLs of propynal is very similar to the energy dependent behavior of the SVL photochemistry of cyclobutanone (S_1) studied earlier by Lee and co-workers (Hemminger and Lee, 1972, and references therein; Tang and Lee, 1976b) and that of perfluorocyclobutanone (S_1) studied by Lewis and Lee (1975). Brief reviews of the SVL photochemistry of cyclobutanone have appeared recently (Rice, 1975; Lee, 1977) and therefore no review is necessary here. If one examines the photoluminescence excitation spectra of propynal, it is clear that there is considerable congestion of the peaks of luminescence intensity above $27{,}000\ \mathrm{cm}^{-1}$ and therefore the integrity of the SVL excitation is definitely compromised. Thayer *et al.* (1975) suggest that the ν_2 modes (aldehyde C—H stretch) and to some extent combination modes with ν_2 are quite inefficient with regards to IC (small frequency and displacement change), but that they can couple effectively to a predissociative continuum.

C. Biacetyl

1. Electronic Spectrum

The wavelength disposition of the $\tilde{A}\,^1A_u \leftarrow \tilde{X}\,^1A_g$ and $\tilde{a}\,^3A_u \leftarrow \tilde{X}\,^1A_g$ absorption bands in biacetyl are similar to those observed in glyoxal. The band origins of the $\tilde{A}\,^1A_u$ and $\tilde{a}\,^3A_u$ states are thought to lie at ~4575Å (~21,850 cm^{-1}) and ~5000 Å (~20,000 cm^{-1}), respectively (Drent *et al.*, 1973). In the wavelength region 4700–4000 Å, although SVL resolution of biacetyl absorption is not possible, there is an extensive system of fairly sharp absorption bands which is analogous to the $\tilde{A}$–$\tilde{X}$ system in glyoxal. There are six band groups, which must be assigned to upper state frequencies, that are separated by 1060 and 620 cm^{-1} (Herzberg, 1966). At higher pressure a continuous absorption joins on to these bands and extends to approximately 3500 Å.

The positions of the ^1B_g and ^3B_g states are uncertain and could lie energetically near the corresponding A_u states (McClelland and Yardley, 1973). Calculations, however, place the ^1B_g and ^3B_g states considerably higher in energy (Swenson and Hoffman, 1970). By analogy to glyoxal, electronic energy levels of the *cis* form may be energetically near the ^1A_u and ^3A_u states (Parmenter and Poland, 1969). In addition, biacetyl would be expected to reach a dissociative state at wavelengths shorter than ~4080 Å (Sidebottom *et al.*, 1972).

2. Collision Independence of Intersystem Crossing Following Excitation at 4358 Å

In an early study, Parmenter and Poland (1969) examined the $S_1 \rightarrow T$ intersystem crossing process in biacetyl vapor following excitation by 4358 Å light by directly observing emission from both the initial and final states. Under these conditions, they found that (1) the singlet emission yield from 1.0 torr of biacetyl is constant over the pressure range of 40–0.1 torr of added cyclohexane, (2) the triplet-emission yield from 0.1 torr of biacetyl is also constant over this same pressure range of added cyclohexane provided the experiments are performed in a large volume cell, and (3) the triplet-emission yield decreases at pressures below about 0.1 torr. This decrease is believed not to be due to a decrease in triplet formation but instead to result from wall deactivation processes.

Since isolated molecule conditions were believed to be established for singlet relaxation in the lower limit of this pressure range, it was concluded that upon 4358 Å excitation, triplet formation must occur in biacetyl in the absence of collisional perturbations involving the excited singlet state. [An S_1 lifetime of $\tau_F = \Phi_F \tau_R = 2.1 \times 10^{-8}$ sec is calculated from the high-pressure singlet-emission yield of $\Phi_F = 2.5 \times 10^{-3}$ (Okabe and Noyes, 1957) and the radiative lifetime of $\tau_R = 8.5 \times 10^{-6}$ sec (Almy and Anderson, 1940), estimated from the integrated absorption coefficient.] The apparent independence of intersystem crossing to collisional effects, as well as similar gas phase (Anderson and Parmenter, 1970; Okabe and Noyes, 1957) and solution phase (Backström and Sandros, 1960) fluorescence quantum yields were taken to indicate that biacetyl exhibits statistical limit behavior when excited at 4358 Å. The effective density of vibrational states in the triplet manifold that could be reached by an intersystem crossing process following absorption at 4358 Å was estimated to be approximately 100 states/cm^{-1}. Parmenter and Poland concluded that, contrary to observation, biacetyl would not be expected to exhibit statistical limit behavior.

3. Fluorescence Lifetimes

McClelland and Yardley (1973) used a pulsed dye laser to directly measure, with medium resolution (10–30 Å), the fluorescence decay time of biacetyl vapor as a function of excitation wavelength between 4600 and 3700 Å. The dependence of these lifetimes on increasing biacetyl pressure and pressures of added Ar were also determined. This included the use of sufficiently low total pressures that the observed relaxation times were considerably shorter than the mean time between gas kinetic collisions. Under these collision-free conditions (0.2–1.0 torr), the fluorescence lifetimes were observed to vary smoothly from ~14 nsec for excitation near 4450 Å to ~6 nsec for excitation at 3720 Å. Due to the complexity of the biacetyl absorption spectrum and the wide-band excitation used, several different vibrational states were believed to be populated at each excitation wavelength.

McClelland and Yardley looked for, and failed to observe, evidence for quantum beats and other reversible behavior over the excitation wavelength region. Such reversible behavior might have been anticipated in certain excitation regions due to the expected low density of triplet vibronic levels to which energy conserving transitions from the biacetyl $\tilde{A}\,{}^1A_u$ state apparently take place. McClelland and Yardley pointed out that the 10–30 Å broad-band excitation used by them may have been insufficiently selective to observe such phenomena in biacetyl. They calculated the density of the triplet vibronic levels isoenergetic to the vibrationless $\tilde{A}\,{}^1A_u$ state to be equal to or less than 320 states/cm^{-1}, and concluded, as had Parmenter and Poland (1969), that the apparent statistical limit behavior of biacetyl was anomalous.

McClelland and Yardley found that for biacetyl ($\tilde{A}\,{}^1A_u$) molecules prepared by excitation near 3750 Å (an excess $\tilde{A}\,{}^1A_u$ vibrational energy of ~3800 cm^{-1}) each collision with Ar removes approximately 90 cm^{-1} of vibrational energy, while collisions with biacetyl ($\tilde{X}\,{}^1A_g$) remove roughly 730 cm^{-1} of energy. They suggested that, although the transition dipole for the biacetyl $\tilde{A}\,{}^1A_u \leftarrow \tilde{X}\,{}^1A_g$ transition is quite small, the unusual efficiency for vibrational relaxation during self-collisions conceivably could be a result of resonant electronic energy transfer (as proposed to similarly account for efficient vibrational relaxation in glyoxal).

4. Phosphorescence Excitation Spectra and Decay Rates

Drent *et al.* (1973) studied the phosphorescence excitation spectrum of biacetyl at a pressure of 1.0 torr both in the absence and presence of the foreign gases N_2 and cyclohexane. Three regions of the excitation spectrum with different pressure behavior were identified. Between ~5000 and 4650 Å, region (a), the phosphorescence excitation spectrum was independent of the pressure of foreign gas and was believed to correspond to direct excitation to a biacetyl triplet (presumably $\tilde{a}\,^3A_u$). Between ~4650 and 4440 Å, region (b), the spectrum was observed to exhibit structure even though the absorption spectrum was completely diffuse in this wavelength region. In addition the excitation spectrum was reported to be dependent upon the pressure of added foreign gas. Phosphorescence in this region was thought to result from excitation to the $\tilde{A}\,^1A_u$ state. At excitation wavelengths shorter than ~4440 Å, region (c), the excitation spectrum was observed to become pressure independent and structureless. Phosphorescence produced in region (c) was thought to result from excitation to higher vibronic components of the $\tilde{A}\,^1A_u$ state or from excitation to a different electronic state.

The independence of phosphorescence intensity on the pressure of added foreign gas in region (a) was attributed to the very rapid (relative to phosphorescence decay) vibrational relaxation in the triplet manifold at a pressure of 1.0 torr of biacetyl. The different pressure behavior in the regions (b) and (c) was concluded to be a result of a sudden change at ~4440 Å in the character of the radiationless S ⇝ T process. At excitation wavelengths longer than ~4440 Å, biacetyl was thought to be in the resonant limit, while at shorter wavelengths it was thought to be in the statistical limit. The latter conclusion was found to be consistent with the aforementioned statistical limit behavior observed by Parmenter and Poland (1969) upon excitation of biacetyl at 4358 Å.

The sudden transition to statistical limit behavior at wavelengths shorter than ~4440 Å (including the apparently anomalous statistical limit behavior observed by Parmenter and Poland at 4358 Å) was attributed to the presence of a ^3B_g state in this wavelength region. This state was proposed to give rise to statistical limit behavior in biacetyl at wavelengths below ~4440 Å for two reasons. First, the presence of the ^3B_g state was estimated to enhance the coupling between the biacetyl $\tilde{A}^1A_u$ and $\tilde{a}^3A_u$ states by a factor of ~5 (Drent, 1971; Bixon and Jortner, 1968). Second, the presence of the ^3B_g state was believed to increase the effective density of triplet states to which ISC from the $\tilde{A}\,^1A_u$ state could take place. Without the existence of a ^3B_g state only triplet states of vibrational symmetry b_g would be expected to effectively couple to the singlet. The presence of the ^3B_g state was suggested to introduce strong vibronic interactions via b_u vibrations between ^3B_g and $\tilde{a}\,^3A_u$. Under these conditions, $\tilde{a}\,^3A_u$ states of vibrational symmetry b_u were proposed as also contributing to the effective density of states. In addition, the $\tilde{A}\,^1A_u \leftrightarrow {}^3B_g$ interaction was believed to proceed via two a_u promoting modes, instead of only one b_g mode as in the $\tilde{A}\,^1A_u \leftrightarrow \tilde{a}\,^3A_u$ interaction. This was also thought to lead to an increase in the number of effectively coupled states.

At wavelength longer than ~4440 Å, radiationless transitions from $\tilde{A}\,^1A_u$ to $\tilde{a}\,^3A_u$ were proposed to be in the resonant limit due to the small ($\leq 2500\ \mathrm{cm}^{-1}$) energy gap between the vibrationless $\tilde{a}\,^3A_u$ state and the vibronic levels of $\tilde{A}\,^1A_u$ produced. The structured phosphorescence spectrum in this wavelength region (b) was concluded to have a connection with an energy dependence of the effectively coupled triplet vibronic levels. This effect was also concluded to be a characteristic of resonant limit behavior, since the dissipative manifold in the statistical limit was envisioned as being close to a uniform quasicontinuum.

Moss and Yardley (1974) used tunable laser excitation of the $\tilde{A}\,^1A_u$ state of biacetyl to form biacetyl triplets with specific amounts of vibrational excitation as a result of energy conserving ISC transitions. To deduce the fate of the biacetyl triplets prepared in this manner, they monitored the intensity and decay rate of phosphorescence from biacetyl vapor as a function of excitation wavelength and added pressure of cyclohexane or Ar. The phosphorescence decay rate of biacetyl was observed to exhibit no dependence on excitation wavelength or pressure in a series of experiments employing the following conditions: (a) excitation of 14–40 torr biacetyl at 4440 Å, (b) excitation of biacetyl at pressures down to 5×10^{-3} torr in the presence of 10 torr of Ar, (c) excitation of 1 torr biacetyl in up to ~110 torr of Ar, and (d) excitation of 1 torr of biacetyl in up to ~80 torr of cyclohexane at various wavelengths between 3650 and 4500 Å. The constancy in the phosphorescence decay rate under these conditions is expected, since vibrational relaxation of the triplet levels initially produced as a result of ISC should be very rapid relative to phosphorescence decay. In these experiments, $(4.83 \pm 0.05) \times 10^2\ \mathrm{sec}^{-1}$ was obtained as an upper limit value to the correct thermalized phosphorescence decay rate constant at 297 K.

Although the phosphorescence decay rate was observed to be independent of pressure and excitation wavelength, the phosphorescence yield was found to decrease with increasing excitation energy and decreasing pressure. Moss and Yardley concluded that a triplet state radiationless decay process, with a rate that increases exponentially with increasing vibrational energy in the $\tilde{a}\,^3A_u$ state, was necessary to account for the observed loss of phosphorescence intensity with increasing excitation energy and decreasing pressure. This radiationless triplet state decay process was interpreted as being due to an efficient $\tilde{a}\,^3A_u \rightsquigarrow \tilde{X}\,^1A_g$ ISC. Unimolecular rate constants varying between $\sim 10^2\ \mathrm{sec}^{-1}$ for vibrationless $\tilde{a}\,^3A_u$ and $\sim 10^9\ \mathrm{sec}^{-1}$ for $\tilde{a}\,^3A_u$ with roughly $9000\ \mathrm{cm}^{-1}$ of excess vibrational energy were estimated for this nonradiative process. Moss and Yardley indicated that this simple model would not explain all of the previously described experimental observations of Drent, *et al.* (1973), which led the latter authors to propose the existence of a ${}^3B_g$ state energetically near the $\tilde{A}\,^1A_u$

levels reached by 4440 Å excitation. Moss and Yardley further indicated that their simple model would require some revision if the experimental results suggesting this ^3B_g state were correct.

5. Evidence for Reversible ISC

It may be recalled that McClelland and Yardley (1973) found no evidence of reversible behavior for biacetyl ($\sim$0.2–1.0 torr) using broad-band (10–30 Å) tunable dye laser excitation at various wavelengths between 4450 and 3729 Å. This is the same excitation region [below 4440 Å, region (c)] for which Drent *et al.* (1973) observed statistical limit behavior from 1.0 torr of biacetyl at an excitation bandwidth of 5 Å. The study of biacetyl photophysics was extended in this wavelength region by van der Werf *et al.* (1974) to pressures as low as 10^{-3} torr using tunable dye laser excitation bandwidths down to 0.3 Å. Upon excitation at 4200 Å, three distinct kinds of emission were observed. Besides the earlier reported fluorescence with a lifetime of $\sim$10 nsec (McClelland and Yardley, 1973) and phosphorescence with a lifetime of 1.5 msec (Sidebottom *et al.*, 1972b), they observed an emission on the microsecond time scale. This emission exhibited nonexponential decay behavior and had the same spectral distribution as the fast fluorescence. The ratio of the yields of slow to fast fluorescence was 3 at 10^{-3} torr.

The fast fluorescence was found to have the same weak pressure dependence as observed by McClelland and Yardley (1973). Its lifetime increased with pressure. The lifetime of the slow fluorescence, on the other hand, was observed to decrease with pressure and obey Stern–Volmer kinetics. The phosphorescence, as defined by its spectral composition, was found to be collisionally induced and to disappear in the limit of zero pressure.

All emissions were observed to be linear in laser intensity, and the possibility of triplet–triplet annihilation (Badcock *et al.*, 1972) could be excluded.

Other excitation wavelengths than 4200 Å (4600–3700 Å) were also examined and it appeared that there were always two fluorescence decay components. The lifetime of the slow fluorescence was observed to depend upon excitation wavelength. Preliminary results suggested that at higher excitations the lifetime decreased in an exponential manner with increasing energy. The lifetime of the slow decay component at 3500 Å was about 50 nsec.

The fast fluorescence was interpreted as reflecting competition between direct radiative decay from the singlet ($S_1 \xrightarrow{k_R} S_0$), internal conversion to the ground state ($S_1 \overset{k_{IC}}{\rightsquigarrow} S_0$), and intersystem crossing to the triplet ($S_1 \overset{k_{ISC}}{\rightsquigarrow} T$). Slow fluorescence decay was thought to result from triplet molecules that had undergone a reversible transition back to the singlet state ($T \overset{k_{RISC}}{\rightsquigarrow} S_1$). Phosphorescence was suggested to occur only for triplet molecules that were sufficiently relaxed by collisions to be energetically below the vibrationless singlet state. Analysis of the decay data for excitation at 4200 Å, assuming

$k_R \approx 10^5\ \text{sec}^{-1}$ (from integrated absorption data of Almy and Anderson, 1940), gave rate constants for the above processes of $k_{IC} = 2.4 \times 10^7\ \text{sec}^{-1}$, $k_{ISC} = 7.6 \times 10^7\ \text{sec}^{-1}$, and $k_{RISC} = 1.9 \times 10^5\ \text{sec}^{-1}$.

The kinetic treatment of van der Werf *et al.*, when transformed to its quantum mechanical analogue (Lahmani *et al.*, 1974), yielded values for the triplet level density, $\rho_T = 6.3 \times 10^5/\text{cm}^{-1}$, the S_1–T coupling element, $V_{ST} = 1.0 \times 10^{-5}\ \text{cm}^{-1}$, and the number of triplet states coupled to the excited singlet, $N = 400$. It was concluded that biacetyl at 4200 Å fits neither the statistical limit case nor the sparse intermediate case (Bixon and Jortner, 1969) but constitutes a case in between that exhibits features of both limits. At times much shorter than the recurrence time the decay obeys the behavior of the statistical limit (viz., small pressure dependence), while at longer times the decay has features of the sparse intermediate case (viz., pressure dependent nonexponential slow decay, and only collisionally induced phosphorescence).

D. Methylglyoxal

1. Electronic Spectrum

Figure 19 compares the low-resolution absorption spectrum of methylglyoxal (CH_3COCHO) with those of glyoxal and biacetyl. The lowest energy observed methylglyoxal transition, believed to correspond to the ${}^1A''({}^1A_u) \leftarrow {}^1A'({}^1A_g)(\pi^* \leftarrow n)$ band, occurs between ~4600 and 3500 Å (21,739–28,571 cm^{-1}) (Coveleskie and Yardley, 1975a). Some structure can be observed between 4600 and 4300 Å, but the spectrum becomes increasingly diffuse at wavelengths below ~4300 Å. Coveleskie and Yardley assigned the 0–0 band of the ${}^1A''({}^1A_u) \leftarrow {}^1A'({}^1A_g)$ methylglyoxal transition to lie at either 4527 Å (22,090 cm^{-1}) or at 4492 Å (22,260 cm^{-1}). They reasoned that if the potential surfaces of the methylglyoxal ${}^1A''({}^1A_u)$ and ${}^1A'({}^1A_g)$ states are only slightly displaced as in glyoxal (Paldus and Ramsay, 1967), the 0–0 band should be the most intense feature in the absorption spectrum. This suggested a 4492 Å band origin. Various 270–280 cm^{-1} progressions apparently originate at 4492 Å. On the other hand, an absorption and emission mirror image relationship appeared to be best satisfied by the weaker 4527 Å absorption peak which could not be ruled out as the 0–0 band.

The increasing diffuseness in the $S_1 \leftarrow S_0$ absorption spectra for the series of glyoxal, methylglyoxal, and biacetyl was suggested by Byrne and Ross (1971) to result from either or both (1) spectral congestion due to the addition of vibrational degrees of freedom, and (2) CH_3 group perturbations which dilute the oscillator strength over a broader spectral interval.

The onset of a second methylglyoxal absorption band was reported by Coveleskie and Yardley (1975a) to lie near 3300 Å (30,300 cm^{-1}). This band,

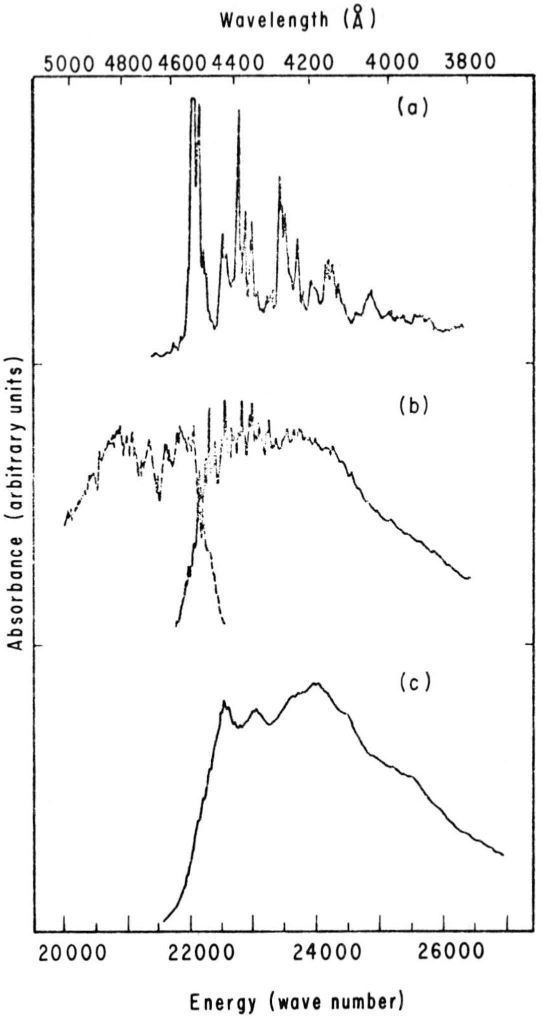

Fig. 19. Low-resolution visible absorption spectra for (a) glyoxal, (b) methylglyoxal, and (c) biacetyl. For comparison, a portion of the observed emission spectrum of methylglyoxal is also displayed (dashed lines). Taken from Coveleskie and Yardley (1975a), with permission of the *Journal of the American Chemical Society*, copyright 1975 by the American Chemical Society.

which has similar oscillator strength to the first, was assigned to a $^1A''(^1B_g) \leftarrow {}^1A(^1A_g)(\pi^* \leftarrow n)$ transition. Due to the small ($<3000\ cm^{-1}$) singlet–triplet separation for $\pi^* \leftarrow n$ excited states, Coveleskie and Yardley estimated the ^3B_g state to lie at $\sim 27{,}000\ cm^{-1}$, and to be energetically inaccessible to excitation wavelengths longer than ~ 3700 Å. They noted that this suggestion was inconsistent with the conclusion of Drent *et al.* (1973) concerning the existence of a

^3B_g state in biacetyl near 22,500 cm^{-1} (4440 Å). The absence of additional absorption features between 6500 and 3500 Å led Coveleskie and Yardley (1975a) to conclude that absorptions involving additional $\pi^* \leftarrow n$ singlet states were located at still higher energies.

2. Emission Spectra and Luminescence Decay

Coveleskie and Yardley (1975a) attributed the vapor phase methylglyoxal emission in the wavelength regions 4400–5200 and 5000–6200 Å to fluorescence and phosphorescence, respectively. The difference in the origins of the fluorescence ($\sim$22,100 cm^{-1}) and phosphorescence ($\sim$19,700 cm^{-1}) spectra suggested that the $^3A''(^3A_u)$ state was approximately 2400 cm^{-1} lower in energy than the $^1A''(^1A_u)$ level.

The fluorescence decay of the $^1A''(^1A_u)$ state of methylglyoxal was studied by Coveleskie and Yardley (1975a, b) in the vapor phase as a function of additive pressure and excitation wavelength. At high pressures (above $\approx$2 torr methylgloxal or $\approx$100 mtorr methylglyoxal with $\geq$5 torr of argon) two exponential decays were observed corresponding to phosphorescence (msec time scale, 5000–6200 Å) and fluorescence (nanosecond time scale, 4400–5200 Å). Under these conditions the initial intensity of the phosphorescence decay was observed to be less than 10^{-3} times that of the fluorescence decay.

Coveleskie and Yardley (1975b) found that the fluorescence decay rates observed for 100 mtorr of methylglyoxal with 5 torr of added argon increased smoothly and linearly over the $\sim$2500 cm^{-1} span in excess vibrational energy studied in the $^1A''(^1A_u)$ state. Under these conditions (bandpass less than 1 Å) the fluorescence lifetime was observed to decrease by a factor of 2, from its value of $\sim$20 nsec at 4492 Å (22,260 cm^{-1}), with an increase in E_{vib} of $\sim$1850 cm^{-1}. The fluorescence decay rates and the variation in decay rate with increasing vibrational energy observed for methylglyoxal under these conditions were similar to those observed by McClelland and Yardley (1973) for biacetyl at low pressures. These observations were suggested to indicate that the dominant process for depopulating the optically excited S_1 state in both methylglyoxal and biacetyl was intersystem crossing ($S_1 \rightsquigarrow T$).

For excitation of lower pressures of methylglyoxal at 4492 Å a more complicated nonexponential fluorescence behavior was observed. Coveleskie and Yardley (1975b) analyzed the fluorescence decay intensity in terms of a simple biexponential decay, i.e., $I(t) = I_1 \exp(-\lambda_1 t) + I_2 \exp(-\lambda_2 t)$, where $\lambda_2 < \lambda_1$. As the pressure was increased, the rate constant λ_1 and the ratio of initial intensities I_1/I_2 remained relatively constant with λ_1 slowly increasing and I_1/I_2 slowly decreasing. The rate constant λ_2, however, was observed to increase by a factor of 2 as the methylglyoxal pressure was increased from 50 to 120 mtorr. The values obtained ($\sigma^{\text{Me-Gly}} \approx 480$ Å^2 and $\sigma^{\text{Ar}} \approx 160$ Å^2) for the cross sections for collisional quenching of the long decay component (λ_2)

were very much larger than gas kinetic and are suggestive of very long range interactions. Both components of the decay were demonstrated to have a similar spectral nature.

To explain this low-pressure behavior Coveleskie and Yardley (1975b) postulated a kinetic scheme similar to that used by van der Werf *et al.* (1974) to account for the nonexponential fluorescence decay observed for biacetyl at low pressures. The unique feature of this kinetic scheme involved the inclusion of a reversible collision-free intersystem crossing process as first proposed by Ashpole *et al.* (1971) and Baba *et al.* (1971) and later discussed by Lahmani *et al.* (1974).

Based on the assumption that S_1–T ISC was the dominant decay process, Coveleskie and Yardley (1975b) employed the theoretical model of Lahmani *et al.* (1974) to establish correspondences between the methylglyoxal kinetic data and the quantum mechanical parameters of the triplet level density (ρ_T), the singlet–triplet coupling element V_{ST}, and the number of triplet states N coupled to the excited singlet. The theoretical model of Lahmani *et al.* (1974) was based on the assumption of intermediate case behavior in the strong coupling limit. A similar transformation had been carried out by van der Werf *et al.* (1974) for biacetyl. From their experimental results for 4492 Å excitation of methylglyoxal, Coveleskie and Yardley obtained apparent values for $\rho_T \approx 1.1 \times 10^4$ states/cm^{-1}, $V_{ST} \approx 6.0 \times 10^{-5}$ cm^{-1}, and $N \approx 5$. Using the Haarhoff (1963) approximation, however, they calculated that no more than 50–80 states/cm^{-1} would be expected in the triplet state $\sim$2500 cm^{-1} above its vibrationless level. The greater than two orders of magnitude larger values for ρ_T derived from the experimental results led Coveleskie and Yardley to suggest that nonradiative decay must involve IC to higher vibrational levels of the ground state as well as ISC to the triplet. The density of triplet levels for biacetyl was calculated to be similarly smaller (by a factor of $\sim$50) than the corresponding value derived from experimental data obtained upon 4200 Å excitation (van der Werf *et al.* 1974). Coveleskie and Yardley (1975b) suggested that IC processes to S_0 might have to be postulated in the case of biacetyl as well as for methylglyoxal.

Coveleskie and Yardley did not observe quantum beats in their methylglyoxal study. Quantum beats might have been expected on the basis of predictions of the quantum mechanical formulation of Lahmani *et al.* (1974). They concluded that such behavior in methylglyoxal had been averaged out or was beyond their experimental resolution. The authors indicated that due to the comparable intensity of both fluorescence decay components, however, methylglyoxal clearly exhibited reversible decay behavior.

Additional data on methylgyloxal has recently been reported in the doctoral dissertation of van der Werf (1976). Coveleskie and Yardley (1975b) had studied the decay rates of methylglyoxal up to $\sim$2500 cm^{-1} of excess vibrational energy in the S_1 state. This study was extended by van der Werf to $\sim$5000 cm^{-1}

of excess energy. In comparable energy regions these studies agree quite well. An extensive comparison of the photophysical behavior of glyoxal, methylglyoxal, and biacetyl was also carried out by van der Werf (1976). These results have been extensively reviewed by Avouris *et al.* (1977). The intermediate case strong coupling limit was also used by van der Werf to interpret the dicarbonyl decay data. Avouris *et al.* (1977) showed that it is possible to similarly analyze the dicarbonyl emissions in terms of an intermediate case weak coupling model. They concluded that both coupling limits could account qualitatively for the apparent reversible kinetics of the decay processes. They suggested that neither of the two coupling limits should be assumed to have quantitative significance until independent information is obtained concerning the dicarbonyl level spacings, interaction energies, and decay widths. Avouris *et al.* (1977) suggested that it is conceivable that both coupling limits may be simultaneously operative in the dicarbonyl excited states.

E. K. C. Lee wishes to acknowledge financial support for research on the photochemistry of small molecules from the Office of Naval Research, the Department of Energy (Office of Basic Energy Sciences), and the National Science Foundation.

REFERENCES

Abe, K., Meyers, F., McCubbin, T. K., and Polo, S. R. (1971). *J. Mol. Spectrosc.* **38**, 552.
Abe, K., Meyers, F., McCubbin, T. K., and Polo, S. R. (1974). *J. Mol. Spectrosc.* **50**, 413.
Almy, G. M., and Anderson, S. (1940). *J. Chem. Phys.* **8**, 805.
Altmann, J. A., Csizmadia, I. G., Yates, K., and Yates, P. (1977). *J. Chem. Phys.* **66**, 298.
Ambartzumian, R. V., Letokhov, V. S., Markarov, G. N., and Puretzkui, A. A. (1974). *In* "Laser Spectroscopy" (R. G. Brewer and A. Mooradian, eds.), p. 611. Plenum Press, New York.
Anderson, L. G., and Parmenter, C. S. (1970). *J. Chem. Phys.* **52**, 466.
Anderson, L. G., Parmenter, C. S., Poland, H. M., and Rau, J. D. (1971). *Chem. Phys. Lett.* **8**, 232.
Anderson, L. G., Parmenter, C. S., and Poland, H. M. (1973). *Chem. Phys.* **1**, 401.
Ashpole, C. W., Formosinho, S. J., and Porter, G. (1971). *Proc. R. Soc. London Ser. A* **323**, 11.
Atherton, N. M., Dixon, R. H., and Kirby, G. H. (1964). *Trans. Faraday Soc.* **60**, 1688.
Atkinson, G. H., Laufer, A. H., and Kurylo, M. J. (1973). *J. Chem. Phys.* **59**, 350.
Avouris, P., Gelbart, W. M., and El-Sayed, M. A. (1977). *Chem. Rev.* **77**, 793.
Baba, H., Nakajima, A., Aoi, M., and Chihara, K. (1971). *J. Chem. Phys.* **55**, 2433.
Back, R. A., and Koda, S. (1977). *Can J. Chem.* **55**, 1387.
Backström, H. L. J., and Sandros, K. (1960). *Acta Chem. Scand.* **14**, 48.
Badcock, C. C., Sidebottom, H. W., Calvert, J. G., Rabe, R. E., and Damon, E. K. (1972). *J. Am. Chem. Soc.* **94**, 19.
Baronavski, A. P. (1975). Ph.D. Dissertation, Univ. of California, Berkeley, California.
Baronavski, A. P., Hartford, A., Jr., and Moore, C. B. (1976). *J. Mol. Spectrosc.* **60**, 111.
Becker, K. H., Lippmann, H., and Schurath, U. (1977). *Ber. Bunsenges. Phys. Chem.* **81**, 567.
Beyer, R. A., and Lineberger, W. C. (1975). *J. Chem. Phys.* **62**, 4024.
Beyer, R. A., Zittel, P. F., and Lineberger, W. C. (1975). *J. Chem. Phys.* **62**, 4016.
Bird, G. R., and Marsden, M. J. (1974). *J. Mol. Spectrosc.* **50**, 403.
Birss, F. W., Dong, R. Y., and Ramsay, R. A. (1973). *Chem. Phys. Lett.* **18**, 111.
Birss, F. W., Ramsay, D. A., and Till, S. M. (1978). *Chem. Phys. Lett.* **53**, 14.

Bixon, M., and Jortner, J. (1968). *J. Chem. Phys.* **48**, 715.
Bixon, M., and Jortner, J. (1969). *J. Chem. Phys.* **50**, 3284.
Brand, J. C. D. (1954). *Trans. Faraday Soc.* **50**, 431.
Brand, J. C. D. (1956). *J. Chem. Soc.* 858.
Brand, J. C. D., Callomon, J. H., and Watson, J. K. G. (1963). *Dis. Faraday Soc.* **35**, 175.
Brand, J. C. D., and Nanes, R. (1973). *J. Mol. Spectrosc.* **46**, 194.
Brand, J. C. D., and Srikameswaran, K. (1972). *Chem. Phys. Lett.* **15**, 130.
Brand, J. C. D., and Stevens, C. (1973). *J. Chem. Phys.* **58**, 3331.
Brand, J. C. D., Jones, V. T., and di Lauro, C. (1971). *J. Mol. Spectrosc.* **40**, 616.
Brand, J. C. D., Hardwick, J. L. Pirkle, P. J., and Seliskar (1973a). *Can. J. Phys.* **51**, 2184.
Brand, J. C. D., Humphrey, D. R., Douglas, A. E., and Zanon, I. (1973b). *Can. J. Phys.* **51**, 530.
Brand, J. C. D., Jones, V. T., and di Lauro, C. (1973c). *J. Mol. Spectrosc.* **45**, 404.
Brand, J. C. D., Hardwick, J. L., Humphreys, D. R., Hamada, Y., and Merer, A. J. (1976a). *Can. J. Phys.* **54**, 186.
Brand, J. C. D., Chiu, P. H., Hoy, A. R., and Bist, H. D. (1976b). *J. Mol. Spectrosc.* **60**, 43.
Briggs, J. P., Caton, R. B., and Smith, M. J. (1975). *Can. J. Chem.* **53**, 2133.
Brus, L. E., and McDonald, J. R. (1973). *Chem. Phys. Lett.* **21**, 283.
Brus, L. E., and McDonald, J. R. (1974). *J. Chem. Phys.* **61**, 97.
Busch, G. E., and Wilson, K. R. (1972). *J. Chem. Phys.* **56**, 3638.
Butler, S., and Levy, D. H. (1977). *J. Chem. Phys.* **66**, 3538.
Butler, S., Kahler, C., and Levy, D. H. (1975). *J. Chem. Phys.* **62**, 815.
Byrne, J. P., and Ross, I. G. (1971). *Aust. J. Chem.* **24**, 1107.
Callomon, J. H., and Innes, K. K. (1963). *J. Mol. Spectrosc.* **10**, 166.
Calvert, J. G. (1973). *Chem. Phys. Lett.* **20**, 484.
Calvert, J. G., and Pitts, J. N. (1966). "Photochemistry." Wiley, New York.
Calvert, J. G., Kerr, J. A., Demerjian, K. L., and McQuigg, R. D. (1972). *Science* **175**, 752.
Campbell, J. D., Hancock, G., Halpern, J. B., and Welge, K. H. (1976). *Chem. Phys. Lett.* **44**, 404.
Clark, J. H., Moore, C. B., and Reilly, J. P. (1978). *Int. J. Chem. Kinet.* **10**, 427.
Clements, J. H. (1935). *Phys. Rev.* **47**, 224.
Coon, J. B., Cesani, F. A., and Huberman, F. P. (1970). *J. Chem. Phys.* **52**, 1647.
Cossart-Magos, C., Frad, A., and Tramer, A. (1978). *Spectrochim Acta.* **34A**, 195.
Coveleskie, R. A., and Yardley, J. T. (1975a). *J. Am. Chem. Soc.* **97**, 1667.
Coveleskie, R. A., and Yardley, J. T. (1975b). *Chem. Phys.* **9**, 275.
DiGiorgio, V. E., and Robinson, G. W. (1959). *J. Chem. Phys.* **31**, 1678.
Dixon, R. N., and Halle, M. (1973). *Chem. Phys. Lett.* **22**, 450.
Dong, R. Y., and Ramsay, D. A. (1973). *Can. J. Phys.* **51**, 1491.
Donnelly, V. M., and Kaufman, F. (1977a). *J. Chem. Phys.* **66**, 4100.
Donnelly, V. M., and Kaufman, F. (1977b). *J. Chem. Phys.* **67**, 4768.
Douglas, A. E. (1963). *Disc. Faraday Soc.* **35**, 158.
Douglas, A. E. (1966). *J. Chem. Phys.* **45**, 1007 (1966).
Douglas, A. E., and Huber, K. P. (1965). *Can J. Phys.* **43**, 74.
Drent, E. (1971). Thesis, Univ. of Groningen.
Drent, E., van der Werf, R. P., and Kommandeur, J. (1973). *J. Chem. Phys.* **59**, 2061.
Duchesne, J., and Rosen, B. (1947). *J. Chem. Phys.* **15**, 631.
Englman, R., and Jortner, J. (1970). *Mol. Phys.* **18**, 145.
Evans, K., and Rice, S. A. (1972). *Chem. Phys. Lett.* **14**, 8.
Evans, K., Heller, D., Rice, S. A., and Scheps, R. (1973). *J. Chem. Soc. Faraday Trans. II* **69**, 856.
Fleming, G. R., Gijzeman, O. L. J., and Lin, S. H. (1973). *Chem. Phys. Lett.* **21**, 527.
Forst, W. (1973). "Theory of Unimolecular Reactions." Academic Press, New York.
Frad, A., and Tramer, A. (1973). *Chem. Phys. Lett.* **23**, 297.
Freed, K. F. (1976a). *Topics Appl. Phys.* **15**, 1.

Freed, K. F. (1976b). *Chem. Phys. Lett.* **37**, 47.
Freed, K. F. (1976c). *J. Chem. Phys.* **64**, 1604.
Freeman, D. E., and Klemperer, W. (1966). *J. Chem. Phys.* **45**, 52.
Gelbart, W. M. (1977). *Ann. Rev. Phys. Chem.* **28**, 323.
Gelbart, W. M., and Freed, K. F. (1973). *Chem. Phys. Lett.* **18**, 470.
Gelbart, W. M., Heller, D. F., and Elert, M. L. (1975). *Chem. Phys.* **7**, 116.
Gillespie, G. D., and Khan, A. U. (1976). *J. Chem. Phys.* **65**, 1624.
Gillespie, G. D., Khan, A. U., Wahl, A. C., Hosteny, R. P., and Krauss, M. (1975). *J. Chem. Phys.* **63**, 3425.
Gordon, R. G., and Chiu, Y. N. (1971). *J. Chem. Phys.* **55**, 1469.
Haarhoff, P. C. (1963). *Mol. Phys.* **7**, 101.
Haas, Y., Houston, P. L., Clark, J. H., and Moore, C. B., Rosen, H., and Robrish, P. (1975). *J. Chem. Phys.* **63**, 4195.
Hackett, P. A., Back, R. A., and Koda, S. (1976). *J. Chem. Phys.* **65**, 5103.
Hallin, K. E. J., and Merer, A. J. (1976). *Can. J. Phys.* **54**, 1157.
Hallin, K. E. J., Hamada, Y., and Merer, A. J. (1976). *Can. J. Phys.* **54**, 2118.
Hamada, Y., and Merer, A. J. (1974). *Can. J. Phys.* **52**, 1443.
Hamada, Y., and Merer, A. J. (1975). *Can. J. Phys.* **53**, 2555.
Hardwick, J. L. and Brand, J. C. D. (1973). *Chem. Phys. Lett.* **21**, 458.
Hardwick, J. L., and Till, S. M. (1979). *J. Chem. Phys.* **70**, 2340.
Hase, W. L., and Feng, D. F. (1976). *J. Chem. Phys.* **64**, 651.
Hayes, D. M., and Morokuma, K. (1972). *Chem. Phys. Lett.* **12**, 539.
Heller, D. F., and Freed, K. F. (1972). *Int. J. Quant. Chem.* **6**, 267.
Heller, D. F., Freed, K. F., and Gelbart, W. M. (1972). *J. Chem. Phys.* **56**, 2309.
Heller, D. F., Elert, M. L., and Gelbart, W. M. (1978). *J. Chem. Phys.* **69**, 4061 (1978).
Hemminger, J. C., and Lee, E. K. C. (1972). *J. Chem. Phys.* **56**, 5284.
Herzberg, G. (1945). "Infrared and Raman Spectra of Polyatomic Molecules." Van Nostrand-Reinhold, Princeton, New Jersey.
Herzberg, G. (1950). "Spectra of Diatomic Molecules," 2nd ed. Van Nostrand-Reinhold, Princeton, New Jersey.
Herzberg, G. (1966). "Electronic Spectra and Electronic Structure of Polyatomic Molecules." Van Nostrand-Reinhold, Princeton, New Jersey.
Hillier, I. H., and Sanders, V. R. (1971). *Mol. Phys.* **22**, 193.
Hochstrasser, R. M., and Marchetti, A. P. (1970). *J. Mol. Spectrosc.* **35**, 335.
Holzer, W., and Ramsay, D. A. (1970). *Can. J. Phys.* **48**, 1759.
Horowitz, A., and Calvert, J. G. (1978). *Int. J. Chem. Kinet.* **10**, 713, 805.
Houston, P. L., and Moore, C. B. (1976). *J. Chem. Phys.* **65**, 757.
Howard, W. E., and Schlag, E. W. (1978). *J. Chem. Phys.* **68**, 2679.
Howe, J. A., and Goldstein, J. H. (1958). *J. Am. Chem. Soc.* **80**, 4846.
Huber, K. P. (1966). As presented by G. Herzberg, "Electronic Spectra and Electronic Structure of Polyatomic Molecules," p. 483. Van Nostrand-Reinhold, Princeton, New Jersey.
Huber, J. R., and Kumar, D. (1977). *Ber. Bunsenges. Phys. Chem.* **81**, 215.
Hui, M. H., and Rice, S. A. (1972). *Chem. Phys. Lett.* **17**, 474.
Ingold, C. K., and King, G. W. (1953). *J. Chem. Soc.*, 2072.
Jaffe, R. L., and Morokuma, K. (1976). *J. Chem. Phys.* **64**, 4881.
Job, V. A., Sethuraman, V., and Innes, K. K. (1969). *J. Mol. Spectrosc.* **30**, 365.
Jortner, J., Rice, S. A., and Hochstrasser, R. M. (1969). *Adv. Photochem.* **7**, 149.
Jungen, C., and Merer, A. J. (1976). "Molecular Spectroscopy: Modern Research" (K. N. Rao, ed.), Vol. 2, p. 127. Academic Press, New York.
Kawasaki, M., Lee, S. J., and Bersohn, R. (1975). *J. Chem. Phys.* **63**, 809.
Keyser, L. F., Levine, S. Z., and Kaufman, F. (1971). *J. Chem. Phys.* **54**, 355.

Koda, S., and Back, R. A. (1977). *Can. J. Chem.* **55**, 1380.
Koda, S., Hackett, P. A., and Back, R. A. (1974). *Chem. Phys. Lett.* **28**, 532.
Kühn, I., Heller, D. F., and Gelbart, W. M. (1977). Cited in Avouris *et al.* (1977).
Kumar, D., and Huber, J. R. (1976). *Chem. Phys. Lett.* **38**, 537.
Kusch, P., and Loomis, F. W. (1939). *Phys. Rev.* **55**, 850.
Küttner, H. G., Selzle, H. L., and Schlag, E. W. (1977a). *Chem. Phys. Lett.* **48**, 207.
Küttner, H. G., Selzle, H. L., and Schlag, E. W. (1977b). *Israel J. Chem.* **176**, 264.
Lahmani, A., Tramer, A., and Tric, C. (1974). *J. Chem. Phys.* **60**, 4431.
Lee, E. K. C. (1977). *Accounts Chem. Res.* **10**, 319.
Lee, E. K. C., and Uselman, W. M. (1972). *Faraday Disc. Chem. Soc.* **53**, 125.
Letokhov, V. S. (1977). *Ann. Rev. Phys. Chem.* **28**, 133.
Lewis, R. S., and Lee, E. K. C. (1975). *J. Phys. Chem.* **79**, 187.
Lewis, R. S., and Lee, E. K. C. (1978). *J. Phys. Chem.* **82**, 250.
Lewis, R. S., Tang, K. Y., and Lee, E. K. C. (1976). *J. Chem. Phys.* **65**, 2910.
Lin, S. H. (1966). *J. Chem. Phys.* **44**, 3759.
Lin, S. H. (1976). *Proc. R. Soc. London Ser. A* **352**, 57.
Lin, S. H., and Bersohn, R. (1968). *J. Chem. Phys.* **48**, 2732.
Lin, C. T., and Moule, D. C. (1971a). *J. Mol. Spectrosc.* **37**, 280.
Lin, C. T., and Moule, D. C. (1971b). *J. Mol. Spectrosc.* **38**, 136.
Liu, D. S., and Brand, J. C. D. (1974). *J. Phys. Chem.* **78**, 2270.
Loper, G. L., and Lee, E. K. C. (1972). *Chem. Phys. Lett.* **13**, 140.
Lucchese, R. R., and Schaefer, H. F. (1978). *J. Am. Chem. Soc.* **100**, 298.
Luntz, A. C. (1975). Private communication.
Luntz, A. C. (1978). *J. Chem. Phys.* **69**, 3436.
Luntz, A. C., and Maxon, V. T. (1974). *Chem. Phys. Lett.* **26**, 553.
Marcus, R. A. (1952). *J. Chem. Phys.* **20**, 355, 359.
Marcus, R. A. (1965). *J. Chem. Phys.* **43**, 2658.
Marling, J. B. (1977). *J. Chem. Phys.* **66**, 4200.
Matsuzaki, A., and Nagakura, S. (1976). *Chem. Phys. Lett.* **37**, 204.
McClelland, G. M., and Yardley, J. T. (1973). *J. Chem. Phys.* **58**, 4368.
McMurry, H. L. (1941). *J. Chem. Phys.* **9**, 231, 241.
McQuigg, R. D., and Calvert, J. G. (1969). *J. Am. Chem. Soc.* **91**, 1590.
Melton, L. A., and Klemperer, W. (1971). *J. Chem. Phys.* **55**, 1468.
Merer, A. J., and Hallin, K. E. J. (1978). *Can. J. Phys.* **56**, 838.
Mettee, H. D. (1968). *J. Chem. Phys.* **49**, 1784.
Mettee, H. D. (1969). *J. Phys. Chem.* **73**, 1071.
Miller, R. G., and Lee, E. K. C. (1974). *Chem. Phys. Lett.* **27**, 475.
Miller, R. G., and Lee, E. K. C. (1975). *Chem. Phys. Lett.* **33**, 104.
Miller, R. G., and Lee, E. K. C. (1976). *Chem. Phys. Lett.* **41**, 52.
Miller, R. G., and Lee, E. K. C. (1978). *J. Chem. Phys.* **68**, 4448.
Monts, D. L., Soep, B., and Zare, R. N. (1978). "Nitrogen Dioxide," Wiley, New York.
Moore, C. B. (1973). *Accounts Chem. Res.* **6**, 323.
Moortgat, G. K., Slemr, F., Seiler, W., and Warneck, P. (1978). Paper presented at the *Informal Conf. Photochem., 13th, Clearwater Beach, Florida, January 4–7.*
Moss, A. Z., and Yardley, J. T. (1974). *J. Chem. Phys.* **61**, 2883.
Moule, D. C., and Walsh, A. D. (1975). *Chem. Rev.* **75**, 67.
Nicolet, M. (1975). *Rev. Geophys. Space Sci.* **13**, 593.
Oka, T. (1973). *Adv. At. Mol. Phys.* **9**, 127.
Okabe, H. (1971). *J. Amer. Chem. Soc.* **93**, 7095.
Okabe, H., and Noyes, W. A. (1957). *J. Am. Chem. Soc.* **79**, 801.
Otsuka, K., and Calvert, J. G. (1971). *J. Am. Chem. Soc.* **93**, 2581.
Paech, F., Schmieldle, R., and Demtröder, W. (1975). *J. Chem. Phys.* **63**, 4369.

Paldus, J., and Ramsay, D. A. (1967). *Can. J. Phys.* **45**, 1389.
Parkin, J. E., Poole, H. G., and Raynes, W. T. (1962). *Proc. Chem. Soc.* (*London*) 248.
Parmenter, C. S. (1972a). *Adv. Chem. Phys.* **22**, 365.
Parmenter, C. S. (1972b). *In* "MTP International Review of Science" (D. A. Ramsay, ed.), Ser. One, Vol. 3, p. 297. University Park Press, Baltimore, Maryland.
Parmenter, C. S., and Poland, H. M. (1969). *J. Chem. Phys.* **51**, 1551.
Parmenter, C. S., and Rordorf, R. F. (1978). *Chem. Phys.* **27**, 1.
Parmenter, C. S., and Tang, K. Y. (1978). *Chem. Phys.* **27**, 127.
Photos, E., and Atkinson, G. H. (1975). *Chem. Phys. Lett.* **36**, 34.
Prais, M. G., Heller, D. F., and Freed, K. F. (1974). *Chem. Phys.* **6**, 331.
Rao, T. N., and Calvert, J. G. (1970). *J. Phys. Chem.* **74**, 681.
Rettschnick, R. P. H., Ten Brink, H. M., and Langelaar, J. (1978). *J. Mol. Struct.* **47**, 261.
Rice, O. K. (1929). *Phys. Rev.* **34**, 1451.
Rice, S. A. (1975). *In* "Excited States" (E. C. Lim, ed.), Vol. 2, p. 111. Academic Press, New York.
Riley, S. J., Sanders, R. K., and Wilson, K. R. (1974). *In* "Laser Spectroscopy" (R. G. Brewer and A. Mooradian, eds.), p. 597. Plenum Press, New York.
Robinson, G. W. (1956). *Can. J. Phys.* **34**, 699.
Robinson, G. W. (1967). *J. Chem. Phys.* **47**, 1967.
Robinson, G. W. (1974). *In* "Excited States" (E. C. Lim, ed.), Vol. 1, p. 1. Academic Press, New York.
Robinson, G. W., and DiGiorgio, V. E. (1958). *Can. J. Phys.* **36**, 31.
Rordorf, R. F., Knight, A. E. W., and Parmenter, C. S. (1978). *Chem. Phys.* **27**, 31.
Sackett, P. B., and Yardley, J. T. (1971). *Chem. Phys. Lett.* **9**, 612.
Sackett, P. B., and Yardley, J. T. (1972). *J. Chem. Phys.* **57**, 152.
Sakurai, K., and Broida, H. P. (1969). *J. Chem. Phys.* **50**, 2404.
Schwartz, S. E., and Johnston, H. S. (1969). *J. Chem. Phys.* **51**, 1286.
Schwartz, S. E., and Senum, G. I. (1975). *Chem. Phys. Lett.* **32**, 569 (1975).
Senum, G. I., and Schwartz, S. E. (1977). *J. Mol. Spectrosc.* **64**, 75.
Sethuraman, V., Job, V. A., and Innes, K. K. (1970). *J. Mol. Spectrosc.* **33**, 189.
Shaw, R. J., Kent, J. E., and O'Dwyer, M. F. (1976a). *Chem. Phys.* **8**, 155.
Shaw, R. J., Kent, J. E., and O'Dwyer, M. F. (1976b). *Chem. Phys.* **8**, 165.
Shibuya, K., and Lee, E. K. C. (1978). *J. Chem. Phys.* **69**, 758.
Sidebottom, H. W., Badcock, C. C., Calvert, J. G., Reinhardt, G. W., Rabe, B. R., and Damon, E. K. (1971). *J. Am. Chem. Soc.* **93**, 2587.
Sidebottom, H. W. Otsuka, K., Horowitz, A., Calvert, J. G., Rabe, B. R., and Damon, E. K. (1972a). *Chem. Phys. Lett.* **13**, 337.
Sidebottom, H. W., Badcock, C. C., Calvert, J. G., Rabe, B. R., and Damon, E. K. (1972b). *J. Am. Chem. Soc.* **94**, 13.
Simons, J. P., and Tasker, P. W. (1973). *Mol. Phys.* **26**, 1267.
Sloane, C. S., and Hase, W. L. (1977). *J. Chem. Phys.* **66**, 1523.
Smalley, R. E., Ramakrishna, B. L., Levy, D. H., and Wharton, L. (1974). *J. Chem. Phys.* **61**, 4363.
Smalley, R. E., Wharton, L., and Levy, D. H. (1975). *J. Chem. Phys.* **63**, 4977.
Sodeau, J. R., and Lee, E. K. C. (1978). *Chem. Phys. Lett.* **57**, 71.
Solarz, R., and Levy, D. H. (1974). *J. Chem. Phys.* **60**, 842.
Solarz, R., Butler, S., and Levy, D. H. (1973). *J. Chem. Phys.* **58**, 5172.
Stevens, C. G., and Brand, J. C. D. (1973). *J. Chem. Phys.* **58**, 3324.
Stevens, C. G., and Zare, R. N. (1975). *J. Mol. Spectrosc.* **56**, 167.
Stevens, C. G., Swagel, M. W., Wallace, R., and Zare, R. N. (1973). *Chem. Phys. Lett.* **18**, 465.
Strickler, S. J., and Howell, D. B. (1968). *J. Chem. Phys.* **49**, 1947.
Strickler, S. J., Vikesland, J. P., and Bier, H. D. (1976). *J. Chem. Phys.* **60**, 664.
Su, F., Bottenheim, J. W., Thorsell, D. L., Calvert, J. G., Damon, E. K. (1977). *Chem. Phys. Lett.* **49**, 305.

Su, F., Bottenheim, J. W., Sidebottom, H. W., Calvert, J. G., and Damon, E. K. (1978). *Int. J. Chem. Kinet.* **10**, 125.
Swenson, J. R., and Hoffman, R. (1970). *Helv. Chem. Acta* **53**, 2331.
Tang, K. Y., and Lee, E. K. C. (1976a). *Chem. Phys. Lett.* **43**, 232.
Tang, K. Y., and Lee, E. K. C. (1976b). *J. Phys. Chem.* **80**, 1833.
Tang, K. Y., Fairchild, P. W., and Lee, E. K. C. (1977). *J. Chem. Phys.* **66**, 3303.
Thayer, C. A., and Yardley, J. T. (1972). *J. Chem. Phys.* **57**, 3992.
Thayer, C. A., and Yardley, J. T. (1974). *J. Chem. Phys.* **61**, 2487.
Thayer, C. A., Pocius, A. V., and Yardley, J. T. (1975). *J. Chem. Phys.* **62**, 3712.
Thomson, R., and Warsop, P. A. (1969). *Trans. Faraday Soc.* **65**, 2806.
Thomson, R., and Warsop, P. A. (1970). *Trans. Faraday Soc.* **66**, 1871.
Uselman, W. M., and Lee, E. K. C. (1976a). *J. Chem. Phys.* **65**, 1948.
Uselman, W. M., and Lee, E. K. C. (1976b). *J. Chem. Phys.* **64**, 3457.
van der Werf, R. (1976). Doctoral Dissertation, Gröningen.
van der Werf, R., Zevenhuijzen, D., and Kommandeur, J. (1974). *Chem. Phys. Lett.* **27**, 325.
van der Werf, R., Schutten, E. and Kommandeur, J. (1975). *Chem. Phys.* **11**, 281.
Volk, L. J., and Lee, E. K. C. (1977). *J. Chem. Phys.* **67**, 238.
Walsh, R., and Benson, S. W. (1966). *J. Am. Chem. Soc.* **88**, 4570.
Walsh, A. D., and Warsop, P. A. (1961). *Trans. Faraday Soc.* **57**, 345.
Wampler, F. B., Otsuka, K., Calvert, J. G., and Damon, E. K. (1973). *Int. J. Chem. Kinet.* **5**, 669.
Ware, W. R. (1971). *In* "Creation and Detection of the Excited State," Vol. 1, Part A (W. R. Ware, ed.). Dekker, New York, p. 213.
Weisshaar, J., Baronavski, A., Cabello-Albala, A., and Moore, C. B. (1978). *J. Chem. Phys.* **69**, 4720.
Wilson, E. B., Decius, J. C., and Cross, P. C. (1955). "Molecular Vibrations." McGraw-Hill, New York.
Yardley, J. T. (1972a). *J. Chem. Phys.* **56**, 6192.
Yardley, J. T. (1972b). *J. Am. Chem. Soc.* **94**, 7283.
Yardley, J. T., Holleman, G. W., and Steinfeld, J. I. (1971). *Chem. Phys. Lett.* **10**, 266.
Yeung, E. S., and Moore, C. B. (1973). *J. Chem. Phys.* **58**, 3988.
Yeung, E. S., and Moore, C. B. (1974). *J. Chem. Phys.* **60**, 2139.
Zittel, P. F., and Lineberger, W. C. (1977). *J. Chem. Phys.* **66**, 2972.

2

Rotational Fine Structure in Radiationless Transitions

W. E. Howard and E. W. Schlag*
Institut für Physikalische Chemie
Technische Universität München
Garching, Germany

I. INTRODUCTION

A transition between two electronic states of a molecule may be radiative or nonradiative. In the case of radiative transitions it is known that the transition probabilities depend on the nature of the electronic, vibrational, and rotational states. If the two states differ in multiplicity, the transition probability also depends on the coupling of the spin states.

In the case of radiationless transitions, the situation is not well understood. Early solid and liquid phase experiments showed that the electronic states and

* Present address: Department of Chemistry, University of California, Irvine, California.

ISBN 0-12-450650-X

electronic energy gap (total vibrational energy of the final state) were important for a description of these processes. Early gas phase experiments demonstrated that the nature of the vibrational states was also of importance.

However, only recently have experiments shown that the rotational states may also take an active part in the nonradiative process. If this is the case, one must consider the nature of all molecular states in a description of nonradiative processes, as one must in the radiative case.

In this paper we review some of the more important experimental and theoretical evidence for the role of the rotational states in nonradiative processes. In Section II, the formation of a rotational ensemble in an excited electronic state by optical absorption is discussed. In Section III, processes which can alter the distribution are considered.

Given a particular ensemble of rotational levels, the question still remains as to whether or not different levels are characterized by different nonradiative rate constants.

In Section IV, the experimental evidence relating to this question is presented. In Sections V–VIII, detailed expressions for the nonradiative rate constant are given which incorporate the rotational matrix elements.

In Section IX, a stochastic model is given by which ensemble observables can be calculated from individual level populations and rate constants. Collision effects and finite width pulse excitation are included in the model. In Section X, numerical calculations based on this model are presented.

Thus, we discuss the creation, coupling, decay, and observation of an ensemble of rotational levels which can undergo nonradiative transitions. We do not attempt to give a comprehensive review of the theory of radiationless transitions. We have restricted our discussion to only those topics connected with the role of the rotational states in radiationless transitions. Many topics such as irreversibility, the small molecule or statistical limit, Green's function techniques, etc., have already received much attention in the literature and are not discussed.

II. THE EXCITED STATE ROTATIONAL DISTRIBUTION

Through optical absorption, a molecule may be prepared in a state which can decay via radiative or radiationless processes. Generally, broadband excitation creates an ensemble of states. The ensemble is characterized by an excited state rotational and vibrational distribution. The limit of this distribution is the single rovibronic level. However, it is important to note that even an infinitely narrow bandwidth does not necessarily result in the excitation of a single rovibronic level. However, if such excitation can be achieved, one may unambiguously determine the radiative and nonradiative rate constants. This is the ideal experimental situation.

Excitation of a level which is uniquely characterized by a single set of a quantum numbers allows the extraction of the maximum amount of information about the system. Otherwise, it is necessary to use averaging or statistical methods. This is often the case in optical absorption for which band contour analysis is used to reproduce the shape of an observed rotational contour instead of the position and intensity of individual spectral lines.

Doppler broadening and a finite excitation bandwidth generally result in the creation of an ensemble of rotational and vibrational levels in the excited electronic state. Each of these levels may be characterized by different rate constants. This complication necessitates the explicit consideration of level populations within the ensemble as the first step in an analysis of excited state decay processes. We will, for simplicity, assume that the resolution of the excitation light is such that only one vibrational level is populated. Later in this section, we will relax this assumption and consider its implications. Thus, the various rotational states of a single vibronic level are taken as the components of the ensemble. In order to describe the excited state ensemble, it is necessary to begin with the ground state ensemble of rotational levels.

For any given temperature T, the ground state rotational distribution may be described by the equation

$$N(J, \tau) = (2J + 1)g_{\mathrm{N}} \exp[-E(J, \tau)/kT], \tag{2.1}$$

where (J, τ) denotes the rotational state of an asymmetric rotor, g_{N} is the nuclear statistical weight, and the exponential gives the Boltzmann factor. For naphthalene-h_8 the nuclear statistical weights are 136 for states of Wang symmetry E^+ and E^- and 120 for O^+ and O^-. Using Eq. (2.1) and a program to calculate asymmetric rotor energies, the ground state rotational distribution for this molecule has been calculated as a function of rotational energy. The results are shown in Fig. 1. An expanded scale would show only discrete peaks for each rotational level with a mean energy separation of about 0.04 cm^{-1}. Equation (2.1) and the corresponding shape of a Boltzmann distribution are, of course, well known. They have been presented here primarily for the sake of comparison with excited state distributions which are not as well documented.

Optical excitation maps the ground state distribution into the excited state. This mapping is a function of the excitation wavelength and resolution, temperature, and the quantum mechanical transition probabilities. The contour intensity as a function of energy (wavelength) is given by

$$C(E) = \sum_{i,f} C_{fi}(E), \tag{2.2}$$

where

$$C_{fi}(E) = N(J, K)(2J + 1)^{-1} I_{fi}\, \Delta(E - \Delta E_{fi}),$$

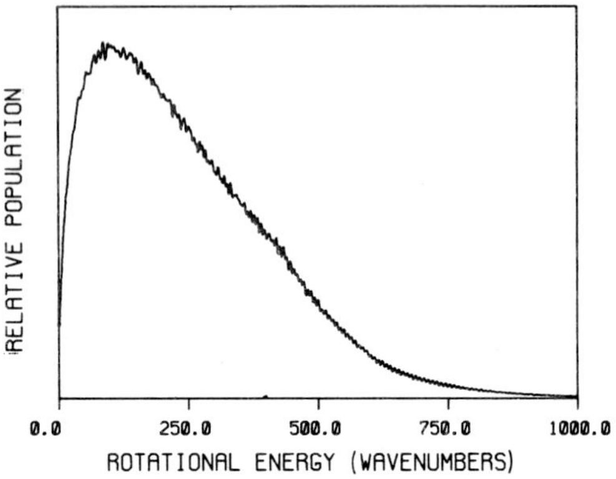

Fig. 1. The ground state rotational distribution of naphthalene h_8 at 298 K.

and where I_{fi} is the line strength of the transition and $\Delta(E - \Delta E_{fi})$ is the resolution function. The form of this function is often Gaussian for laser excitation and triangular if monochromators are used. The line strength of a symmetric rotor is given by

$$I_{J'K' \leftarrow JK} = (2J + 1)(2J' + 1)\tfrac{1}{3} \left| M_{\Delta K} \begin{pmatrix} J & J' & 1 \\ K & -K' & \Delta K \end{pmatrix} \right|^2 . \tag{2.3}$$

Terms of the type

$$\begin{pmatrix} J & J' & 1 \\ K & -K' & \Delta K \end{pmatrix}$$

are the Wigner 3-J symbols. They are discussed by Edmonds (1959) in great detail. $M_{\Delta K}$ is the electronic–vibrational integral of the dipole operator expressed in spherical tensor coordinates. Methods for transforming Eq. (2.3) to an asymmetric rotor line strength are given, for example, by Metz *et al.* (1978).

Rotational energies may be obtained using standard methods such as given by Allen and Cross (1963). A plot of $C(E)$ as a function of the energy E is the rotational band contour. Standard methods for computation of it have been described by Brand (1972).

In Fig. 2 we give a calculated band contour for the naphthalene-h_8 $8(b_{1g})_0^1$ optical transition. This is a vibronically induced type B contour. It is often the case that one wants to describe the structure underlying a contour. The most common method for doing this has been the use of a Fortrat plot. The Fortrat analysis shows the substructure of a band as a function of J and K. Such an analysis has been performed of the naphthalene contour by Hollas and Thakur (1971). However, as stated earlier we wish to concentrate not on the band contour but on the energy distribution of the excited state. To obtain this information, one requires only a trivial modification of the Fortrat code of a

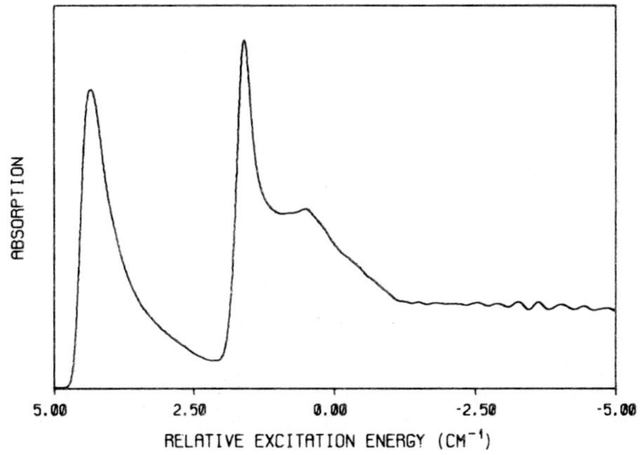

Fig. 2. The rotational band contour for the $8(b_{1g})_0^1$ transition in naphthalene h_8 at 298 K.

band contour program. Instead of storing J and K, one must store the rotational energy and transition probability as a function of excitation energy. From this data, the excited state rotational distribution may be obtained as a function of rotational energy for a given excitation energy and resolution.

However, there is an important case for which a calculation is not necessary. Consider the symmetric rotor with inertial constants which are equal in both electronic states, i.e., $A = A'$, $B = B'$, and $C = C'$. One finds that all Q branch transitions occur at exactly the same excitation energy.

This means that even infinite resolution will produce a Boltzmann distribution in the excited state. In real molecules this is usually not the case because the inertial constants in the electronic states generally differ. Thus, there may be an excitation bandwidth small enough to generate non-Boltzmann distributions. The limit of a non-Boltzmann distribution is single rovibronic level excitation. Resolution of ΔJ transitions generally requires an excitation bandwidth smaller than ΔC while resolution of ΔJ and ΔK transitions demands that the bandwidth be smaller than $\Delta A - \Delta C$.

Thus, it is usually the case that one must calculate the excited state distribution. Using the procedure outlined above, various distributions of rotational levels in the first excited singlet of naphthalene-h_8 were calculated. The calculations and relative wavelengths refer to Fig. 2. The resolution was fixed at 0.2 cm^{-1}. The calculated excited state rotational distributions are shown for relative excitation energies of 1.5 (red peak), 2.5, 3.5, and 4.23 cm^{-1} (blue peak) on Figs. 3–6, respectively. In Figs. 3 and 6 the distributions are bunched within a roughly 250 cm^{-1} energy region. However, Figs. 4 and 5 show bifurcated distributions, i.e., two primary groups centered at different rotational energies. These calculations show that the excited state distributions vary strongly with excitation energy. Variations of over 200–300 cm^{-1} seem to be typical. All

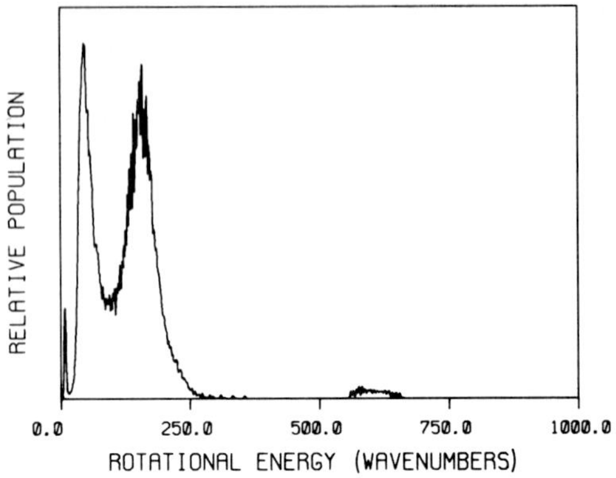

Fig. 3. The excited state rotational distribution of naphthalene h_8 resulting from excitation of the $8(b_{1g})_0^1$ band at $+1.5$ cm^{-1} from the band origin (red peak) with an excitation bandwidth of 0.2 cm^{-1}.

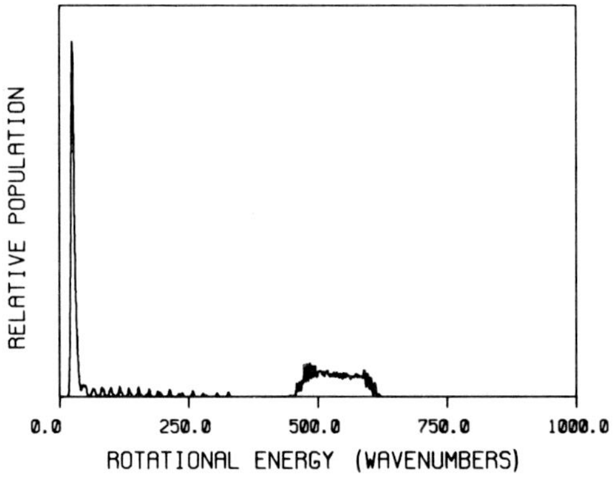

Fig. 4. The excited state rotational distribution for excitation at $+2.5$ cm^{-1} from the band origin. (Same data as Fig. 3.)

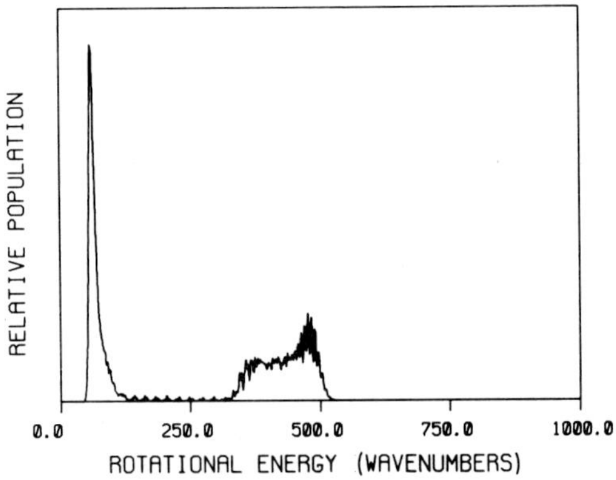

Fig. 5. The excited state rotational distribution for excitation $+3.5$ cm^{-1} from the band origin. (Same data as Fig. 3.)

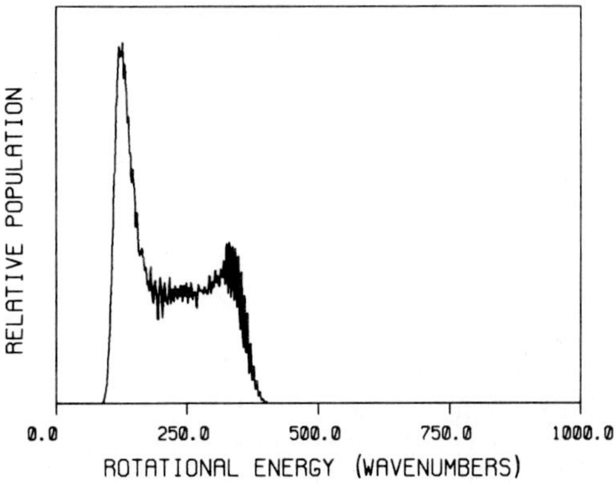

Fig. 6. The excited state rotational distribution for excitation at $+4.23\ \mathrm{cm}^{-1}$ from the band origin (blue peak). (Same data as Fig. 3.)

distributions are markedly non-Boltzmann and none are similar to Fig. 1. Note that the discrepancy between excitation bandwidth and the width of the excited distribution is about two orders of magnitude.

The effect of excitation bandwidth also has an important influence on these distributions. In Figs. 7 and 8, calculated excited state distributions are shown for excitation of the origin of the $8(b_{1g})_0^1$ band of naphthalene-h_8 (see Fig. 2) with bandwidths of 0.2 and $0.05\ \mathrm{cm}^{-1}$, respectively. A smaller bandwidth is seen to make the distribution more linelike. Conversely, an increase in the excitation bandwidth will lead to a broadening of the excited state rotational distribution. In the limit of an excitation bandwidth which is broader than the rotational contour, the ground state Boltzmann distribution will be mirrored in the excited state.

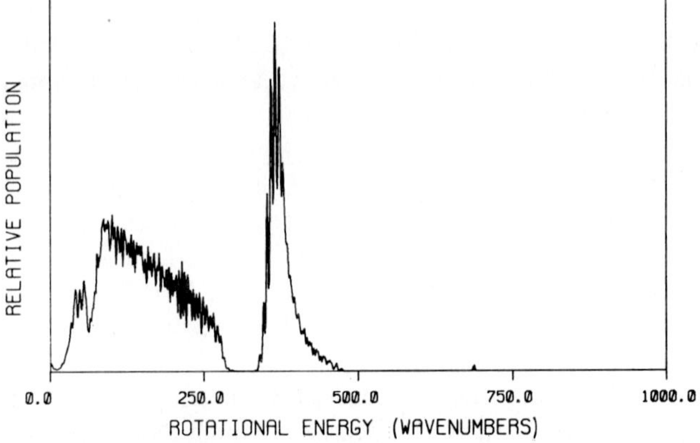

Fig. 7. The excited state rotational distribution of naphthalene h_8 for excitation at the origin of the $8(b_{1g})_0^1$ band with a resolution of $0.2\ \mathrm{cm}^{-1}$.

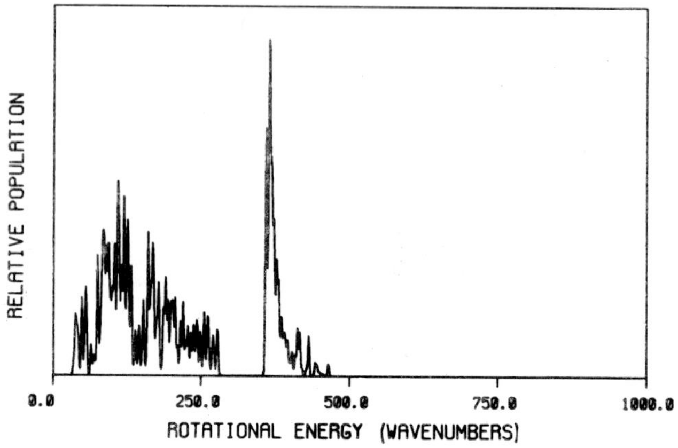

Fig. 8. The excited state rotational distribution of naphthalene h_8 for excitation at the origin of the $8(b_{1g})_0^1$ band with a resolution of 0.05 cm^{-1}.

The population of the excited state $|J'K'\rangle$ will be given by

$$N'(J', K', E) = \sum_i C_{fi}(E), \tag{2.4}$$

where E indicates that N' depends explicitly on the excitation bandwidth and wavelength. If the excitation bandwidth is large relative to the range of ΔE_{fi} and if the Boltzmann factor changes very slowly over the range $\Delta J = +1, 0, -1$ for each J, one finds that

$$N'(J', K', E) = N(J, K)[(2J' + 1)/(2J + 1)]\tfrac{1}{3}|M_Z|^2. \tag{2.5}$$

The relative M state degeneracies will only slightly modify the Boltzmann distribution for $\Delta J = \pm 1$ and not at all for $\Delta J = 0$. Note that broad bandwidth excitation makes the right-hand side of Eq. (2.5) independent of E. For a sufficiently large bandwidth, the excited state rotational distribution will always assume a near Boltzmann form.

We note that for a P or R branch, ΔE_{fi} will generally scan the entire range of the rotational contour. Thus, the excitation bandwidth Δ must cover the entire contour to generate an excited state Boltzmann distribution via $\Delta J = \pm 1$. For a Q branch, one finds that the transition energies from the various J levels are closely spaced. As discussed above, for $A = A'$, $B = B'$, and $C = C'$ the Q branch transitions of a symmetric rotor will all occur at the same transition energy.

We note that the ideal situation is excitation of single rovibronic levels. It is convenient to have a measure of the degree to which rotational purity may be achieved as a function of excitation resolution and wavelength. For this purpose we introduce the rotational purity index (RPI). It assumes a value of zero if the

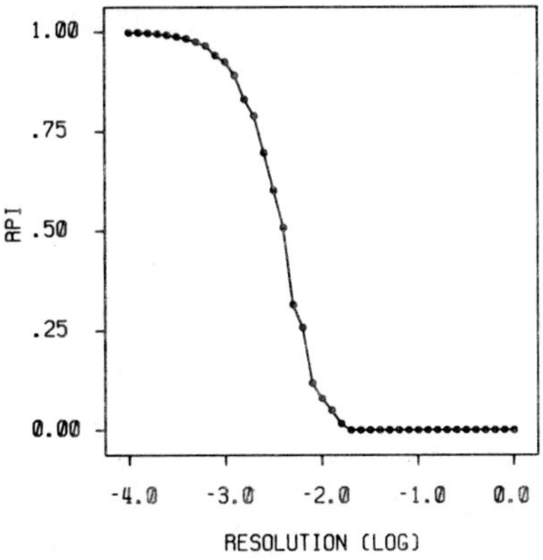

Fig. 9. The rotational purity index of naphthalene h_8 for excitation at the origin of the $8(b_{1g})_0^1$ band.

excitation always results in the population of two or more rotational levels and of unity if it never does.

In order to evaluate the RPI, transition energy space is divided into intervals of fixed width ε. Each interval is then subdivided into intervals of width Δ, corresponding to an excitation width. Within each interval $n\Delta$, for $n = 1$ to ε/Δ, the number of rotational states which have a nonzero population $N'(J', K', E)$ are counted. The RPI is then given by the ratio of the number of intervals having a unity count to those having a nonzero count. This is then repeated for various Δ. A plot of the RPI as a function of Δ then shows the probability of single rotational level excitation as a function of bandwidth within the interval ε.

With reference to Fig. 9, we have calculated the RPI for naphthalene-h_8 with $\varepsilon = 1\ \text{cm}^{-1}$ centered at $\Delta E = 0$ (the band origin). It is seen that multiple rotational level excitation always occurs at resolutions down to $\Delta \cong 10^{-2}\ \text{cm}^{-1}$. However, the Doppler limit occurs around 0.05 cm^{-1}. This means that no Doppler-limited experiments can result in single rotational level excitation at this wavelength, whatever the resolution may be. For Doppler-free experiments, single rotational level excitation will almost always be achieved at bandwidths below $10^{-3}\ \text{cm}^{-1}$. These values will vary with ΔE, temperature, and, of course, the molecule.

These excited state distributions define the initial ensemble of levels which may undergo radiative and nonradiative decay processes. In general these distributions vary as a function of excitation energy, resolution, and temperature. We note that they are also influenced by collisional processes, which tend to restore any distribution to the Boltzmann form. Collisional processes are discussed in Section III.

We have considered the rotational distributions associated with the excitation of a single vibronic level. We note that the vibrational purity of excitation can only be partially determined by the experimentalist. A very important effect is that of sequence congestion. It results from the fact that the vibrational levels in the electronic ground state are also populated according to the Boltzmann equation. Hot bands can thus contribute to the absorption spectrum. If the Franck–Condon factors for these modes are nearly degenerate in the vibrational quantum numbers, the primary intensity from sequence bands will be at an offset of $\Delta\omega = \omega - \tilde{\omega}$ from the main absorption band. ω and $\tilde{\omega}$ denote the excited and ground state vibrational frequencies, respectively. The modes of most importance have large Boltzmann factors and thus small frequencies. In naphthalene-h_8 the two observable sequences have ground state frequencies of 176 and 195 cm^{-1} for offsets of -10 and -55 cm^{-1}, respectively. The negative sign indicates that both occur to the red of the main absorption band. In this molecule one is thus limited to a region of roughly 10 cm^{-1} in which vibrational purity is relatively high. The regions of negligible sequence congestion will, of course, be different for each molecule. This effect compounds the difficulty of describing the excited state ensemble. In any experimental situation, it must be explicitly considered for a meaningful interpretation of the data.

There is a second source of vibrational impurity in the excited ensemble. This comes from the fact that at higher energies above the origin of an electronic state, the spacing of vibrational levels becomes increasingly smaller. (This is much different from rotational level spacings which are roughly equal to the inertial constant B at all energies.)

The increase in the vibrational density of states with total vibrational energy makes the excitation of only one mode very difficult at high energies. In fact, at high enough energies vibrational levels will overlap within their natural line-widths. The energy at which this sets in is known as the turnover point and has been discussed by Schlag and Howard (1976). It is given by the equation

$$1/\tau_{\text{obs}} = 2\pi c/\rho, \tag{2.6}$$

where τ_{obs} and ρ refer to the observed lifetime and the vibronic density of states, respectively. A plot of the left- and right-hand sides of Eq. (2.6) on the same figure as a function of total vibrational or excess energy allows a graphical determination of the turnover point from the point of intersection. This is illustrated in Fig. 10 for β-naphthylamine. Above the turnover point a quasi-continuum is said to exist. In the sense of the Heisenberg uncertainty principle, one may regard it as a true continuum. For many large, polyatomic molecules such as β-naphthylamine, napththalene, and aniline, the turnover point occurs at excess energies in the region of 4000–5000 cm^{-1}.

The turnover point, which results from lifetime broadening, is an upper limit, above which single vibronic level excitation can not be achieved. It is possible that vibrational impurity may dominate at even lower energies due to such effects as the Doppler limit (translational temperature) and sequence conjestion (vibrational temperature).

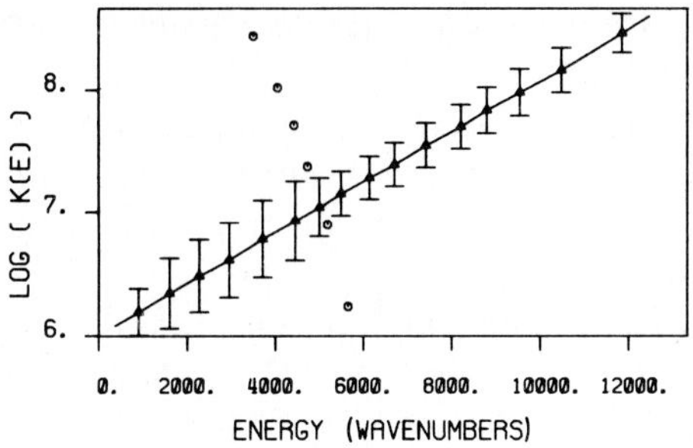

Fig. 10. The turnover point for β-naphthylamine. (From Schlag and Howard, 1976.)

In the discussion which follows we will assume that the excitation is well below the turnover point and that the resolution is sufficient to excite only single vibronic levels. Poor resolution is a trivial complication in a description of the ensemble. Excitation above the turnover point introduces new theoretical problems related to intramolecular energy transfer. Some of these problems have been discussed by Rice (1975).

III. ISOLATED MOLECULE CONDITIONS AND COLLISIONAL ENERGY REDISTRIBUTION

For a molecule to be isolated, the effect of the environment must be such that neither the energies of the molecular states are shifted on the relevant molecular energy scale nor are transitions induced between states on the relevant experimental time scale. These two requirements describe the maximum perturbation which can come from the environment. In the following we will assume that energy shifts are negligible and consider only collision induced state transitions. In particular, we are interested in any transition which alters the initially prepared distribution of rovibronic levels. In general, one requires that the time between collisions be long enough to allow the molecule to decay from the initially prepared level or ensemble of levels.

In the gas phase, collisional effects come from two sources: the cell wall and other molecules. Wall collisions are the most drastic. As a minimum, the initial rovibronic distribution is completely altered. Usually, the molecule will be returned to its electronic ground state. This effect is potentially a very great source of error in lifetime and quantum yield measurements. This is clearly seen, for example, in the work on SO_2 by Brus and McDonald (1973). They reported lifetimes which were longer than those observed in any previous experiment. They attributed this to the larger cell employed in their experiments which reduced the number of wall collisions. This problem has been analyzed in detail by Sackett (1972), and it is generally always possible to select a cell size large enough to eliminate this effect.

Much more important are molecule–molecule collisions. At high enough pressures, any rotational distribution assumes a thermalized form similar to Fig. 1 on a time scale short relative to radiative decay. This also applies to the vibrational distribution.

A dipole–dipole interaction which produces electronic energy transfer at large distances can occur. This effect has been postulated as operative in NO by Melton and Klemperer (1973) and in glyoxal by Anderson *et al.* (1973). However, for the singlet states of molecules the size of benzene, or larger, no experiments have indicated the presence of this effect. We note that the most common examples of electronic–electronic energy transfer processes are for systems with nonzero spin.

However, for all molecules, there are efficient processes which lead to transitions in the vibrational and rotational states. These processes are thus the most important for our consideration of effects which can alter the prepared rovibronic ensemble of levels. It is necessary to note that there is no finite pressure at which collisional effects are entirely absent. However, a decrease in pressure will generally decrease the probability of a collisionally induced transition. The collisionless limit as discussed in the literature is that pressure regime in which the probability of a collision is negligible on the time scale of the measurement. Careful reading of the discussions on the collisionless limit shows that experimentalists generally consider only processes which can change the vibronic level. Rotational transitions are usually ignored.

There is often some justification for this procedure. Let us assume that the excitation bandwidth is broad enough to excite a near Boltzmann distribution of rotational levels. Complete rotational thermalization in the excited state will then result in a negligible change in the population of each excited rotational level. Experimentally, one will not observe any effects due to these very minor changes in the excited distribution. The observable collisional effects will then be related to vibrational level redistribution. Hence, one is concerned with a collisionless limit correlated only with V–V transitions. Rotational transitions may be occurring but the rotational distribution is left essentially unchanged. The pressure regime in which V–V transitions are negligible on the experimental time scale is referred to here as the first collisionless limit.

Let us now consider the case of an excited state rotational distribution which is very non-Boltzmann, e.g., such as those shown in Figs. 3–8. In this case rotational thermalization results in a large change in the rotational distribution. If any observables depend on this distribution, collisional processes leading towards rotational thermalization will be important.

Generally, rotational cross sections are much larger than those of V–V transfer. There will thus exist a pressure regime for which rotational redistribution occurs rapidly on a time scale for which vibrational transitions may be neglected. This implies that there is a second collisionless limit which occurs at a lower pressure than the first collisionless limit and for which rotational redistribution is negligible. Thus, retention of the excited vibrational distribution requires one to measure at pressures below the first collisionless limit, while retention of the original rotational distribution requires one to measure below the second collisionless limit.

These collisionless limits are generally different for each molecule. However, we would like to give an order of magnitude estimate for them. For large molecules the V–V transfer cross section is often on the order of hard sphere cross sections. This is a result of the short range forces which induce the transitions. For molecules larger and heavier than benzene with lifetimes less than several hundred nsec, the first collisionless limit is usually in the range of 0.1–1 torr.

For example, Brown *et al.* (1975) have measured quenching rate constants for *p*-fluorotoluene. They obtained the value of $k_Q = 2.5 \times 10^{-11}$ molecules/cm^3/sec. With a lifetime of roughly 10 nsec, quenching will account for only 1% of the total vibronic state decay at 1 torr. The actual numerical values will change as a function of wavelength and molecule. However, the range of 0.1–1 torr seems to be typical for most large molecules.

For small molecules, the range may be different. In the case of glyoxal, Rordorf *et al.* (1978) have observed rotational cross sections of roughly 300 $Å^2$ while Beyer *et al.* (1976) have observed vibrational cross sections on the order of 14 $Å^2$. Thus, the pressure onset of the two collisionless limits will differ by about one order of magnitude. They found that vibrational relaxation occurs rapidly above 25 mtorr, which we may take as the pressure onset of the first collisionless limit.

If vibrational relaxation is negligible for glyoxal at 25 mtorr, a further decrease in pressure corresponds to a decrease in rotational redistribution. In the pressure region of 0.2–20 mtorr, Kuettner *et al.* (1978) have measured glyoxal phosphorescence. Their most important results are shown in Fig. 11. In contrast to earlier work, the phosphorescence of glyoxal-d_2 became constant in the region of 0–3 mtorr. This latter value is thus the pressure onset of the second collisionless region. This indicates that no collisional rearrangement is occurring in this region. A further increase in pressure above 3 mtorr is seen to increase the phosphorescence. This was attributed to collisional population of rotational levels which had a stronger coupling strength to the final triplet state. For glyoxal-h_2 the ISC is purely collision induced and it is seen from Fig. 11 that the phosphorescence decreases to zero in the limit of zero pressure.

For larger molecules, the rotational cross sections have been measured by microwave pressure broadening and are also on the order of 100–300 $Å^2$. This will put the second collisionless region in the pressure range of 1–10 mtorr for many molecules. Measurement below the second collisionless limit assures that the rovibronic levels which are observed to decay are the same levels populated by absorption. To have these measurements be distinguished from

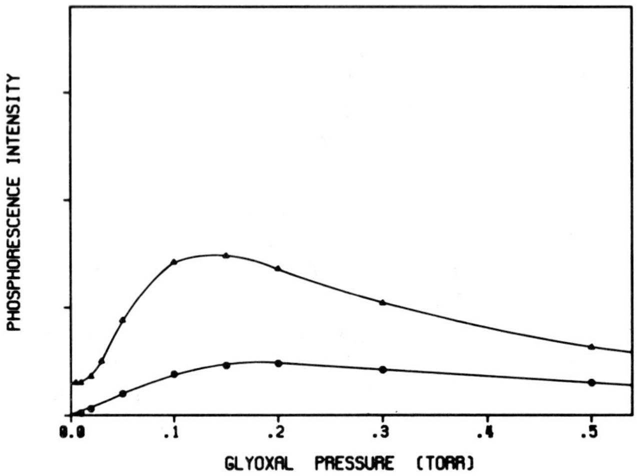

Fig. 11. The pressure dependence of the phosphorescence of glyoxal h_2 (circles) and d_2 (triangles). (From Kuettner *et al.*, 1978.)

those above this second collisionless limit one must prepare levels with a high degree of purity, i.e., nearly single rovibronic levels. In the case of many rotational levels, the initially prepared ensemble may well give results indistinguishable from a rotationally thermalized distribution above this limit. Between the first and second collisionless limits, one then has collisional redistribution of the initially prepared rotational levels and hence measures the properties of a rotational ensemble. In such a situation it is still possible to speak of single vibronic level excitation and decay. Only the rotational levels are scrambled. At pressures above the first collisionless limit both rotational and vibrational redistribution occur on the time scale of radiative decay and hence form a single Boltzmann ensemble. Measurement above either of the collisionless limits requires that the bimolecular rate constants for energy transfer be included in the rate equations describing the time development of the excited ensemble. This will be discussed in more detail in Section IX.

IV. EXPERIMENTAL EVIDENCE FOR A ROTATIONAL EFFECT IN NONRADIATIVE PROCESSES

Up to this point we have followed the excitation process from the ground state to an excited electronic state. We have shown that the Boltzmann ground state distribution is mapped into a non-Boltzmann excited state distribution and that various collisional processes tend to restore the excited ensemble to the Boltzmann form.

In the theory of nonradiative processes, it has long been assumed that every rotational level belonging to the same vibronic state is characterized by the same radiative and nonradiative rate constants. If this is the case, it is only necessary to excite a single vibronic level below the first collisionless limit in order to measure everything that can be measured. The nature of the excited state rotational distribution would be of no importance. It is thus necessary to question this assumption by consideration of the experimental observations.

The first lifetime spectrum of a molecular system in the gas phase was made by Schlag *et al.* (1967) and Schlag and von Weyssenhoff (1969) on β-naphthlyamine. The measured spectrum showed that the lifetime monotonically decreased with increasing excitation energy. However, no indication of the dependence of the nonradiative rate constant on vibrational or rotational excitation could be obtained, because of the broad bandwidth excitation.

Single vibronic level measurements of quantum yields and/or lifetimes have been made on benzene by Spears and Rice (1971) and on naphthalene by Schlag *et al.* (1971a) and Knight *et al.* (1973). Since these initial investigations, many other molecules have been studied under conditions of single vibronic level excitation. Without giving any details, we only wish to note that the nonradiative rate was observed to fluctuate in a nonmonotonic manner as a function of vibronic level energy or as a function of excitation energy. These experiments have demonstrated that the nature of the excited vibronic level plays an important role in nonradiative transitions. The increased resolution thus lead to conclusions which could not be reached by analysis of earlier data based on broad bandwidth excitation. At the same time, such experiments are not capable of giving information on the nonresolved rotational states. The important point is that only those experiments having at least partial rotational state resolution can indicate whether or not the rotational states play an important role in nonradiative transitions.

The first investigation of rotational state selection on intramolecular processes was made by Parmenter and Schuh (1972). Fluorescence quantum yields were measured at various positions along the rotational contours of four different vibronic levels of benzene-h_6. To within the stated accuracy of 10%, no dependence of the yield on the rotational distribution was found.

In contrast to these results, measurements by Boesl *et al.* (1975) along the 0–0 band of naphthalene-h_8 revealed strong structure in the lifetime spectrum. This is shown in Fig. 12. The maximum lifetime of 310 nsec was observed to coincide with the maximum of the absorption spectrum. This was the first observation of a rotational effect on the nonradiative process of ISC in a large molecule.

Fluorescence quantum yield measurements along the rotational contours of various vibronic levels of naphthalene-h_8 by Howard and Schlag (1976) have shown strong rotational state dependence. The results for excitation along the $8(b_{1g})_0^1$ band are shown in Fig. 13. As may be seen, the fluorescence quantum

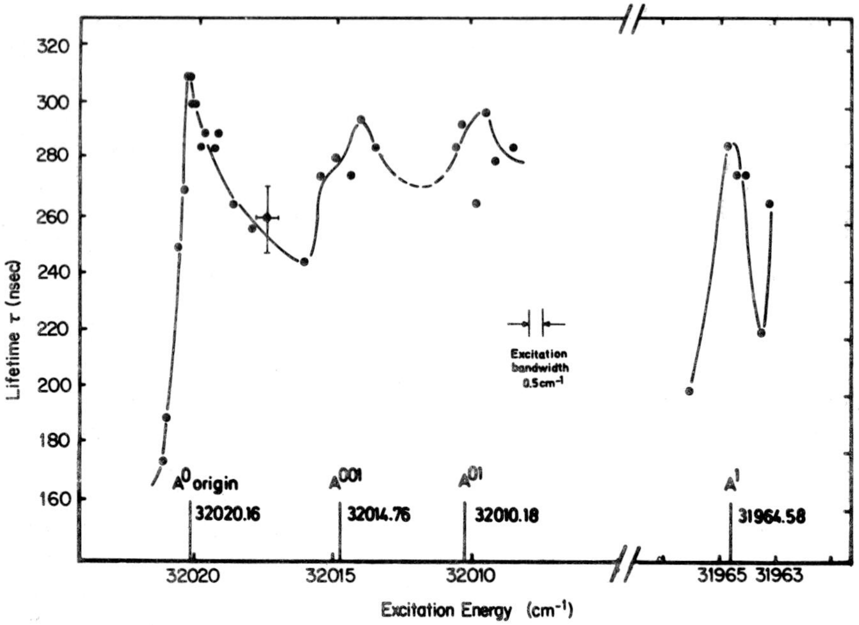

Fig. 12. Measured fluorescence lifetimes along the 0–0 band of naphthalene h_8. (From Boesl *et al.*, 1975.)

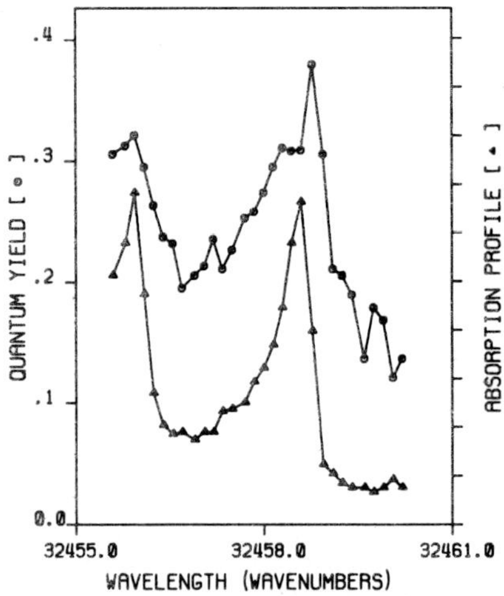

Fig. 13. Measured fluorescence quantum yields along the $8(b_{1g})_0^1$ band of naphthalene h_8. (From Howard and Schlag, 1976.)

yield shows large fluctuations as a function of position along the band contour and thus of the excited state rotational distribution.

These two sets of measurements indicate that ISC in naphthalene is strongly dependent upon the excited rotational distribution. Another demonstration of a dependence of the nonradiative rate constant on the rotational states would come from the observation of the decay properties of single rovibronic levels. Such measurements have been made on formaldehyde by Tang *et al.* (1977). Single rotational levels of the $2_0^3 4_0^1$ vibronic transition were populated using laser excitation. The results are shown in Fig. 14. Comparison of the fluorescence excitation spectrum (top of figure) with the absorption spectrum (bottom of figure) indicates that the relative fluorescence quantum yields (ratio of the curves) shows a large variation with rotational level excitation. The fluctuations extend over a factor of 4 and show no simple dependence on J and K. The nonradiative decay was explained via a mechanism of internal conversion.

We note that the experiments on naphthalene measured rotational ensembles while those on formaldehyde measured single rovibronic levels. With regard to rotational purity of excitation, the experiments on formaldehyde must be regarded as a better example of rotational effects in nonradiative processes. However, formaldehyde is an intermediate size molecule and the role of collisions is not completely clear. The formaldehyde experiments were made at a pressure of 200 mtorr, while those on naphthalene were made at 70 mtorr. Thus, the experiments on naphthalene probably had less collisional rearrangement of the rotational levels. To date no experiments have observed the decay of large polyatomic molecules excited below the second collisionless limit with single rovibronic level excitation. However, if the rotational distribution in formaldehyde or naphthalene had been complete, no rotational dependence in the observables could have been measured. We can thus conclude that the experiments which could partially or completely resolve the rotational levels demonstrate that a rotational effect exists in nonradiative transitions, that the effect can be very large and that it can not be ignored in any theoretical model. The excited state rotational distribution as discussed in Section II thus assumes a role of utmost importance. Each of the levels will be characterized by different nonradiative rate constants and the observed lifetimes and fluorescence quantum yields of the ensemble will not be related to the microscopic rate constants in a simple manner. The next step in our analysis of nonradiative processes will be a consideration of the coupling between a single rovibronic level in an excited singlet state with the final states of the nonradiative transition.

V. ROTATIONAL STATE MATRIX ELEMENTS AND SELECTION RULES

The problem to be considered in this section and the next is how the rotational states can influence the nonradiative rate constant. In this section we discuss the determination of the matrix elements coupling the initial state to

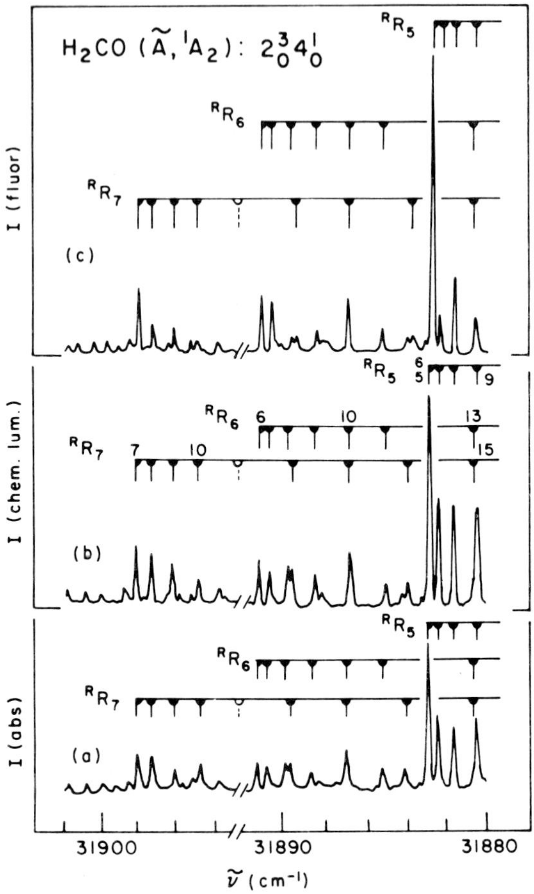

Fig. 14. A portion of H_2CO spectra of the $2_0^3 4_0^1$ transition:
(a) absorption spectrum with ordinate in absorbance unit;
(b) red HNO* chemiluminescence excitation spectrum;
(c) blue H_2CO^* fluorescence excitation spectrum. The rotational sub-branch assignment as a near prolate top is ${}^{\Delta K}\Delta J_{K''}$ with J'' shown on top of the line. The line for $J'' = 11$ of ${}^R R_7$ is not shown. (From Tang *et al.*, 1977.)

the final state. This requires a knowledge of the wavefunctions as well as the operators coupling the states. Let us first consider the wavefunctions.

To describe the wavefunctions, we first introduce the quantum numbers which we will use. J, N, and S denote the quantum numbers for total angular momentum including spin, rotational angular momentum, and electron spin angular momentum, respectively. The projection of J, N, and S on the space fixed axis system is denoted by M, M_N and M_S, respectively. Their projection on the molecular fixed axis system is denoted by Ω, K, and Σ, respectively. Other quantum numbers, such as those of nuclear spin angular momentum, will not be considered here.

We must now choose a set of basis functions for the molecular rotation problem. In accordance with the phase choice of Hougen (1970) and Brown and Howard (1976) we set

$$|JM\Omega\rangle = [(2J + 1)/8\pi^2]^{1/2} D^{J*}_{M\Omega}(\omega), \tag{5.1}$$

where $D^J_{M\Omega}(\omega)$ is the Wigner function. It is discussed in detail by Edmonds (1959). For the electron spin basis functions we choose $|S\Sigma\rangle$. This choice of basis functions has the advantage that for purely intramolecular processes the selection rules $\Delta J = 0$ and $\Delta\Omega = 0$ apply. This does not imply that J and Ω are good quantum numbers, but only that their associated operators commute with the operators responsible for inducing the state transitions. For the cases of interest here we will find that J is a good quantum number and Ω is not.

The basis set we have chosen may be designated by $|JM\Omega; S\Sigma\rangle$. This corresponds to a Hund's coupling case a. However, the molecules with which we will be concerned are generally characterized by a Hund's coupling case b. This case will be appropriate when spin splittings are small relative to rotational energies.

This will apply to most light, polyatomic molecules. It may not apply if the molecule contains very heavy atoms. We note that no molecule will be adequately described in Hund's case a if it is highly rotationally excited. The good quantum numbers in case b are J, S, N, K, and M with wavefunctions denoted by $|NKSJM\rangle$. The coupling of momenta in states described by these quantum numbers is given by

$$\tilde{J} - \tilde{S} = \tilde{N}, \tag{5.2}$$

and has been discussed by Hougen (1964). The relation between the two basis sets has been given by Brown and Howard (1976) and Howard and Schlag (1976) as

$$|\gamma; NKSJM\rangle = \sum_{\Omega,\Sigma} (-1)^{N-S+\Omega}(2N + 1)^{1/2} \begin{pmatrix} J & S & N \\ \Omega & -\Sigma & -K \end{pmatrix} |\gamma; JM\Omega; S\Sigma\rangle, \tag{5.3}$$

where γ denotes all other quantum numbers such as those of the vibronic states. We let $|\gamma\rangle$ denote adiabatic Born–Oppenhiemer states and assume that the electronic–vibrational problem has been solved. Terms of the form

$$\begin{pmatrix} J & S & N \\ \Omega & -\Sigma & -K \end{pmatrix}$$

are the Wigner 3-J symbols.

The distinction between case a and case b coupling disappears for $S = 0$, or for singlet states. For these states $J = N$, $\Omega = K$, and $\Sigma = 0$. Thus,

$$|\gamma; NK0NM\rangle = (-1)^{N-K}(2N+1)^{1/2}\begin{pmatrix} N & 0 & N \\ K & 0 & -K \end{pmatrix}|\gamma; NMK; 00\rangle$$

$$= |\gamma; NMK; 00\rangle. \tag{5.4}$$

It is now necessary to consider the operators which couple these states. The various Born–Oppenheimer approximations have been discussed by Schlag *et al.* (1971b). For that chosen here the Hamiltonian may be written as

$$\tilde{H} = \tilde{H}_0 + \tilde{T}_{\mathrm{N}} + \tilde{H}_{\mathrm{so}} \tag{5.5}$$

where $\tilde{H}_0$ is the zero-order Hamiltonian, $\tilde{T}_{\mathrm{N}}$ is the nuclear kinetic energy operator and $\tilde{H}_{\mathrm{so}}$ is the spin-orbit operator. The initially excited singlet state is coupled to another singlet state by $\tilde{T}_{\mathrm{N}}$ and to a triplet by $\tilde{H}_{\mathrm{so}}$. Using these two operators we now wish to obtain the rotational matrix elements. Hougen (1964) has discussed these matrix elements with respect to singlet–triplet absorption. Brand and Stevens (1973) have given matrix elements for the spin-orbit interaction. Howard and Schlag (1978b) have given the matrix elements of $\tilde{H}_{\mathrm{so}}$ and $\tilde{T}_{\mathrm{N}}$ for intersystems crossing between a singlet and a triplet and for internal conversion between two singlets or two triplets.

For internal conversion (IC), two singlets are coupled via

$$\langle\gamma'; N'M'K'; 00|\tilde{T}_N|\gamma; NMK; 00\rangle = \langle\gamma'|\tilde{T}_N|\gamma\rangle\delta_{N,N'}\delta_{M,M'}\delta_{K,K'}. \tag{5.6}$$

Thus, $\tilde{T}_N$ operates only on the electronic-vibrational wavefunctions and the rotational selection rules are given by

$$\Delta N = 0, \qquad \Delta K = 0, \qquad \text{and} \qquad \Delta M = 0.$$

No mention was made of the inertial constants of the two states. In general, they are different for the initial and final states. This results in a shift in the rotational energies. However, it does not affect the wavefunctions or the selection rules given above.

For intersystems crossing (ISC), a singlet and a triplet are coupled via

$$\langle\gamma'; N'K'S'J'M'|\tilde{H}_{\mathrm{so}}|\gamma; NK0NM\rangle$$

$$= \sum_{\Omega',\Sigma'}(-1)^{N'-S'+\Omega'}(2N'+1)^{1/2}\begin{pmatrix} J' & S' & N' \\ \Omega' & -\Sigma' & -K' \end{pmatrix}$$

$$\times \langle\gamma'; J'M'\Omega'; S'\Sigma'|\tilde{H}_{\mathrm{so}}|\gamma; NMK; 00\rangle. \tag{5.7}$$

The matrix element on the right-hand side of Eq. (5.7) may be simplified due to the requirements

$$\Delta J = 0, \qquad \Delta M = 0, \qquad \text{and} \qquad \Delta\Omega = 0.$$

This leaves only the matrix element

$$\langle \gamma'; S'\Sigma' | \tilde{H}_{so} | \gamma; 00 \rangle.$$

For those molecules of interest here the spin–orbit operator is given by a sum of two one-electron operators as

$$\tilde{H}_{so} = \tilde{h}_{so,1} + \tilde{h}_{so,2}. \tag{5.8}$$

The one electron operators may be written in spherical tensor coordinates to give

$$\tilde{H}_{so} = \lambda_1 \sum_{\alpha=\pm 1,0} (-1)^{\alpha} \tilde{l}_1^{\alpha} \tilde{s}_1^{-\alpha} + \lambda_2 \sum_{\alpha=\pm 1,0} (-1)^{\alpha} \tilde{l}_2^{\alpha} \tilde{s}_2^{-\alpha}, \tag{5.9}$$

where λ_i, $\tilde{l}_i^{\alpha}$, and $\tilde{s}_i^{\alpha}$ refer to the spin-orbit coupling constant, electron orbital angular momentum operator, and electron spin operator, respectively, for the ith electron.

Every tensor may be decomposed into a sum of symmetric and antisymmetric parts. For $\tilde{H}_{so}$ this is given by

$$\tilde{H}_{so} = \sum_{\alpha=\pm 1,0} \tilde{l}_-^{\alpha} \tilde{s}_-^{\alpha} + \sum_{\alpha=\pm 1,0} \tilde{l}_+^{\alpha} \tilde{s}_+^{\alpha}, \tag{5.10}$$

where

$$\tilde{l}_{\pm}^{\alpha} = \tfrac{1}{2}(\lambda_1 \tilde{l}_1^{\alpha} \pm \lambda_2 \tilde{l}_2^{\alpha}), \tag{5.11}$$

and

$$\tilde{s}_{\pm}^{\alpha} = \tfrac{1}{2}(-1)^{\alpha}(\tilde{s}_1^{-\alpha} \pm \tilde{s}_2^{-\alpha}). \tag{5.12}$$

For ISC only the antisymmetric component induces transitions because $\langle 1\Sigma | \tilde{s}_+^{\alpha} | 00 \rangle$ is identically zero for every choice of Σ and α. The matrix element of the antisymmetric spin operator is given by

$$\langle 1\Sigma | \tilde{s}_-^{\alpha} | 00 \rangle = \delta_{-\Sigma,\alpha}. \tag{5.13}$$

Thus, the final matrix element is

$$\langle \gamma'; N'K'S'J'M' | \tilde{H}_{so} | \gamma; NK0NM \rangle$$
$$= \sum_{\alpha} (-1)^{N+K} (2N'+1)^{1/2} \begin{pmatrix} N & N' & 1 \\ K & -K' & \alpha \end{pmatrix} \langle \gamma' | \tilde{l}_-^{\alpha} | \gamma \rangle. \tag{5.14}$$

The rotational selection rules may be obtained from the Wigner 3-J symbol and are

$$\Delta N = \pm 1, 0 \qquad \text{and} \qquad \Delta K = \pm 1, 0.$$

In a case b coupling the projection of S on the molecular axis Σ is not a good quantum number. However, spin–spin interactions are not eliminated. For a given $|NK\rangle$ state this interaction gives rise to three closely spaced sublevels.

These are denoted by F_1, F_2, and F_3 and are associated with $J = N + 1$, N, and $N - 1$, respectively. For a given J, the F_1, F_2, and F_3 sublevels lie in order of increasing energy. For a given N this is not necessarily the case. For a given rotational branch in ISC (ΔN and ΔK), only one of these three levels can be populated because of the $\Delta J = 0$ selection rule. This gives rise to additional selection rules which determine the spin sublevel which can be reached through a particular ΔN transition. These selection rules are as follows.

ΔN	spin sublevel
$+1$	F_3
0	F_2
-1	F_1

Note that these are not the three closely spaced spin sublevels belonging to the same $|NK\rangle$ state of a triplet. Their energy difference is on the order of rotational state energy spacings and all are characterized by the same value of J.

These matrix elements determine the contribution of the rotational states to the amplitude of the nonradiative transition rate. However, there are additional rotational effects which emerge by consideration of the full rate constant expression.

VI. THE RATE CONSTANT FOR NONRADIATIVE TRANSITIONS

It is not a trivial matter to write the quantum mechanical expression for the nonradiative rate constant. For a given perturbation Hamiltonian $\tilde{H}'$ which can induce a transition from initial state $|i\rangle$ to a final state $|f\rangle$, a simple solution of the time dependent Schrödinger equation gives

$$k_{\mathrm{nr}} = 2\pi \sum_f |\langle f|\tilde{H}'|i\rangle|^2 \left[\frac{\sin(\Delta E_{fi}t)}{\pi\, \Delta E_{fi}}\right] \equiv 2\pi \sum_f |\langle f|\tilde{H}'|i\rangle|^2\, \Delta_t(\Delta E_{fi}), \quad (6.1)$$

which reduces to

$$k_{\mathrm{nr}} = 2\pi \sum_f |\langle f|\tilde{H}'|i\rangle|^2\, \delta(\Delta E_{fi}) \quad (6.2)$$

in the limit of $t \to \infty$. Goldberger and Watson (1964) have given a higher-order solution valid to all orders of perturbation theory. Their equation is

$$k_{\mathrm{nr}} = 2\pi \sum_f |\langle f|\tilde{R}(E)|i\rangle|^2\, \delta(\Delta E_{fi}), \quad (6.3)$$

where $\tilde{R}(E)$ is the level shift operator. The lowest order solution of Eq. (6.3) is Eq. (6.2).

A closed form solution of Eq. (6.2) was given by Bixon and Jortner (1968). It was based on the following assumptions:

(1) All matrix elements are equal, i.e., $\langle f|\tilde{H}'|i\rangle = V$.
(2) All final states are equally spaced with a spacing of $\varepsilon = 1/\rho$ where ρ is the density of states.

Their final equation was

$$k_{\mathrm{nr}} = 2\pi|V|^2\rho. \tag{6.4}$$

This equation has the form of the Fermi golden rule and gives the rate constant as the product of a coupling strength and density of states. This equation can not be used for realistic calculations because neither of these two basic assumptions are fulfilled for real molecular systems. The basic problem with this and other density of states models such as those given by Siebrand (1971) is the use of the preaveraging assumption. It assumes the separability of ρ and $|V|$ such that

$$\langle V^2\rho\rangle = \langle V^2\rangle\langle\rho\rangle. \tag{6.5}$$

From Eq. (6.2) the assumption is that $|\langle f|\tilde{H}'|i\rangle|^2$ can be factored in front of the sum over delta functions. This is particularly bad if the matrix elements exhibit large fluctuations. As will be discussed in Section VIII, this assumption completely breaks down with respect to the vibrational overlap factors between two bound electronic states.

Christie and Craig (1972b) have introduced the concept of an effective density of states. From Eq. (6.5) $|V|$ was allowed to assume two different values: zero and a constant. The density of states was then calculated only for the nonzero $|V|$. Equation (6.4) is thus seen to be valid if $|V|$ is either constant or partitioned into zero and nonzero (but constant) blocks.

Most models perform the calculation of the rate constant in the time representation. This is accomplished by a Fourier or Laplace transform of Eq. (6.2). Such models are discussed in great detail in the review of Freed (1976). These models thus take into account the variation in the vibrational matrix elements under the sum of Eq. (6.1) or Eq. (6.2). Thus, the full Franck–Condon weighed density of states (FCWDOS) is explicitly considered. As discussed in the previous section, the rotational matrix elements can also vary as a function of the final states. It is thus also desirable to explicitly incorporate these rotational matrix elements into a rate expression.

The time representation does not easily allow the addition of these terms. Thus, let us reconsider Eq. (6.2). The formulation of this equation requires that initial and final states be degenerate. However, from the Heisenberg uncertainty principle, energy matching is required only within the natural linewidths of the states. Freed (1970) has noted that it is more realistic to replace $\Delta_f(E)$ in Eq. (6.1) with a Lorentzian of width γ, the linewidth of the state, instead of with a delta function. We then have

$$\Delta_t(E) \to \Delta_\gamma(E) = \gamma[E^2 + (\gamma/2)^2]^{-1}. \tag{6.6}$$

Such rate equations have been considered by Nitzan and Jortner (1971, 1973) and by Medvedev *et al.* (1977) within a time representation. However, we will find it more convenient to work directly in the energy representation. This method has been discussed by Metz (1976). Making the substitution of Eq. (6.6) in Eq. (6.1) gives

$$k_{\mathrm{nr}} = 2\pi \sum_f |\langle f|\tilde{H}'|i\rangle|^2 \Delta_\gamma(\Delta E_{fi}). \tag{6.7}$$

It is assumed that γ is the same for all final states. We will take Eq. (6.7) to be the basic rate equation for nonradiative transitions in all that follows.

At this point we can substitute the matrix elements of the rotational states into Eq. (6.7) and obtain an equation for the rate constant. We will denote the electronic and vibrational wavefunctions by $|\phi\rangle$ and $|\chi\rangle$, respectively. In addition, we will assume that the product $|\phi\chi\rangle$ adequately describes the BO states. For IC we have $\tilde{H}' = \tilde{T}_{\mathrm{N}}$ and

$$k_{\mathrm{IC}} = 2\pi \sum_{\mathrm{all}'} |\langle\phi'\chi'|\tilde{T}_{\mathrm{N}}|\phi\chi\rangle|^2 \delta_{N,N'}\delta_{K,K'}\delta_{M,M'}\,\Delta_\gamma(\Delta E_{fi}). \tag{6.8}$$

For ISC, $\tilde{H}' = \tilde{H}_{\mathrm{so}}$ and

$$k_{\mathrm{ISC}} = 2\pi \sum_{\mathrm{all}'} \left| \sum_\alpha (-1)^{N+K}(2N'+1)^{1/2} \begin{pmatrix} N & N' & 1 \\ K & -K' & \alpha \end{pmatrix} \langle\phi'\chi'|\tilde{l}^{\alpha}_{-}|\phi\chi\rangle \right|^2 \Delta_\gamma(\Delta E_{fi})$$

$$= 2\pi \sum_{\mathrm{all}'} \sum_\alpha (2N'+1) \begin{pmatrix} N & N' & 1 \\ K & -K' & \alpha \end{pmatrix}^2 |\langle\phi'\chi'|\tilde{l}^{\alpha}_{-}|\phi\chi\rangle|^2 \Delta_\gamma(\Delta E_{fi}). \tag{6.9}$$

It is possible to further separate out the vibrational terms. Let us now define the function

$$D(E) = \sum_{\chi'} |\langle\phi'\chi'|\tilde{H}'|\phi\chi\rangle|^2 \Delta_\gamma(E + \Delta E^{\mathrm{v}}_{fi} + \Delta E^{\mathrm{e}}_{fi}), \tag{6.10}$$

where $\tilde{H}'$ equals $\tilde{T}_{\mathrm{N}}$ or $\tilde{l}^{\alpha}_{-}$ for Eq. (6.8) or (6.9), respectively. We assume that the vibrational wavefunctions $|\chi\rangle$ can be represented by the product of single oscillator wavefunctions corresponding to the normal modes of the molecule. Thus,

$$|\chi\rangle = \prod_{i=1} |n_i\rangle, \tag{6.11}$$

where n_i is the vibrational quantum number of the ith normal mode. Equation (6.10) can now be rewritten as

$$D(E) = |\langle\phi'|\tilde{H}'|\phi\rangle|^2 \sum_{\{n'\}} \prod_{i=1}^{3N-6} |\langle n_i'|n_i\rangle|^2 \Delta_\gamma(E + \Delta E^{\mathrm{e}} + \Delta E^{\mathrm{v}}). \tag{6.12}$$

The summation is over all permutations of final state vibrational quantum numbers. We note that if $\tilde{H}'$ operates on a single vibrational mode (a promoting mode), then the mode can be factored in front of the product, which will then only have $3N - 7$ terms. The constant promoting mode energy difference may then be subsumed under the electronic state energy difference ΔE^{e} and taken out of the vibrational level energy difference ΔE^{v}. We now define

$$D(E) = |\langle\phi'|\tilde{H}'|\phi\rangle|^2 F_\gamma(E + \Delta E^{\mathrm{e}}), \tag{6.13}$$

where

$$F_\gamma(E) = \sum_{\{n'\}} \prod_{i=1}^{3N-6} |\langle n_i'|n_i\rangle|^2 \, \Delta_\gamma(E + \Delta E^{\mathrm{v}}). \tag{6.14}$$

The function $F_\gamma(E)$ is the Franck–Condon weighed density of states (FCWDOS).

The final rate constant expressions become

$$k_{\mathrm{IC}} = 2\pi|\langle\phi'|\tilde{T}_{\mathrm{N}}|\phi\rangle|^2 \delta_{N,N'}\delta_{K,K'}\delta_{M,M'} F_\gamma(\Delta E^{\mathrm{e}} + \Delta E^{\mathrm{r}}), \tag{6.15}$$

and

$$k_{\mathrm{ISC}} = 2\pi \sum_{\alpha=\pm 1,0} |\langle\phi'|\tilde{l}^{\alpha}_{-}|\phi\rangle|^2 (2N' + 1)\begin{pmatrix} N & N' & 1 \\ K & -K' & \alpha \end{pmatrix}^2 F_\gamma(\Delta E^{\mathrm{e}} + \Delta E^{\mathrm{r}}). \tag{6.16}$$

Equations (6.15) and (6.16) are in a convenient form for calculations. The electronic, vibrational, and rotational contributions are clearly separated. As mentioned at the end of Section V, new rotational effects can appear in these equations which do not directly come from the rotational matrix elements. This is discussed in the next section.

VII. ENERGY MATCHING IN ISOLATED MOLECULES

The equations for the nonradiative rate constants contain a linewidth function which requires that initial and final states be isoenergetic within their natural linewidths. This is explicitly described by the $\Delta_\gamma(E)$ function in Eqs. (6.8) and (6.9) and has important consequences for the rotational states. Careful consideration of these two equations shows that the rotational energies must also be included in the energy matching function $\Delta_\gamma(E)$.

Writing the arguments of the linewidth function in full gives

$$\Delta_\gamma(\Delta E_{fi}) = \Delta_\gamma(E_f - E_i) = \Delta_\gamma[(E_f^{\mathrm{e}} + E_f^{\mathrm{v}} + E_f^{\mathrm{r}}) - (E_i^{\mathrm{e}} + E_i^{\mathrm{v}} + E_i^{\mathrm{r}})], \tag{7.1}$$

where superscripts e, v, and r refer to electronic, vibrational and rotational energies, respectively. Isoenergetic conditions are given by

$$\Delta E^{\mathrm{e}} + E_i^{\mathrm{v}} + E_i^{\mathrm{r}} = E_f^{\mathrm{v}} + E_f^{\mathrm{r}}, \tag{7.2}$$

where ΔE^{e} is the constant electronic energy gap including vibrational zero-point energies and possibly promoting mode energies. Equation (7.2) may be rearranged to give

$$\Delta E^{\mathrm{e}} + (E_i^{\mathrm{r}} - E_f^{\mathrm{r}}) = (E_f^{\mathrm{v}} - E_i^{\mathrm{v}}). \tag{7.3}$$

The right-hand side of this equation is the vibrational energy difference of initial and final states. In all models discussed in the literature it has been equated with ΔE^{e}. Inclusion of the rotational energy difference may be viewed as a change in the effective energy gap. The new energy gap must then be filled by the vibrational levels.

Another, equivalent view is obtained by rewriting Eq. (7.2) as

$$\Delta E^{\mathrm{e}} = (E_f^{\mathrm{v}} - E_i^{\mathrm{v}}) + (E_f^{\mathrm{r}} - E_i^{\mathrm{r}}) \equiv -(\Delta E^{\mathrm{v}} + \Delta E^{\mathrm{r}}). \tag{7.4}$$

For a given initial vibrational level, it may happen that a particular final vibrational level is isoenergetic for $\Delta N = 0$ and $\Delta K = 0$. This is illustrated in the following diagram.

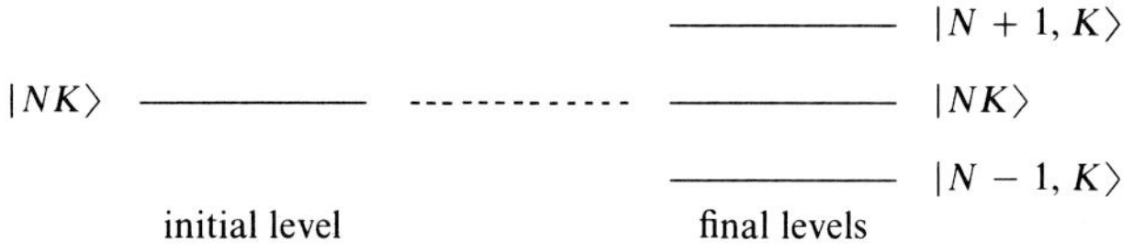

The other two rotational levels, i.e., $|N+1, K\rangle$ and $|N-1, K\rangle$, which belong to the same final vibrational level, do not fulfill the isoenergetic condition even though they are symmetry allowed for ISC.

We note that it is possible to find final rotational states $|N+1, K\rangle$ and $|N-1, K\rangle$ which are isoenergetic with the initial rovibronic level by changing the vibrational energy of the final state via ΔE^{v}. In the following diagram the arrows indicate the direction in which ΔE^{v} must be changed.

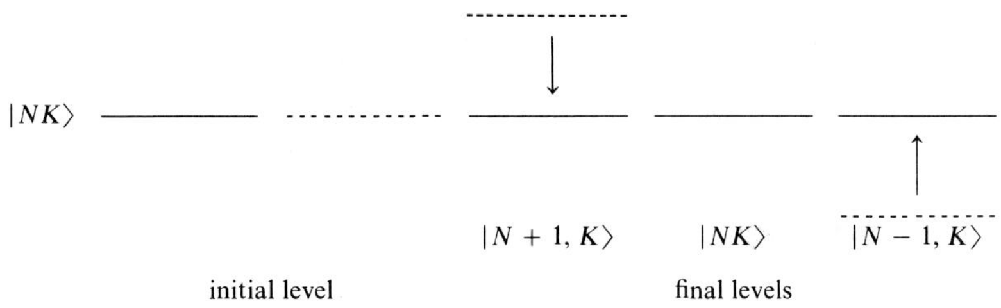

Thus, vibrational energy in the final state must be decreased for higher rotational levels and increased for lower rotational levels. This is the essential meaning of Eq. (7.4).

The conclusion of this analysis is that each of the nine rotational branches $(\Delta N, \Delta K)$ of a given initial rotational state $|NK\rangle$ is associated with a different set of final vibrational levels. This is true, of course, only if the energy difference ΔE^{r} is larger than the linewidth γ.

It is important to note that ΔE^{r} is not necessarily small. Such quantities are directly observable in high resolution optical spectra. They correspond to the transition energy offset from the rotational origin along an absorption contour. Thus, the rotational contour is an effective mapping of ΔE^{r} between two electronic states. It is not uncommon for ΔE^{r} in optical spectra to exceed 50–80 cm^{-1}. Since optical absorption is also characterized by the selection rules $\Delta N = \pm 1, 0$ and $\Delta K = \pm 1, 0$, shifts in this range can also be expected for ISC.

The analogy to optical spectroscopy can help clarify this point by consideration of the requirement

$$h\nu = \Delta E^{\mathrm{e}} + \Delta E^{\mathrm{v}} + \Delta E^{r}, \tag{7.5}$$

where ΔE refers to the upper state energy minus the lower state energy. This requirement states that the photon energy $h\nu$ must equal the transition energy. For the isoenergetic, nonradiative process one must set $h\nu = 0$. For optical absorption, a change in ΔE^{r} is reflected in a change in the photon energy $h\nu$. For the nonradiative process there are no photons and a change in ΔE^{r} must be accounted for by a change in ΔE^{v}. The rotational branches (P, Q, and R for $\Delta N = -1$, 0, and $+1$, respectively) in the optical process generally scan different energy regions of photon energy space. For the isoenergetic process, the rotational branches scan different energy regions of the final vibrational states. We can refer to this effect as energy gap branching. It arises from the fact that a given rotational branch in an isoenergetic process can effectively alter the energy gap as viewed by the vibrational states.

Although internal conversion, in contrast to intersystems crossing, is restricted to $\Delta N = 0$ and $\Delta K = 0$, energy gap branching can also play an important role. To illustrate this, consider the equation for the energy of a prolate symmetric rotor. This is given by

$$E_{N,K} = C[N(N + 1)] + (A - C)K^2. \tag{7.6}$$

The transition energy between $|NK\rangle$ and $|N'K'\rangle$ is given by

$$\Delta E^{\mathrm{r}} = E_{N',K'} - E_{N,K},$$

which for $\Delta N = 0$ and $\Delta K = 0$ is

$$\Delta E^{\mathrm{r}} = \Delta C[N(N + 1)] + (\Delta A - \Delta C)K^2. \tag{7.7}$$

Thus, only when the inertial constants in the two states are equal, i.e., $\Delta A = \Delta C = 0$, do all rotational transitions in IC occur at the same energy. In real molecules this is virtually never the case. A Q branch may be narrow but hardly ever transform limited. By analogy, one expects that the vibrational energy region scanned by the IC transitions will be narrower than that scanned by the ISC transitions. However, ΔE^{r} in both cases should be on the order of tens of wavenumbers.

Up to this point we have primarily considered the rotational energies in the final state of a nonradiative process. We have shown that different rotational states often require coupling with different vibrational states. However, we have not yet shown that this can have an effect on the rate constant. In particular, if the Franck–Condon factors between initial and final states are constant over the energy region scanned by ΔE^{r}, one may employ the sum rule over the Wigner 3-J symbols in Eq. (6.9):

$$\sum_{N', K'} (2N' + 1) \begin{pmatrix} N & N' & 1 \\ K & -K' & \alpha \end{pmatrix}^2 = 1. \tag{7.8}$$

Equation (6.16) then assumes the form

$$k_{\mathrm{ISC}} = 2\pi \sum_{\alpha} |\langle \phi' | \tilde{l}^{\alpha}_{-} | \phi \rangle|^2 F(\Delta E^{\mathrm{e}}). \tag{7.9}$$

This equation predicts that there are no rotational effects. This leads to the important conclusion that rotational structure will be absent in a nonradiative transition if the vibrational levels are characterized by an $F(E)$ function which varies slowly and smoothly over the energy region scanned by ΔE^{r}. The positive form of this statement is that structure in $F(E)$ over the range of ΔE^{r} induces rotational structure in ISC or IC. In ISC the induced structure is not just a function of $F(E)$ but also of the rotational matrix elements. It is now apparent that the exact form of $F(E)$ has important consequences for nonradiative transitions. In the next section we consider the vibrational states in detail.

VIII. VIBRATIONAL EFFECTS IN NONRADIATIVE TRANSITIONS

In the theory of nonradiative transitions, the models may be divided into two classes on the basis of their treatment of the vibrational modes. The first consists of density of states models. The only parameters required are the final state vibrational frequencies. The final density of states $\rho(E)$ gives the number of vibrational levels per unit of total vibrational energy. The second class of models considers the full Franck–Condon weighed density of states $F(E)$. A calculation of $F(E)$ requires knowledge of initial and final state vibrational frequencies and of normal mode displacements.

Our primary interest in this section is with the form of the function $F(E)$. However, the density of states $\rho(E)$ is so closely related to $F(E)$ that a discussion of both is necessary. We wish to note that some of the more important recent advances in this area stem from an algorithm of Bayer and Swinehart and the subsequent analysis of $\rho(E)$ by Stein and Rabinovitch (1972) and of $F_{\gamma}(E)$ by Metz (1976).

Let us first consider the density of states $\rho(E)$. The energy in vibrational degrees of freedom above the vibrationless or spectroscopic origin of an electronic state is given by the formula

$$E = \sum_{\mu} m_{\mu} \omega_{\mu}, \tag{8.1}$$

where E is in units of $\hbar$. In general, many different sets of vibrational quantum numbers $\{m\}$ may generate a state with the same total energy E. The number of sets $\{m\}$ per unit of vibrational energy is the density of states.

Early attempts to evaluate $\rho(E)$ with direct count methods proved to be very time consuming. Other methods focused on the classical equation for the density of states. Both of these techniques have been well reviewed by Schlag *et al.* (1965) and Sandsmark (1967). Much attention was also directed to the fact that the vibrational partition function could be written as a Laplace transform of the density of states. This is given by

$$Q_v(s) = \int_0^\infty \rho(E) \exp(-sE)\, dE. \tag{8.2}$$

The density of states can thus be represented as the inverse Laplace transform of the partition function. This yields

$$\rho(E) = \frac{1}{2\pi} \int_{-\infty}^{\infty} Q_v(s) \exp(sE)\, ds. \tag{8.3}$$

The solution to this inversion is a finite series of roughly $s/2$ terms (for s oscillators). The first terms were given by Schlag and Sandsmark (1962) and Haarhoff (1963). The complete solution was given by Thiele (1963).

The vibrational partition function for a harmonic oscillator can be written in closed form as

$$Q_v(s) = \prod_\mu [2 \sinh(\omega_\mu s/2)]^{-1}. \tag{8.4}$$

Haarhoff expanded Eq. (8.4) in a power series and then took the Laplace transform. The result is the well-known Haarhoff formula:

$$\rho(E) = E^{3N-7}\left[(3N-7)! \prod_\mu \omega_\mu\right]^{-1} (1 + \text{higher order terms}). \tag{8.5}$$

An alternative to the power series expansion has been the use of the method of steepest descents as described by Hoare and Ruijrok (1970). The saddle point s^* is obtained from the equation

$$\frac{\partial}{\partial s}(sE + \ln Q_v(s))|_{s=s^*} = 0. \tag{8.6}$$

The integral of Eq. (8.3) was then reduced to a Gaussian, which could be solved exactly. This method, like the series expansion, forces a certain smoothing which is justifiable only in the limit of high energies.

A most powerful technique is a direct count algorithm developed by Bayer and Swinehart and discussed by Stein and Rabinovitch (1972). Instead of using the permutations of all vibrational quantum numbers, the BS algorithm incorporates a method of iterative convolution for the calculation of $\rho(E)$. Before giving their results, we must first introduce several definitions and the notation which will be used.

The density of states is often defined by

$$\rho(E) = \sum_f \delta(E - E_f). \tag{8.7}$$

However, this formula is somewhat misleading, since the definition is the number of states per unit vibrational energy. Any direct count will partition final energies into finite intervals of width Δ. The total count within any interval is not in general the density of states, since it is the number of states per unit Δ. The actual density of states is given by the count in an interval divided by the interval width Δ. A calculation which directly counts the states is thus related to the formula

$$\rho^{\Delta}[n\Delta \leq E < (n+1)\Delta] = \sum_f \delta^{\Delta}(n\Delta - E_f) \equiv \rho^{\Delta}(n), \tag{8.8}$$

where the δ^{Δ} function assumes the values

$$\begin{aligned} \delta^{\Delta}(n\Delta - E_f) &= 1 \qquad \text{if} \quad n \leq E_f/\Delta < n + 1, \\ \delta^{\Delta}(n\Delta - E_f) &= 0 \qquad \text{otherwise.} \end{aligned} \tag{8.9}$$

The correct density of states is given by

$$\rho(E) = \rho^{\Delta}(n)\Delta^{-1}. \tag{8.10}$$

Note that this insures that the equation

$$\int_0^E \rho(E')\, dE' = \sum_{i=0}^{n=E/\Delta} \rho^{\Delta}(i) \tag{8.11}$$

is independent of Δ.

As the next step, all oscillator energies ω_μ are reduced to integers ε_μ such that

$$\omega_\mu/\Delta \leq \varepsilon_\mu < \omega_\mu/\Delta + 1.$$

The total vibrational energy becomes

$$E_f = \sum_\mu m_\mu \varepsilon_\mu \Delta, \tag{8.12}$$

which yields

$$\rho^{\Delta}(n) = \sum_{\{m\}} \delta\left(n - \sum_\mu m_\mu \varepsilon_\mu\right). \tag{8.13}$$

This is the basic permutational equation for density of states calculations. As written, it is valid only for harmonic oscillators since $m_\mu \varepsilon_\mu$ is taken to be the energy of the mth level. It may be applied to anharmonic oscillators if $m_\mu \varepsilon_\mu$ is replaced by the reduced anharmonic energy ε_{μ, m_μ} of the mth level.

The addition of a single oscillator μ to a set of $\mu - 1$ oscillators will change the density of states of the system. If the initial density of states is denoted $\rho_{\mu-1}^{\Delta}(n)$, then a new density of states can be generated via iteration of the equation

$$\rho_{\mu}^{\Delta}(n) = \sum_{m_\mu} \rho_{\mu-1}^{\Delta}(n - m_\mu \varepsilon_\mu) \tag{8.14}$$

from $n = \varepsilon_{\max}$ to ε_μ where $\varepsilon_{\max}$ is the maximum energy to be considered in the calculation. This is the basic equation for an iterative convolution solution for the density of states, As for Eq. (8.13). it may be used for anharmonic oscillators if ε_{μ, m_μ} is used as the oscillator energy.

For a single oscillator, there is only one level at $n = 0$ to be convoluted. Thus,

$$\rho_0^{\Delta}(n) = \delta(n), \tag{8.15}$$

and the single oscillator density of states function becomes

$$\rho_{\mu}^{\Delta}(n) = \delta(n - m_\mu \varepsilon_\mu). \tag{8.16}$$

Iteration of Eq. (8.14) for $n = \varepsilon_{\max}$ to ε_μ and for all modes $\mu = 1$ to $3N - 6$ generates the total density of states. Metz (1976) has noted that for a calculation employing this technique, the quantities $\rho_{\mu}^{\Delta}(n)$ and $\rho_{\mu-1}^{\Delta}(n)$ of Eq. (8.14) may use the same storage vector since counting from the top never destroys data which is later required. This is important if the levels are not harmonic.

The BS algorithm for harmonic oscillators is based on the fact that harmonic oscillator energies come in equally spaced intervals. The density of states is given by iteration of the equation

$$\rho_{\mu}^{\Delta}(n) = \rho_{\mu-1}^{\Delta}(n) + \rho_{\mu}^{\Delta}(n - \varepsilon_\mu) \tag{8.17}$$

from $n = \varepsilon_\mu$ to $\varepsilon_{\max}$ and for all modes $\mu = 1$ to $3N - 6$. This algorithm requires much less computer time than Eq. (8.14) since it eliminates the need to perform an extra sum over the quantum numbers m_μ. In addition, it can be very efficiently programmed since $\rho_{\mu}^{\Delta}(n)$ and $\rho_{\mu-1}^{\Delta}(n)$ can occupy the same location in core. We will refer to this algorithm as the harmonic iterative convolution technique.

Stein and Rabinovitch (1972) have also given an algorithm based on Eq. (8.17) for anharmonic oscillators. Although it is somewhat faster than the iterative convolution algorithm of Eq. (8.14), the SR anharmonic algorithm requires twice the storage requirements of all those discussed. For detailed information the reader is referred to the original reference.

Using Eq. (8.17) we have calculated the density of states for naphthalene-h_8 with a set of estimated T_1 frequencies. The results are shown in Fig. 15. The frequency scale gives the energy gap relative to the spectroscopic origin of S_1. With 48 normal modes, a nearly exponential increase in $\rho(E)$ is observed around the energy gap of 10,626 cm^{-1} (relative zero). The scale is linear and the value at the relative zero is 3×10^{11} states/cm^{-1}.

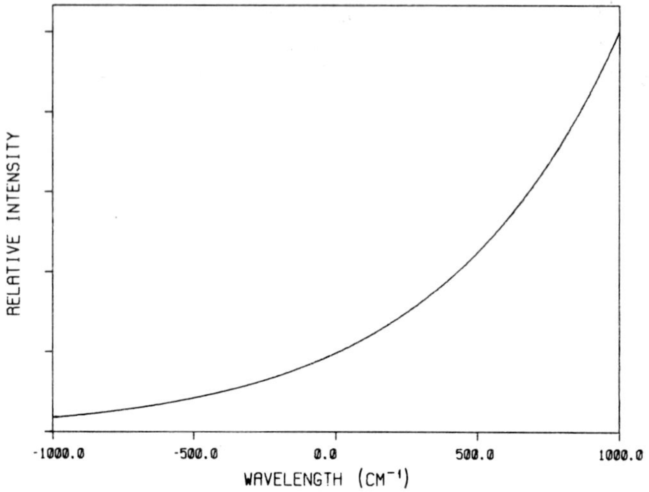

Fig. 15. The density of T_1 vibrational states for naphthalene h_8 relative to the S_0–S_1 energy gap.

Equation (8.7) is the basic equation for density of states calculations. The analog of this equation for a Franck–Condon weighed density of states calculation is

$$F(E) = \sum_f |\langle \chi_i | \chi_f \rangle|^2 \, \delta(E - \Delta E^v). \tag{8.18}$$

Comparison with Eq. (8.7) shows that the difference in $\rho(E)$ and $F(E)$ is that the latter sums the Franck–Condon factors at a given energy E (or energy interval) while $\rho(E)$ simply counts the states. They may be made numerically equal by setting $|\langle \chi_i | \chi_f \rangle|^2 = 1$ for all i and f.

Christie and Craig (1972b) have introduced the concept of an effective density of states. They noted that the Franck–Condon factor of a nontotally symmetric mode was identically zero if the change in vibrational quantum number was odd. This comes from the general selection rule that the overlap of two wavefunctions $\langle \psi_1 | \psi_2 \rangle$ is identically zero if the product of the representations of ψ_1 and ψ_2 does not contain the totally symmetric representation. For two non totally symmetric harmonic oscillators, the product of the representations is nontotally symmetric if the change in the quantum number is odd. Thus, the states could not couple and were not included in the density of states calculation. On the basis of Eq. (8.18), their criteria would be

$$|\langle \chi_i | \chi_f \rangle|^2 = 0 \qquad \text{if identically zero by symmetry, and}$$
$$|\langle \chi_i | \chi_f \rangle|^2 = 1 \qquad \text{otherwise.}$$

Thus, their $\rho_{\text{eff}}(E)$ gave an indication of the number of modes coupled at any given energy but no indication of their coupling strength.

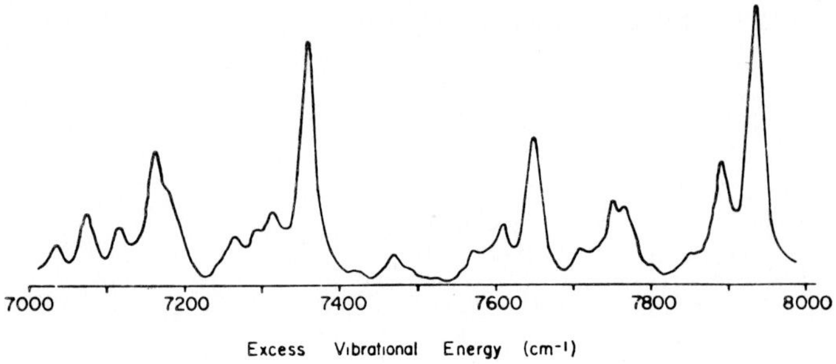

Fig. 16. Franck–Condon weighed density of states for the $S_1 \leftarrow S_0$ transition in benzene h_6 as a function of the electronic gap. (From Christie and Craig, 1972b.)

In order to get the proper vibrational coupling between the initial and final states, one must explicitly evaluate the Franck–Condon factors $|\langle \chi_i | \chi_f \rangle|^2$ and not just assign them a value of zero or unity. Expanding the vibrational wavefunctions via Eq. (6.10), we may rewrite Eq. (8.18) as

$$F(E) = \sum_{\{m\}} \left(\prod_{\mu} |\langle n_\mu | m_\mu \rangle|^2 \right) \delta\left(E - \sum_{\mu} m_\mu \omega_\mu \right). \tag{8.19}$$

This is an analog of Eq. (8.13) (but without pregraining) and is the basic equation for a permutational evaluation of $F(E)$. The first calculation of $F(E)$ using this method was by Christie and Craig (1972b). They evaluated $F(E)$ with a Lorentzian linewidth [$F_\gamma(E)$ given by Eq. (6.12)] of $\gamma = 10\ \mathrm{cm}^{-1}$ for the $S_0 \leftarrow S_1$ IC transition in benzene-h_6. Because a permutational approach was used to evaluate $F(E)$, the calculation was limited to only about 150 of the over 10^8 levels in the benzene ground state.

However, their results are very interesting. They are shown in Fig. 16. The figure shows a very structured coupling strength between the two electronic states as a function of the electronic energy gap. They reported that this structure depended on the width γ but that no γ gave a smooth curve.

As in calculations of $\rho(E)$, Laplace and Fourier transform techniques have been used to evaluate $F_\gamma(E)$. The vast majority of these techniques have employed saddle point procedures. We will briefly discuss them later but the interested reader is referred to Freed (1976) for a comprehensive review of such techniques. We wish to concentrate on direct count methods for evaluating $F_\gamma(E)$, since they are not only facile but exact.

Equation (8.19) can be simplified by reducing E and ω_μ to the integers n and ε_μ by a graining factor Δ as previously described. This yields the equation

$$F(n) = \sum_{\{m\}} \left(\prod_{\mu} |\langle n_\mu | m_\mu \rangle|^2 \right) \delta\left(n - \sum_{\mu} m_\mu \varepsilon_\mu \right). \tag{8.20}$$

This equation is the direct analog of Eq. (8.13) including pregraining for evaluation of $F(n)$.

The correct $F(E)$ is given by

$$F(E) = F(n)\,\Delta^{-1} \tag{8.21}$$

so that the relation

$$\int_0^E F(E')\,dE' = \sum_{i=0}^{n=E/\Delta} F(i) \tag{8.22}$$

is independent of Δ.

Metz (1976) has shown that Eq. (8.20) may be written as the $(3N-6)$th iteration of

$$F_\mu(n) = \sum_{m_\mu} F_{\mu-1}(n - m_\mu \varepsilon_\mu)|\langle n_\mu|m_\mu\rangle|^2, \tag{8.23}$$

from $n = \varepsilon_{\max}$ to ε_μ, where $F_0 = \delta(n)$ and $F_i(n)$ is the ith iteration. This is the analog of Eq. (8.14) for calculation of $\rho(E)$ and is the method of convolutive iteration. This algorithm may be programmed very efficiently since $F^\mu(n)$ and $F_{\mu-1}(n)$ are equivalent in terms of data storage. The last iteration of Eq. (8.23) gives

$$F_{3N-6}(n) = F(n).$$

This may be substituted into Eq. (8.21) to get $F(E)$.

We note that in Eq. (8.18) we have defined $F(E)$ in terms of a delta function while $F_\gamma(E)$ in Eq. (6.14) is defined in terms of the linewidth function. This can be easily rectified if we set

$$F_\gamma(E) = \sum_{n=0}^{\infty} F(n)\left\{\frac{(\gamma/2\pi)}{(E-n\Delta)^2 + (\gamma/2)^2}\right\}. \tag{8.24}$$

We would like to know under which conditions the formula corresponding to Eq. (8.22) will hold for $F_\gamma(E)$. To determine this we integrate both sides of Eq. (8.24) from $E' = 0$ to E. This gives

$$\begin{aligned}\int_0^E F_\gamma(E')\,dE' &= \sum_{n=0}^{\infty} F(n) \int_0^E dE' \left\{\frac{\gamma/2\pi}{(E'-n\Delta)^2 + (\gamma/2)^2}\right\} \\ &= \sum_{n=0}^{\infty} F(n)\pi^{-1}\left[\arctan\left(\frac{-2\Delta n + 2E}{\gamma}\right) + \arctan\left(\frac{2n\Delta}{\gamma}\right)\right] \\ &\equiv \sum_{n=0}^{\infty} F(n) G(n, \Delta, \gamma, E). \end{aligned} \tag{8.25}$$

We require that

$$F(E) = \lim_{\gamma\to 0} F_\gamma(E),$$

since the normalized Lorentzian becomes a delta function in this limit. Let us thus consider

$$\lim_{\gamma\to 0} G(n, \Delta, \gamma, E).$$

In this limit the arguments of the arctan function approach either $+\infty$ or $-\infty$ for which the arctan assumes values of $\pi/2$ or $-\pi/2$, respectively. We find that

$$\lim_{\gamma\to 0} G(n, \Delta, \gamma, E) = 1 \qquad \text{if} \quad n\Delta < E,$$
$$= 0 \qquad \text{if} \quad n\Delta > E. \tag{8.26}$$

Thus,

$$\lim_{\gamma\to 0} \int_0^E F_\gamma(E')\, dE' = \sum_{n=0}^{E/\Delta} F(n), \tag{8.27}$$

in correspondence with Eq. (8.22).

The only problem remaining for an evaluation of $F_\gamma(E)$ is the determination of the single oscillator Franck–Condon factors $\langle n_\mu | m_\mu \rangle$. We will consider only displaced, distorted harmonic oscillators. They are characterized by frequencies $\tilde{\omega}_\mu$ and ω for the initial and final states, respectively. In addition, the reduced displacements of the totally symmetric modes are given by

$$\tilde{\Delta}_\mu = \tilde{\omega}_\mu^{1/2}(\tilde{Q}_\mu - Q_\mu), \qquad \text{where} \qquad Q_\mu = (M_\mu/2)^{1/2} R_\mu$$

and R_μ is the equilibrium position of mode μ. At the low vibrational energies to be considered, anharmonicities should not be very important. However, closed form solutions for the Franck–Condon factors of Morse oscillators have been given by Kuehn *et al.* (1977).

One of the first analytic formulas for the Franck–Condon factors of displaced, distorted harmonic oscillators was given by Katriel (1970). Nitzan and Jortner have given expressions for displaced, identical surfaces (1971) and for undistorted oscillators (1972). Other formulas have been given by Christie and Craig (1972a) and by Heller *et al.* (1972). We have found that the formula of Metz (1975) is very convenient since it accommodates all possibilities. It may be programmed to give a compact routine which uses harmonic oscillator recursion relations to compute rapidly the matrix elements $\langle n | m \rangle$ for all m, given initial quantum number n for mode μ.

Defining

$$\beta = (\tilde{\omega} - \omega)/2(\omega\tilde{\omega})^{1/2} \qquad \text{and} \qquad \alpha = (1 + \beta^2)^{1/2},$$

the integrals are given by

$$\langle n | m \rangle = [m!\,n!\,\alpha^{-1} \exp(-\tilde{\Delta}^2(\alpha - \beta)/\alpha)]^{1/2}$$
$$\times \sum_{k=0}^{\min(n,m)} \left[R_{m-k}\left(-\frac{\beta}{\alpha}, -\frac{\tilde{\Delta}}{\alpha}\right) R_{n-k}\left(\frac{\beta}{\alpha}, \frac{\alpha-\beta}{\alpha}\tilde{\Delta}\right) \left(\frac{\alpha^{-k}}{k!}\right) \right], \tag{8.28}$$

where $R_1(x, y)$ satisfies the recursion relation

$$R_l(x, y) = [2xR_{l-2}(x, y) + yR_{l-1}(x, y)]l^{-1}$$

with $R_0(x, y) = 1$ and $R_{-1}(x, y) = 0$.

In Fig. 16 we have shown one calculation of $F_{\gamma}(E)$ which used the permutational method. Metz (1976) has performed calculations on the $T_1 \leftarrow S_1$ transition in benzene using all Franck–Condon factors in the iterative convolution scheme just discussed. Calculations were made with grain sizes of $\Delta = 50, 10$, and $1\ cm^{-1}$. A decrease in Δ resulted in an increase in the average fluctuations in $F_{\gamma}(E)$. At $\Delta = 50\ cm^{-1}$ fluctuations were about one order of magnitude, while at $\Delta = 1\ cm^{-1}$ they were around five orders of magnitude. However, using just an iterative convolution scheme, it is never necessary to neglect certain modes on an *a priori* basis as done by Christie and Craig to keep the computation manageable.

It was of interest to investigate whether such a severe truncation in the counting problem can be justified. For this purpose, in a given energy region, contributions to $F_{\gamma}(E)$ as well as $\rho(E)$ were considered only for those levels above a preset minimum vibrational overlap factor. The calculations were repeated for lower and lower minimum values until all states were counted. The numerical results are shown in Fig. 17. The ordinate gives the ratio of the integrated $F_{\gamma}(E)$ considering only contributions in excess of a given value to the integrated $F_{\gamma}(E)$ with all states included in the calculation. The integration region was always $1000\ cm^{-1}$. The abscissa gives the log of the ratio of the number of states included in the calculation to the total number of states in the energy region.

One of the most interesting features of this figure is that the integrated $F_{\gamma}(E)$ reaches over 95% of its final value when only a miniscule fraction of all states, i.e. 10^{-6}, have been counted. This result is seen to be the same for vastly differing

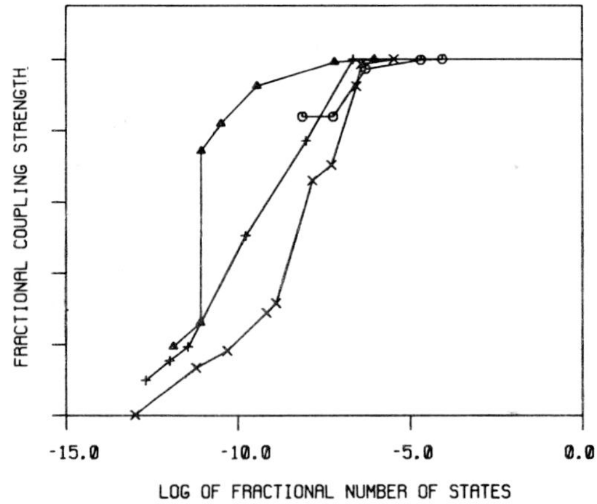

Fig. 17. The ratio of the integrated Franck–Condon weighed density of states $F_{\gamma}(E)$ with limited contributions to the integrated $F_{\gamma}(E)$ with all states included is plotted as a function of the $\log_{10}$ of the ratio of the number of states contributing to $F_{\gamma}(E)$ to the total number of states in the energy region. Four different energy regions for benzene h_6 are considered. (○) 7000–8000, (△) 15,000–16,000, (+) 24,000–25,000, and (×) 39,000–40,000.

energy ranges. For the energy region 7000–8000 cm^{-1}, we find that $10^{-6} \times 10^{8} = 100$ states are responsible for over 95% of the total coupling strength.

This implies that the Christie–Craig calculation shown in Fig. 16 (which considered 150 states) should not differ considerably from a complete calculation based on Eq. (8.23). The small number of states contributing to $F_\gamma(E)$ in this low energy region makes a selective, permutational approach feasible here. However, for the energy region 39,000–40,000 cm^{-1}, the 95% point is reached at $10^{-6} \times 10^{19} = 10^{13}$ states. Such a large number of states makes any permutational method much too time consuming. It is thus necessary to use iterative techniques such as given by Eq. (8.23) at moderate to high energy gaps.

On the other hand, if one employs a saddle point calculation, this completely washes out all structure in $F_\gamma(E)$. The time representation or generating function approach has been discussed by Freed (1976). We note only that it corresponds to $F_\gamma(E)$ in the limit of very large grains. In fact, Nitzan and Jortner (1971) have used grains as large as $\Delta = 1000\ cm^{-1}$ to smooth out the excess energy dependence in their rate constant calculations.

One may, of course, choose Δ as small as one likes and then introduce smoothing through the linewidth γ as shown by Eq. (8.24). In fact, it is this method which gives the most realistic model for averaging. Considered from this view, one is very limited in the choice of γ because it must correspond to the linewidth given by

$$\gamma = \hbar/\tau$$

where τ is on the order of the lifetime of the final states of the nonradiative process.

In the solid phase, phonon coupling may rapidly depopulate the initial vibrational level reached by the nonradiative process. Very short lifetimes result in large linewidths γ. In this case we may expect that saddle point methods are applicable.

In the gas phase the situation is quite different. The primary process for triplet state depopulation is through collision. Measurements by Schroeder *et al.* (1977) on naphthalene have shown that triplet state vibrational thermalization occurs on a timescale of tens of microseconds. Although this timescale does not reflect rotational transitions, gas phase triplet state, vibrational level lifetimes should generally be much longer than those in the solid phase. The linewidth becomes much narrower and saddle point procedures are expected to be invalid.

It is interesting to compare $\rho(E)$ with $F_\gamma(E)$. In Fig. 15, a calculation of $\rho(E)$ for triplet vibrational levels of naphthalene-h_8 has been given. We have also a calculated $F_\gamma(E)$ for the $T_1 \leftarrow S_1$ transition in this molecule using a set of vibrational parameters corresponding to a weak coupling case. The final state linewidth was set at 25 cm^{-1}. The results of this calculation are shown in Fig. 18. The energy scale is the energy gap relative to T_1 and the intensity scale is linear.

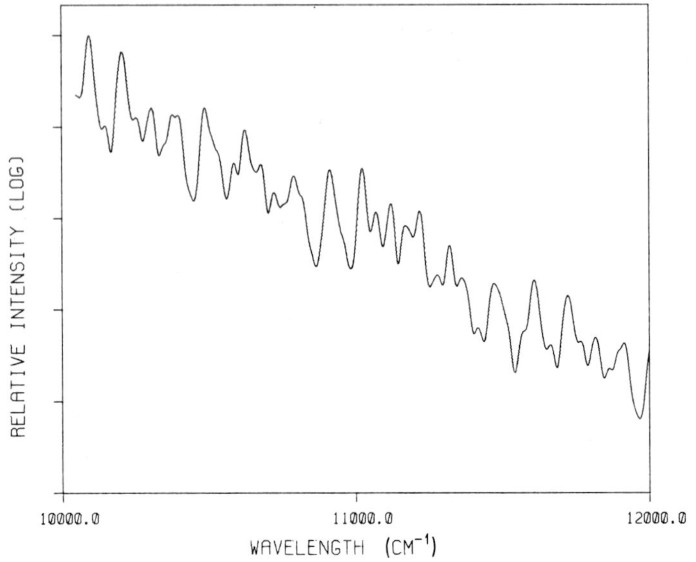

Fig. 18. The Franck–Condon weighed density of states F$\gamma(E)$ for the $T_1 \leftarrow S_1$ transition in naphthalene h_8 with a linewidth of $\gamma = 25\ \text{cm}^{-1}$. The energy scale is relative to T_1.

In contrast to $\rho(E)$, the $F_\gamma(E)$ shows a general decrease with increasing energy gap.

We note that the increase in $\rho(E)$ with E and the decrease in $F_\gamma(E)$ with E is an important difference. The energy gap law states that the rate of radiationless decay out of the vibrationless level of an electronic state to another electronic state decreases exponentially with increasing energy gap between the states. This behavior is predicted by virtually all models incorporating $F_\gamma(E)$ in the time or energy representation. The pure density of states is known to increase exponentially with energy, from which the decrease demanded by the energy gap law requires a very strongly decreasing function $\langle V^2 \rangle$ to be superimposed. This alone makes any separability problematical.

We have also performed the calculation on napthalene-h_8 with a linewidth of 2.5 cm^{-1}. This is shown of in Fig. 19. The fluctuations are about one order of magnitude higher than those seen in Fig. 18. Over the very narrow energy range shown in Fig. 19, very little exponential decrease can be seen since $F_\gamma(E)$ is dominated by large fluctuations. However, this is approximately the energy range scanned by the rotational energy difference ΔE^{r}.

As noted in Section VII, structure in $F_\gamma(E)$ is a prerequisite for rotational structure to appear in the final rate constant expression. As shown in Figs. 17–19, this structure exists for molecules in the gas phase, given a suitable choice of γ.

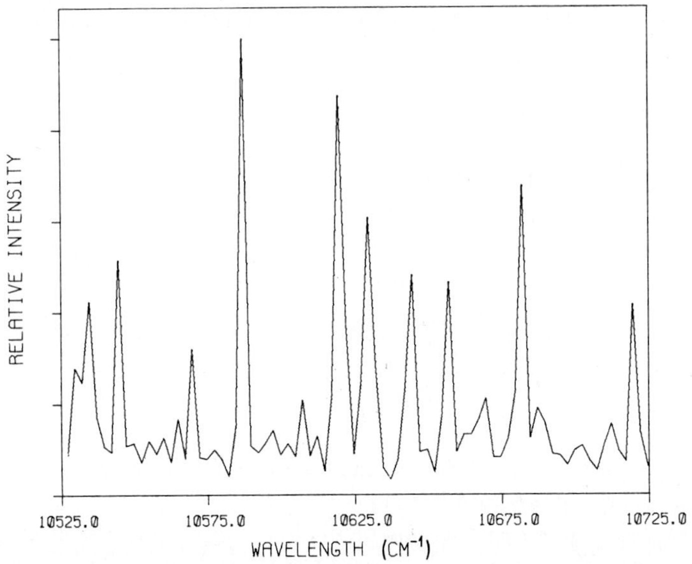

Fig. 19. The Franck–Condon weighed density of states for the $T_1 \leftarrow S_1$ transition in naphthalene h_8 with $\gamma = 2.5\ \text{cm}^{-1}$. (From Howard and Schlag, 1978a.)

We conclude that rotational structure is to be expected in nonradiative rate constants for molecules in the low pressure gas phase. In addition, saddle point solutions to the time representation of $F_\gamma(E)$ should be generally inapplicable to isolated molecules, at least in the lower vibrational levels of the lower electronic states.

IX. THE DECAY OF A MULTILEVEL SYSTEM

Methods have been described by which one can calculate initial rotational level populations and their nonradiative rate constants including the rotational and vibrational matrix elements. The final question which we wish to discuss is how one describes the decay of this ensemble in terms of the microscopic parameters. In general, the observation of fluorescence from an ensemble of electronically excited molecules allows the determination of the ensemble lifetime τ and fluorescence quantum yield ϕ. The relations most commonly employed to relate these observables to the microscopic rate constants are

$$\phi = k_r(k_r + k_{nr})^{-1} \tag{9.1}$$

and

$$\tau = (k_r + k_{nr})^{-1}. \tag{9.2}$$

In Eqs. (9.1) and (9.2), k_r is the radiative rate constant for fluorescence (or phosphorescence) and k_{nr} the nonradiative rate constant for such processes as ISC or IC.

In a multilevel system, these equations are not always valid. The relation between the observables and microscopic rate constants then becomes much more complicated. In this section we discuss the conditions under which Eqs. (9.1) and (9.2) are valid and give the relations which apply when the above equations are not valid.

For a single excited level which decays according to the equation

$$dn(t)/dt = -(k_r + k_{nr})n(t), \tag{9.3}$$

the time dependence of the level is given by

$$n(t) = n(0)\exp(-(k_r + k_{nr})t). \tag{9.4}$$

In Eq. (9.4), $n(0)$ is the population of the level resulting from excitation with a pulse which is a delta function in time at $t = 0$.

The observed signal for collection of all photons with unit efficiency is given by

$$S(t) = k_r n(t) = k_r n(0)\exp(-(k_r + k_{nr})t). \tag{9.5}$$

The quantum yield is given by

$$\phi = \int_0^\infty S(t)\,dt/n(0) = k_r(k_r + k_{nr})^{-1}, \tag{9.6}$$

which is just that given by Eq. (9.1). The lifetime is most generally characterized by the mean first-passage time of the signal function. We note that only this definition of a lifetime allows any random decay curve to be characterized by a unique time constant. For the signal given by Eq. (9.5) one has

$$\tau = \int_0^\infty S(t)t\,dt \bigg/ \int_0^\infty S(t)\,dt = (k_r + k_{nr})^{-1}, \tag{9.7}$$

which is the value given by Eq. (9.2). However, from Eq. (9.5) the signal is seen to decay exponentially according to $\exp(-t/\tau)$ where τ is given by Eq. (9.7). For a single level, the mean first-passage time and the exponential decay time are equal.

The assumptions used in the derivation of these equations are as follows:

(1) A single level is excited by the light source.
(2) The excitation source is a delta function in time.
(3) There are no collisional processes.

The fact that many experiments do not even approximately fulfill the above requirements makes it necessary to explicitly consider their effect on the observables.

A weaker form of condition (1) is that all levels of an ensemble be characterized by the same microscopic rate constants k_r and k_{nr}. This was long considered to be the case for different rotational levels belonging to the same electronic–vibrational state. However, as discussed in previous sections, different rotational levels may be characterized by different nonradiative rate constants. Thus, the weaker form of condition (1) does not hold in general. Gas phase excitation must result in single rovibronic level excitation for Eqs. (9.1) and (9.2) to be applied.

Stochastic theories describing multilevel relaxation have been widely discussed in the literature. Oppenheim *et al.* (1967) have reviewed the general theory of multistate relaxation. Particular applications to chemistry have been discussed by McQuarrie (1967). Molecular decay following sinusoidal excitation has been described by Schlag *et al.* (1969). Multilevel relaxation following pulsed excitation has been discussed in the cases of chemical reactions by Valance (1967) and of fluorescence decay by Howard and Schlag (1976). The latter work, however, assumed that conditions (1) and (2) were fulfilled.

The influence of a finite width excitation pulse on the signal has also received much attention in the literature. In general, the observed signal $S'(t)$ is given by the convolution integral of the signal following delta function excitation $S(t)$ with the excitation waveform $R(t)$. This is given by

$$S'(t) = \int_0^t dt'\, S(t - t')R(t'). \tag{9.8}$$

A solution for $S(t)$ may be obtained by the methods of Laplace transform or iterative convolution.

The effect of collisional relaxation of the excited ensemble has been treated by Valance (1967), Carrington (1961), Williamson (1969), and Lin (1972). Collision-induced transition rates are also the subject of the extensive literature on gas phase energy transfer. Several aspects of collisional processes have been discussed in Section III.

Thus, failure to meet any one of the three conditions has been discussed in the literature. We now wish to discuss a general theory which encompasses all three conditions. The emphasis will be on the effect of these conditions on the experimental observables of fluorescence quantum yield, lifetime, and the fluorescence decay curve.

For simplicity we will define the microscopic parameters with reference to a two-level ensemble. With respect to Fig. 20, the following decay processes for any level i will be considered:

(1) Radiative decay of level i with rate constant $k_{r,i}$.
(2) Collision free nonradiative decay of level i with rate constant $k_{nr,i}$.
(3) Collisional quenching of level i with bimolecular rate constant $k_{Q,i}$.
The final states of this process are considered to be outside the ensemble and hence not observable.

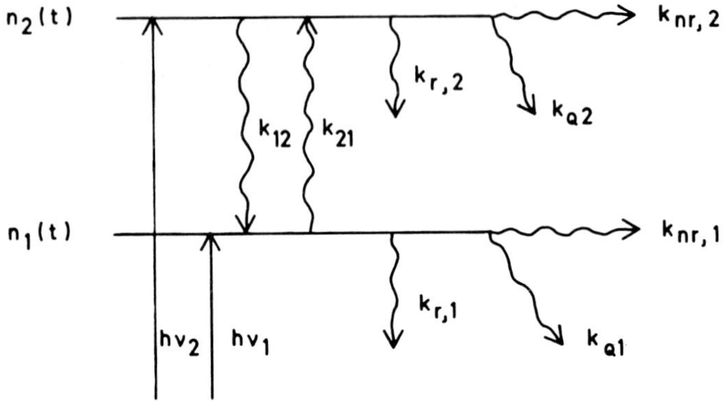

Fig. 20. Microscopic decay processes defined for a two level system.

(4) Collisional induced transitions between levels within the ensemble with bimolecular rate constant k_{ji} for the process $|i\rangle \xrightarrow{M} |j\rangle$ where $[M]$ is the concentration of the collision partner M.

The Pauli master equation for the decay of the ith level of an N level ensemble is given by

$$\begin{aligned}\frac{dn_i(t)}{dt} &= -(k_{r,i} + k_{nr,i} + k_{Q,i}[M])n_i(t) \\ &\quad - \sum_{j\neq i} k_{ji}[M]n_i(t) + \sum_{j\neq i} k_{ij}[M]n_j(t) \\ &\equiv -\sum_j J_{ij}n_j(t),\end{aligned} \tag{9.9}$$

for $i = 1$ to N and where J_{ji} are the elements of the transport matrix $\mathbf{J}$. These are defined by

$$J_{ii} = (k_{r,i} + k_{nr,i} + k_{Q,i}[M]) + \sum_{j\neq i} k_{ji}[M],$$

and

$$J_{ij} = -k_{ij}[M] \qquad \text{for} \quad i \neq j.$$

For a two level ensemble the $\mathbf{J}$ matrix has the form

$$\begin{pmatrix} k_{r,1} + k_{nr,1} + (k_{Q,1} + k_{21})[M] & -k_{12}[M] \\ -k_{21}[M] & k_{r,2} + k_{nr,2} + (k_{Q,2} + k_{12})[M] \end{pmatrix}.$$

Written in matrix form, Eq. (9.9) is given by

$$\dot{\mathbf{n}}(t) = -\mathbf{J}\mathbf{n}(t), \tag{9.10}$$

where $\mathbf{n}(t)$ is the column vector of level populations at time t.

In order to obtain a closed form expression for the individual level populations as a function of time, it is easiest to start with Eq. (9.9). Integrating both sides from $t' = 0$ to t gives

$$\int_0^t dn_i(t') = -\int_0^t \sum_{j=1}^{N} J_{ij} n_j(t')\, dt', \tag{9.11}$$

or

$$n_i(t) - n_i(0) = -\sum_{j=1}^{N} J_{ij} \int_0^t n_j(t')\, dt'. \tag{9.12}$$

This equation gives the population of level i as

$$n_i(t) = n_i(0) - \sum_{j=1}^{N} J_{ij} \int_0^t n_j(t')\, dt', \tag{9.13}$$

where $n_i(0)$ is identified with the population of level i which results from a delta function excitation pulse.

Equation (9.13) gives an expression for $n_i(t)$ in terms of an integral containing $n_j(t)$ for $j = 1$ to N. Iterative substitution of the right-hand side of Eq. (9.13) into the integral of Eq. (9.13) for $n_j(t)$ yields an infinite series. This is given by

$$n_i(t) = n_i(0) + \sum_{p=1}^{\infty} \frac{(-t)^p}{p!} \sum_{m_1=1}^{N} \cdots \sum_{m_p=1}^{N} J_{im_1} J_{m_1 m_2} \cdots J_{m_{p-1} m_p} n_{m_p}(0). \tag{9.14}$$

We note that the summations over m_k define a matrix multiplication. Thus, Eq. (9.14) can be written in matrix form to give

$$\mathbf{n}(t) = \left(\mathbf{I} + \sum_{p=1}^{\infty} \frac{(-t)^p}{p!} \mathbf{J}^p\right)\mathbf{n}(0) = \exp(-t\mathbf{J})\mathbf{n}(0). \tag{9.15}$$

This equation is the final solution for Eq. (9.10).

The vector of fluorescence emission is given by

$$\mathbf{S}(t) = \mathbf{K}_r \mathbf{n}(t), \tag{9.16}$$

where $\mathbf{K}_r$ is the matrix diagonal in the radiative rate constants. If fluorescence is observed equally from all levels, the signal measured by a photomultiplier is given by

$$S(t) = \mathbf{u}\mathbf{K}_r \mathbf{n}(t) \tag{9.17}$$

where $\mathbf{u}$ is a row vector of 1's. The vector $\mathbf{u}$ sums the signal from all levels.

In the case that different levels are observed unequally, e.g., due to the presence of a filter, the vector $\mathbf{u}$ must be replaced by the vector $\mathbf{g}$. The elements of $\mathbf{g}$ give the fraction of radiation observed from each level. This gives

$$S(t) = \mathbf{g}\mathbf{K}_r \mathbf{n}(t). \tag{9.18}$$

However, in that which follows we will assume that the more simple form of Eq. (9.17) is applicable.

The quantum yield is defined as the ratio of the number of photons emitted to that absorbed. The number absorbed is given by the total population at $t = 0$. This is given by $C(0)$ where

$$C(t) = \mathbf{un}(t). \tag{9.19}$$

The number emitted is related to the vector of fluorescence emission and given by the integral of Eq. (9.17) for all time. Thus,

$$\{\phi\} = \int_0^\infty S(t)\,dt/C(0) = \mathbf{uK}_\mathrm{r} \int_0^\infty \mathbf{n}(t)\,dt/C(0). \tag{9.20}$$

Note that we let curled brackets denote the ensemble observable. It is easily shown that

$$\int_0^\infty \mathbf{n}(t)\,dt = \int_0^\infty \exp(-t\mathbf{J})\mathbf{n}(0)\,dt = \mathbf{J}^{-1}\mathbf{n}(0) \tag{9.21}$$

where $\mathbf{J}^{-1}$ is the inverse of $\mathbf{J}$. It follows that

$$\{\phi\} = \mathbf{uK}_\mathrm{r}\mathbf{J}^{-1}\mathbf{n}(0)/C(0). \tag{9.22}$$

It is useful to simplify the notation used in the last equation. We now define two matrices as

$$\boldsymbol{\phi} \equiv \mathbf{K}_\mathrm{r}\mathbf{J}^{-1} \qquad \text{and} \qquad \boldsymbol{\tau} \equiv \mathbf{J}^{-1}.$$

We will also find it convenient to define

$$[\mathbf{X}] \equiv \mathbf{uXn}(0) \qquad \text{and} \qquad \langle X \rangle \equiv [\mathbf{X}]/C(0),$$

for any matrix $\mathbf{X}$. Using this notation, Eq. (9.22) is given by

$$\{\phi\} = [\boldsymbol{\phi}]/C(0) = \langle \phi \rangle. \tag{9.23}$$

This equation gives the correct quantum yield in terms of all microscopic parameters and can be directly used for comparison with experimental results. The assumption that condition (2) is satisfied will be discussed later.

Another observable is the normalized decay curve. This is given by

$$D(t) = S(t)\Big/\int_0^\infty S(t')\,dt' = \frac{\mathbf{uK}_\mathrm{r}\exp(-t\mathbf{J})\mathbf{n}(0)}{\mathbf{uK}_\mathrm{r}\mathbf{J}^{-1}\mathbf{n}(0)} = \frac{[\mathbf{K}_\mathrm{r}\exp(-t\mathbf{J})]}{[\boldsymbol{\phi}]}. \tag{9.24}$$

It has the property

$$\int_0^\infty D(t)\,dt = 1. \tag{9.25}$$

It is often convenient to characterize a decay curve by a single time constant. In the case of a single exponential, for which

$$D(t) = (1/\tau)\exp(-t/\tau) \tag{9.26}$$

applies, the lifetime τ is unambiguous. In the more general case of multilevel decay Eq. (9.26) does not apply. One must then use the mean first-passage time. This is given by

$$\{\tau\} = \int_0^\infty D(t)t\,dt. \tag{9.27}$$

It is easily shown that if $D(t)$ is given by Eq. (9.26), $\{\tau\} = \tau$ [see Eq. (9.7)]. In the general case,

$$\begin{aligned}\{\tau\} &= \int_0^\infty \mathbf{uK}_r\mathbf{n}(t)t\,dt/\mathbf{uK}_r\mathbf{J}^{-1}\mathbf{n}(0)\\ &= \mathbf{uK}_r\left(\int_0^\infty \exp(-t\mathbf{J})t\,dt\right)\mathbf{n}(0)/[\boldsymbol{\phi}].\end{aligned} \tag{9.28}$$

From the identity

$$\int_0^\infty \exp(-t\mathbf{J})t\,dt = \mathbf{J}^{-2}, \tag{9.29}$$

it follows that

$$\{\tau\} = \mathbf{uK}_r\mathbf{J}^{-2}\mathbf{n}(0)/[\boldsymbol{\phi}] = [\boldsymbol{\phi\tau}]/[\boldsymbol{\phi}]. \tag{9.30}$$

Thus, there are two quantities $\{\phi\}$ and $\{\tau\}$ as well as the function $D(t)$ which may be compared with experimental results. In the case of no collisional processes, Eqs. (9.23) and (9.30) reduce to the formulas given by Howard and Schlag (1976). In the case of single level excitation, Eqs. (9.23) and (9.30) reduce to Eqs. (9.6) and (9.7), respectively. The case of finite excitation pulse width will be discussed shortly.

For a calculation of $\{\phi\}$ or $\{\tau\}$ it is not necessary to solve the matrix exponential $\exp(-t\mathbf{J})$ but only the matrix inverse $\mathbf{J}^{-1}$. Thus, for any given set of parameters the calculation of $\{\phi\}$ and $\{\tau\}$ can be very rapid. For a calculation of the decay curve $D(t)$, it is necessary to solve the exponential. A direct power series expansion requires very much computer time before conversion. Faster techniques have been discussed by Oppenheim *et al.* (1967) and Valance (1967).

From Eqs. (9.1) and (9.2) it is possible to obtain

$$k_r = \phi/\tau, \tag{9.31}$$

and

$$k_{nr} = (1 - \phi)/\tau. \tag{9.32}$$

These equations apply only for single level excitation. However, for multilevel excitation we may formally define

$$\{k_r\} = \{\phi\}/\{\tau\}, \tag{9.33}$$

and

$$\{k_{nr}\} = (1 - \{\phi\})/\{\tau\}. \tag{9.34}$$

These formal definitions refer to experimental observables because the quantities on the right-hand side of both equations are observables. However, for the multilevel system the quantities $\{k_r\}$ and $\{k_{nr}\}$ do not correspond to microscopic rate constants.

It is often the case that the microscopic radiative rate constant k_r^m is a constant for all rotational levels of the same vibronic state. This allows a simplification of Eqs. (9.33) and (9.34). The new equations become

$$\{k_r\} = [\boldsymbol{\phi}]^2/[\boldsymbol{\phi\tau}]C(0) = k_r^m\langle\tau\rangle^2/\langle\tau^2\rangle \equiv k_r^m L(\tau), \tag{9.35}$$

and

$$\{k_{nr}\} = \langle\tau\rangle/\langle\tau^2\rangle - k_r^m\langle\tau\rangle^2/\langle\tau^2\rangle = \{1/\langle\tau\rangle - k_r^m\}L(\tau). \tag{9.36}$$

The function $L(\tau)$ may be considered as a lifetime dispersion function. It assumes values from zero to unity. It is unity when all states have the same lifetime. If variations in $\{k_r\}$ are observed along a rotational contour, the maximum value will be the lower limit to the true k_r^m. The variations in $\{k_r\}$ will reflect the dispersion of lifetimes within the ensemble.

Up to this point, our discussion has been based on the assumption of a delta function excitation pulse. For a pulse of finite width, the problem becomes somewhat more complicated. Let us denote the excitation function by $R(t)$ and normalize it so that

$$\int_0^\infty R(t)\,dt = 1. \tag{9.37}$$

The normalization will not affect any observables because any proportionality constant will cancel in all numerators and denominators.

The excited state will be populated via absorption through terms of the form

$$R(t)P_i N_g \equiv R(t)f_i,$$

where N_g is the number of ground state molecules and P_i the probability of absorption to level i. Note that all coherence effects and changes in the ground state population have been ignored. The total number of levels which are excited for all time is given by

$$N_t = \sum_i \int_0^\infty R(t) f_i\,dt = \mathbf{uf}.$$

For a delta function pulse, all levels are excited at $t = 0$, from which it follows that $\mathbf{n}(0) = \mathbf{f}$. If $R(t)$ is not a delta function, excitation is spread out in time and the population at $t = 0$ is not given by **uf**. However, the integrated absorption is still **uf**.

The response of the system to any random function $R(t)$ will be given by the convolution of Eq. (9.8). For the signal defined by Eq. (9.17) we get

$$\begin{aligned} S'(t) &= S(t) * R(t) \\ &= \mathbf{u}\mathbf{K}_f(\mathbf{n}(t) * R(t)) \\ &= \mathbf{u}\mathbf{K}_f \int_0^t dt' \exp[(t' - t)\mathbf{J}]R(t')\mathbf{f} \\ &\equiv \mathbf{u}\mathbf{K}_f \mathbf{n}'(t). \end{aligned} \tag{9.38}$$

We may now define the observables in terms of the new signal function. They are given by

$$\{\phi\}' = \mathbf{u}\mathbf{K}_r \int_0^\infty \mathbf{n}'(t)\, dt/\mathbf{u}\mathbf{f}, \tag{9.39}$$

$$D'(t) = \mathbf{u}\mathbf{K}_r\mathbf{n}'(t)/\mathbf{u}\mathbf{K}_r \int_0^\infty \mathbf{n}'(t)\, dt, \tag{9.40}$$

and

$$\{\tau\}' = \mathbf{u}\mathbf{K}_r \int_0^\infty t n'(t)\, dt/\mathbf{u}\mathbf{K}_r \int_0^\infty n'(t)\, dt. \tag{9.41}$$

Note that the primed variables refer to those with a finite width pulse excitation function, while unprimed variables refer to delta function excitation.

Let us first consider the quantum yield. From Eq. (9.39) it is necessary to solve the integral

$$\int_0^\infty \mathbf{n}'(t)\, dt.$$

This is given by

$$\begin{aligned} \int_0^\infty \mathbf{n}'(t)\, dt &= \int_0^\infty dt \int_0^t dt' \exp[(t' - t)\mathbf{J}]R(t')\mathbf{f} \\ &= \mathscr{L}\left[\int_0^t dt' \exp(t'\mathbf{J})R(t')\right]_{s=\mathbf{J}} \mathbf{f}, \end{aligned} \tag{9.42}$$

where $\mathscr{L}$ is the Laplace transform operator. Using well-known Laplace identities gives

$$\int_0^\infty \mathbf{n}'(t)\, dt = [s^{-1}R(s - \mathbf{J})]_{s=\mathbf{J}}\mathbf{f} = \mathbf{J}^{-1}R(s = 0)\mathbf{f}. \tag{9.43}$$

However,

$$R(s = 0) = \int_0^\infty R(t)\, dt = 1,$$

so that

$$\int_0^\infty \mathbf{n}'(t)\, dt = \mathbf{J}^{-1}\mathbf{f}, \tag{9.44}$$

and

$$\{\phi\}' = \mathbf{u}\mathbf{K}_f\mathbf{J}^{-1}\mathbf{f}/\mathbf{u}\mathbf{f} = \{\phi\}. \tag{9.45}$$

From this equation it follows that the excitation pulse width has no effect on the observable quantum yields.

The decay function $D'(t)$ is given by the convolution

$$D'(t) = D(t) * R(t). \tag{9.46}$$

No simpler form may be obtained. In contrast to the result for quantum yields, $D'(t)$ and $D(t)$ are generally always different.

The final observable to be considered is the mean first-passage time. From Eq. (9.41) the integral involving t must be solved. This is given by

$$\begin{aligned}\int_0^\infty \mathbf{n}'(t)t\, dt &= \int_0^\infty dt\, t \int_0^t dt' \exp[(t' - t)\mathbf{J}]R(t')\mathbf{f} \\ &= \mathscr{L}\left[t \int_0^t dt' \exp(t'\mathbf{J})R(t')\right]_{s=\mathbf{J}} \mathbf{f}.\end{aligned} \tag{9.47}$$

Using the Laplace identity

$$\mathscr{L}\{tF(t)\} = -f'(s)$$

yields

$$\begin{aligned}\int_0^\infty \mathbf{n}'(t)t\, dt &= \left[-\frac{d}{ds}(s^{-1}R(s - \mathbf{J}))\right]_{s=\mathbf{J}} \mathbf{f} \\ &= \mathbf{J}^{-2}\mathbf{f} + \mathbf{J}^{-1}\mathbf{f}\int_0^t dt'\, R(t')t' \\ &\equiv \mathbf{J}^{-2}\mathbf{f} + \mathbf{J}^{-1}\mathbf{f}\tau_R,\end{aligned} \tag{9.48}$$

where τ_R is the mean first-passage time of the excitation pulse. From Eq. (9.41) we find

$$\{\tau\}' = \mathbf{u}\mathbf{K}_r(\mathbf{J}^{-2} + \mathbf{J}^{-1}\tau_R)\mathbf{f}/\mathbf{u}\mathbf{K}_r\mathbf{J}^{-1}\mathbf{f} = \{\tau\} + \tau_R. \tag{9.49}$$

Thus, the mean first-passage time for a finite-width excitation pulse $\{\tau\}'$ is given by that for a delta function excitation pulse plus that of the pulse itself. This leads to the very simple relation

$$\{\tau\} = \{\tau\}' - \tau_R. \tag{9.50}$$

It is often the case that mean first-passage times are not measured. Rather, an attempt is made to fit the observed decay curve to an exponential. If $D(t)$ is given by Eq. (9.26), the observed curve $D'(t)$ will always be exponential for times after $R(t)$ has reached zero. The single lifetime τ can then be obtained from the exponential portion of the curve by any of several mathematical techniques. It will be equal to $\{\tau\}$ given by Eq. (9.50) only if Eq. (9.26) applies.

If the ensemble consists of a mixture of exponentials, only Eq. (9.50) will give a unique time constant which is characteristic of the ensemble. An exponential best fit to any portion of the curve will lead, in general, to values which are dependent on the start and end points of the calculation.

To summarize this section we note that a finite-excitation pulse width results in the following changes in the observables:

(1) Quantum yields are not changed.

(2) $\{\tau\}$ is increased by the mean first-passage time of the excitation waveform $R(t)$.

(3) $D(t)$ is convoluted with the excitation waveform.

The final equations are applicable to the observables of any fluorescent ensemble of molecular states which obey Eq. (9.9) and which are excited by any finite pulse. Equation (9.9) excludes bimolecular processes between two molecules if the concentrations of both are time dependent. It does include collisional effects between molecules which are in the ensemble and those which are not.

Effects such as absorption from the excited state, discussed by Speiser *et al.* (1973), have not been included here. However, such minor modifications are easily made to the theory presented here. In general, the methods discussed in this section allow one to calculate the observables of a fluorescent ensemble of states for a large range of experimental conditions. In the next section numerical calculations based on this procedure are discussed.

X. MODEL CALCULATIONS

Howard and Schlag (1978a) have performed numerical calculations for the $T_1 \leftarrow S_1$ transition in naphthalene-h_8. They assumed that collisional processes could be neglected. The vector of excited state rotational level populations was obtained from band contour methods as discussed in Section II. Two different sets were considered:

(1) Type A contours as exhibited by the 0–0 band and

(2) Type B contours as exhibited by the $8(b_{1g})_0^1$ band.

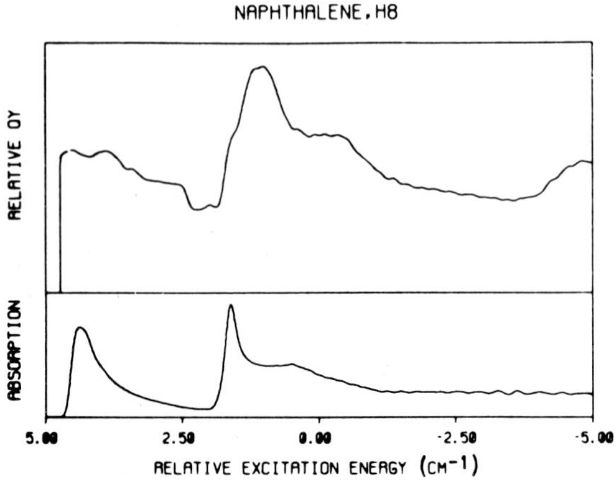

Fig. 21. The calculated ensemble fluorescence quantum yield $\{\phi\}$ as a function of excitation wavelength along the $8(b_{1g})_0^1$ band contour of naphthalene h_8. (From Howard and Schlag, 1978a.)

Nonradiative rate constants were calculated by Eq. (6.16) including all vibrational and rotational contributions to the matrix elements and energy gap. Ensemble quantum yields were obtained by solution of Eq. (9.23).

In Fig. 21 a calculation of $\{\phi\}$ over the type B contour is shown. This may be compared with the experimental data of Fig. 13. Although the calculated fluorescence quantum yield peak to the blue is too small, the major features of the measured data are produced.

In Fig. 22, a calculation of $\{\phi\}$ over the 0–0 band is shown. The peak in the quantum yield near the peak in absorption is also in qualitative agreement with experimental data.

These calculations represent the first attempt to include rotational effects in a theoretical description of nonradiative rate constants and observed quantum yields. The correlation with experiment is only qualitative. Even though triplet parameters were realistic in that they were on the correct order of magnitude, most of them could only be estimated. In spite of these problems, several important conclusions were reached:

(1) Rotational structure is expected in the observables along a rotational contour, given a reasonably narrow excitation bandwidth.

(2) The structure is dependent on the electronic matrix elements $\langle \phi_T | \tilde{l}_-^{\alpha} | \phi_s \rangle$ which weight the various ΔK transitions.

(3) Calculations with $F(E) = 1$ for all E showed no structure in agreement with Eq. (7.9). Thus, structure in $F(E)$ is necessary for structure in k_{nr} and hence in the observables.

(4) Calculations with a constant rotational contribution to the amplitude but with rotational energies in the energy gap produced a structure different

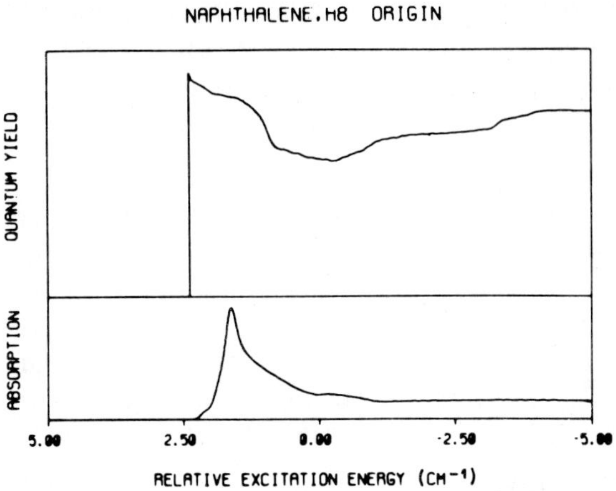

Fig. 22. The calculated ensemble fluorescence quantum $\{\phi\}$ yield as a function of excitation wavelength along the 0–0 band contour of naphthalene h_8. (From Howard and Schlag, 1978a.)

from that including the Wigner 3-*J* symbol. Hence the nature of the angular momentum coupling strongly influences k_{nr}.

(5) Calculated structure was a strong function of the promoting mode energy in the final state. A shift in this energy resulted in a constant shift in $F(E)$. Hence the fine structure of $F(E)$ is seen strongly to influence k_{nr}.

These calculations show that rotational effects can be as important to the nonradiative rate as vibrational effects. The calculations cannot be expected to become quantitative until more information exists on the triplet rotational and vibrational states. Such data are almost totally lacking in the literature except on some very small molecules.

XI. CONCLUSION

Experimental fluorescence quantum yield and lifetime data on naphthalene and formaldehyde show that the nonradiative rate constant depends upon rotational state selection. This requires an extension of the models of nonradiative decay beyond the vibronic level.

In this work we have discussed the creation and decay of an ensemble of rovibronic levels in S_1. A detailed test of the theory presented here—or of any theory of nonradiative transitions—requires experimental data obtained under the conditions of single rovibronic level excitation and at pressures below the second collisionless limit, but for large molecules. To date, no experiment has fulfilled these conditions, but nevertheless, as in optical spectroscopy, much information can be gleaned from contour analysis.

Particular emphasis was placed on the nature of the coupling of the angular momenta in the molecule and the consequent rotational selection rules for nonradiative processes. It was shown that for ISC the terms

$$(2N' + 1)\begin{pmatrix} N & N' & 1 \\ K & -K' & \alpha \end{pmatrix}^2$$

must be included in the rate constant expression. For ISC and IC, the rotational energies must be included in the energy matching of initial and final states. Although rotational wavefunctions were given in the symmetric rotor basis, transformation to an asymmetric rotor basis is straightforward. For a numerical calculation, the problem is that the inertial constants are virtually unknown for the triplet states of large, polyatomic molecules. In principle, the inertial constants can be obtained via band contour analysis of high resolution S–T absorption spectra. However, such data as this is lacking except for some small molecules.

It has been shown that any structure from rotational states depends on structure in the vibrational overlap factors. In fact, a proper evaluation of the Franck–Condon weighed density of states $F_\gamma(E)$ is the crucial step in the calculation of nonradiative rate constants. However, there are several problems to be overcome. Our assumption that the modes are described by harmonic oscillators is certainly not valid at large electronic energy gaps. However, if one wishes to evaluate the vibrational factors, one must know the vibrational frequencies including the anharmonic parameters. Most vibrational frequencies and normal coordinate displacements are unknown for molecular excited states. It is thus difficult to obtain reasonable numerical values even within the harmonic approximation. Only recently, Robey and Schlag (1977) have succeeded in performing an accurate excited state normal mode analysis on the S_1 state of benzene-h_6 and -d_6. Such calculations are also needed on triplet states. For smaller molecules such as glyoxal and formaldehyde, the vibrational frequencies in S_1 and T_1 have been determined. Further experimental data are needed on the vibrational levels of the triplet states of large polyatomic molecules.

An additional problem is that it is not generally valid to describe all final states with the same linewidth γ. In general, each final level will be characterized by a unique width γ_i. Moreover, at large density of final states in molecules such as naphthalene and benzene, there will be processes which correspond to energy randomization among the isoenergetic levels as well as inelastic collision processes. One expects that both of these processes must enter into an evaluation of γ. Only a very few experiments, such as that of Schroeder *et al.* (1977), have even given an indication of the lifetime of the final states of the nonradiative transition. It is evident that further experiments on the states prepared by the nonradiative transition are necessary.

Like the radiative process, a nonradiative process depends on the nature of the electronic, vibrational, rotational, and spin states which are coupled by an appropriate operator. The theory which we have outlined above explicitly considers all of these states and is able to explain the major features of the data on high resolution nonradiative transitions, but it remains for the future to characterize the relevant states accurately by spectroscopy.

REFERENCES

Allen, H. C., and Cross, P. C. (1963). "Molecular Vib-Rotors." Wiley, New York.

Anderson, L. G., Parmenter, C. S., and Poland, H. M. (1973). *Chem. Phys.* **1**, 401.

Beyer, R. A., Zittel, P. F., and Lineberger, W. C. (1975). *J. Phys.* **62**, 4016.

Bixon, M., and Jortner, J. (1968). *J. Chem. Phys.* **48**, 715.

Boesl, U., Neusser, H. J., and Schlag, E. W. (1975). *Chem. Phys. Lett.* **31**, 1.

Brand, J. C. D. (1972). "MTP International Review of Science: Spectroscopy" (D. A. Ramsey, ed.). Butterworths, London.

Brand, J. C. D., and Stevens, C. G. (1973). *J. Chem. Phys.* **58**, 3331.

Brown, J. M., and Howard, B. J. (1976). *Mol. Phys.* **31**, 1517.

Brown, R. G., Rockley, M. G., and Phillips, D. (1975). *Chem. Phys.* **7**, 41.

Brus, L. E., and McDonald, J. R. (1973). *Chem. Phys. Lett.* **21**, 283.

Carrington, T. (1961). *J. Chem. Phys.* **35**, 807.

Christie, J. R., and Craig, D. P. (1972a). *Mol. Phys.* **23**, 345.

Christie, J. R., and Craig, D. P. (1972b). *Mol. Phys.* **23**, 352.

Edmonds, A. R. (1959). "Angular Momentum in Quantum Mechanics." Princeton Univ. Press, Princeton, New Jersey.

Freed, K. F. (1970). *J. Chem. Phys.* **52**, 1345.

Freed, K. F. (1976). *Topics Appl. Phys.* **15**, Chapter 2.

Goldberger, M., and Watson, K. (1964). "Collision Theory." Wiley, New York.

Haarhoff, P. C. (1963). *Mol. Phys.* **7**, 101.

Heller, D. F., Freed, K. F., and Gelbart, W. M. (1972). *J. Chem. Phys.* **56**, 2309.

Hoare, M. R., and Ruijrok, T. W. (1970). *J. Chem. Phys.* **52**, 113.

Hollas, J. M., and Thakur, S. N. (1971). *Mol. Phys.* **22**, 203.

Hougen, J. T. (1964). *Can. J. Phys.* **42**, 433.

Hougen, J. T. (1970). National Bureau of Standards Monograph 115. U.S. Govt. Printing Office, Washington, D.C.

Howard, W. E., and Schlag, E. W. (1976). *Chem. Phys.* **17**, 123.

Howard, W. E., and Schlag, E. W. (1978a). *Chem. Phys.* **29**, 1.

Howard, W. E., and Schlag, E. W. (1978b). *J. Chem. Phys.* (in press).

Katriel, J. (1970). *J. Phys. B At. Mol. Phys.* **3**, 1315.

Knight, A. E. W., Selinger, B. K., and Ross, I. G. (1973). *Aust. J. Chem.* **26**, 1159.

Kuehn, I. H., Heller, D. F., and Gelbart, W. M. (1977). *Chem. Phys.* **22**, 435.

Kuettner, H. G., Selzle, H. L., and Schlag, E. W. (1978). *Chem. Phys.* (in press).

Lin. S. H. (1972). *J. Chem. Phys.* **56**, 4155.

McQuarrie, D. A. (1967). "Stoichastic Approach to Chemical Kinetics." Methuen, London.

Medvedev, E. S., Osherov, V. I., and Pschenichnikov, V. M. (1977). *Chem. Phys.* **23**, 397.

Melton, L. A., and Klemperer, W. (1973). *J. Chem. Phys.* **59**, 1099.

Metz, F. (1975). *Chem. Phys.* **9**, 121.

Metz, F. (1976). *Chem. Phys.* **18**, 385.

Metz, F., Howard, W. E., Wunsch, L., Neusser, H. J., and Schlag, E. W. (1978). *Proc. R. Soc. London Ser. A.* **363**, 381.

Nitzan, A., and Jortner, J. (1971). *J. Chem. Phys.* **55**, 1355.

Nitzan, A., and Jortner, J. (1972). *J. Chem. Phys.* **56**, 3360.
Nitzan, A., and Jortner, J. (1973). *Theor. Chim. Acta* **30**, 217.
Oppenheim, I., Schuler, K. E., and Weis, G. H. (1967). *Adv. Mol. Relaxation Processes* **1**, 13.
Parmenter, C. S., and Schuh, M. D. (1972). *Chem. Phys. Lett.* **13**, 120.
Rice, S. A. (1975). "Excited States" (E. C. Lim, ed.), Vol. 2, Chapter 3. Academic Press, New York.
Robey, M. J., and Schlag, E. W. (1977). *J. Chem. Phys.* **67**, 2775.
Rordorf, B. F., Knight, A. E. W., and Parmenter, C. S. (1978). *Chem. Phys.* **27**, 11.
Sackett, P. B. (1972). *Appl. Opt.* **11**, 2181.
Sandsmark, R. A. (1967). PhD thesis, Northwestern Univ.
Schlag, E. W., and Howard, W. E. (1976). "Molecular Energy Transfer" (R. Levine and J. Jortner, eds.). Wiley, New York.
Schlag, E. W., and Sandsmark, R. A. (1962). *J. Chem. Phys.* **37**, 168.
Schlag, E. W., and von Weyssenhoff, H. (1969). *J. Chem. Phys.* **51**, 2508.
Schlag, E. W., Sandsmark, R. A., and Valance, W. G. (1965). *J. Phys. Chem.* **69**, 1431.
Schlag, E. W., von Weyssenhoff, H., and Starzak, M. (1967). *Int. Conference Photochem. Munich, Germany* Preprints, Part I, p. 68.
Schlag, E. W., Yao, S. J., and von Weyssenhoff, H. (1969). *J. Chem. Phys.* **50**, 732.
Schlag, E. W., Schneider, S., and Chandler, D. W. (1971a). *Chem. Phys. Lett.* **11**, 474.
Schlag, E. W., Schneider, S., and Fischer, S. F. (1971b). *Ann. Rev. Phys. Chem.* **22**, 405.
Schroeder, H., Neusser, H. J., and Schlag, E. W. (1977). *Chem. Phys. Lett.* **48**, 12.
Siebrand, W. (1971). *J. Chem. Phys.* **54**, 363.
Spears, K. G., and Rice, S. A. (1971). *J. Chem. Phys.* **55**, 5561.
Speiser, S., van der Werf, R., and Kommandeur, J. (1973). *Chem. Phys.* **1**, 297.
Stein, S. E., and Rabinovitch, B. S. (1972). *J. Chem. Phys.* **58**, 2438.
Tang, K. Y., Fairchild, P. W., and Lee, E. K. C. (1977). *J. Chem. Phys.* **66**, 3303.
Thiele, E. (1963). *J. Chem. Phys.* **39**, 3258.
Valance, W. G. (1967). PhD thesis, Northwestern Univ.
Williamson, J. H. (1969). *J. Chem. Phys.* **50**, 2719.

3

On Rotational Effects in Radiationless Processes in Polyatomic Molecules

Frank A. Novak, Stuart A. Rice, Michael D. Morse, and Karl F. Freed*

The Department of Chemistry
and
The James Franck Institute
The University of Chicago
Chicago, Illinois

* Fannie and John Hertz Foundation Fellow.

ISBN 0-12-450650-X

I. INTRODUCTION

All physical processes are subject to constraints imposed by the requirements of conservation of energy, of linear momentum, and of angular momentum. The still developing theory of elementary photophysical and photochemical processes in polyatomic molecules has often explicitly elucidated the roles of the first two of these conditions, but relatively little is known of the details of the role of the conservation of angular momentum. In part this disparity in level of knowledge is a consequence of lack of stimulus from experimental findings, and in part it is a consequence of the very great extra complication introduced into the theoretical analysis when molecular rotation is included in the Hamiltonian. However, new developments in laser and molecular beam technology have opened the way to the study of radiationless processes in selected rovibronic states of an isolated polyatomic molecule, and even the very few data already available challenge our theoretical understanding of the observations. This paper reviews the basic ideas in the analysis of the influence of molecular rotation on radiationless processes in polyatomic molecules, paying attention to the interaction of the rotational, vibrational, and electronic degrees of freedom. A reduction of the theory to limiting cases will also be described.

II. THE DIATOMIC MOLECULE CASE

For a diatomic molecule the only significant radiationless process is fragmentation into atoms or ions. The effect of molecular rotation on this fragmentation is well understood, and a detailed account of the theory can be found in Herzberg (1950) and the review article by Child (1974). Here we merely quote several of the results for the purpose of emphasizing the obvious but important point that the effect of angular momentum selection rules is to reduce the possible final states accessible from the prepared state, but these rules are in themselves insufficient to determine which of the allowed processes are fast or slow. To understand the latter we must also account for Franck–Condon effects, where this term is used in its most general sense to mean evaluation of a coupling matrix element with due regard for the phases of the initial and final state wavefunctions and to variation of coupling strength over a region of space.

In Herzberg's notation, a type I predissociation of a diatomic molecule arises from the overlap (in energy) of the rovibronic levels of one electronic state with the dissociation continuum of some lower electronic state, and a type III predissociation arises from the overlap of rovibronic and continuum levels of the same electronic state (the case of a rotational barrier to dissociation). The selection rules are, in the usual notation,

$$\Delta J = 0, \tag{2.1}$$

$$+ \not\leftrightarrow -, \tag{2.2}$$

$$\mathrm{g} \not\leftrightarrow \mathrm{u}, \tag{2.3}$$

for all electronic coupling schemes. In Hund's cases (a) and (b) and neglecting spin–orbit coupling, these selection rules are augmented by

$$\Delta S = 0, \tag{2.4}$$

$$\Delta\Lambda = 0, \pm 1, \tag{2.5}$$

which can be somewhat further simplified if both electronic states belong to the same coupling case, either (a) or (b); if case (a) then $\Delta\Sigma = 0$, if case (b) then $\Delta K = 0$. In general, the selection rules for a radiationless transition in a diatomic molecule are the same as for radiative transitions except that g $\leftrightarrow$ u in the radiative case is replaced by g $\not\leftrightarrow$ u in the nonradiative case. When spin–orbit coupling is important, condition (2.4) is no longer a restriction on the predissociation.

Child has published an extensive discussion of the influence of predissociation on the linewidths of transitions in diatomic molecules, a paper to which we refer the reader for details which flesh out the following sketch of the important results. Consider the case of predissociation by rotation (Herzberg type III). The wavefunction must satisfy the radial Schrödinger equation,

$$\left[\frac{d^2}{dr^2} + k^2 - U(r) - \frac{J(J+1)}{r^2}\right]\chi_{EJ}(r) = 0,$$
$$k^2 \equiv 2\mu E/\hbar^2, \qquad U(r) \equiv 2\mu V(r)/\hbar^2, \tag{2.6}$$

where, as usual, μ is the reduced mass and $V(r)$ the potential energy curve. When the wavefunction χ_{EJ} is normalized to a delta function of the energy and vanishes at the origin, it has the asymptotic form

$$\chi_{EJ} \underset{r\to\infty}{\longrightarrow} \left(\frac{2\mu}{\pi\hbar^2 k}\right)^{1/2} \sin\left[kr - \frac{J\pi}{2} + \eta_J(E)\right], \tag{2.7}$$

where $\eta_J(E)$ is the phase shift due to the potential $V(r)$. In this case, since only one electronic state is involved, the selection rules (2.1) and (2.4) are automatically satisfied.

In the energy region where resonances occur it is convenient to consider the energy to be complex

$$E = E_n - i\tfrac{1}{2}\Gamma_n, \qquad \Gamma_n > 0, \tag{2.8}$$

so that an initial nonstationary state with amplitude concentrated inside the well formed by the rotational barrier decays as $\exp(-\Gamma_n t/\hbar)$. The lifetime τ_n of the level n is related to the phase shifts by

$$\eta_J(E) = \arg F(E) + \tfrac{1}{2}J\pi + \arctan(\Gamma_n/2(E_n - E)). \tag{2.9}$$

The term $F(E)$ in (2.9) is determined as follows. When the wavevector k is regarded as complex, (2.7) can be rewritten in the form

$$\chi_{kJ} \underset{r\to\infty}{\longrightarrow} (1/2i)[f_J(k)e^{ikr} - f_J(-k)e^{-ikr}] \tag{2.10}$$

TABLE 1

Widths (Lifetimes) of Resonances for the H_2 Ground State[a]

v	J	$\Gamma_{nJ}(\text{cm}^{-1})$
0	38	98.1
0	37	5.98
9	19	66.2
9	18	0.53
11	14	18.5
11	13	0.005
12	12	11.6
12	11	2.62
13	9	89.6
13	8	1.89
14	6	79.0
14	5	26.4
14	4	0.007

[a] From LeRoy and Bernstein (1971).

with $f_J(k)$ and $f_J(-k)$ Jost functions, having the property $f_J(-k) = [f_J(k)]^*$. The complex energies characteristic of the resonant states are located at the zeros of $f_J(-k)$, and $F(E)$ is the coefficient of the linear term in the expansion in energy about such a zero:

$$f_J(k) = [f_J(-k)]^* = F(E)(E - E_n - \tfrac{1}{2}i\Gamma_n). \tag{2.11}$$

Practical algorithms relating the E_n and Γ_n to the form of the potential have been developed. By virtue of (2.9), these give the explicit dependence of the lifetime of the resonant state on the angular momentum of the system. Because the rotational barrier height is a quadratic function of J it is to be expected that the lifetime of the resonant state varies strongly with J. That this is so is shown by the entries in Table 1.

Consider now predissociation involving two electronic states (Herzberg type I). The coupling of the zeroth-order bound state to the continuum leads to level shifts and interference effects, which are manifest in the absorption line shape, the so-called Fano lineshape. Let n denote a particular vibrational, total angular momentum level. It is commonly argued that the matrix element which couples the continuum and bound vibrational states (represented by χ_{1E} and $\chi_{2n'}$ respectively),

$$V_{nE} = \int_0^\infty \chi_{2n}(r) H_{21}(r) \chi_{1E}(r)\, dr, \tag{2.12}$$

TABLE 2

Values of Γ_n *for* $A^{2}\,{}^{+}OD$ (cm^{-1})[a]

	v					
N	4	5	6	7	8	9
0	0.00	0.21	1.47	0.15	0.00	0.02
4	0.00	0.26	1.40	0.28	0.02	0.01
8	0.00	0.49	1.15	0.54	0.12	0.02
12	0.02	0.80	0.56	0.63	0.23	—
16	0.11	1.31	0.01	0.23	0.00	—
20	0.43	1.19	0.00	—	—	—

[a] From Czarny *et al.* (1971).

is nonvanishing only in a small region near the crossing point of the two potential energy curves. Then

$$\Gamma_n = 2\pi|V_{nE}|^2 = 2\pi|H_{12}^0|^2 \left| \int_0^\infty \chi_{2n}(r)\chi_{1E}(r)\,dr \right|^2, \tag{2.13}$$

with H_{12}^0 the value of $H_{21}(r)$ at the crossing point. Note that the influence of electronic and rotational angular momentum are incorporated in $H_{21}(r)$, which is

$$H_{21}(r) = \langle P_2|H(q, r)|P_1\rangle, \tag{2.14}$$

with $|P_1\rangle$ and $|P_2\rangle$ the appropriate electron and rotational angular momentum states, and the expectation value is with respect to electron coordinates and the angle variables of the nuclear position coordinate r. Provided that the linewidth is small relative to the level spacing, this Franck–Condon approximation is rather good. Instances when this condition fails, and the Franck–Condon approximation needs to be improved, are discussed by Child.

Our interest is in the rotational state dependence of Γ_n. The calculated dependence of Γ_n on rotational quantum number for the $A^2\Sigma^+$ state of OD is shown in Table 2. Note the regularity of the variation of Γ_n with $N = J - \Sigma$, in particular the oscillatory behavior. It is clear that in this case vibration–rotation interaction leads to a marked dependence of Γ_n on N, which is also manifest as a nonmonotone dependence of Γ_n on energy (see Fig. 1).

The results quoted in Table 2 were obtained by numerical integration of the overlap integrals. Very useful insight into the nature of the results can be obtained from analytical approximations to the overlap integral. The two most important factors affecting the resonance lifetime as a function of J are the contribution to the energy by rotational motion and the change of wavefunction phase with J. The first of these determines, by energy difference, the particular vibrational state–continuum overlap that enters (2.13), and hence can lead to a

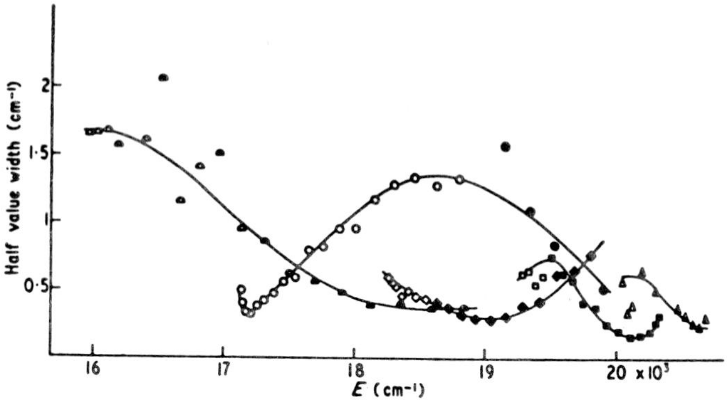

Fig. 1. The full width at half maximum of rotational levels for several vibrational levels of the OD $A^2\Sigma^+$ state. The full curves are smooth curves through experimental points for each ϑ. Observed values deduced from (semicircles) 0 and 8; (circles) 0, 9, and 1, 9; (diamonds) 0 and 10; (squares) 0 and 11; and (triangles) 1 and 12 bands. Full symbols are resolved, open symbols unresolved doublets. This figure from J. Czarny *et al.* (1971).

nonmonotone dependence of Γ_n on energy. The second of these provides a direct modulation of the overlap integral, and hence of Γ_n. For a detailed discussion of these matters the reader is referred to the review by Child.

We shall see in the following sections that the features characteristic of the rotational state dependence of radiationless processes in diatomic molecules have one-to-one counterparts in the rotational state dependence of radiationless processes in polyatomic molecules. In addition, there are several new features in these processes that are unique to polyatomic molecules.

III. THE POLYATOMIC MOLECULE CASE

A. The Eckart Frame and the Nuclear Kinetic Energy Operator

The present theory of intramolecular radiationless relaxation among bound states of polyatomic molecules concentrates on the interactions between the vibrational modes and the electronic motion in the molecule. This description is used in the case of both internal conversion and intersystem crossing. The relaxation is thereby considered to arise from homogeneous perturbations, i.e., couplings which can occur in a nonrotating molecule. The fact that rotational structure exists can, in the first approximation, alter the relaxation rates by modification of the energy gap and by possibly giving an angular momentum dependence to the homogeneous perturbation. These rotational modifications have been discussed at length by Howard and Schlag (1976, 1978a, b). A treatment of singlet–triplet interactions in polyatomic molecules where a heterogeneous perturbation can exist due to the interaction of rigid body and electronic angular momenta has been given by Stevens and Brand (1973).

Because of the complexity of the separation of the nuclear kinetic energy into rotational, translational, and for a nonlinear molecule, the remaining $3N - 6$ nuclear coordinate components, it is important to consider carefully the nature of the couplings between states involved in the radiationless relaxation process. In the general case this would surely be extremely difficult starting from first principles. It is possible, however, to make a first principles calculation by assuming that the molecule under consideration has the same symmetry in all of its electronic states. This means, more explicitly, that it is possible to move from any equilibrium molecular configuration to any other by displacing the nuclei in such a way such that the center of mass remains fixed and the molecule has not rotated as a whole. These are precisely the conditions that allow the rotations and translation of the molecular frame to be separated from the normal vibrational modes, i.e., the Eckart conditions (Wilson *et al.*, 1955; Kroto, 1975). The above requirements insure that the Eckart frame remains unique throughout any relaxation process. On physical grounds it is seen that only totally symmetric normal modes are displaced from one electronic state to another.

If $\mathbf{d}_n$ denotes the displacement of nucleus n from a reference position $\mathbf{r}_n^e$, the nuclear kinetic energy may be written classically as (Wilson *et al.*, 1955; Kroto, 1975)

$$T_N = \tfrac{1}{2} I_{\alpha\beta}\omega_\alpha\omega_\beta + \sum_n m_n\omega_\alpha e_{\alpha\beta\gamma} d_{n\beta}\dot{d}_{n\gamma} + \tfrac{1}{2}\sum_n m_n \dot{d}_{n\alpha}\dot{d}_{n\alpha} \tag{3.1}$$

under the constraints (Eckart conditions)

$$\sum_n m_n \mathbf{d}_n = 0, \qquad \sum_n m_n \mathbf{r}_n^e \times \mathbf{d}_n = 0. \tag{3.2}$$

The ω_α represent angular velocities and $I_{\alpha\beta}$ moments of inertia and $e_{\alpha\beta\gamma}$ is the completely antisymmetric third rank tensor. In the usual construction of a Hamiltonian operator using (3.1) a set of normal modes is now introduced which describes the nuclear motion in a particular electronic state. Because the normal modes in one electronic state are in general different from those in another, we will work in terms of displacements and momenta conjugate to these positions instead of normal modes, for the purpose of developing the problem in a more general manner.

The angular momentum is given by

$$J_\alpha = \partial T/\partial\omega_\alpha = I_{\alpha\beta}\omega_\beta + \sum_n m_n e_{\alpha\beta\gamma} d_{n\beta}\dot{d}_{n\gamma} \tag{3.3}$$

and the momenta conjugate to the displacements are

$$P_{n\gamma} = \partial T/\partial\dot{d}_{n\gamma} = m_n\omega_\alpha e_{\alpha\beta\gamma} d_{n\beta} + m_n\dot{d}_{n\gamma}. \tag{3.4}$$

These relations lead to the classical nuclear kinetic energy

$$T_N = \tfrac{1}{2}(J_\alpha - M_\alpha)\mathscr{M}_{\alpha\beta}(J_\beta - M_\beta) + \tfrac{1}{2}\sum_{n,\alpha} P_{n\alpha}^2, \tag{3.5}$$

where

$$M_\alpha = \sum_n e_{\alpha\beta\gamma} d_{n\beta} P_{n\gamma} \qquad \text{and} \qquad \mathscr{M} = \mathscr{I}^{-1}, \tag{3.6}$$

with

$$\mathscr{I}_{\alpha\beta} = I_{\alpha\beta} - \sum_n m_n e_{\alpha\alpha'\gamma} e_{\beta\beta'\gamma} d_{n\alpha'} d_{n\beta'} \tag{3.7}$$

The above expressions are very similar to those containing normal modes. They are, however, more general since they are independent of electronic state. Quantum mechanically we may write the total Hamiltonian as

$$H = T_e + V(x, \{d\}) + \tfrac{1}{2}\sum_{n,\alpha} P_{n\alpha}^2 + \tfrac{1}{2}\mathscr{M}_{\alpha\beta}(J_\alpha - M_\alpha)(J_\beta - M_\beta) - \tfrac{1}{8}\sum_\alpha \mathscr{M}_{\alpha\alpha}, \tag{3.8}$$

$$H_e = T_v + V(x, \{d\}), \qquad H_R = \tfrac{1}{2}\mathscr{M}_{\alpha\beta}(J_\alpha - M_\alpha)(J_\beta - M_\beta), \tag{3.9}$$

where Watson's (1968) simplification has been used.

Proceeding as usual, assume that the electronic Schrödinger equation may be solved as

$$H_e \phi_n(x, \{d\}) = \varepsilon_n(\{d\})\phi_n(x, \{d\}), \tag{3.10}$$

subject to conditions (3.2). Then it is assumed that the eigenstates of (3.8) may be expanded as

$$\Psi(x, \{d\}, \Omega) = \sum_n \psi_n(\{d\}, \Omega)\phi_n(x, \{d\}), \tag{3.11}$$

where Ω labels the Euler angle describing the orientation of the molecular axes in space. In the absence of an external field the electronic wavefunction cannot depend upon Ω. A prediagonalization of (3.8) in the basis (3.10) assuming no coupling yields adiabatic rovibronic states. If one expands the potential felt by the nuclei keeping only terms corresponding to the harmonic approximation, a set of normal modes can be constructed for each electronic state by choosing a proper combination of mass weighted displacements; i.e., one can write

$$\sqrt{m_n}\, d_{n\alpha} = \sum_r l_{nr}^{(\alpha)} Q_r. \tag{3.12}$$

Coriolis constants can also be introduced via the definition

$$\zeta_{rr'}^{(\alpha)} = \sum_n e_{\alpha\beta\gamma} l_{nr}^{(\beta)} l_{nr'}^{(\gamma)}. \tag{3.13}$$

These procedures lead to the usual rovibronic energy levels and wave functions, which can be written as

$$\Psi_{nvR}(x, Q^n, \Omega) = \phi_n(x, Q^n)\psi_{vR}(Q^n, \Omega), \tag{3.14}$$

where n, v, and R label electronic, vibrational, and rotational quantum numbers, respectively.

We will be primarily concerned with the coupling between electronic manifolds. The displacements $\{d\}$ or the mass weighted displacements as in (3.11) provide the proper set of coordinates to put all states on equal footing. We may also choose a set of normal coordinates belonging to one electronic state since there is a one-to-one correspondence between mass weighted displacements and normal modes. The Jacobian for the transformation is (within a sign) unity.

We select as basis the set of normal modes belonging to the originally prepared state, say m, and express all operators and wavefunctions in terms of the $\{Q_r^m\}$. In terms of these normal modes the coupling between rovibronic states is given by

$$\langle\psi_{mvR}(Q^m, \Omega)|\langle\phi_m(x, Q^m)|\left[-\frac{1}{2}\sum_r \frac{\partial^2}{\partial Q_r^2} + H_{\mathrm{R}}\right]|\phi_n(x, Q^m)\rangle|\psi_{nv'R'}(Q^m, \Omega)\rangle, \tag{3.15}$$

where the operator in brackets acts on everything to its right. H_{rot} takes on the familiar normal mode representation (Wilson *et al.*, 1955; Kroto, 1975),

$$H_R = \tfrac{1}{2}\mu_{\alpha\beta}(J_\alpha - \pi_\alpha)(J_\beta - \pi_\beta) \tag{3.16}$$

with

$$\pi_\alpha = \sum_r \sum_{r'} \zeta_{rr'}^{(\alpha)} Q_r P_{r'}. \tag{3.17}$$

For simplicity the vibronic part of the coupling will be taken as

$$\sum_r c_{mn}^{(r)}\langle\psi_{mvR}(Q^m, \Omega)|P_r|\psi_{nv'R'}(Q^m, \Omega)\rangle, \tag{3.18}$$

with

$$c_{mn}^{(r)} = -i\langle\phi_m(x, Q^m)|(\partial/\partial Q_r)|\phi_n(x, Q)\rangle_{x, Q^m=0}. \tag{3.19}$$

In the rotational part of the coupling the principal inverse moment of inertia axis will be assumed the same in both electronic states; we also assume that its coordinate dependence can be taken into account by writing in the coupling operator the average of its values in the two electronic states. With these assumptions the rotational coupling simplifies to

$$\begin{aligned}&\tfrac{1}{2}\langle\psi_{mvR}(Q^m, \Omega)|\langle\phi_m(x, Q^m)|\bar{\mu}_{\alpha\alpha}(J_\alpha^2 - 2J_\alpha\pi_\alpha + \pi_\alpha^2)\\ &\quad\times|\phi_n(x, Q^m)\rangle|\psi_{nv'R'}(Q^m, \Omega)\rangle.\end{aligned} \tag{3.20}$$

Because of the complexity of the π operators the coupling will be left in this form until it is used in the consideration of specific models. Note that the term in

J_α^2 is not taken to be zero despite the orthogonality of the electronic wavefunction, because **J** is the rigid body angular momentum [usually denoted by **N** in Hund's case (b)]. Since

$$\mathbf{J} = \mathbf{J}_{\text{tot}} - \mathbf{L} - \mathbf{S}, \tag{3.21}$$

there is the possibility that the electronic angular momentum can cause electronic coupling between states of the same parity. This is the type of interaction that underlies the lambda doubling in diatomic molecules.

Of course the states appearing in (3.20) are of the same multiplicity. In order to describe intersystem crossing, spin–orbit interactions must be added to the list of couplings.

The lesson to be learned from the considerations in this section is that one *cannot* write the commutation relation

$$[J_\alpha, T_N] = 0 \tag{3.22}$$

in a rotating frame. The mechanics of the rotation–vibration coupling complicates the dynamical properties of polyatomic molecules even more than it complicates their static spectral properties. In Section IV a model will be developed for a simple case considering the coupling given by (3.20).

B. Rational Selection Rules for Nonradiative Processes

The arguments presented in this section are very similar to those recently presented by Howard and Schlag (1976) in a discussion of the effect of rotation on radiationless processes.

Consider first the process of internal conversion between two singlet levels in a symmetric top molecule. In this case, let us take the angular momentum J_α as the rigid body angular momentum of the molecule and assume that the electronic manifolds couple only through

$$T_N' = -\frac{1}{2}\sum_{r=1}^{3N-6}\frac{\partial^2}{\partial Q_r^2}. \tag{3.23}$$

Now T_N' and the angular momentum operators satisfy the commutation relations

$$[J^2, T_N'] = 0, \qquad [J_z, T_N'] = 0. \tag{3.24}$$

These relations imply that $\Delta J = 0$, which is just the condition of conservation of angular momentum. They also imply that $\Delta K = 0$, or in words, that the projection of **J** along the top (z) axis is also conserved. Denoting vibronic quantum numbers by γ_n one can then write the intersystem crossing rate between states m and n as

$$k_{m\to n} = 2\pi\sum_n |\langle\gamma_n|T_N'|\gamma_m\rangle|^2\delta_{J_m,J_n}\delta_{K_m,K_n}\delta(E_n - E_m). \tag{3.25}$$

Examination of (3.25) shows that there are two mechanisms which can lead to $k_{m\to n}$ being a function of the angular momentum quantum numbers. The first is simply a modification of the energy difference $E_n - E_m$, since the molecule will have different rotational constants in the two electronic states. The second is the effect of centrifugal distortion and Coriolis coupling within a vibronic manifold. Centrifugal distortion causes the Franck–Condon factors appearing in (3.25) to become angular momentum dependent, while Coriolis coupling causes the vibrations to become mixed. Just this type of Coriolis interaction has been proposed to explain the variation in the internal conversion rate as a function of rotational state in formaldehyde (Lee, 1977). Note, however, that this Coriolis coupling induced mixing does not include the effects of the coupling given by (3.20). Equation (3.20) describes a dynamic Coriolis interaction leading to electronic coupling between states m and n due to the variation in the Coriolis coupling during the relaxation process.

Intersystem crossing between a singlet and a triplet state is another relaxation process of great importance in the physics of polyatomic molecules. Spin–orbit coupling is the important interaction in this process, and since it explicitly depends upon spin angular momentum, its rotation dependence is more complicated than the internal conversion case treated above.

Since in nonlinear symmetric top polyatomic molecules the electronic angular momentum is effectively quenched, Hund's case (b) is a natural choice for the angular momentum coupling scheme in such molecules. In this case the good quantum numbers are the total angular momentum J, the spin angular momentum S, and the difference, given by the eigenvalues of the operator $\mathbf{N}^2 = (-\mathbf{S} + \mathbf{J})^2$. Since the spin–orbit interaction is usually written in terms of molecular frame operators, i.e., case (a) operators, it is useful to transform from a case (a) basis to a case (b) basis. In terms of the Wigner 3-J symbol (Edmonds, 1960) the transformation is given by

$$|SJNK\rangle = \sum_{P,S_z} (-1)^{J+P}(2N+1)^{1/2} \begin{pmatrix} J & N & S \\ P & -K & -S_z \end{pmatrix} |SS_zJP\rangle, \tag{3.26}$$

where P is the projection of J on the top axis and S_z is the projection of S on the top axis. Transformation (3.26) is much discussed in the literature (Stevens and Brand, 1973; Hougen, 1964) and may be looked upon as an application of Van Vleck's reversed angular momentum technique (van Vleck, 1951; Freed, 1966).

As a simplification it is useful to consider only a two-electron system and assume that all of the other electrons in the molecule act simply as a core without major influence on the outer two electrons. The singlet spin state is given by

$$|S_S\rangle = |0,0\rangle = 2^{-1/2}(\alpha_1\beta_2 - \alpha_2\beta_1), \tag{3.27}$$

and the members of the triplet by

$$|1S_z\rangle = \begin{cases} |1,1\rangle = \alpha_1\alpha_2, \\ |1,0\rangle = 2^{-1/2}(\alpha_1\beta_2 + \alpha_2\beta_1), \\ |1,-1\rangle = \beta_1\beta_2. \end{cases} \tag{3.28}$$

Using (3.26) along with the identities $J_T = N_S, P_T = K_S, S_T = 1$, it is found that

$$\langle \Psi_T | H_{so} | \Psi_S \rangle = \sum_{S_z} (-1)^{N_S + K_S} (2N_T + 1)^{1/2}$$

$$\times \begin{pmatrix} N_S & N_T & 1 \\ K_S & -K_T & -S_z \end{pmatrix} \langle \gamma_T, 1S_z | H_{so} | \gamma_S S_S \rangle. \tag{3.29}$$

Now writing the spin–orbit coupling as the sum of one electron operators in spherical form (Edmonds, 1960; Rose, 1950)

$$H_{so} = \sum_{j=1}^{2} \lambda_j \sum_{\alpha = \pm 1, 0} (-1)^{\alpha} l_j^{(\alpha)} S_j^{(-\alpha)}, \tag{3.30}$$

the matrix element (3.29) becomes

$$\langle \Psi_T | H_{so} | \Psi_S \rangle = \sum_{\alpha = \pm 1, 0} (-1)^{N_S + K_S} (2N_T + 1)^{1/2}$$

$$\times \begin{pmatrix} N_S & N_T & 1 \\ K_S & -K_T & \alpha \end{pmatrix} \langle \gamma_T | l_-^{(\alpha)} | \gamma_S \rangle, \tag{3.31}$$

where $l_-^{(\alpha)} = \frac{1}{2}(\lambda_1 l_1^{(\alpha)} - \lambda_2 l_2^{(\alpha)})$. From the properties of the 3-J symbols the selection rules for intersystem crossing via direct spin–orbit coupling are

$$\Delta N = \pm 1, 0, \qquad \Delta K = \pm 1, 0. \tag{3.32}$$

The discussion of intrastate rotationally dependent couplings in the case of internal conversion is valid in intersystem crossing as well. The rotational dependence of intersystem crossing is thus more complicated than internal conversion because there exist specific rotational prefactors in the rates, not just simple delta functions.

Howard and Schlag also point out that because of the sum rule

$$\sum_{N_T, K_T} (2N_T + 1) \begin{pmatrix} N_S & N_T & 1 \\ K_S & -K_T & \alpha \end{pmatrix}^2 = 1 \tag{3.33}$$

it is necessary that the Franck–Condon weighted density of states in the triplet manifold is not constant over the range of J states swept out. If it is constant the rotational dependence is again only of the type occurring in internal conversion.

Since the coupling (3.20) cannot mix states of different multiplicity, any rotational dependence of the radiationless rate arising from this term can only occur through a second-order interaction similar to the situation of mixed spin–orbit–vibronic interaction. We will return to this point in a later section.

IV. MODELING THE EFFECT OF INTERMANIFOLD CORIOLIS COUPLING

Let us consider the effect of the coupling (3.20) on the internal conversion process. Because of the complexity of the π operators it is more useful to specialize to a particular model than to attempt a general analysis. Consider, for example, the benzene molecule; we shall assume it has D_{6h} symmetry in all of its states. This means that one electronic state differs from another only by displacements along totally symmetric normal modes and by frequency shifts. We also assume, for simplicity, that all vibrations are harmonic and that the vibrational frequencies are the same in the states considered.

The in-plane vibrations in benzene have symmetry species

$$2A_{1g},\quad A_{2g},\quad 4E_{2g},\quad 2B_{1u},\quad 3E_{1u}.$$

Suppose that one of the A_{1g} modes is a promoting mode. Call this mode 2. Then the only in plane mode to which 2 may couple is the A_{2g} mode, labeled 1, because of the symmetry property of the Coriolis constants,

$$\zeta_{rr'}^{(\alpha)} \neq 0. \tag{4.1}$$

Again, in the interest of simplicity, we disregard out-of-plane modes and neglect any coupling due to electronic angular momentum described by (3.10). With these assumptions it is necessary to consider only the operator

$$\pi_z = \zeta_{21}^{(z)}(P_2 Q_1 - P_1 Q_2) \tag{4.2}$$

in the matrix element (3.20). For a planar symmetric top $\mu_{zz} = \frac{1}{2}\mu_{xx} = \frac{1}{2}\mu_{yy} = \frac{1}{2}A$, and therefore (3.20) becomes

$$\begin{aligned} &\langle\psi_{mvR}(Q,\Omega)|\langle\phi_m(x,Q)|[-\tfrac{1}{4}J_z\zeta_{21}^{(z)}(A_m + A_n)(P_2Q_1 - P_1Q_2) \\ &\quad + \tfrac{1}{8}(A_m + A_n)\zeta_{21}^{2(z)}(P_2Q_1 - P_1Q_2)^2]|\phi_n(x,Q)\rangle|\psi_{nv'R'}(Q,\Omega)\rangle. \end{aligned} \tag{4.3}$$

Since P_r is represented by $i\partial/\partial Q_r$ and since $r = 2$ is the promoting mode, (4.3) reduces to

$$\begin{aligned} &-\tfrac{1}{4}C_{mn}^{(2)}(A_m + A_n)\zeta_{21}^{(z)}\langle\psi_{mvR}(Q,\Omega)|Q_1 J_z|\psi_{nv'R'}(Q,\Omega)\rangle \\ &\quad - \tfrac{1}{8}C_{mn}^{(2)}(A_m + A_n)\zeta_{21}^{2(z)}\langle\psi_{mvR}(Q,\Omega)|(P_1Q_1 + Q_1P_1)Q_2|\psi_{nv'R'}(Q,\Omega)\rangle \\ &\quad + \tfrac{1}{4}C_{mn}^{(2)}(A_m + A_n)\zeta_{21}^{2(z)}\langle\psi_{mvR}(Q,\Omega)|Q_1^2 P_2|\psi_{nv'R'}(Q,\Omega)\rangle. \end{aligned} \tag{4.4}$$

To (4.4) one must add the vibronic part of the coupling

$$C_{mn}^{(2)}\langle\psi_{mvR}(Q,\Omega)|P_2|\psi_{nv'R'}(Q,\Omega)\rangle. \tag{4.5}$$

Consider (4.4) and (4.5) to cause the coupling between electronic states m and n due to the presence of a bath of oscillators. This is possible because of the assumption of equal vibrational frequencies in both states. In order to formulate this precisely note that the energy of a particular rovibronic state in a symmetric top molecule can be approximated by

$$\omega_{nvJK} = \omega_n + \omega_{vJK} = \omega_n + \sum_s \omega_s(n_s + \tfrac{1}{2}) + \sum_t \omega_t(n_t + 1) + A_n J(J+1) - \tfrac{1}{2}A_n\left[K - \sum_t \zeta_t l_t\right]^2, \tag{4.6}$$

where s labels nondegenerate vibration, t labels degenerate modes, and l_t is the vibrational angular momentum quantum number for mode t.

Since the frequencies ω_s and ω_t are assumed equal in all electronic states, a Hamiltonian having diagonal elements equal to (4.6) can be written in the form

$$H_0 = \sum_n \omega_n |n\rangle\langle n| + \sum_s \omega_s(a_s^\dagger a_s + \tfrac{1}{2}) + \sum_t \omega_t(N_t + 1) + \sum_n \left\{A_n \mathbf{J}^2 - \tfrac{1}{2}A_n\left[J_z - \sum_t \zeta_t L_t\right]^2\right\}|n\rangle\langle n|. \tag{4.7}$$

For explicit forms for the degenerate oscillator number operator N_t and angular momentum operator L_t in terms of ladder operators see Messiah (1958). The Hamiltonian (4.7) acts on states of the form

$$|m\rangle \prod_s |n_s\rangle \prod_t |n_t, l_t\rangle |JK\mathscr{M}\rangle. \tag{4.8}$$

The coupling (4.4) and (4.5) can be written as

$$\tfrac{1}{4}C_{mn}^{(2)}(A_m + A_n)[-\zeta_{21}^{(z)}Q_1 J_z - \tfrac{1}{2}\zeta_{21}^{2(z)}(P_1Q_1 + Q_1P_1)Q_2 + \zeta_{21}^{2(z)}Q_1^2P_2]|m\rangle\langle n| \prod_{s\in \mathrm{A_{1g}}} D(\Delta_s) + C_{mn}^{(2)}P_2 \prod_{s\in \mathrm{A_{1g}}} D(\Delta_s) + \text{h.c.} \tag{4.9}$$

The $D(\Delta_s)$ in (4.9) are displacement operators which make the Franck–Condon overlaps work out correctly (Nitzan and Jortner, 1973). Explicitly,

$$D(\Delta_s) = \exp[-(\Delta_s/\sqrt{2})(a_s^\dagger - a_s)]. \tag{4.10}$$

Since we are considering essentially a two-level system of states m and n, it is useful to put the Hamiltonian into a form that shows explicitly the energy gap $\omega_m - \omega_n = \omega_0$. Writing H_0 in terms of σ_3, the z Pauli matrix, and choosing the zero of energy halfway between ω_m and ω_n, one finds

$$H_0 = \tfrac{1}{2}\omega_0\sigma_3 + \sum_s \omega_s(a_s^\dagger a_s + \tfrac{1}{2}) + \sum_t \omega_t(N_t + 1) + \left\{\tfrac{1}{2}(A_m - A_n)\mathbf{J}^2 - \tfrac{1}{4}(A_m - A_n)\left[J_z - \sum_t \zeta_t L_t\right]^2\right\}\sigma_3 + \left\{\tfrac{1}{2}(A_m + A_n)\mathbf{J}^2 - \tfrac{1}{4}(A_m + A_n)\left[J_z - \sum_t \zeta_t L_t\right]^2\right\}. \tag{4.11}$$

Since no degenerate modes appear explicitly in (4.9) the last terms in braces in (4.11) may be dropped; it commutes with both H_0 and the coupling (4.9); i.e., it generates no dynamics in the model. The zeroth-order Hamiltonian may then be taken as

$$H_0 = \tfrac{1}{2}\left\{\omega_0 + (A_m - A_n)\mathbf{J}^2 - \tfrac{1}{2}A_m\left[J_z - \sum_t \zeta_{t,m}L_t\right]^2 + \tfrac{1}{2}A_n\left[J_z - \sum_t \zeta_{t,n}L_t\right]^2\right\}\sigma_3 + \sum_s \omega_s(a_s^\dagger a_s + \tfrac{1}{2}) + \sum_t \omega_t(N_t + 1). \quad (4.12)$$

If out-of-plane modes are considered—and in general they must be—the A_{1g} promoting mode can Coriolis couple to the out of plane E_{1g} degenerate vibrational mode yielding a coupling dependent upon J_x and J_y. Because the angular momentum operators do not commute amongst themselves, the last term in (4.11) must be retained if considering the coriolis coupling to out-of-plane modes.

In order to calculate the rate of transition from electronic state m to state n consider the golden rule expression

$$k_{m\to n} = 2\pi \sum_n |\langle n|V|m\rangle|^2 \delta(E_m - E_n). \quad (4.13)$$

If in (4.13) the δ function is replaced by its Fourier representation, one finds in a standard manner

$$k_{m\to n} = \int_{-\infty}^{\infty} d\tau\, \langle m|V(\tau)V(0)|m\rangle, \quad (4.14)$$

where $V(\tau)$ is the interaction picture coupling:

$$V(\tau) = e^{iH_0\tau}Ve^{-iH_0\tau}. \quad (4.15)$$

In the two level system of electronic states m and n, where m is the initially prepared state, (4.9) may be written as

$$V = c_{mn}^{(2)}\{P_2 - \tfrac{1}{4}\xi_{21}^{(z)}Q_1J_z - \tfrac{1}{8}\zeta_{21}^{(z)}\xi_{21}^{(z)}(P_1Q_1 + Q_1P_1)Q_2 + \tfrac{1}{4}\zeta_{21}^{(z)}\xi_{21}^{(z)}Q_1^2P_2\}\sigma_+ \prod_{s\in A_{1g}} D^\dagger(\Delta_s) + \text{h.c.}, \quad (4.16)$$

where

$$\xi_{21}^{(z)} = \zeta_{21}^{(z)}(A_m + A_n). \quad (4.17)$$

Using (4.15) and (4.12) yields

$$V(\tau) = c_{mn}^{(2)}\{P_2(\tau) - \tfrac{1}{4}\xi_{21}^{(z)}Q_1(\tau)J_z - \tfrac{1}{8}\zeta_{21}^{(z)}\xi_{21}^{(z)}[P_1(\tau)Q_1(\tau) + Q_1(\tau)P_1(\tau)]Q_2(\tau) + \tfrac{1}{4}\xi_{21}^{(z)}\zeta_{21}^{(z)}Q_1^2(\tau)P_2(\tau)\} \prod_{s\in A_{1g}} D^\dagger(\Delta_s,\tau)\sigma_+ \exp\{i(\omega_0 + \Delta H_{\text{rot}}^0)t\} + \text{h.c.} \quad (4.18)$$

$\Delta H^0_{\mathrm{rot}}$ is simply the difference in the rotational Hamiltonian in states m and n. Note that this term modifies the energy gap. The time dependence of the P and Q operators comes only from the vibrational Hamiltonian. J_z contains no time dependence since it is a constant of the motion. From (4.14), (4.16), and (4.18) the rate constant becomes

$$\begin{aligned} k_{m\to n} = |c^{(2)}_{mn}|^2 \int_{-\infty}^{\infty} d\tau\, \langle \{V\}, JKM | \{P_2(\tau) - \tfrac{1}{4}\xi^{(z)}_{21} Q_1(\tau) K \\ - \tfrac{1}{8}\zeta^{(z)}_{21}\xi^{(z)}_{21}[P_1(\tau)Q_1(\tau) + Q_1(\tau)P_1(\tau)]Q_2(\tau) \\ + \tfrac{1}{4}\xi^{(z)}_{21}\zeta^{(z)}_{21} Q_1^2(\tau)P_2(\tau)\} \prod_{s\in \mathrm{A_{1g}}} D^\dagger(\Delta_s, \tau) \prod_{s\in \mathrm{A_{1g}}} D(\Delta_s)\{P_2 - \tfrac{1}{4}\xi^{(z)}_{21}Q_1 K \\ - \tfrac{1}{8}\zeta^{(z)}_{21}\xi^{(z)}_{21}[P_1Q_1 + Q_1P_1]Q_2 + \tfrac{1}{4}\xi^{(z)}_{21}\zeta^{(z)}_{21}Q_1^2P_2\} | \{V\}, JKM\rangle \\ \times e^{i(\omega_0 + \Delta E_{\mathrm{rot}})\tau}. \end{aligned} \tag{4.19}$$

Since the Q_1 normal mode is not displaced (it is not a totally symmetric mode), (4.9) becomes (abbreviating the product of displacement operator by $\mathscr{D}$)

$$\begin{aligned} k_{m\to n} = |c^{(2)}_{mn}|^2 \int_{-\infty}^{\infty} d\tau\, e^{i(\omega_0 + \Delta E_{\mathrm{rot}})\tau}\{\langle P_2(\tau)\mathscr{D}^\dagger(\tau)\mathscr{D}(0)P_2(0)\rangle \\ + \tfrac{1}{16}\xi^{2(z)}_{21}K^2\langle Q_1(\tau)\mathscr{D}^\dagger(\tau)\mathscr{D}(0)Q_1(0)\rangle \\ + \tfrac{1}{64}\xi^{2(z)}_{21}\zeta^{2(z)}_{21}\langle [P_1(\tau)Q_1(\tau) + Q_1(\tau)P_1(\tau)]Q_2(\tau)\mathscr{D}^\dagger(\tau)\mathscr{D}(0)Q_2(0) \\ \times [P_1(0)Q_1(0) + Q_1(0)P_1(0)]\rangle \\ + \tfrac{1}{16}\xi^{2(z)}_{21}\zeta^{2(z)}_{21}\langle Q_1^2(\tau)P_2(\tau)\mathscr{D}^\dagger(\tau)\mathscr{D}(0)Q_1^2(0)P_2(0)\rangle \\ - \tfrac{1}{8}\zeta^{(z)}_{21}\xi^{(z)}_{21}\langle P_2(\tau)\mathscr{D}^\dagger(\tau)\mathscr{D}(0)[P_1(0)Q_1(0) + Q_1(0)P_1(0)]Q_2(0)\rangle \\ - \tfrac{1}{8}\zeta^{(z)}_{21}\xi^{(z)}_{21}\langle [P_1(\tau)Q_1(\tau) + Q_1(\tau)P_1(\tau)]Q_2(\tau)\mathscr{D}^\dagger(\tau)\mathscr{D}(0)P_2(0)\rangle \\ + \tfrac{1}{4}\xi^{(z)}_{21}\zeta^{(z)}_{21}\langle P_2(\tau)\mathscr{D}^\dagger(\tau)\mathscr{D}(0)Q_1^2(0)P_2(0)\rangle \\ + \tfrac{1}{4}\xi^{(z)}_{21}\zeta^{(z)}_{21}\langle Q_1^2(\tau)P_2(\tau)\mathscr{D}^\dagger(\tau)\mathscr{D}(0)P_2(0)\rangle \\ - \tfrac{1}{32}\zeta^{2(z)}_{21}\xi^{2(z)}_{21}\langle [P_1(\tau)Q_1(\tau) + Q_1(\tau)P_1(\tau)]Q_2(\tau)\mathscr{D}^\dagger(\tau)\mathscr{D}(0)Q_1^2(0)P_2(0)\rangle \\ - \tfrac{1}{32}\zeta^{2(z)}_{21}\xi^{2(z)}_{21}\langle Q_1^2(\tau)P_2(\tau)\mathscr{D}^\dagger(\tau)\mathscr{D}(0)[P_1(0)Q_1(0) + Q_1(0)P_1(0)]Q_2(0)\rangle\}. \end{aligned} \tag{4.20}$$

Only the second term in (4.20) depends explicitly on the angular momentum quantum number K. The first term is the usual vibronic coupling. The remaining terms arise from the electronic coupling between the states m and n. The second term produces a Coriolis coupling between vibrations even when the molecule is in a $J = 0$ state. Physically this is due to the fact that the $\mathrm{A_{2g}}$ mode in the molecule behaves in a similar manner to a rotation, an effect similar to that which corrects the vibrational energy in a molecule with degenerate modes excited, and is due to the presence of vibrational angular momentum.

For simplicity let us keep only the first two terms in (4.19). Then

$$k_{m\to n} = |c_{mn}^{(2)}|^2 \int_{-\infty}^{\infty} d\tau\, e^{i(\omega_0+\Delta E_{\rm rot})\tau}$$

$$\times \left\{ \langle \{v\}_m | \prod_{s\in A_{1g}} D^\dagger(\Delta_s, \tau) P_2(\tau) \prod_{s\in A_{1g}} P_2(0) D(\Delta_s) | \{v\}_m \rangle \right.$$

$$\left. + \tfrac{1}{16}\zeta_{21}^{2(z)} K^2 \langle \{V\}_m | Q_1(\tau) \prod_{s\in A_{1g}} D^\dagger(\Delta_s, \tau) \prod_{s\in A_{1g}} D(\Delta_s) Q_1(0) | \{v\}_m \rangle \right\}. \tag{4.21}$$

Writing all operators in the form of ladder operators, we have

$$Q_r = (2\omega_r)^{-1/2}(a_r^\dagger + a_r), \qquad Q_r(\tau) = (2\omega_r)^{-1/2}(a_r^\dagger e^{i\omega_r\tau} + a_r e^{-i\omega_r\tau});$$

$$P_r = i(\omega_r/2)^{1/2}(a_r^\dagger - a_r), \qquad P_r(\tau) = i(\omega_r/2)^{1/2}(a_r^\dagger e^{i\omega_r\tau} - a_r e^{-i\omega_r\tau}).$$

The displacement operators have the property of producing coherent states (Merzbacker, 1970) from the zero-point levels of the two totally symmetric modes which are displaced. It is assumed that initially there are no quanta in these states (as would be the case for absorption into the first singlet of benzene, where an E_{2g} mode is excited). Performing the required algebra yields

$$k_{m\to n} = |c_{mn}^{(2)}|^2 \exp\left[-\frac{1}{2}\sum_s \Delta_s^2\right] \int_{-\infty}^{\infty} d\tau \exp\left[i(\omega_0 + \Delta E_{\rm rot})\tau + \frac{1}{2}\sum_s \Delta_s^2 e^{-i\omega_s\tau}\right]$$

$$\times \left[\tfrac{1}{2}\omega_2 e^{-i\omega_2\tau} + (1/32\omega_1)K^2\zeta_{21}^{2(z)} e^{-i\omega_1\tau}\right]. \tag{4.22}$$

The integral appearing in (4.22) has been considered previously for the case of large energy gaps since an asymptotic result may be obtained in this case using a steepest descent method (Freed, 1976). It is not necessary to evaluate (4.22) explicitly in order to understand its consequences. The first term in brackets in the integrand in (4.22) is the vibronic contribution to the rate constant. This part of the rate is rotationally dependent through a simple energy gap modification

$$\omega_0 - \omega_2 \to \omega_0 - \omega_2 + \Delta E_{\rm rot}. \tag{4.23}$$

The second term in brackets is the interstate Coriolis coupling term. Here the energy gap is also modified by the addition of a rotational energy, and furthermore the promoting mode frequency ω_2 is replaced by the frequency of the mode to which the promoting mode is Coriolis coupled. This term also contains an explicit angular momentum dependence through the factor K^2.

For small or moderate values of K the rotational effect will be small since $\zeta_{21}^{(z)}$ is on the order of a rotational energy separation. When K^2 becomes large it is necessary to consider the intrastate coupling also, since (4.12) may no longer be a reasonable choice for H_0 because of the presence of centrifugal distortion effects which will modify the harmonic vibrations and hence the Franck–Condon factors describing the relaxation process. The result (4.22) is intended simply to demonstrate that interstate coupling due to the nuclear kinetic energy contains effects due to rotations in addition to effects attributed to normal vibrational modes.

V. ROTATIONAL EFFECTS IN SPIN–ORBIT COUPLING

A. Vibronic Spin–Orbit Coupling

The preceding section dealt with the situation in which relaxation takes place between singlet states. Because of the importance of the singlet–triplet intersystem crossing processes in many experimentally well studied prototype molecules, e.g., aromatic hydrocarbons, it is important to consider how interstate Coriolis coupling may influence these spin dependent processes.

The angular momentum dependence of the direct spin–orbit coupling has been considered by Howard and Schlag and discussed in Section IIIB. It has long been recognized that direct spin–orbit coupling is not the only important mechanism leading to intersystem crossing. One must also consider the effect of a mixed spin–orbit and vibronic, or nuclear kinetic energy, coupling. This type of mechanism was considered by Siebrand (1970) and Henry and Siebrand (1971) for the case of a norotating molecule, in which case the nuclear kinetic energy operator is given by (3.23). In the treatment given here the nuclear kinetic energy operator will include the rotational kinetic energy as well as the normal mode kinetic energy (3.23), and the rovibronic coupling will by given by (3.15). Because of the algebraic complexity in the case no explicit situation will be considered. We will instead concentrate on general formulas and models.

Following Siebrand we consider the matrix element of the spin–orbit coupling operator in a basis where the T_{N} coupling has been taken account of in first order perturbation theory. That is, we write for the initially excited singlet state

$$ {}^1\tilde{\Psi}_m = {}^1\Psi_m + \sum_{i \neq m} \frac{\langle {}^1\Psi_i | [T_{\mathrm{N}}^1 + H_{\mathrm{rot}}] | {}^1\Psi_m \rangle}{E_m - E_i} \, {}^1\Psi_i . \tag{5.1} $$

A similar relation will be taken to hold for the triplet states ${}^3\tilde{\Psi}_n$ into which the singlet will decay. Equation (5.1) differs from Siebrand's in that the rotational coupling is included. The matrix element responsible for the intersystem crossing process is given by

$$ \begin{aligned} \langle {}^1\tilde{\Psi}_m | H_{\mathrm{so}} | {}^3\tilde{\Psi}_n \rangle = {} & \langle {}^1\Psi_m | H_{\mathrm{so}} | {}^3\Psi_n \rangle \\ & + \sum_i \frac{\langle {}^1\Psi_m | [T_N^1 + H_{\mathrm{rot}}\rangle | {}^1\Psi_i \rangle \langle {}^1\Psi_i | H_{\mathrm{so}} | {}^3\Psi_n \rangle}{{}^1E_m - {}^1E_i} \\ & + \sum_j \frac{\langle \Psi_m | H_{\mathrm{so}} | {}^3\Psi_j \rangle \langle {}^3\Psi_j | [T_N^1 + H_{\mathrm{rot}}] | {}^3\Psi_n \rangle}{{}^3E_n - {}^3E_i} . \end{aligned} \tag{5.2} $$

Recalling that

$$^1\Psi_m = {}^1\phi_m(x, Q)\psi_{mvR}(Q, \Omega) \tag{5.3}$$

and furthermore assuming

$$^1\phi_m(x, Q) = {}^1\phi_m(x, 0) + [\partial\, {}^1\phi_m(x, Q)/\partial Q]_{Q=0} Q + \cdots \tag{5.4}$$

allows us to construct an effective Hamiltonian describing the relaxation process when rotational coupling is present. Let us recall that in the two-electron approximation of Section III the spin–orbit operator is taken to be

$$H_{\text{so}} = \sum_{j=1}^{2} \lambda_j \sum_{\alpha=\pm1,0} l_j^{(\alpha)} S_j^{(-\alpha)}. \tag{5.5}$$

Equation (5.2) may now be written as the sum of three terms:

$$V_{mn} = V_{mn}^{(1)} + V_{mn}^{(2)} + V_{mn}^{(3)}. \tag{5.6}$$

The first two terms in (5.6) are given by Siebrand as

$$V_{mn}^{(1)} = \langle {}^1\phi_m(x, 0) | H_{\text{so}}(0) | {}^3\phi_n(x, 0)\rangle_x \langle \psi_{nvR}(Q, \Omega) | \psi_{mv'R'}(Q, \Omega)\rangle, \tag{5.7}$$

$$V_{mn}^{(2)} = [(\partial/\partial Q)\langle {}^1\phi_m(x, Q) | H_{\text{so}}(Q) | {}^3\phi_n(x, Q)\rangle]_{Q=0} \langle \psi_{mvR}(Q, \Omega) | Q | \psi_{nv'R'}(Q, \Omega)\rangle. \tag{5.8}$$

Equation (5.7) describes the direct coupling as discussed in Section III. The Q in (5.8) is meant as a collective symbol, and for the case of aromatic hydrocarbons this term is dominated by in-plane vibrations. The angular momentum dependence of (5.8) will be the same as that of (5.7) since the spin operators determine this. The magnitudes of the two terms will of course be different.

The third term in (5.6) comes about from the sums in the matrix element (5.2). This term consists of essentially two parts, one vibronic and the other rotational. Assuming that the electronic energy separations are much greater than any rotational or vibrational level separations, the vibronic part of $V_{mn}^{(3)}$ is given by

$$V_{mn,\,\text{vib}}^{(3)} = \sum_\alpha \sum_q \sum_{j=1}^{2} \left\{ \sum_i \frac{{}^1c_{mi}^{(q)} \langle {}^1\phi_i(x, 0) | \lambda_j(x) l_j^{(\alpha)} | {}^3\phi_n(x, 0)\rangle_x}{{}^1\bar{E}_m - {}^1\bar{E}_i} + \sum_i \frac{\langle {}^1\phi_m(x, 0) | \lambda_j(x) l_j^{(\alpha)} | {}^3\phi_i(x, 0)\rangle\, {}^3c_{in}^{(q)}}{{}^3\bar{E}_n - {}^3\bar{E}_i} \right\} S_j^{(-\alpha)} P_q \sigma_+ + \text{h.c.}, \tag{5.9}$$

with

$$^1c_{mi}^{(q)} = \langle {}^1\phi_m(x, Q) | i(\partial/\partial Q_q) | {}^1\phi_i(x, Q)\rangle_{Q=0}. \tag{5.10}$$

Equation (5.9) is to be interpreted as an effective operator coupling electronic states m and n. It is to be evaluated between the crude adiabatic rovibronic states $|mVR\rangle$ and $|nV'R'\rangle$. Note that the rotational dependence of (5.9) is the same as that of (5.7) and (5.8); the magnitude of the effect is different of course. We also note that Henry and Siebrand point out that the important modes q in (5.10) are out of plane vibrations.

Judging from the complexity of (4.4), which is the direct rotational coupling, one expects the indirect coupling to be even more complicated, especially since it is necessary to consider out-of-plane modes. The rotational coupling now contains contributions from the x and y components of the angular momentum as well as the z component considered in Section IV. The part of the rotational coupling depending explicitly on the angular momentum operator is given by an expression similar to (5.9):

$$V^{(3)}_{mn,\,\mathrm{rot}} = -\sum_{\beta}\sum_{\alpha}\sum_{q}\sum_{r}\sum_{j=1}^{2}\left\{\sum_{i}\frac{\bar{\mu}^{mi}_{\beta\beta}\,{}^{1}c^{(q)}_{mi}\langle {}^{1}\phi_i(x,0)|\lambda_j(x)l^{(\alpha)}_j|{}^{3}\phi_n(x,0)\rangle_x}{{}^{1}\bar{E}_m - {}^{1}\bar{E}_i} + \sum_{i}\frac{\langle {}^{1}\phi_m(x,0)|\lambda_j(x)l^{(\alpha)}_j|{}^{3}\phi_i(x,0)\rangle_x\,{}^{3}c^{(q)}_{in}\bar{\mu}^{in}_{\beta\beta}}{{}^{3}\bar{E}_n - {}^{3}\bar{E}_i}\right\}\zeta^{(\beta)}_{rq}Q_r N_\beta s^{(-\alpha)}_j\sigma_+ + \mathrm{h.c.} \tag{5.11}$$

In (5.10) N_β is the βth component of the rigid body angular momentum

$$\mathbf{N} = \mathbf{J} - \mathbf{S}. \tag{5.12}$$

Thus missing from (5.11) is any type of electronic angular momentum coupling. This type of interaction is contained in the Stevens and Brand model. It could be included here in a simple manner by noting that the interstate interaction from H_{rot} may be written in Hund's case (b) [see (3.20)] as

$$\langle\psi_{mvR}(Q,\Omega)|\langle\phi_m(x,Q)|\tfrac{1}{2}\bar{\mu}_{\alpha\alpha}\{(N_\alpha - L_\alpha)^2 - 2(N_\alpha - L_\alpha)\pi_\alpha + \pi_\alpha\pi_\alpha\} \times |\phi_n(x,Q)\rangle|\psi_{nv'R'}(Q,\Omega)\rangle, \tag{5.13}$$

where α is here the αth Cartesian component of the vector operators. The operator L_α is the sum of all of the one-electron orbital angular momentum operators. The matrix elements (5.9) and (5.11) have been written so that the terms in curly brackets can be treated as effective coupling constants, which makes the expression much less formidable looking.

It should be emphasized that the sum of (5.9) and (5.11) is not the total rotational coupling. We must also consider the $\boldsymbol{\pi}\cdot\boldsymbol{\pi}$ terms in (5.13) even though this term has the same rotational dependence as do all other terms except (5.11), [and its modification due to the $\mathbf{N}\cdot\mathbf{L}$ term in (5.13)]. This is especially the case if one wishes to use isotope effects in trying to unravel any rotationally dependent effects in both internal conversion and intersystem crossing. Of course, one must also try to take into account the possible symmetry differences when isotopic substitution is made.

B. Intersystem Crossing in a Two-Level System: A Displaced Oscillator Model

Let us reconsider the effect of rotation on the intersystem crossing rate constant. This was previously discussed in Section IIIB, where the results of Howard and Schlag were reviewed. The discussion in this section will be based upon the displaced oscillator model, and methods similar to those used in Section IV will be applied.

Consider the Hamiltonian

$$H = \tfrac{1}{2}\omega_0 \sigma_3 + \sum_j \omega_j (a_j^\dagger a_j + \tfrac{1}{2}) + A(Q)\mathbf{N}^2 + B(Q)N_z^2 + H_{so} + H_{cor}. \quad (5.14)$$

This operator describes a two-level symmetric top molecule for which the upper level is now a singlet and the lower level a triplet. Degenerate vibrations are not treated explicitly because we assume that the important modes are of A_{1g} symmetry. Thus our model assumes that the molecule remains a symmetric top throughout the relaxation process. $A(Q)$ and $B(Q)$ are inverse moments of inertia depending explicitly on the positions of the nuclei. As before we assume that the vibrational frequencies in the singlet and triplet are the same but the equilibrium positions are shifted. H_{cor} is the Coriolis interaction and H_{so} the spin–orbit coupling Hamiltonian, which we take as

$$H_{so} = \sigma_- \prod_j{}' D(\Delta_j) l_-^0 s_-^0 + \text{h.c.} \quad (5.15)$$

This expression for H_{so} is similar to (3.30), where we have summed over the two electrons and absorbed the λ coefficients into l. Only one component is kept because we use nondegenerate orbital functions. The operator S_- is defined by

$$S_-^\alpha = S_1^\alpha - S_2^\alpha. \quad (5.16)$$

The products of displacement operators take into account the Franck–Condon overlap between vibrations in the singlet and triplet. Equation (15.15) is the part of the spin–orbit coupling which mixes the singlet and triplet. The part coupling triplet sublevels has been ignored. The intersystem crossing rate constant will be calculated as in (4.14) via

$$k_{isc} = \int_{-\infty}^{\infty} dt \, \langle H_{so}(t) H_{so}(0) \rangle, \quad (5.17)$$

where the average in (5.17) is taken in the initial rovibronic singlet state, denoted $|2; \{v\}; N_S K_S 00\rangle$. In this state 2 labels the upper electronic singlet, $\{v\}$ labels the vibrational quantum number, N_S labels the angular momentum associated with the molecular tumbling, K_S its component on the top axis, and 0, 0 labels the spin states S and M_S.

Consider first the case where $A(Q)$ and $B(Q)$ are constant and H_{cor} is neglected. Substituting (5.18) into (5.17) one finds that (5.17) factors into a rotational and a vibrational part. Explicit calculation of the rotational factor shows it to be independent of N_S and K_S because of the appearance of sums such as (3.33). Thus there is no rotational dependence of k_{isc} in this case. The only way that rotational effects occur is through the interaction of vibration and rotation.

In the case of small displacements the effect of the change of moment of inertia can be taken into account by adding a term

$$\sum_k \left[\left(\frac{\partial A}{\partial Q_k}\right)_0 Q_k \mathbf{N}^2 + \left(\frac{\partial B}{\partial Q_k}\right) Q_k N_z^2\right] \tag{5.18}$$

to the zeroth-order Hamiltonian (5.14). Carrying through the calculation for k_{isc} yields

$$k_{isc} = 2[I_-^0]^2 \exp\left\{-\tfrac{1}{2}\sum_j{}' \Delta_j^2\right\} \int_{-\infty}^{\infty} dt\, e^{i\omega_0 t} \exp\left\{\tfrac{1}{2}\sum_j{}' \Delta_j^2 \cos \omega_j t - \tfrac{1}{2} i \sum_j{}' [\Delta_j^2 - 2N_S(N_S+1)\Delta_j \xi_j - 2K_S^2 \Delta_j \eta_j] \sin \omega_j t\right\}, \tag{5.19}$$

where

$$\xi_k = (\hbar/2m_k\omega_k^3)^{1/2}(\partial A/\partial Q_k)_0 \tag{5.20}$$

and

$$\eta_k = (\hbar/2m_k\omega_k^3)^{1/2}(\partial B/\partial Q)_0. \tag{5.21}$$

For $N_S = K_S = 50$ and a displacement of 0.1 Å the ratio of vibration to rotational coefficients of $\sin \omega_j t$ is only about 1% for $\omega_k \sim 10^3$ cm^{-1}. This effect is clearly small.

Coriolis interaction and the effect of vibration–rotation interaction in the initial singlet state are also easily calculated; these also lead to small effects. If any significant rotational effects exist in experimental data, then, they are probably due to resonances particular to specific molecules.

VI. ANGULAR MOMENTUM CONSERVATION IN POLYATOMIC FRAGMENTATION REACTIONS

We now examine how the conservation of angular momentum affects the rate of fragmentation and the distribution of energy amongst the products of dissociation of a polyatomic molecule. Although the general theory of polyatomic fragmentation reactions can be readily formulated, reduction of the complex analysis to useful algorithms and instructive physical models is only just beginning, so very little is known about the detailed dynamics of the elementary processes involved.

We begin with a description of a crude model, now superseded, which displays very clearly all but one aspect of the influence of angular momentum constraints on the rate of formation of products from a photofragmentation, and the distribution of energy between them. The omitted feature, which is of very great importance for the understanding of real reactions, will be discussed later.

A. A Crude Model

Florida and Rice (1975) introduced a model designed to provide an explicit representation of the role of angular momentum conservation in defining the reaction dynamics when the geometries of reactant and products are different. Specifically, the model neglects the evolution of reactant normal modes into a recoil trajectory and product normal modes by assuming that the bond which breaks is independent of other bonds (the latter can be coupled intramolecularly). This assumption is removed in the more general theory proposed by Morse *et al.* (1979), and the effects induced thereby are shown to be very important. Nevertheless, for the limited purpose for which it was designed, the Florida–Rice model is still very useful.

The physical idea underlying the Florida–Rice model is very simple. If a molecule is nonlinear in the initial state, it must be expected that the line defined by the bond that breaks, which also almost always will define the line along which the fragments of the molecule separate, will not pass through the center of mass. But then as the molecule separates orbital angular momentum is generated, and this must be compensated by a change in molecular angular momentum of the fragments. It is the interplay between these contributions to the angular momentum and the constraint of conservation of energy, and their influence on the photodissociation dynamics, that we seek to understand.

Consider a polyatomic molecule which, in the zeroth-order representation, has a bound adiabatic electronic surface V_i and a repulsive adiabatic electronic surface V_f. Predissociation occurs via coupling of the corresponding zeroth-order adiabatic electronic states ϕ_i and ϕ_f. The full molecular Hamiltonian is, in an obvious notation,

$$H(r, R) = T(\mathbf{R}) + H^{\mathrm{BO}}(\mathbf{r}, \mathbf{R}), \tag{6.1}$$

where r and R denote, respectively, the electronic and nuclear position vectors. Now, the electronic wavefunctions ϕ_i and ϕ_f are eigenfunctions of the full electronic Hamiltonian with parametric dependence on the nuclear coordinates, and the adiabatic potential surfaces $V_i(R)$ and $V_f(R)$ are defined by

$$V_i(\mathbf{R}) = \langle \phi_i | H^{\mathrm{BO}} | \phi_i \rangle, \qquad V_f(\mathbf{R}) = \langle \phi_f | H^{\mathrm{BO}} | \phi_i \rangle. \tag{6.2}$$

The Florida–Rice model assumes the following:

(i) The adiabatic potential $V_f(R)$ is separable in the sense

$$V_f(\mathbf{R}) = V_f^{(1)}(\mathbf{X}) + V_f^{(2)}(\mathbf{Q}), \tag{6.3}$$

where **X** refers to some reaction coordinate and **Q** to the remaining coordinates of the energy surface ($\mathbf{R} \equiv \mathbf{X} \oplus \mathbf{Q}$).

(ii) The reaction coordinate **X** is taken to be the local coordinate vector of the dissociating bond.

(iii) The interaction between the bond represented by **X** and the remainder of the coordinates represented by **Q** is neglected. Note that this assumption does not preclude interactions between modes in the set **Q**; it does imply that the nuclear kinetic energy operator is diagonal with respect to **X**.

(iv) The resonance widths are much smaller than their separations.

The use of assumptions (i)–(iii) requires that we consider three wavefunctions for nuclear motion. These are defined by the equations

$$[T(\mathbf{R}) + E_{in} - V_{\mathrm{i}}(\mathbf{R})]\chi_{in}(\mathbf{R}) = 0, \tag{6.4}$$

$$[T^{(1)}(\mathbf{X}) + E - E_{\mathrm{f}m}^{(2)} - V_{\mathrm{f}}^{(1)}(\mathbf{X})]\xi_{\mathrm{f}\beta}(\mathbf{X}) = 0, \tag{6.5}$$

$$[T^{(2)}(\mathbf{Q}) + E_{\mathrm{f}m}^{(2)} - V_{\mathrm{f}m}^{(2)}(\mathbf{Q})]\chi_{\mathrm{f}m}(\mathbf{Q}) = 0, \tag{6.6}$$

where only $\xi_{\mathrm{f}\beta}^{(1)}$ is a continuum function. The superscript notation identifies the group of coordinates which form the argument of a function: $F^{(1)}$ is a function of only the reaction coordinate (bond coordinate) X, and $F^{(2)}$ is a function only of the remaining coordinates Q. We treat the case where coupling between ϕ_{i} and ϕ^{f} derives from $T(\mathbf{R})$, i.e., internal conversion. Then the resonance widths take the form

$$\Gamma_{in} = 2\pi \sum_{m,\,\beta} |\langle \chi_{in}(\mathbf{R}) U_{\mathrm{if}}(\mathbf{R}) \chi_{\mathrm{f}m}^{(2)}(\mathbf{Q}) \xi_{\mathrm{f}\beta}^{(1)}(\mathbf{X}) \rangle|^2, \tag{6.7}$$

with

$$U_{\mathrm{if}}(\mathbf{R}) = \langle \phi_{\mathrm{i}} | T(\mathbf{R}) | \phi_{\mathrm{f}} \rangle, \tag{6.8}$$

and with conservation of angular momentum an implicit constraint on the sum over m and β. By virtue of assumptions (ii) and (iii), the vibrational motion in ϕ_{i} is described by the product function

$$\chi_{in}(\mathbf{R}) = \chi_{in}^{(2)}(\mathbf{Q}) \Xi_{in}^{(1)}(\mathbf{X}), \tag{6.9}$$

which greatly simplifies evaluation of (6.7).

It is now convenient to introduce cylindrical coordinates in the space spanned by **X** using the bond axis as the axis of symmetry. In this coordinate system Eq. (6.6) assumes the form

$$\left[\frac{\partial^2}{\partial x^2} + \frac{1}{\rho}\frac{\partial}{\partial \rho}\left(\rho \frac{\partial}{\partial \rho}\right) + \frac{1}{\rho^2}\frac{\partial^2}{\partial \phi^2} + \left(\frac{2\mu}{\hbar^2}\right)\{E - E_{\mathrm{f}m}^{(2)} - V_{\mathrm{f}}^{(1)}(\rho, \phi, x)\}\right]\xi_{\mathrm{f}\beta}^{(1)} = 0, \tag{6.10}$$

with x the distance along the bond axis, ρ the distance perpendicular to the bond axis, and ϕ the azimuthal angle. Continuing, the coupling term $U_{\text{if}}(\mathbf{R})$ can now be expressed in the form

$$U_{\text{if}}(\mathbf{R}) = U_{\text{if}}^{(1)}(\mathbf{X}) + U_{\text{if}}^{(2)}(\mathbf{Q}), \tag{6.11}$$

$$U_{\text{if}}^{(1)}(\mathbf{X}) = \frac{\hbar^2}{\mu}\left[I_x \frac{\partial}{\partial x} + I_\rho\left(\frac{\partial}{\partial \rho} + \frac{1}{2\rho}\right) + \frac{1}{\rho^2} I_\phi \frac{\partial}{\partial \phi}\right], \tag{6.12}$$

$$U_{\text{if}}^{(2)}(\mathbf{Q}) = \sum_j \left(\frac{\hbar^2}{\mu_j}\right)\frac{I_j \partial}{\partial R_j}, \tag{6.13}$$

where

$$I_j = \langle \phi_{\text{i}} | \partial/\partial R_j | \phi_{\text{f}} \rangle, \tag{6.14}$$

with the integration taken over the electronic coordinates. To make any further progress we must know some of the properties of the potential surfaces $V_{\text{i}}(\mathbf{R})$ and $V_{\text{f}}(\mathbf{R})$.

Florida and Rice illustrate the explicit influence of angular momentum conservation on the rate of fragmentation for the case that the prepared state has nonlinear molecular geometry and the state in which bond breaking occurs has linear molecular geometry. They assume that

$$V_{\text{f}}^{(1)}(\mathbf{X}) = \tfrac{1}{2}\mu\omega_{\text{f}\rho}\,\rho^2 + \vartheta_{\text{f}}^{(1)}(x), \tag{6.15}$$

$$V_{\text{i}}^{(1)}(\mathbf{X}) = \tfrac{1}{2}\mu\omega_{\text{i}x}x^2 + \tfrac{1}{2}\mu\omega_{\text{i}\rho}\rho^2 + Ae^{-\gamma\rho^2}, \tag{6.16}$$

which corresponds to treating the ρ dependence of the vibrational energy as that of a double-well potential in state ϕ_{i} and as that of an isotropic oscillator in state ϕ_{f}: A and γ are the double-well potential constants, μ the reduced mass, and $\omega_{\text{i}x}$, $\omega_{\text{i}\rho}$, and $\omega_{\text{f}\rho}$ the vibration frequencies associated with coordinates x and ρ in states ϕ_{i} and ϕ_{f} as indicated. The term $\vartheta_{\text{f}}^{(1)}(x)$ represents a repulsive potential curve along the x direction in state ϕ_{f}. The electronic wavefunctions ϕ_{i} and ϕ_{f} are referred to the linear geometry in both states.

The solution of Eq. (6.10) using the potential (6.15) is

$$\xi_{\text{f}\beta}^{(1)}(\mathbf{X}) = \eta_{\text{f}\beta}^{\sigma l}(x) G_{\text{f}\beta}^{\sigma l}(\rho, \phi), \tag{6.17}$$

with

$$G_{\text{f}\beta}^{\sigma l}(\rho, \phi) = N_{\sigma l} e^{\iota l \phi} e^{-\alpha_{\text{f}}\rho^2/2} (\alpha_{\text{f}}\rho)^{l/2} L_\sigma^l(\alpha_{\text{f}}\rho^2). \tag{6.18}$$

In Eqs. (6.17) and (6.18), $N_{\sigma l}$ is a normalization constant, $\alpha_{\text{f}} \equiv \mu\omega_{\text{f}\rho}/\hbar$, L_σ^l is the associated Laguerre polynomial, and $\eta_{\text{f}\beta}^{\sigma l}$ is a continuum wavefunction satisfying

$$\{d^2/d^2x + (2\mu/\hbar^2)[E - E_{\text{f}m}^{(2)} - \hbar\omega_{\text{f}\rho}(2\sigma + l + 1) - \vartheta_{\text{f}}^{(1)}(x)]\}\eta_{\text{f}\beta}^{\sigma l}(x) = 0. \tag{6.19}$$

Moreover, it can be shown that

$$\Xi_{\text{in}}^{(1)}(\mathbf{X}) = N_{\text{i}n} H_{\text{i}n}(x) F_{\text{in}}^{jk}(\rho, \phi), \tag{6.20}$$

where $H_{\text{i}n}(x)$ is the nth harmonic oscillator wavefunction, $N_{\text{i}n}$ is a normalization constant, and

$$F_{\text{in}}^{jk}(\rho, \phi) = \sum_{\sigma'} a_{\sigma'}^{jk} G_{\text{in}}^{\sigma' k}(\rho, \phi), \tag{6.21}$$

$$G_{\text{in}}^{\sigma' k}(\rho, \phi) = N_{\sigma' k} e^{ik\phi} e^{-\alpha_{\text{i}}\rho^2/2} (\alpha_{\text{i}} \rho^2)^{k/2} L_{\sigma'}^{k}(\alpha_{\text{i}} \rho^2), \tag{6.22}$$

$$(n + \tfrac{1}{2})\hbar\omega_{\text{i}x} = E - E_{\text{in}}^{(2)} - \varepsilon_{jk}, \tag{6.23}$$

with ε_{jk} and F_{in}^{jk} the eigenvalues and eigenfunctions, respectively, of the double-well potential Hamiltonian expressed as expansions in the set of isotropic harmonic oscillator functions and energies. As before, $\alpha_{\text{i}} \equiv \mu\omega_{\text{i}}/\hbar$ and $N_{\sigma' k}$ is a normalization constant.

Consider the case when (6.12) reduces to

$$U_{\text{if}}^{(1)}(\mathbf{X}) = (\hbar^2/\mu)[I_\rho(\partial/\partial\rho + \tfrac{1}{2}\rho)] \tag{6.24}$$

as, e.g., in the photodissociation of HCCCl. Suppose predissociation is induced by a conical intersection of $V_{\text{i}}^{(1)}$ and $V_{\text{f}}^{(1)}$ so that there is a critical value of x in the linear geometry where $V_{\text{i}}^{(1)}(x = x^*) = V_{\text{f}}^{(1)}(x = x^*)$. The coupling should be negligibly small except in the immediate neighborhood of $x = x^*$, $\rho = 0$. We expect, then, that it is satisfactory to take $V_{\text{if}}^{(1)} = \varepsilon\rho$ with ε a constant. It follows that the partial level width $\Gamma_{in,\,fm\beta}^{jk,\,\sigma l}$ can be written in the form

$$\Gamma_{in,\,fm\beta}^{jk,\,\sigma l} = \varepsilon^2 |\langle \chi_{\text{in}}^{(2)}(\mathbf{Q}) | \chi_{\text{f}m}^{(2)}(\mathbf{Q}) \rangle|^2$$
$$\times |\langle H_{\text{i}n}(x) | \eta_{\text{f}\beta}^{\sigma l}(x) \rangle|^2 |\langle F_{\text{in}}^{jk}(\rho, \phi) | \rho | G_{\text{f}\beta}^{\sigma l}(\rho, \phi) \rangle|^2. \tag{6.25}$$

Integration over ϕ yields for the last term in (6.25) the relation

$$\langle F_{\text{in}}^{jk}(\rho, \phi) | \rho | G_{\text{f}\beta}^{\sigma l}(\rho, \phi) \rangle = \delta_{kl} \langle F_{\text{in}}^{jl}(\rho) | \rho | G_{\text{f}\beta}^{\sigma l}(\rho) \rangle. \tag{6.26}$$

This mathematical result can be viewed as a statement of conservation of angular momentum about the molecular axis.

The overlap factor in the bond extension x can be evaluated using the method of Caplan and Child (1972). Let the pseudocrossing point velocity $\mathbf{W}$ and the classical forces $\mathbf{F}_{\text{i}}$, $\mathbf{F}$ be defined by

$$E - E_{\text{in}}^{(2)} - \varepsilon_{jl} - V_{\text{i}}^{(1)}(x^*) = E - E_{\text{f}m}^{(2)} - \hbar\omega_{\text{i}\rho}(2\sigma + l + 1) - \vartheta_{\text{f}}^{(1)}(x^*)$$
$$= \tfrac{1}{2}\mu\mathbf{W}^2, \tag{6.27}$$

$$\mathbf{F}_{\text{i}} = -(\partial V_{\text{i}}^{(1)}/\partial x)_{x=x^*}, \qquad \mathbf{F}_{\text{f}} = -(\partial V_{\text{f}}^{(1)}/\partial x)_{x=x^*}. \tag{6.28}$$

Then we have

$$\langle H_{\text{i}n} | \eta_{\text{f}\beta}^{\sigma l} \rangle \approx \frac{4\pi\omega_{\text{i}x}}{W^* |\Delta\mathbf{F}|} [\text{Ai}(t(E))]^2, \tag{6.29}$$

with

$$|\Delta\mathbf{F}| = |\mathbf{F}_f - \mathbf{F}_i|, \tag{6.30}$$

$$W^* = 2(\hbar|\mathbf{F}_i||\mathbf{F}_j|/4\mu^2|\Delta\mathbf{F}|)^{1/3}, \tag{6.31}$$

$$t(E) = -(\mathbf{W}/W^*)^2 = -\{(2/\mu W^{*2})[E - E_{in}^{(2)} - \varepsilon_{jl} - \vartheta_i^{(1)}(x^*)]\}, \tag{6.32}$$

and Ai(t) the Airy function. Equation (6.29) is a refinement of the semiclassical transition probability formula derived by Landau and Lifshitz (1958). It includes the effect of redistribution of energy between the bending mode and the other modes of vibration. Suppose the time required for the atom to leave the neighborhood of $x = x^*$ is very short compared to the period of the bending vibration in ϕ_f. Then the impact parameter can be taken as the average value of ρ in the state labeled σl, i.e., the coupling of motions along ρ and x in ϕ_f can be neglected. Furthermore, if the energy gap ΔE_{elec} between vibrationless levels of ϕ_i and ϕ_f is large we expect that $\vartheta_f^{(1)}(x)$ is very steep near $x = x^*$, whereupon the value of ρ which defines the impact parameter is the value at $x = x^*$, i.e., once the molecule is on the surface $V_f^{(1)}$ it dissociates very rapidly, hence with ρ frozen at its initial value. Then, via the relation between the average value of ρ and σ, l, the values of the latter are fixed. Note that the value of l in ϕ_i can be (crudely) thought of as the quantum number for the initial rotational excitation of the bent molecule about the x axis.

Using the interpretation outlined above, the orbital angular momentum of the recoiling fragments is, approximately,

$$\mathbf{L} = \mu\bar{\rho}_{\sigma l}\mathbf{W}_{\sigma l}, \tag{6.33}$$

with $\bar{\rho}_{\sigma l}$ the average value of ρ in the state labeled by σl and $\mathbf{W}_{\sigma l}$ the final velocity of the leaving atom. From the recursion relation for the Laguerre polynomials we find

$$\bar{\rho}_{\sigma l} = [(2\sigma + l + 1)/\alpha_f]^{1/2} \tag{6.34}$$

using the root mean square average for $\bar{\rho}_{\sigma l}$. The rotational kinetic energy of the molecular fragment which is consistent with (6.33) is

$$L^2/2I = [(2\sigma + l + 1)/2\alpha_f I]\mu^2 W_{\sigma l}^2, \tag{6.35}$$

and hence the partial level width corresponding to interaction of the states $x_{in}^{(2)}(\mathbf{Q})F_{in}^{jl}(\rho)H_{in}^{(x)}$ and $x_{fm}^{(2)}(Q)\xi_{f\beta}^{(1)}(\mathbf{X})$ is

$$\Gamma_{in,fm\beta}^{jl} = \varepsilon^2|\langle\chi_{in}^{(2)}(\mathbf{Q})|\chi_{fm}^{(2)}(\mathbf{Q})\rangle|^2 K_{jl}(\mathscr{E}), \tag{6.36}$$

where

$$K_{jl}(\mathscr{E}) = \sum_{\sigma} (4\pi^2\omega_{ix}|W^*|\Delta\mathbf{F}|)[\text{Ai}(t(E))]^2|\langle F_{in}^{jl}(\rho)|\rho|G_{f\beta}^{\sigma l}(\rho)\rangle|^2\delta(\mathscr{E}), \tag{6.37}$$

and

$$\mathscr{E} = (E_{in}^{(2)} - E_{fm}^{(2)}) + \left[\Delta E_{\text{elec}} + (n + \tfrac{1}{2})\hbar\omega_{ix} - \tfrac{1}{2}\mu \mathbf{W}_{\sigma l}^2 - \lim_{x\to\infty} \vartheta_f^{(1)}(x)\right]$$

$$+ [\varepsilon_{jl} - (2\sigma + l + 1/2\alpha_f I)\mu^2 \mathbf{W}_{\sigma l}^2]. \tag{6.38}$$

The form of Eq. (6.36) with (6.37) and (6.38) displays explicitly the consequences of conservation of angular momentum, as desired.

The Florida–Rice model suggests that conservation of angular momentum can have a subtle but important influence on the rate of a predissociation. Aside from the obvious coupling to the energy conservation condition, evident in (6.38), there is an implied dynamical constraint. The Airy function, which plays the role of a generalized Franck–Condon factor for energies near the intersection point of the two potential surfaces, oscillates in amplitude as the energy is changed. Consequently the rate of fragmentation depends on the precise partitioning between product rotational and translational kinetic energy, and hence on the product angular momentum.

The consequence of assumptions (i)–(iii) of the Florida–Rice model is that the overlap in x is treated as in predissociation of a diatomic molecule. Consequently, transfer of energy between recoiling product molecules and generation of internal excitation of polyatomic species is not accounted for. Also, but mainly of technical importance, the model represents the potential surface of the final state as having a quadratic resistance to bending near the crossing point of the potential surfaces. Clearly, the bending force constant must decrease rapidly as x increases; hence this representation can only be valid in the immediate vicinity of $x = x^*$ and not elsewhere along x.

B. The Morse–Freed–Band Theory: Photodissociation of a Linear Triatomic Molecule

The Florida–Rice model displays the ubiquitous features of radiationless processes, namely a dependence on selection rules to limit the possible final channels, and on generalized Franck–Condon factors to order the importance of the allowed final channels. However, as noted, it neglects the change in bonding on dissociation and the evolution of reactant coordinates into product coordinates, hence omitting important aspects of the dynamics of decay of the prepared state. In contrast, the Morse–Freed–Band theory properly treats the generalized Franck–Condon factors which emerge from use of the Condon or Q centroid approximations applied to the rate expression

$$\Gamma_{\text{fi}} = \left| \int \Psi_f(\mathbf{Q}) V(\mathbf{Q}) \Psi_i(\mathbf{Q})\, d\mathbf{Q} \right|^2 = \left| \bar{V} \int \Psi_f(\mathbf{Q}) \Psi_i(Q)\, d\mathbf{Q} \right|^2. \tag{6.39}$$

In all that follows $\Psi_f(\mathbf{Q})$ represents the exact nuclear wavefunction on the dissociative potential surface; it is characterized by quantum numbers f in the limit of infinite product separation and is normalized to 2π times a delta function in energy. Similarly, $\Psi_i(\mathbf{Q})$ represents the exact nuclear wave function on the initial potential surface; it is characterized by quantum numbers i representing the prepared rotational and vibrational state of the molecule. Then $V(\mathbf{Q})$ is given by

$$V(\mathbf{Q}) = \langle \phi_f(\mathbf{X}, \mathbf{Q}) | V | \phi_i(\mathbf{X}, \mathbf{Q}) \rangle_{\mathbf{X}}, \tag{6.40}$$

where ϕ_i, ϕ_f are the initial and final state Born–Oppenheimer electronic wavefunctions at the nuclear positions $\mathbf{Q}$, the electronic coordinates x refer to the molecule fixed frame (r in the space-fixed frame), and $\langle \cdot \rangle_{\mathbf{X}}$, as usual, denotes an expectation value with respect to the electronic coordinates.

In the case of predissociations from a well-defined initially prepared predissociating state the coupling element $V(\mathbf{Q})$ is a zero-rank tensor; there is then no preferred direction in space. The only spatial anisotropy that can occur in this case is the result of the method of preparation of the predissociating state. For direct photodissociation, however, the direction of polarization of the absorbed photon determines a preferred direction in space. It is usually sufficient to make a dipole approximation to $V(\mathbf{Q})$, resulting in a first-rank tensor effective coupling $V(\mathbf{Q})$ and the familiar dipole selection rules for total angular momentum J. Since dipole selection rules limit ΔJ to 0, ± 1, $V(\mathbf{Q})$ can sometimes be approximated as a zero-rank tensor, even in the case of direct photodissociation, so long as the calculated distribution of fragment rotational quantum states produced is wide compared to $\Delta J = 0, \pm 1$. The relevant tensor algebra, retaining the full dipole selection rules, is presented in the Appendix. In this review we consider the simple case when $V(\mathbf{Q})$ is a zero-rank tensor.

We now examine fragmentation of a triatomic molecule, following the analysis of Morse, Freed, and Band. In general, repulsive potential energy surfaces do not yield a separable Schrödinger equation, thereby giving rise to inelastic scattering. Thus the wavefunction Ψ_f must be written as the sum of terms

$$\Psi_f(\mathbf{Q}') = \sum_n \sum_{\mu, m} \langle J'M' | lj\mu m \rangle Y_{l\mu}(\theta_{SF}, \phi_{SF}) Y_{jm}(\beta', \alpha') \Psi_n(Q_2') \Psi_{Eln}(Q_1') \tag{6.41}$$

if vibrationally inelastic scattering processes can occur on the repulsive surface. Here $Y_{l\mu}(\theta_{SF}, \phi_{SF})$ and $Y_{jm}(\beta', \alpha')$ describe the orbital motion of the fragments (in the space-fixed frame) and the rotational motion of the diatomic fragment, respectively, $\Psi_n(Q_2')$ the vibrations of the diatom, and $\Psi_{Eln}(Q_1')$ the translational motion as the fragments separate. As indicated in Fig. 2, the coordinates (θ_{SF}, ϕ_{SF}) and (β', α') are the polar and azimuthal angles, respectively, describing the orientation of the vector from the atom to the diatom center of mass with respect to the axis of the diatomic molecule. The vibrational coordinate of the diatom is Q_2' and the distance between the diatom center of mass and the atom

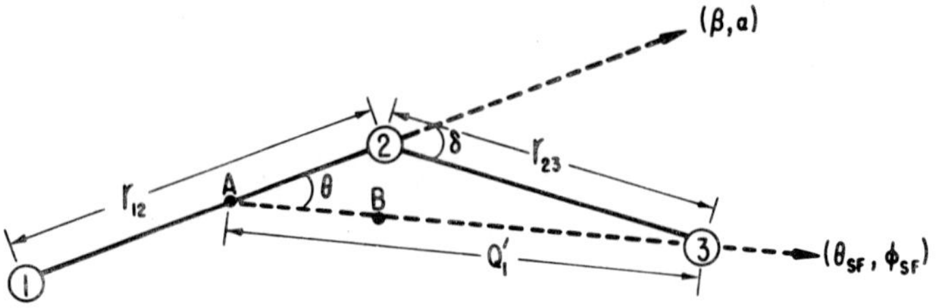

Fig. 2. Coordinates used in the theory. Bond 2–3 breaks in the photodissociation process.

is Q'_1. The functions in (6.41) comprise a natural basis set, which is appropriate to the asymptotic region of separated fragments. The sum over n indicates that the vibrational and translational motions are nonseparable. We require as boundary condition that as Q'_1 goes to infinity only one term in the sum over n remain, so that a well-defined vibrational state of the fragments is produced. The functions $\Psi_{Eln}(Q'_1)$ are solutions of the coupled equations which result from Schrödinger's equation for Ψ_f with this boundary condition.

The techniques developed by Morse, Freed, and Band, to be described below, can be applied term by term to the individual contributions in (6.41). They also permit the consideration of rotationally inelastic scattering by a form analogous to (6.41) with a sum over j and $\Psi_{Eln}(Q'_1)$ also dependent on j. Such a scheme, however, requires an enormous number of coupled equations for the solution of $\Psi(\mathbf{Q}')$ and is presently unnecessary because of lack of knowledge concerning the dissociative potential energy surfaces of typical molecules, e.g., ICN.

In the evaluation of the Franck–Condon factors appearing in (6.39) special care must be taken because $\Psi_i(\mathbf{Q})$ and $\Psi_f(\mathbf{Q}')$ are generally expressed in different coordinate systems (Podolsky, 1928). The final state wavefunction $\Psi_f(\mathbf{Q})$ is expressed in terms of the asymptotic coordinates appropriate for the fragments, $\mathbf{Q}'$, while $\Psi_i(\mathbf{Q})$ is expressed in terms of the normal coordinates (or local modes) $\mathbf{Q}$ of the initial bound state of the molecule. As usual, Ψ_f and Ψ_i are normalized with respect to their volume elements $d\mathbf{Q}'$ and $d\mathbf{Q}$, respectively:

$$\int \Psi_f^*(\mathbf{Q}')\Psi_f(\mathbf{Q}')\, d\mathbf{Q}' = 2\pi\delta(E_f - E_i), \tag{6.42}$$

$$\int \Psi_i^*(\mathbf{Q})\Psi_i(\mathbf{Q})\, d\mathbf{Q} = 1. \tag{6.43}$$

A transformation between $\mathbf{Q}$ and $\mathbf{Q}'$ is necessary in order to evaluate $\langle f|i\rangle$. The Jacobian J^t of the transformation is introduced, through the definition

$$d\mathbf{Q}' = |\partial\mathbf{Q}'/\partial\mathbf{Q}|\, d\mathbf{Q} \equiv J^t\, d\mathbf{Q}, \tag{6.44}$$

so that (6.42) can be rewritten as

$$\int \Psi_{\mathrm{f}}^{*}(\mathbf{Q}')\Psi_{\mathrm{f}'}(\mathbf{Q}')|J^{\mathrm{t}}|\, d\mathbf{Q} = 2\pi\delta(E_{\mathrm{f}} - E_{\mathrm{f}'}), \tag{6.45}$$

where $|J^{\mathrm{t}}|$ is the magnitude of the Jacobian. Thus, when the integration is with respect to the variables $\mathbf{Q}$, each factor of $\Psi_{\mathrm{f}}(\mathbf{Q}')$ must be multiplied by $|J^{\mathrm{t}}|^{1/2}$. We then obtain

$$\langle \mathrm{f}|\mathrm{i}\rangle = \int \Psi_{\mathrm{f}}^{*}(\mathbf{Q}')|J^{\mathrm{t}}|^{1/2}\Psi_{\mathrm{i}}(\mathbf{Q})\, d\mathbf{Q} \tag{6.46}$$

or

$$\langle \mathrm{f}|\mathrm{i}\rangle = \int \Psi_{\mathrm{f}}^{*}(\mathbf{Q}')|J^{\mathrm{t}}|^{-1/2}\Psi_{\mathrm{i}}(\mathbf{Q})\, d\mathbf{Q}'. \tag{6.47}$$

The inclusion of the square root of the Jacobian becomes especially important in studying the photodissociation of a bent molecule (Morse *et al.*, 1979) since in this case J^{t} depends strongly on the coordinates even in the simplest models.

In order to discuss the photodissociation of any molecule, the relevant potential energy surfaces must be known. At present we have extremely little information about the structure of excited state dissociative surfaces. Therefore, Morse, Freed, and Band adopt a model dissociative surface with a few parameters, adjust these parameters to reproduce available experimental data at one (or a few) photon energies, and then predict the product energy distributions at other energies. The result is a semiempirical potential energy surface which may be tested by further experiments or by *ab initio* calculations in special cases. By inverting the procedure, photodissociation experiments can provide a means of determining the properties of dissociative potential energy surfaces.

For the zeroth-order model surface, one which is qualitatively correct and has the minimum of adjustable parameters, the diatomic oscillator is described by a harmonic potential characterized by the force constant of the isolated diatom, and an exponential repulsion is taken to act between atoms 2 and 3, $V = A\exp(-DR_{23})$. For simplicity, Morse, Freed, and Band use basis functions appropriate to two simple potentials. One is $V = V_0 \exp(-DQ_1')$ where $V_0 = \exp\{[m_1/(m_1 + m_2)]r_{12}^0\}$, and r_{12}^0 is the equilibrium interatomic distance in the diatomic fragment. Some residual attractive interactions are incorporated into the model surface by using the potential $V(Q_1') = V_0 \exp(-DQ_1') - C$, where C is a function which is constant at small Q_1', and asymptotically goes to zero as, for example, $Q_1'^{-6}$. The exact asymptotic form taken by C is unimportant since only the behavior of $V(Q_1')$ in the Franck–Condon region matters. This model potential yields a separable Schrödinger equation and thus removes the sum over n in (6.41); it is chosen solely for convenience and illustrative purposes. There are no limitations intrinsic to the Morse–Freed–Band theory which require such a simple, separable wavefunction.

The obvious coordinates for the initially bound triatomic molecule are the normal or local modes of the molecule. The initial wavefunction is then given as a product of wavefunctions for these normal modes, or more generally as a linear combination of terms of this form. For a linear molecule these include two normal coordinates for the stretching vibrations, Q_1 and Q_2, two angles describing the orientation of the axis, α and β, and two angles describing the bending vibration, γ and δ. (The three center of mass degrees of freedom are simply removed.) As indicated in Fig. 2, β and α are polar and azimuthal angles, respectively, and γ is an azimuthal angle describing the orientation of the instantaneously bent molecule relative to the equilibrium axis. With this coordinate system we may express the rotational–vibrational wavefunction as

$$\Psi_{\mathrm{i}}(\mathbf{Q}) = [(2J + 1)/8\pi^2]^{1/2} D^{J*}_{MK}(\alpha\beta\gamma)\Psi_{n_1}(Q_1)\Psi_{n_2}(Q_2)\Psi^k_v(\delta), \tag{6.48}$$

where a harmonic potential has been invoked for the bending and stretching vibrations and rotation–vibration interactions are ignored. Then $\Psi_{n_1}(Q_1)$ and $\Psi_{n_2}(Q_2)$ are ordinary harmonic-oscillator wavefunctions (Merzbacher, 1970, p. 61),

$$\Psi_n(Q) = 2^{-n/2}(n!)^{-1/2}(\mu\omega/\hbar\pi)^{1/4} \exp[-(\mu\omega/2\hbar)Q^2]H_n[\sqrt{\mu\omega/\hbar}\,Q], \tag{6.49}$$

and $\Psi^k_v(\delta)$ is a two-dimensional isotropic harmonic oscillator function with vibrational angular momentum k along the axis (Townes and Schalow, 1975; Gradshteyn and Ryzhik (1965),

$$\Psi^k_v = \kappa\left[2\left(\frac{v - |k|}{2}\right)! \Big/ \left(\frac{v + |k|}{2}\right)!\right]^{1/2} (\kappa\delta)^{|k|} \exp\left[-\frac{\kappa^2\delta^2}{2}\right] L^{|k|}_{(v-|k|)/2}(\kappa^2\delta^2), \tag{6.50}$$

where

$$\kappa = \left[\frac{(r^0_{12})^2(r^0_{23})^2}{(r^0_{12})^2/m_3 + (r^0_{23})^2/m_1 + (r^0_{12} + r^0_{23})^2/m_2}\left(\frac{\omega_{\mathrm{bend}}}{\hbar}\right)\right]^{1/2}. \tag{6.51}$$

Any anharmonicities characteristic of the initial potential surface can be represented by a linear superposition of the functions (6.48) or any other convenient set of bond vibration functions (Carney *et al.*, 1977). Likewise, any vibration–rotation couplings can be included by taking a sum of terms of the form of (6.48). Since each of the individual terms in such a sum is treated separately as described below, we continue with (6.48) for notational simplicity.

The final state wavefunction is described in terms of the coordinates appropriate in the asymptotic region of separated fragments. Here the angles β' and α' denote the polar and azimuthal angles, respectively, of the diatomic axis, while θ_{SF} and ϕ_{SF} denote the corresponding angles of the vector from the diatom center of mass to the third atom. In the asymptotic limit the relative motion of the fragments separates from the diatomic vibration, so the coordinates are Q'_1 and Q'_2, with Q'_1 the distance from atom 3 to the diatom center of mass and Q'_2 the normal coordinate for the diatomic stretching vibration. Thus, the asymptotic wavefunction has the form

$$\Psi_f = \sum_{\mu m} \langle J'M' | lj\mu m \rangle Y_{jm}(\beta', \alpha') Y_{l\mu}(\theta_{SF}, \phi_{SF}) \Psi_n(Q'_2) \Psi_{El}(Q'_1), \qquad (6.52)$$

with $\langle J'M' | lj\mu m \rangle$ a vector coupling coefficient to provide eigenstates of total angular momentum quantum number J' and space-fixed z projection M'. This is the exact final state wavefunction for the simple model potential that Morse, Freed, and Band have chosen. For a more complicated, realistic, potential surface a linear combination of terms of this type would be necessary, or approximations must be introduced to account for the scattering.

For a linear molecule in its equilibrium configuration the axis of the molecule is equivalent to the axis of the resulting diatomic fragment. Consequently no rotation or axis-switching transformation (Hougen and Watson, 1965) is required to change the angles (α, β) to (α', β') if $\delta = 0$. In the case of the dissociation of a bent molecule, however, the axis of the diatomic fragment is generally not a principal axis of inertia for the molecule, and a static axis switching transformation must be made to relate the axis of the diatomic to the principal axes for the equilibrium configuration of the molecule. For both linear and bent molecules one must account for dynamical axis switching, arising because the molecule is vibrating about its equilibrium configuration. For a linear molecule, given any instantaneous displacement from the equilibrium axis, the axis of the diatom is not equivalent to the equilibrium axis of the molecule, so the angles (β', α') differ from (β, α). Similarly, for the case of a bent molecule, given any instantaneous deviation from the equilibrium angle, the static axis switching transformation does not rotate the principal axis of inertia into the diatomic axis as it does at equilibrium. In principle, the effect of dynamical axis switching should be incorporated in the analysis, but to do so requires that the rigid rotor approximation giving (6.48) be dropped. The result is that one needs a linear combination of terms like (6.48). However, as there is little reason to expect significant contributions from other terms, Morse, Freed, and Band use (6.48). In the rigid rotor approximation, then, the rotational wavefunction is obtained assuming a rigid molecule fixed at its equilibrium position. Hence, the angles (β, α) of (6.48) are identical to (β', α') of (6.52).

By using expressions (6.48) and (6.52), noting that a rotation by the Euler angles (α, β, γ) takes the angles (θ_{SF}, ϕ_{SF}) to $(\theta, 0)$, where θ is the angle between the diatomic axis and the vector from atom 3 to the center of mass of the diatom, and using the orthogonality properties of the D^J_{MK} functions and Clebsch–Gordon coefficients (Rose, 1957), it is found that (see appendix)

$$\begin{aligned} \langle f | i \rangle = {} & [2\pi(2j + 1)/(2J + 1)]^{1/2} \langle Jk | ljk0 \rangle \delta_{JJ'} \delta_{MM'} \\ & \times \langle Y_{lk}(\theta, 0) \Psi_{n'_2}(Q'_2) \Psi_{El}(Q'_1) | \Psi_{n_1}(Q_1) \Psi_{n_2}(Q_2) \Psi^k_2(\delta) \rangle. \end{aligned} \qquad (6.53)$$

The integral remaining in (6.53) is three dimensional and nonseparable; it is not readily evaluated without use of approximations. The transformation between (Q_1, Q_2) and (Q'_1, Q'_2) has been derived for the case of a collinear dissociation (Band and Freed, 1975; Freed and Band, 1977). With the inclusion of the bending degree of freedom, this transformation depends explicitly on the bending angle δ, although the dependence is slight for small amplitude bends. Similarly, the transformation between θ and δ retains some dependence on Q'_1 and Q'_2, although this too is small, provided Q'_1 and Q'_2 deviate only slightly from a fixed point in Ψ_i.

The Jacobian for the transformation from (Q_1, Q_2, δ) to (Q'_1, Q'_2, θ) also varies with coordinates; again, the variation is small for small amplitude vibrations in Ψ_i. For instance, if the δ dependence of the transformation between (Q_1, Q_2) and (Q'_1, Q'_2) were ignored, and θ were approximated by the first term in its Taylor series expansion in δ, we would obtain

$$|J| = (\det C')(1 - A)^2, \tag{6.54}$$

where C' is the transformation matrix which takes (Q'_1, Q'_2) to (Q_1, Q_2), and

$$A = \frac{m_1}{m_1 + m_2} \frac{r_{12}}{r_{23} + m_1 r_{12}/(m_1 + m_2)},$$

where r_{12} and r_{23} are instantaneous interatomic atomic distances. Equation (6.54) is really the first term in a Taylor series expansion of $|J^t|$, about the point given by r_{12}, r_{23} and $\delta = 0$.

In order to evaluate (6.53) more precisely, define

$$F(Q'_1, Q'_2, \delta) = \Psi^*_{n_i}(Q'_2)\Psi^*_{El}(Q'_1)\Psi_{n_1}(Q_1)\Psi_{n_2}(Q_2), \tag{6.55a}$$

$$G(Q'_1, Q'_2, \delta) = Y^*_{lk}(\theta', 0)\Psi^k_v(\delta), \tag{6.56a}$$

and

$$J^{1/2} = |J^t(Q'_1, Q'_2, \delta)|^{1/2}. \tag{6.57a}$$

The functions F and G exhibit weak dependence on δ and (Q'_1, Q'_2), respectively, and $J^{1/2}$ depends only weakly on all three variables. We therefore expand these functions in Taylor series in their respective slowly varying arguments to obtain

$$F(Q'_1, Q'_2, \delta) = \sum_i \mathrm{f}_i(Q'_1, Q'_2)[(\delta - \delta_0)^i/i!], \tag{6.55b}$$

$$G(Q'_1, Q'_2, \delta) = \sum_{jk} g_{jk}[(Q'_1 - Q^0_1)^j/j!][(Q'_2 - Q^0_2)^k/k!], \tag{6.56b}$$

$$J^{1/2}(Q'_1, Q'_2, \delta) = \sum_{lmn} J^{1/2}_{lmn}[(Q'_1 - Q^0_1)^l(Q'_2 - Q^0_2)^m(\delta - \delta_0)^n/l!m!n!]. \tag{6.57b}$$

Inserting these expressions in (6.53) we obtain

$$\langle \mathrm{f}|\mathrm{i}\rangle = \delta_{JJ'}\delta_{MM'}\left[\frac{2\pi(2j+1)}{2J+1}\right]^{1/2}\langle Jk|ljk0\rangle \sum_{i}\sum_{jk}\sum_{lmn}\frac{J^{1/2}_{lmn}}{i!j!k!l!m!n!}$$
$$\times \int_{-\infty}^{\infty} dQ'_1 \int_{-\infty}^{\infty} dQ'_2\, \mathrm{f}_i(Q'_1, Q'_2)(Q'_1 - Q^0_1)^{j+l}(Q'_2 - Q^0_2)^{k+m}$$
$$\times \int_0^{\infty} d\left(\frac{\delta^2}{2}\right)(\delta - \delta_0)^{i+n} g_{jk}(\delta). \tag{6.58}$$

The series (6.55b), (6.56b), and (6.57b) converge very rapidly, and by prudent choice of Q^0_1, Q^0_2, and δ_0 the summation (6.58) converges rapidly as well. The essence of the Q centroid method (Freed and Lin, 1975) is to choose values of Q^0_1, Q^0_2, and δ_0 which insure the rapid convergence of this sum.

Since δ is constrained to be near zero for the bound state of the triatomic, $g_{jk}(\delta)$ is peaked near zero and decays exponentially at large δ. For convenience, we therefore choose $\delta_0 = 0$. Furthermore, since the functions $F(Q'_1, Q'_2, \delta)$ and $J^{1/2}$ depend so weakly on δ near $\delta = 0$, we truncate the sums over i and n after their first (constant) term (Morse, Freed, and Band have also evaluated higher-order terms). This leads to the approximation

$$\langle \mathrm{f}|\mathrm{i}\rangle = \delta_{JJ'}\delta_{MM'}\left[\frac{2\pi(2j+1)}{2J+1}\right]^{1/2}\langle Jk|ljk0\rangle \sum_{jklm}\frac{J^{1/2}_{lm0}}{j!k!l!m!}$$
$$\times \int_{-\infty}^{\infty} dQ'_1 \int_{-\infty}^{\infty} dQ'_2\, \mathrm{f}_0(Q'_1, Q'_2)(Q'_1 - Q^0_1)^{j+l}(Q'_2 - Q^0_2)^{k+m}$$
$$\times \int_0^{\infty} d\left(\frac{\delta^2}{2}\right) g_{jk}(\delta). \tag{6.59}$$

Similarly, the remaining sum should converge rapidly, so we truncate after terms linear in $(Q'_1 - Q^0_1)$ and $(Q'_2 - Q^0_2)$. Q^0_1 and Q^0_2 are then chosen so that

$$\begin{aligned}&\int_{-\infty}^{\infty} dQ'_1 \int_{-\infty}^{\infty} dQ'_2\, \mathrm{f}_0(Q'_1, Q'_2)(Q'_1 - Q^0_1) = 0,\\ &\int_{-\infty}^{\infty} dQ'_1 \int_{-\infty}^{\infty} dQ'_2\, \mathrm{f}_0(Q'_1, Q'_2)(Q'_2 - Q^0_2) = 0,\end{aligned} \tag{6.60}$$

thereby leaving an integral with the single term,

$$\langle \mathrm{f}|\mathrm{i}\rangle = \delta_{JJ'}\delta_{MM'}\left[\frac{2\pi(2j+1)}{2J+1}\right]^{1/2}\langle Jk|ljk0\rangle J^{1/2}_{000}$$
$$\times \int_{-\infty}^{\infty} dQ'_1 \int_{-\infty}^{\infty} dQ'_2\, \mathrm{f}_0(Q'_1, Q'_2)\int_0^{\infty} d\left(\frac{\delta^2}{2}\right) g_{00}(\delta). \tag{6.61}$$

The remaining integral over δ may also be analytically evaluated if appropriate small angle approximations are made. If necessary, correction terms can be included, but for initially linear molecules in low energy bending states these have not been found necessary in the numerical calculations performed to date. With the results of the last paragraph, $\langle \mathrm{f}|\mathrm{i}\rangle$ becomes

$$\langle \mathrm{f}|\mathrm{i}\rangle = F_{El} G(J, v, k, l, j)\delta_{JJ'}\delta_{MM'}(\det C')^{1/2}, \tag{6.62}$$

where

$$F_{El} = \int_{-\infty}^{\infty} dQ_1' \int_{-\infty}^{\infty} dQ_2'\, \mathrm{f}_0(Q_1', Q_2'), \tag{6.63}$$

$$G(J, v, k, l, j) = \left[\frac{2\pi(2j + 1)}{2J + 1}\right]^{1/2} \langle Jk|ljk0\rangle \int_0^{\infty} d\delta\, \delta Y_{lk}^*(\theta, 0)\Psi_v^k(\delta)(1 - A). \tag{6.64}$$

The integral (6.63) for the stretches and relative motions is of the same form as arises in collinear models of dissociation processes and can be evaluated by the methods given by Band and Freed. Again, Morse, Freed, and Band give explicit formulas for higher-order terms.

The spherical harmonic $Y_{lk}^*(\theta, 0)$ is readily expressed as an associated Legendre function. Because the initial state bending wavefunction is negligible except for small angles, it may be approximated by its small angle Bessel function limit:

$$P_l^k(\cos\theta) = [(l + |k|)!/(l - |k|)!](l + \tfrac{1}{2})^{-|k|} J_{|k|}[(l + \tfrac{1}{2})\theta]. \tag{6.65}$$

Using a small angle approximation in the argument of the Bessel function $[\theta = (1 - A)\delta]$ and a standard integral, an analytic result for G is readily obtained:

$$\begin{aligned} G(J, v, k, l, j) = &\left\{\frac{(2j + 1)(2l + 1)(l + |k|)![(v - |k|)/2]!}{(2J + 1)(l - |k|)![(v + |k|)/2]!}\right\}^{1/2} \left(\frac{1 - A}{\kappa}\right)^{1+|k|} \\ &\times (-1)^{(|k|+v)/2} \langle Jk|ljk0\rangle \exp[-(l + \tfrac{1}{2})^2(1 - A)^2/2\kappa^2] \\ &\times L_{(v-|k|)/2}^{|k|}[(l + \tfrac{1}{2})^2(1 - A)^2/\kappa^2]. \end{aligned} \tag{6.66}$$

Correction terms from the δ centroid expansion and corrections to the Bessel approximation can be retained if needed. However, for A and κ appropriate to ICN, and $v = k = 0$, the validity of the Bessel approximation resulting in (6.66) has been checked by direct numerical integration of (6.64), and it is found to be accurate to within 0.3%. Unless the prepared state of the triatomic molecule involves a highly excited bending mode, (6.66) is sufficiently accurate.

C. Calculations and Interpretations

Given the Morse–Freed–Band expression for $\langle f|i\rangle$, it is instructive to consider the distribution of fragment states which is obtained when scattering on the final repulsive surface is negligible. In this case, the product distribution is governed by the Franck–Condon factor $|\langle f|i\rangle|^2$ alone. Using (6.62) and (6.66),

$$|\langle f|i\rangle|^2 = N\langle Jk|ljk0\rangle^2 \exp[-\rho(l+\tfrac{1}{2})^2]\{L^{|k|}_{(v-|k|)/2}[\rho(l+\tfrac{1}{2})^2]\}|F_{El}|^2,$$

where

$$N = \frac{(2j+1)(2l+1)(l+|k|)!}{(2J+1)(l-|k|)!}\frac{[(v-|k|)/2]!}{[(v+|k|)/2]!}\rho^{|k|+1}\delta_{JJ'}\delta_{MM'} \qquad (6.67)$$

and $\rho = (1-A)^2/\kappa^2$.

The exponential term in (4.1) insures that only small values of the orbital momentum l are significantly populated; then the conservation of angular momentum, expressed by the Clebsch–Gordon coefficients and Kronecker delta functions, leads to the expectation that on the average most of the angular momentum initially present in the triatomic molecule goes into diatomic fragment rotation, specified by the quantum number j. Calculations using (6.67) show that this conclusion is correct, with the peak in the distribution of j falling close to the value of total angular momentum, $j \sim J$, provided the dependence of the factor $|F_{El}|^2$ on j and l is temporarily ignored. This factor includes the constraint of overall energy conservation and often depends strongly on j. It, therefore, has an important influence on the partitioning of energy between translational, vibrational, and rotational degrees of freedom (Morse *et al.*, 1976; Florida and Rice, 1975).

Let us momentarily ignore $|F_{El}|^2$, however, and examine the conditional rotational distributions $P_J(l, j) = |\langle f|i\rangle|^2/|F_{El}|^2$, as these are universally valid, the only molecular parameter being $\rho \equiv (1-A)^2/\kappa^2$. The conditional probability of producing orbital angular momentum l (neglecting $|F_{El}|^2$) is simply

$$P_J(l) = \sum_j P_J(l, j) = \frac{[(v-|k|)/2]!(2l+1)(l+|k|)!}{[(v+|k|)/2]!(l-|k|)!}\rho^{|k|+1}$$

$$\times \{L^{|k|}_{(v-|k|)/2}[\rho(l+\tfrac{1}{2})]\}^2 \exp[-\rho(l+\tfrac{1}{2})^2], \qquad (6.68)$$

where the relation $\sum_j (2j+1)\langle Jk|ljk0\rangle^2 = 2J+1$ has been utilized. This conditional distribution of values of l does not depend on the total angular momentum J, so once again we are led to expect that as we increase J the distribution in diatomic rotational quantum number j must shift to higher values, with $j_{\text{average}} \approx J$.

We may similarly obtain the conditional distributions for diatomic rotational quantum number j, $P_J(j)$, but the sum has been analytically evaluated only for $J = 0$; the result for $J = 0$ is

$$P_{J=0}(j) = (2j + 1)[(1 - A)/\kappa]^2 \exp[-\rho(j + \tfrac{1}{2})^2]\{L^0_{(v/2)}[\rho(j + \tfrac{1}{2})^2]\}^2. \quad (6.69)$$

A comparison of (6.69) and (6.68) when $J = k = 0$ shows that the distributions in j and l are identical. Note that expressions (6.68) and (6.69) are normalized probability distributions, provided sums over j and l are replaced by integrals over $(-\frac{1}{2}, \infty)$.

Consider again the case $J = 0$ for which we can separate the features of the distribution of j that arise from bending vibrations alone from those features due to overall rotation of the molecule. A convenient procedure is to determine moments of $P_{J=0}(j)$, thereby obtaining expressions for the average energy in diatomic rotation $\langle E_j \rangle$, and the root mean square deviation from this average, $\sigma(E_j) = [\langle E_j^2 \rangle - \langle E_j \rangle^2]^{1/2}$. These are

$$\langle E_j \rangle = (B/\rho)(v + 1), \quad (6.70)$$

$$\sigma(E_j) = (B/\rho)(\tfrac{1}{2}v^2 + v + 1)^{1/2}, \quad (6.71)$$

where B is the rotational constant for the diatomic fragment. For $J = k = 0$, for which these formulas apply, we see that $\langle E_j \rangle$is proportional to the energy initially in the bending mode, given by $\hbar\omega(v + 1)$. A factor of ω is embedded in $1/\rho = \kappa^2/(1 - A)^2$ [see (6.51)]. Equation (6.71) shows that $\sigma(E_j)$ is also proportional to the energy initially in the bending mode, provided v is large, in which case $\sigma(E_j) \sim (B/\rho\sqrt{2})(v + 1)$. Although these results provide the actual Franck–Condon distributions only if the dependence of $|F_{El}|^2$ on j and l is ignored, they have general qualitative value, since it is intuitively clear that the average energy partitioned into rotation in the dissociation of a nonrotating molecule is proportional to the energy of the bending motion to which the rotation correlates.

The factor of B/ρ which enters into both (6.70) and (6.71) gives particular insight into the dissociation process. In expression (6.50) for the bending wave function, $\sqrt{2}/\kappa$ plays the role of a range parameter, specifying the amplitude of the zero point bending motion. Since $\theta = (1 - A)\delta$ in the small angle approximation, $(1 - A)(\sqrt{2}/\kappa)$ is the amplitude of the zero-point bending motion measured in terms of θ. Hence, for floppy bending modes, where κ is small, the amount of energy going into fragment rotation is predicted by (6.70) to be small, provided the molecule is initially not rotating. Similarly, for stiff bending modes with small bending amplitudes, a substantial excitation of the fragment rotational states is predicted. (Recall, however, the neglect of the energy conservation constraints which enter through the factor $|F_{El}|^2$.)

The general result described can also be understood in terms of the Heisenberg uncertainty principle, which requires $\Delta q \, \Delta P \geq \hbar$ for any generalized coordinate q and its conjugate momentum p. The available region in θ space is specified by $(1 - A)(2/\kappa)$, so the range in its conjugate angular momentum is roughly $\hbar\kappa/\sqrt{2}(1 - A)$. Thus for large $\kappa/(1 - A)$ the range in rotational quantum numbers j and l is large, and for small $\kappa/(1 - A)$ this range is small. Rotational energy is proportional to the square of angular momentum, so we expect $E_j \propto [\kappa^2/(1 - A)^2] = 1/\rho$ as found in (6.70).

Figures 3–5 present plots of the distributions $P_J(j)$ for various values of total angular momentum J and various initial bending quantum numbers v and k, taken from Morse, Freed, and Band. The abscissa is chosen to be $j - J$, to emphasize the peaking of $P_J(j)$ about $j = J$, especially for large values of J. These plots ignore the variation of F_{El}^2 with j and l, so the probability distribution $P_J(l)$ is identical to that of $P_{J=0}(j)$. All curves correspond to $\rho = 1.51 \times 10^{-2}$, which is appropriate for the ground electronic state of ICN. Other calculations show that comparable values are found for many other molecules. On the scale of these graphs the differences between the results assuming scalar coupling or parallel or perpendicular transitions cannot be detected; these differences are seldom more than 3% of the peak in the probability distribution.

In the case of either parallel or perpendicular transitions from excited bending states, Morse, Freed, and Band find considerable structure in the rotational distributions. This arises from the presence of nodes in the bending wavefunction. Such structure might be difficult to observe experimentally, since it can be washed out by the variation of $|F_{El}|^2$ with j and l, and by the effects of thermal averaging over the total angular momentum, J, in the initial bound electronic state.

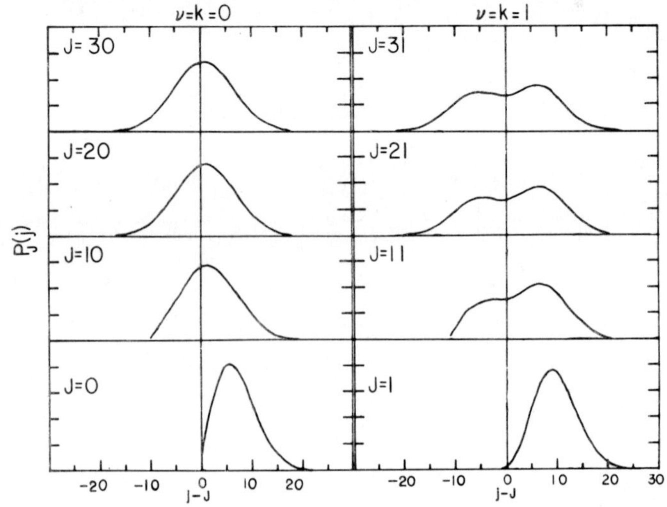

Fig. 3. Probabilities of producing fragment rotational state j from initial triatomic rotational state J for $\rho = 0.00151$ and bending quantum numbers v and k, ignoring energy conservation and final state interaction effects.

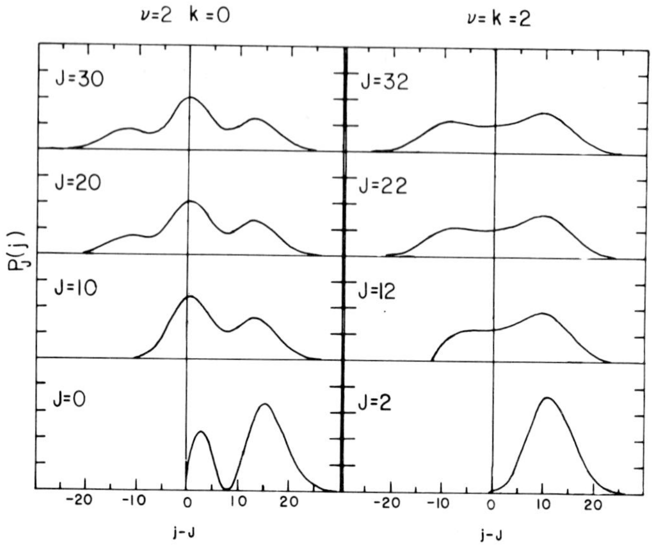

Fig. 4. Probabilities of producing fragment rotational state j from an initial quantum state specified by J, v, and k, ignoring energy conservation and final state interaction effects.

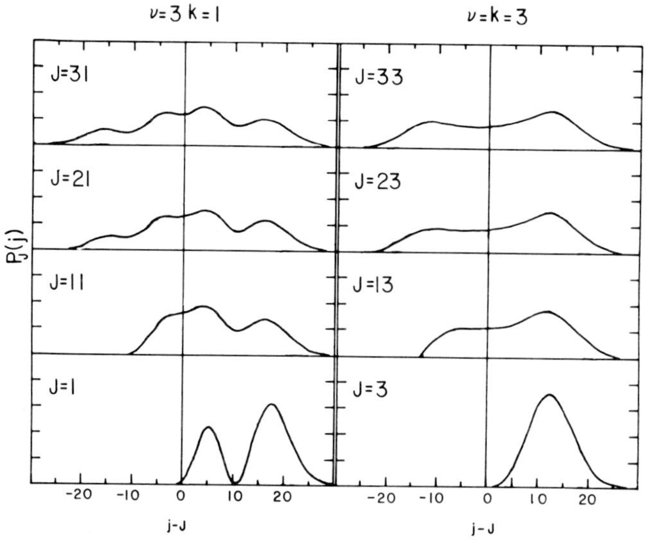

Fig. 5. Probabilities of producing fragment rotational state j from an initial quantum state specified by J, v, and k, ignoring energy conservation and final state interaction effects. These correspond to dissociations of molecules with excited bending modes ($v = 3$).

Finally, we consider the effects of $|F_{El}|^2$ on the distribution of angular momentum. The factor $|F_{El}|^2$ is identical to that entering into the theory of the collinear dissociation of triatomic molecules. Hence, its influence on the rotational distribution of the photofragments is readily calculated by the methods of Band and Freed. Qualitatively, these effects are easily understood in terms of the effective oscillator and the repulsive potential surface (Band and Freed, 1975; Freed and Band, 1977), schematically shown in Fig. 6. For the example given in Fig. 6, with a photon of sufficiently low energy, the Franck–Condon overlap between the interfragment relative translation wavefunction and the effective oscillator wavefunction is small, with the major contribution occurring in the classically forbidden region of both curves. This resembles a tunneling process, and the rate of dissociation will depend strongly on the translational energy. Such a dissociation process is most efficient if a high fraction of the available energy goes into interfragment translation. Hence, the energy conservation constraint in the factor $|F_{El}|^2$ favors low energy rotational states. At higher photon energies this effect becomes less pronounced. At still higher photon energies, $|F_{El}|^2$ favors high energy rotational states (just as it favors high energy vibrational states), although this may be modified by some oscillatory behavior.

In general, the structure of $|F_{El}|^2$ can be guessed from the observed vibrational distribution. For example, if the vibrational distribution varies roughly as $\exp[-C(n_2' + \frac{1}{2})]$, then $|F_{El}|^2$ is roughly an exponential decreasing function of the rotational energy difference $E_f(j) - E_i(J)$. When experimental vibrational distributions are peaked about some value $n_2' \neq 0$, $|F_{El}|^2$ favors some conversion of translational energy into rotation, and is peaked about some nonzero value of the rotational energy difference.

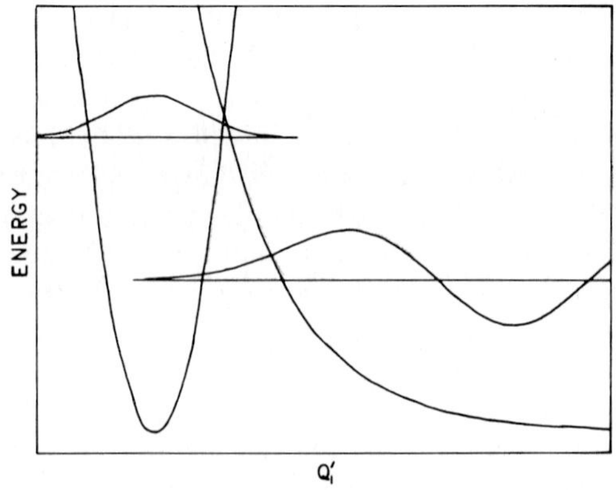

Fig. 6. One-dimensional representation obtained by evaluating all vibrational Franck–Condon factors except the one involving the reaction coordinate Q_1'. The harmonic well is the effective oscillator potential. Its equilibrium position and force constant depend on both stretching vibrations in the bound initial electronic state as well as on the fragment vibration. The relative energies of the effective oscillator and the repulsive curves have no physical significance.

The remaining source of rotational excitation is the final state interaction, which arises when the real interfragment potential does not yield the separable wavefunction (6.52). An effective treatment of this problem with assured accuracy and ease of computation remains to be devised. The calculation of these effects is particularly difficult in the case of rotation, where so many final states are available.

VII. CONCLUDING REMARKS

This review has been concerned with several manifestations of the conservation of angular momentum in radiationless processes. Comparison of the ideas used to describe predissociation of diatomic molecules, photodissociation of polyatomic molecules, and intersystem crossing and internal conversion in polyatomic molecules reveals several common themes. In each case the conservation of angular momentum generates selection rules that restrict the final states which are accessible from the prepared state. Furthermore, in each case the rate of an allowed process depends on a variety of factors that are either directly or indirectly influenced by molecular rotation. The most obvious of these factors is the partitioning of energy between rotation and other degrees of freedom; this leads to a modification of the energy gap between the coupled states. Rather more subtle, but nonetheless ubiquitous, the phase of the rotational wavefunction modulates the rate of decay of the prepared state. Finally, in each case the rate of decay can be written so as to display a dependence on generalized Franck–Condon factors.

In the case of predissociation of a diatomic molecule these rotational effects combine to give a nonmonotone but simple dependence of lifetime on energy.

The situation is more complex in polyatomic molecules. In the case of internal conversion centrifugal distortion induces a dependence of Franck–Condon factors on angular momentum and Coriolis coupling mixes the molecular vibrations, both of which effect the rate. In addition, there is a dynamic interaction between the two electronic states by virtue of the variation of the Coriolis coupling during relaxation. A simple two-state symmetric top model shows that the rate of the radiationless process is altered from the value characteristic of a nonrotating molecule; specifically the rate contains an explicit K^2 dependence and a term in which the frequency of the promoting mode is replaced by that of the mode to which it is Coriolis coupled.

Similarly, the photodissociation of polyatomic molecules is rich in effects traceable to the conservation of angular momentum. In the models of polyatomic photodissociation thus far considered all of these effects are located via the behavior of the Franck–Condon factors. The rotational energy of the products is found to depend sensitively on the nature of the repulsive interaction between separating fragments and on the details of the evolution of reactant into product coordinates.

We have seen that inclusion of molecular rotation in the system Hamiltonian very greatly complicates the analysis of decay processes. Indeed, the theory has been reduced to practical algorithms only for very simple models; much remains to be done. As of this date it appears that a very fruitful interplay between experimental and theoretical developments is in the offing.

APPENDIX: TENSOR AND ANGULAR MOMENTUM ALGEBRA

A. Direct Photodissociation Using a First-Rank Tensor Coupling Operator

Using Born–Oppenheimer wavefunctions for the initial and final states, the golden rule of time-dependent perturbation theory, and a dipole approximation to the matter–radiation interaction Hamiltonian, the required matrix element $\langle \mathrm{f}|V|\mathrm{i}\rangle$ for direct photodissociation is

$$\langle \mathrm{f}|V|\mathrm{i}\rangle = \langle \Psi_\mathrm{f}(\mathbf{Q}')\hat{\mathbf{e}}\cdot\langle \phi_\mathrm{f}(\mathbf{x},\mathbf{Q}')|\sum_j \mathbf{r}_j|\phi_\mathrm{i}(\mathbf{x},\mathbf{Q})\rangle_\mathbf{x}|\Psi_\mathrm{i}(\mathbf{Q})\rangle, \tag{A.1}$$

where $\hat{\mathbf{e}}$ is the polarization of the photon, $\mathbf{x}$ are the electronic coordinates in the body-fixed system, and $\mathbf{r}$ are the electronic coordinates in the space-fixed system. By expressing $\hat{\mathbf{e}}$ and $\mathbf{r}_j$ in spherical components with basis vectors $\hat{\mathbf{u}}_{\pm 1} = \mp(\hat{\mathbf{u}}_x \pm i\hat{\mathbf{u}}_y)/\sqrt{2}$, $\hat{\mathbf{u}}_0 = \hat{\mathbf{u}}_z$ ($\hat{\mathbf{u}}_x$, $\hat{\mathbf{u}}_y$, $\hat{\mathbf{u}}_z$ denote unit vectors) we obtain

$$\langle \mathrm{f}|V|\mathrm{i}\rangle = \sum_{q=-1}^{1}(-1)^q\hat{\mathbf{e}}_q\langle\Psi_\mathrm{f}(\mathbf{Q}')|\langle\phi_\mathrm{f}(\mathbf{x},\mathbf{Q})|\sum_j r_{jq}|\phi_\mathrm{i}(\mathbf{x},\mathbf{Q})\rangle_\mathbf{x}|\Psi_\mathrm{i}(\mathbf{Q})\rangle. \tag{A.2}$$

Now, r_{jq} is a spherical tensor of the first rank. It transforms according to well-known rules under rotations [Brink and Satchler, 1975, Eq. (4.5)]. Under the conventions of Brink and Satchler a rotation by the Eulerian angles $(-\gamma, -\beta, -\alpha)$ takes the space-fixed coordinates into body-fixed coordinates, so we obtain

$$\langle \mathrm{f}|V|\mathrm{i}\rangle = \sum_{q=-1}^{1}(-1)^q\hat{\mathbf{e}}_q\sum_{M''}\langle\Psi_\mathrm{f}(\mathbf{Q})|D^1_{M''q}(-\gamma,-\beta,-\alpha)$$
$$\times\langle\phi_\mathrm{f}(\mathbf{x},\mathbf{Q}')|\sum_j x_{jM''}|\phi_\mathrm{i}(\mathbf{x},\mathbf{Q})\rangle_\mathbf{x}|\Psi_\mathrm{i}(\mathbf{Q})\rangle, \tag{A.3}$$

where D^J_{MK} are the well-known rotation matrices. Invoking a Condon approximation yields

$$\langle \mathrm{f}|V|\mathrm{i}\rangle = \sum_{q=-1}^{1}(-1)^q\hat{\mathbf{e}}_q\sum_{M''}\langle\Psi_\mathrm{f}|D^1_{M''q}(-\gamma,-\beta,-\alpha)|\Psi_\mathrm{i}\rangle\langle\phi_\mathrm{f}|\sum_j x_{jM''}|\phi_\mathrm{i}\rangle_x. \tag{A.4}$$

The subscript q indicates the polarization of the incoming photon. Circularly polarized light in the x-y plane has $q = \pm 1$, while $q = 0$ corresponds to light polarized in the z direction. Similarly $M'' = 0$ for a parallel electronic transition, and $M'' = \pm 1$ for a perpendicular transition. We adopt the notation

$$X_{M''} = \langle \phi_f | \sum_j x_{jM''} | \phi_i \rangle_x, \tag{A.5}$$

$$V_{fi}^{M''q} = \langle \Psi_f | D^1_{M''q}(-\gamma, -\beta, -\alpha) | \Psi_i \rangle \tag{A.6}$$

and use generalizations, appropriate for the presence of electronic angular momentum, of expressions (6.48) and (6.52) for Ψ_i and Ψ_f

$$\Psi_i = \sqrt{(2J+1)/8\pi^2}\, D^{J*}_{Mk+\Lambda_i}(\alpha, \beta, \gamma)\Psi_{n_1}(Q_1)\Psi_{n_2}(Q_2)\Psi^k_v(\delta), \tag{6.48a}$$

$$\Psi_f = \sum_{\mu, m} \langle J'M' | lj\mu m \rangle [(2j+1)(2l+1)/16\pi^2]^{1/2} \times D^{j*}_{m\Lambda_d}(\alpha, \beta, \gamma) D^{l*}_{\mu, \Lambda_a}(\phi_{SF}, \theta_{SF}, 0)\Psi_n(Q_2')\Psi_{E,l}(Q_1'). \tag{6.52a}$$

In these expressions the axis projections of the electronic angular momenta of the initial state of triatomic, the state of fragment diatomic, and that of the fragment atom (quantized along the atom–diatom center of mass axis) are denoted by Λ_i, Λ_d, and Λ_a, respectively. Noting [Rose, 1957, Eq. (4.21)] that

$$D^1_{M''q}(-\gamma, -\beta, -\alpha) = D^{1*}_{qM''}(\alpha, \beta, \gamma), \tag{A.7}$$

$$D^{l*}_{\mu\Lambda_a}(\phi_{SF}, \theta_{SF}, 0) = \sum_{m'} D^{l*}_{\mu m'}(\alpha, \beta, \gamma) D^{l*}_{m'\Lambda_a}(0, \theta, 0), \tag{A.8}$$

where θ_{SF}, ϕ_{SF}, and θ are defined in Section VIB and are indicated in Fig. 2, we obtain

$$\Psi_f = \sum_{\mu, m, m'} \langle J'M' | lj\mu m \rangle [(2j+1)(2l+1)/16\pi^2]^{1/2} \times D^{j*}_{m\Lambda_d}(\alpha, \beta, \gamma) D^{l*}_{\mu m'}(\alpha, \beta, \gamma) D^{l*}_{m'\Lambda_a}(0, \theta, 0)\Psi_n(Q_2')\Psi_{El}(Q_1'). \tag{A.9}$$

The Clebsch–Gordon series for rotation functions may be substituted into (A.9) to yield

$$\Psi_f = \sum_{\mu m, m'J''} \langle J'M' | lj\mu m \rangle \langle J''M' | lj\mu m \rangle \langle J''m' + \Lambda_d | ljm'\Lambda_d \rangle \times [(2j+1)(2l+1)/16\pi^2]^{1/2} D^{J''*}_{M', m'+\Lambda_d}(\alpha, \beta, \gamma) D^{l*}_{m'\Lambda_a}(0, \theta, 0) \times \Psi_n(Q_2')\Psi_{E,l}(Q_1'). \tag{A.10}$$

Thus, use of (A.10) in (A.6) produces

$$V_{fi}^{M''q} = \sum_{m\mu, m'J''} \langle J'M' | lj\mu m \rangle \langle J''M' | lj\mu m \rangle \langle J''m' + \Lambda_d | ljm'\Lambda_d \rangle \times \left[\frac{(2j+1)(2l+1)(2J+1)}{128\pi^4} \right]^{1/2} \int d\Omega\, D^{J*}_{M, k+\Lambda_i}(\Omega) D^{1*}_{q, M''}(\Omega) \times D^{J''}_{M', m'+\Lambda_d}(\Omega) \langle D^{l*}_{m'\Lambda_a}(0, \theta, 0)\Psi_n(Q_2')\Psi_{E,l}(Q_1') | \Psi_{n_1}(Q_1)\Psi_{n_2}(Q_2)\Psi^k_v(\delta) \rangle. \tag{A.11}$$

Rose shows that

$$\int d\Omega\, D^{J*}_{M,k+\Lambda_i}(\Omega) D^{1*}_{q,M''}(\Omega) D^{J''}_{M',m'+\Lambda_d}(\Omega)$$

$$= [8\pi^2/(2J''+1)]\langle J''M'||JqM\rangle\langle J''m'+\Lambda_d||JM''k+\Lambda_i\rangle, \quad (A.12)$$

and that

$$\sum_{m\mu} \langle J'M'|lj\mu m\rangle\langle J''M'|lj\mu m\rangle = \delta_{J',J''}, \quad (A.13)$$

so (A.11) reduces to

$$\begin{aligned} V^{M'q}_{fi} = {} & \langle J'k+\Lambda_i+M''|ljk+M''+\Lambda_i-\Lambda_d\Lambda_d\rangle\langle J'M'|1JqM\rangle \\ & \times \langle J'k+\Lambda_i+M''|1JM''k+\Lambda_i\rangle \\ & \times [(2j+1)(2l+1)(2J+1)/2(2J'+1)^2]^{1/2} \\ & \times \langle D^{l*}_{k+\Lambda_i+M''-\Lambda_d,\Lambda_a}(0,\theta,0)\Psi_n(Q_2')\Psi_{E,l}(Q_1')|\Psi_{n_1}(Q_1)\Psi_{n_2}(Q_2)\Psi^k_v(\delta)\rangle. \end{aligned} \quad (A.14)$$

Electronic selection rules implicit in $X_{M''}$ require that $\Lambda_i + M'' = \Lambda_a + \Lambda_d$, so that D function involved in the bending Franck–Condon factor is reduced to $D_{k+\Lambda_a,\Lambda_a}(0,\theta,0)$. Furthermore, only small values of θ contribute significantly to this integral, so a small angle approximation to $D_{k+\Lambda_a,\Lambda_a}(0,\theta,0)$ is sufficient. Hougen and Watson (1965) have shown that for small angles $D^J_{K,K'}(0,\theta,0) \propto J_{|K-K'|}(J\theta)$, but we have obtained a more accurate and useful approximation,

$$\begin{aligned} D^l_{MK}(0,\theta,0) = {} & (l+\tfrac{1}{2})^{-|M-K|}[(l+K)!(l-M)!/(l-K)!(l+M)!]^{1/2} \\ & \times J_{|K-M|}[\theta(l+\tfrac{1}{2})] \end{aligned} \quad (A.15)$$

when $K \geq M$ or

$$\begin{aligned} D^l_{MK}(0,\theta,0) = {} & (-1)^{M-K}(l+\tfrac{1}{2})^{-|M-K|} \\ & \times [(l-K)!(l+M)!/(l+K)!(l-M)!]^{1/2} J_{|K-M|}[\theta(l+\tfrac{1}{2})] \end{aligned} \quad (A.16)$$

when $M \geq K$. When M or K is zero, this reduces to the Bessel approximation for associated Legendre polynomials.

With this approximation, the bending Franck–Condon overlap integral reduces to a form which is identical to that discussed in Section III for scalar coupling, and the same calculational techniques apply. The resulting joint probability for producing l and j given a value of the total initial angular momentum J, summed over final angular momenta J' and M', and averaged over M is

$$P_J(l, j) = [(2l + 1)(2j + 1)/6] |X_{M''}|^2 |Z_l|^2 \sum_{J'} (2J' + 1)$$
$$\times \begin{pmatrix} j & l & J' \\ \Lambda_d & k + \Lambda_a & -(k + \Lambda_i + M'') \end{pmatrix}^2$$
$$\times \begin{pmatrix} J & 1 & J' \\ k + \Lambda_i & M'' & -(k + \Lambda_i + M'') \end{pmatrix}^2 \quad (A.17)$$

where

$$Z_l = \langle D^{l*}_{k+\Lambda_a, \Lambda_a}(0, \theta, 0) \Psi_n(Q_2') \Psi_{E,l}(Q_1') | \Psi_{n_1}(Q_1) \Psi_{n_2}(Q_2) \Psi_v^k(\delta) \rangle. \quad (A.18)$$

Evaluation of Z_l utilizing the techniques of Section VIB gives a product of a vibrational-translational factor, $F_{E,l}$, which is defined in Eq. (6.63), and a bending factor, yielding

$$Z_l = F_{E,l}(-1)^{(v+|k|)/2} \left[\frac{2[(v - |k|)/2]!(l + k + \Lambda_a)!(l - \Lambda_a)!}{[(v + |k|)/2]!(l - k - \Lambda_a)!(l + \Lambda_a)!} \right]^{1/2}$$
$$\times \left(\frac{1 - A}{\kappa} \right)^{|k|+1} \exp\left[-\frac{(1 - A)^2 (l + \frac{1}{2})^2}{2\kappa^2} \right]$$
$$\times L^{|k|}_{(v-|k|)/2} \left[\frac{(1 - A)^2 (l + \frac{1}{2})^2}{\kappa^2} \right]. \quad (A.19)$$

Summing over either l or j enables us to obtain the probability of obtaining states with definite j or l independent of the values of l or j. Thus we obtain

$$P_J(j) = \frac{2j + 1}{6} |X_{M''}|^2 \sum_{J'l} |Z_l|^2 (2l + 1)(2J' + 1)$$
$$\times \begin{pmatrix} j & l & J' \\ \Lambda_d & k + \Lambda_a & -(k + \Lambda_i + M'') \end{pmatrix}^2$$
$$\times \begin{pmatrix} J & 1 & J' \\ k + \Lambda_i & M'' & -(k + \Lambda_i + M'') \end{pmatrix}^2.$$

Similarly, it is found that

$$P_J(l) = \frac{(2l + 1)}{6} |X_{M''}|^2 \sum_{J'j} |Z_l|^2 (2J' + 1)(2j + 1)$$
$$\times \begin{pmatrix} j & l & J' \\ \Lambda_d & k + \Lambda_a & -(k + \Lambda_i + M'') \end{pmatrix}^2$$
$$\times \begin{pmatrix} J & 1 & J' \\ k + \Lambda_i & M'' & -(k + \Lambda_i + M'') \end{pmatrix}^2. \quad (A.21)$$

In the calculation of Figs. 3–5, as discussed in Section IV, we have ignored the variation of $F_{E,l}$ with l and j. The variation with l will be small for large, heavy molecules like ICN, where the reduced mass of the fragments is large, and the classical turning point on the unbound surface has a value of Q'_1 considerably larger than 0. For smaller, lighter molecules (such as HCN) this variation may well be important and can be easily included in the evaluation of Ψ_{EL}. We ignore the variation with respect to j in order to separate the aspects of the problem due to angular momentum conservation from those due to energy conservation. In this approximation, expression (A.21) for $P_J(l)$ simplifies, giving

$$P_J(l) = [(2l + 1)/6]|X_{M''}|^2|Z_l|^2. \tag{A.22}$$

Thus the only dependence of the distribution of orbital angular momentum l on J arises from the dependence of $F_{E,l}$ on J through energy conservation. Calculations using (A.20) with the assumption that F_{El}^2 is independent of j, J, and l are presented in Figs 3–5, where they are compared to results obtained through the use of a scalar form of the interaction operator.

In Eqs. (A.17), (A.20), and (A.21) we have assumed that only one value of M'' contributes to the transition. For a given choice of the electronic substates i and f, only one value of M'' gives a nonvanishing $X_{M''}$. Of course, energy differences separating the electronic substates with positive and negative values of the axis projection of electronic angular momentum are often very small, so there are usually two overlapping electronic transitions in a perpendicular band. Because the final electronic states are distinct, however, the photodissociation rates sum incoherently, and there is no interference between them. In the treatment presented here we have ignored this possibility and in the interests of simplicity have only considered a single electronic transition with a well-defined value of M''.

B. Photodissociation by Scalar Coupling

Having obtained expressions for $\langle \mathrm{f}|V|\mathrm{i}\rangle$ with V a first-rank tensor, the case when it is a scalar is trivial, formulas (A.8)–(A.11) still apply, giving

$$\begin{aligned}\langle \mathrm{f}|V|\mathrm{i}\rangle = \bar{V} \sum_{\mu m,\, m'J''} &\langle J'M'|lj\mu m\rangle\langle J''M'|lj\mu m\rangle\langle J''m'|ljm'0\rangle \\ &\times \left[\frac{(2J+1)(2j+1)}{32\pi^3}\right]^{1/2} \\ &\times \langle Y_{lm'}(\theta,0)\Psi_n(Q'_2)\Psi_{El}(Q'_1)|\Psi_{n_1}(Q_1)\Psi_{n_2}(Q_2)\Psi_v^k(\delta)\rangle \\ &\times \int d\Omega\, D^{J''}_{M'm'}(\Omega)D^{j*}_{Mk}(\Omega).\end{aligned} \tag{A.23}$$

Evaluation of the sum over μ, m and the integral over Ω yields

$$\langle \mathrm{f}|V|\mathrm{i}\rangle = \bar{V}\langle J'k|ljk0\rangle\delta_{MM'}\delta_{JJ'}[2\pi(2j+1)/(2J+1)]^{1/2} \times \langle Y_{lk}(\theta,0)\Psi_n(Q_2')\Psi_{El}(Q_1')|\Psi_{n_1}(Q_1)\Psi_{n_2}(Q_2)\Psi_v^k(\delta)\rangle, \quad \text{(A.24)}$$

as listed in (6.53). Equivalently, (A.24) can be written as

$$\langle \mathrm{f}|V|\mathrm{i}\rangle = \bar{V}\langle J'k|ljk0\rangle\delta_{MM'}\delta_{JJ'}[(2j+1)(2l+1)/2(2J+1)]^{1/2}Z_l.$$

In this derivation, since the electronic angular momenta are unchanged in the dissociation, we have assumed $\Lambda_\mathrm{i} = \Lambda_\mathrm{d} = \Lambda_\mathrm{a} = 0$ for simplicity.

This research has been supported by the Air Force Office of Scientific Research (F49620-76-C-0017), The National Science Foundation (CHE 78-01573 and CHE 77-24652), and the Fannie and John Hertz Foundation.

REFERENCES

Band, Y. B., and Freed, K. F. (1975). *J. Chem. Phys.* **63**, 3382.
Brink, D. M., and Satchler, G. R. (1975). "Angular Momentum," 2nd ed., Oxford Univ. Press (Clarendon), London and New York.
Caplan, C. E., and Child, M. S. (1972). *Mol. Phys.* 249.
Carney, G. D., Sprandel, L. L., and Kern, C. W. (1977). *Adv. Chem. Phys.* **36**, 305.
Child, M. S. (1974). "Diatomic Predissociation Line Widths," Vol. 2, p. 466. Spec. Rep. Electron Spectrosc. Chem. Soc., London.
Czarny, J., Felenbok, P., Lefebvre-Brion, H. (1971). *J. Phys. B* **4**, 124.
Edmonds, A. R. (1960). "Angular Momentum in Quantum Mechanics." Princeton Univ. Press, Princeton, New Jersey.
Florida, D., and Rice, S. A. (1975). *Chem. Phys. Lett.* **33**, 207.
Freed, K. F. (1966). *J. Chem. Phys.* **45**, 4214.
Freed, K. F. (1976). *in* "Topics in Applied Physics" (K. F. Fong, ed.), Vol. 15, pp. 122–125. Springer Verlag, Berlin and New York.
Freed, K. F., and Band, Y. B. (1977). *Excited States* **3**, 109.
Freed, K. F., and Lin, S. H. (1975). *Chem. Phys.* **11**, 409.
Gradshteyn, I. S., and Ryzhik, I. M. (1965). "Table of Integrals, Series and Products," 4th ed. Academic Press, New York.
Henry, B. R., and Siebrand, W. (1971). *J. Chem. Phys.* **54**, 1072.
Herzberg, G. (1950). "Molecular Spectra and Molecular Structure," Vol. 1, "Spectra of Diatomic Molecules," pp. 413–432. Van Nostrand-Reinhold, Princeton, New Jersey.
Hougen, J. T. (1964). *Can. J. Phys.* **42**, 433.
Hougen, J. T., and Watson, J. K. G. (1965). *Can. J. Phys.* **43**, 298.
Howard, W. E., and Schlag, E. W. (1976). *Chem. Phys.* **18**, 123.
Howard, W. E., and Schlag, E. W. (1978a). *Chem Phys.* **29**, 1.
Howard, W. E., and Schlag, E. W. (1978b). *J. Chem. Phys.* **68**, 2679.
Kroto, W. H. (1975). "Molecular Rotation Spectra." Wiley, New York.
Landau, L. D., and Lifshitz, E. M. (1958). "Quantum Mechanics." Addison-Wesley, Reading, Massachusetts.
Lee, E. K. C. (1977). *Accounts Chem. Res.* **10**, 319.
LeRoy, R. J., and Bernstein, R. B. (1971). *J. Chem. Phys.* **54**, 5114.
Merzbacher, E. (1970). "Quantum Mechanics," p. 369. Wiley, New York.
Messiah, A. (1958). "Quantum Mechanics," pp. 454–456. Wiley, New York.

Morse, M. D., Freed, K. F., and Band, Y. B. (1976). *Chem. Phys. Lett.* **44**, 125.
Morse, M. D., Freed, K. F., and Band, Y. B. (1970). *J. Chem. Phys.* **70**, 3604, 3620.
Morse, M. D., Freed, K. F., and Band, Y. B. (1979). Unpublished results.
Nitzan, A., and Jortner, J. (1973). *J. Chem. Phys.* **58**, 2412.
Podolsky, B. (1928). *Phys. Rev.* **32**, 812.
Rose, M. E. (1957). "Elementary Theory of Angular Momentum." Wiley, New York.
Siebrand, W. (1970). *Chem. Phys. Lett.* **6**, 192.
Stevens, C. G., and Brand, J. C. D. (1973). *J. Chem. Phys.* **58**, 3324.
Townes, C. H., and Schalow, A. L. (1975). "Microwave Spectroscopy," p. 32. Dover, New York.
van Vleck, J. H. (1951). *Rev. Mod. Phys.* **23**, 213.
Waller, I. (1926). *Z. Phys.* **38**, 635.
Watson, J. K. G. (1968). *Mol. Phys.* **15**, 479.
Wilson, E. B., Decius, J. C., and Cross, P.C. (1955). "Molecular Vibrations." McGraw-Hill, New York.

4

Vibrational Relaxation of Isolated Molecules

Rudolf P. H. Rettschnick
Laboratory for Physical Chemistry
University of Amsterdam
Amsterdam, The Netherlands

I. INTRODUCTION

The concept of intramolecular vibrational relaxation dates from the early days of unimolecular reaction theory (Rice and Ramsperger, 1928; Rice, 1930). During a long period of time many experimental data could be explained on the basis of statistical theories. The RRKM theory has been the most successful and commonly used theory of unimolecular dissociation and isomerization reactions. This theory is based on the assumption that the vibrational energy, in excess of the zero-point energy, is very rapidly exchanged among the vibrational modes of the molecule, thus giving rise to an effective energy randomization on a timescale determined by the vibrational periods. In the past few years numerous experimental studies have been made to investigate the limits of validity of this assumption. The success of the statistical theories might be due to the relatively

ISBN 0-12-450650-X

low time resolution of the experiments and, of course, to the fact that the activation step is not very specific in most of the reactions. Recent investigations involving a specific activation of the molecule include photoactivation and chemical activation studies and molecular beam experiments. These investigations show that in many cases the energy is not statistically distributed over the vibrational modes before the reaction takes place. The reported redistribution rates vary within a wide range and therefore the assumption of a rapid randomization of the internal energy is not indisputable.

From these studies it follows that the problem of vibrational redistribution has at least two aspects: (i) the rate of the intermode energy flow in an isolated molecule and (ii) the ergodicity of the redistribution process.

A discussion of the dynamics of unimolecular reactions is beyond the scope of this chapter. We will focus our attention on spectroscopic investigations of intramolecular vibrational relaxation. Such a process seems to be the most elusive nonradiative process in isolated molecules. In recent years our understanding of electronic relaxation processes in polyatomic molecules has improved considerably, but still little is known about intramolecular vibrational relaxation.

A spectroscopic argument in favor of a rapid redistribution of excess vibrational energy has to do with the structure of the emission spectra of large molecules under collision-free conditions. The absence of hot bands in these spectra has been taken as evidence for energy randomization subsequent to the excitation of an optically active vibrational mode. This is in many cases a totally symmetric C–C skeletal mode. The concept of intramolecular vibrational redistribution as a dynamic process is illustrated in Fig. 1. Arrow **a** represents a vibronic transition in which a particular vibronic level $v_s' = 2$ of a totally symmetric mode s is excited. If there were no interaction between this particular vibronic state and other vibronic states the fluorescence spectrum would display resonance emission including hot bands represented by the arrows **b**. The relative intensities of the emission bands would be governed by Franck–Condon factors $|\langle v_s'' | v_s' = 2\rangle|^2$ as in the resonance emission spectra of small molecules.

Interaction between the vibrational modes of the molecule gives rise to the population of many different vibronic states. Vibrational redistribution subsequent to the excitation of the vibronic state $v_s' = 2$ is represented by the wavy arrow at the top of Fig. 1. Such a redistribution process causes a depopulation of the symmetric mode s and the excitation of several optically inactive modes i. The final distributions $\{v_i'\}$ must obey the energy conservation condition

$$(v_s' + \tfrac{1}{2})\hbar\omega_s' = \sum_i (v_i' + \tfrac{1}{2})\hbar\omega_i'.$$

The inactive modes i do not change their vibrational quantum number in the electronic transition $S_0 \leftrightarrow S_1$. For the sake of simplicity it is assumed that the

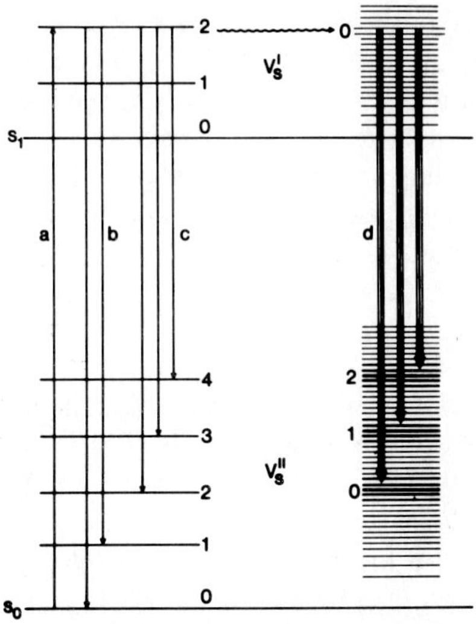

Fig. 1. Schematic diagram used to discuss intramolecular vibrational redistribution in an isolated large molecule. Emission in the hot band and normal fluorescence regions, originating from the optically excited vibronic state ($v_s' = 2$) is represented by arrows **b** and **c**, respectively. The wavy arrow at the top of the diagram represents the deactivation of the optically active mode. The fluorescence originating from the levels populated by the redistribution process is indicated on the right (arrows **d**).

vibrational structure of this electronic transition is determined by only one vibrational mode s. Since most of the vibronic levels, isoenergetic with the optically prepared level, are characterized by $v_s' = 0$; vibrational redistribution causes a drastic reduction of the intensity of the hot bands.

Another characteristic feature of the fluorescence spectra of isolated large molecules is the diffuse appearance of the vibronic bands, which, as a rule, are red shifted as compared with the spectra measured at higher pressures, where the average energy of the molecules is decreased as a result of intermolecular vibrational relaxation. This diffuseness may be due to vibrational redistribution as well. Instead of one single vibronic transition, such as the $v_s' = 2 \rightarrow v_s'' = 2$, which is situated near the electronic origin, many transitions take place, originating from the isoenergetic levels. These transitions are characterized by the vibrational quantum numbers $v_s' = v_s'' = 0$ and $\{v_i' = v_i'' \neq 0\}$. All these transitions together give rise to the so-called quasi 0–0 band. Since generally the vibrational frequencies are smaller in the excited electronic state, the quasi 0–0 band as well as the progression bands $v_s' = 0 \rightarrow v_s'' \neq 0$ and $v_i' = v_i'' \neq 0$ are red shifted with respect to the emission bands originating from the vibrationless S_1 state. The diffuseness is due to the diversity of these frequency shifts.

It should be noted, however, that the diffuse appearance of fluorescence spectra does not provide conclusive evidence of a rapid randomization of vibrational energy in isolated molecules. Also spectral congestion and sequence congestion effects may produce a diffuse background in the emission spectrum (Section III). For this reason studies of intramolecular vibrational relaxation should include time-resolved measurements. As yet, time-resolved spectroscopic experiments dealing with vibrational redistribution are sparse. Only a few of these studies seem to provide reliable information about the dynamics of the redistribution process.

II. THEORETICAL MODELS

In this section we will review very briefly some theoretical models which are concerned with studies of the redistribution of vibrational energy in isolated molecules. Emphasis will be placed on studies dealing with vibrational redistribution subsequent to optical excitation. The models presented in the literature draw heavily on theories of electronic relaxation processes. For this reason we will not go into a detailed discussion of the theoretical models but will just present a basis for the interpretation of spectroscopic studies.

Let us consider the manifold of harmonic adiabatic Born–Oppenheimer states corresponding to the vibronic levels of the lowest excited singlet state S_1. Selection rules for vibronic transitions can be expressed conveniently in terms of harmonic functions; therefore, the vibrational structure of absorption and emission spectra is usually described with harmonic oscillator functions. We will distinguish between vibronic states $\{|s\rangle\}$ which carry oscillator strength from the vibrationless electronic ground state, denoted by $|0\rangle$, and vibronic states $\{|l\rangle\}$ which cannot be excited directly by optical transitions starting from $|0\rangle$ because of very small Franck–Condon factors. The states $\{|s\rangle\}$ and $\{|l\rangle\}$ are coupled by the anharmonic interactions and Coriolis interactions which are neglected in the zero-order effective Hamiltonian. This coupling is quantified by the matrix elements $\{v_{sl}\}$ and $\{v_{ll'}\}$. The choice of a basis set of harmonic functions, all belonging to the same electronic state, implies that on the average the matrix elements v_{sl} and $v_{ll'}$ are of the same order of magnitude.

The zero-order states are characterized by their energies $\{\varepsilon_s\}$ and $\{\varepsilon_l\}$ and widths $\{\gamma_s\}$ and $\{\gamma_l\}$ which arise from radiative and nonradiative decays *not* including intramolecular vibrational redistribution. The effective Hamiltonian defined by this model is diagonalized by quasi-stationary states $\{n\}$, which are characterized by energies $\{\varepsilon_n\}$ and widths $\{\gamma_n\}$.

The distinction between $|s\rangle$ and $|l\rangle$ is based on entirely different values of the transition dipole moments $\boldsymbol{\mu}_{0s}$ and $\boldsymbol{\mu}_{0l}$, but it should be noted that the radiative widths of these zero-order states are not essentially different. The levels s and l are radiatively connected with different subsets of levels in the electronic ground state owing to different Franck–Condon factors; i.e., $|s\rangle$ communicates with $|0\rangle$ and vibrationally excited states $|g_s\rangle$, whereas $|l\rangle$ communicates with $|g_l\rangle$.

Fig. 2. Schematic representation of molecular energy levels. The s character $|\langle s|n\rangle|^2$ of the quasi-stationary states $\{|n\rangle\}$ is indicated by heavy lines. The widths of the levels are not indicated in this figure. It is assumed that the optical transition dipole moment μ_{0s} between the vibrationless vibronic ground state $|0\rangle$ and the harmonic state $|s\rangle$ is nonzero, whereas μ_{0l} is assumed to be zero.

In Fig. 2 a simple molecular energy level scheme is shown. The heavy lines indicate the s character of the quasi-stationary states, which is given by $|\langle s|n\rangle|^2$. This is a measure of the radiative transition probability between these states and the vibrationless ground state $|0\rangle$. In order to simplify this scheme, the submanifold $\{|l''\rangle\}$ which does not interact with $|s\rangle$ owing to symmetry restrictions, is not included in the figure.

We will consider a coherent excitation of a set of quasi-stationary states and the subsequent time evolution. The initially prepared state is described by

$$|\psi(t=0)\rangle = \sum_n c_n|n\rangle, \tag{1}$$

and the state vector at subsequent times is

$$|\psi(t)\rangle = \sum_n c_n|n\rangle \exp(-i\varepsilon_n t/\hbar - \gamma_n t/2\hbar). \tag{2}$$

The time evolution of the molecule in the fluorescent state S_1 may be studied experimentally as the zero-order states $|s\rangle$ and $\{|l\rangle\}$ are associated with different frequency distributions in the emission spectrum or (transient) absorption spectrum. The temporal behavior is given by the projection of the state vector on $|s\rangle$ and $\{|l\rangle\}$:

$$P_s(t) = |\langle s|\psi(t)\rangle|^2, \tag{3a}$$

$$P_l(t) = \sum_l |\langle l|\psi(t)\rangle|^2. \tag{3b}$$

The disappearance of the s character is due to radiative decay as well as electronic relaxation and the build up of the l character. The temporal behavior of $P_l(t)$ and $P_s(t)$ is complicated, even if the harmonic state $|s\rangle$ is the initially prepared state. This problem has been discussed extensively in relation to electronic relaxation processes (Jortner and Mukamel, 1974; Lahmani *et al.*, 1974; Freed, 1976a; see also Chapter 5 of this volume). The time evolution of the system depends on the energy profile of the prepared state $|\psi(t=0)\rangle$. At short enough times the initial state decays exponentially.

In recent years a few studies dealing with the problem of vibrational redistribution in optically excited molecules have been published. In a recent paper on intramolecular vibrational relaxation, Freed (1976b) has discussed a number of experimental results, taken from the literature, in terms of the small and large molecular limits which are characterized by the conditions $\gamma_l \rho_l \ll 1$ and $\gamma_l \rho_l \gg 1$, respectively. In the latter case the l manifold serves as a dissipative quasicontinuum, so that $|s\rangle$, if it could be produced as an initial state, would decay irreversibly into the l manifold on the time scale of the experiment. In the small molecule limit however, the time development of $|s\rangle$ may display recurrences on the time scale determined by the lifetime of the excited state, in addition to oscillations in the decay pattern; in this case the l manifold does not act as an effective dissipative quasi-continuum.

According to Freed the initial decay of a harmonic level s, due to anharmonic coupling with a manifold of levels $\{l\}$ is governed by a rate constant

$$k_{\text{red}} = \frac{2\pi}{\hbar} \sum_l |v_{sl}|^2 \rho_l(\varepsilon_s), \tag{4}$$

where ρ_l^{-1} is the average effective level spacing in the l manifold for those levels which are efficiently coupled to s. Actually, this expression applies to the interaction of $|s\rangle$ with a diagonalized set of states $\{|l\rangle\}$.

Tric (1976) has developed a model in which the anharmonic coupling between the zero-order states $\{|l\rangle\}$ is explicitly taken into account. In this theory it is assumed that coupling constants and energy spacings between zero-order levels are randomly distributed. This model was applied both to redistribution in a single electronic state and to deactivation processes due to vibrational redistribution in mixed electronic states. The latter case refers to an initially prepared harmonic level s which is efficiently coupled to a number of harmonic levels $\{t\}$ belonging to a different electronic state and weakly coupled to the remaining levels. It has been shown by Tric that the width of $|s\rangle$ is governed by the width of the distribution of anharmonic coupling elements, and is independent of their average value, provided that the average level spacing in the l manifold is smaller than the width of this distribution. The width of $|s\rangle$ due to random anharmonic coupling in the strong coupling limit is given by

$$\Delta_s = 2\pi\rho(\varepsilon_s)\langle(\Delta v)^2\rangle, \tag{5a}$$

where $\langle(\Delta v)^2\rangle$ is the variance of the distribution of coupling matrix elements and ρ is the average vibrational level density in the l manifold at the excess energy ε_s. The initial decay rate of $|s\rangle$,

$$k_{\text{red}} = \frac{2\pi}{\hbar} \rho(\varepsilon_s)\langle(\Delta v)^2\rangle, \tag{5b}$$

is determined by the dephasing of the coherently excited quasi-stationary states.

Tric has presented a detailed study of the time evolution of $P_s(t)$ at subsequent times and shown how the variables of the model can be fitted to measured fluorescence decay curves.

A different approach to the problem of vibrational redistribution in large molecules has been presented by Fleming *et al.* (1974). They pointed out the similarity of intramolecular and intermolecular vibrational relaxation if in the latter case the vibrationally excited molecule together with its surroundings is considered as a supermolecule. In both cases vibrational energy is transferred from a particular vibrational mode to the other modes. The authors solved a master equation for the deactivation of a single excited vibrational mode which is linearly coupled by three-phonon interaction to all the other modes. It was assumed that the occupation numbers of all but the relaxing mode preserve a thermal equilibrium distribution during the relaxation process.

The subsequent time development of $P_l(t)$ is connected with the problem of molecular ergodicity. The question whether the redistribution of vibrational energy in an isolated molecule is ergodic or not, has received much attention in recent years. The problem is of importance for the understanding of unimolecular reactions. Therefore, much higher excess energies are considered in the theoretical models devoted to this problem than in the models which refer to spectroscopic studies.

Kay (1974) has discussed the time evolution of an initially prepared harmonic state $|\psi(t = 0)\rangle$ in a relatively large, isolated molecule and the buildup of occupation probability in all molecular states. His study is based on a random coupling assumption and predicts the establishment of a microcanonical distribution due to strong anharmonic interactions between zero-order states if the molecule is so highly excited that multiple quantum exchanges are possible. It is assumed that matrix elements v_{mn} of the anharmonic perturbation V between the zero-order states $|m\rangle$ and $|n\rangle$ do not show any systematic changes when E_m is varied over a range ΔE in which the vibrational level density is sensibly constant.

Numerical studies performed by Nordholm and Rice (1974) on a number of different molecular models predict a prevalence of nonergodic behavior in small molecules even at high energies below the minimum energy required for dissociation. An extensive discussion of the topic of molecular ergodicity has been given by Rice (1975). It seems that prevalence for nonergodic behavior at relatively low excess energies and a tendency to ergodic behavior at higher energies is to be expected, at least in large molecules.

In most theoretical studies the rotational degrees of freedom are left out of consideration. Since we are concerned with molecules in the gas phase, vibrational–rotational interactions may not be neglected. The average values of the coupling matrix elements $\{v_{sl}^{r}, v_{ll'}^{r}\}$ which are involved in the Coriolis coupling of the zero-order vibrational states may differ considerably from those of the anharmonic coupling matrix elements. Unlike anharmonic interaction, Coriolis interaction may couple vibrational levels of different symmetries.

Optical excitation will produce almost pure harmonic levels if the coupling terms are small enough, so that the s character is confined to a narrow energy range. Very little is known about the magnitudes of the off-diagonal matrix elements v and v^{r}. Resolved Fermi resonance effects in electronic absorption and emission spectra of large molecules are rare. The anharmonic coupling terms which have been derived from such spectra have magnitudes up to some tens of wavenumbers. For example, cubic coupling terms on the order of 10 cm^{-1} have been found for benzene (Fischer *et al.*, 1975). Probably, the vast majority of the coupling terms is considerably smaller, since otherwise the electronic spectra of large molecules, measured in low temperature matrices, would be more complex. Also the narrow bandwidths of the vibrational fine structure in these spectra suggest that in general the magnitude of v is rather small. These widths do not exceed values of a few wavenumbers at energies where the density of vibrational states is in the order of $10^4/\mathrm{cm}^{-1}$ or even more, so that according to Eqs. (4) and (5) the average magnitude of v will certainly not exceed a value of $10^{-2}\ \mathrm{cm}^{-1}$. It is to be expected that the magnitudes of the coupling terms decrease at higher energies because of the rapidly increasing number of higher order interactions.

An obvious condition for the observability of vibrational redistribution is that the time constant of the redistribution process exceeds τ_{p}, the duration of the excitation pulse:

$$\hbar\Delta_s^{-1} > \tau_{\mathrm{p}}. \tag{6}$$

A coherent excitation of $|s\rangle$ implies that the widths of the photon wavepackets, Δ_{ph}, exceed Δ_s, the energy spread of the quasi-stationary states among which the s character is distributed. If the excitation is accomplished by means of (nearly) bandwidth-limited pulses from a mode-locked laser, τ_{p} has approximately the same magnitude as the coherence time τ_{coh}, which is related to Δ_{ph} by a Fourier transformation.

Simultaneous measurements of the pulse duration and the frequency distribution of single picosecond pulses (von der Linde, 1972) have provided an empirical relation

$$\tau_{\mathrm{p}}\Delta_{\mathrm{exc}} \approx 2 \times 10^{-11} \quad \mathrm{sec\ cm}^{-1}, \tag{7}$$

where Δ_{exc} is the width of the frequency distribution. This empirical relation may be compared with the relation between the coherence parameters (Rhodes, 1974),

$$\tau_{\mathrm{coh}}\Delta_{\mathrm{ph}} = \hbar \approx 5 \times 10^{-12} \quad \mathrm{sec\ cm}^{-1}. \tag{8}$$

On the other hand, for a conventional light source, τ_{p} may exceed τ_{coh} by several orders of magnitude, and Δ_{ph} may be considerably smaller than Δ_{exc}, which can give rise to an incoherent excitation of harmonic states.

III. EXPERIMENTAL STUDIES

The experimental data available from the literature concerning intramolecular vibrational relaxation are characterized by a great diversity. The time scales which have been reported for the redistribution process vary considerably, As so many of these data are contradictory we will not give a comprehensive survey of the experimental results obtained so far. Instead, we will review some studies which are felt to be illustrative of the present state of the art and which may serve as a starting point for future investigations.

Attention will be paid primarily to spectroscopic investigations in the visible and ultraviolet regions dealing with vibrational redistribution in electronically excited molecules, since many of these compounds can be studied by means of fluorescence techniques at pressures where the molecules can be considered to be collision free on the time scale of their fluorescence lifetimes. We will use the term resonance fluorescence for the emission from molecules which are not perturbed by collisions during their excited state lifetime.

Spectroscopic studies of intramolecular vibrational relaxation imply the excitation of a vibronic level which is part of a dense manifold of levels belonging to a single electronic state, and therefore the investigations are confined to large molecules or to highly excited small polyatomic molecules. In a relatively small molecule like benzene the vibrational level density in the S_1 manifold at energies below the threshold for the so-called "channel 3" decay is not high enough to make probable any redistribution of vibrational energy. At vibrational energies above this threshold, which is situated about 3000 cm^{-1} above the electronic origin of the S_1 manifold, a competitive channel becomes involved in the decay of the S_1 vibronic levels giving rise to a sudden drop of the fluorescence quantum yield to extremely low values (Parmenter, 1972). For this reason benzene is not a suitable molecule for a study of vibrational redistribution.

Slightly larger molecules like toluene and aniline also exhibit a sharp falloff in fluorescence yield beyond a similar threshold energy, but below this threshold the density of vibrational states is already in the order of $10^3/cm^{-1}$, or even more, so that spectroscopic studies of intramolecular vibrational relaxation might be possible. Such studies, however, are plagued by the occurrence of sequence excitations. This problem is much more troublesome when large molecules like anthracene or pentacene are studied.

Before we discuss the experimental results emerging from these studies, it may be useful to consider the problem of sequence congestion which is inherent in spectroscopic investigations of large molecules (Section IIIA). Studies concerned with vibrational redistribution subsequent to optical excitation of isolated aromatic molecules will be reviewed in Section IIIB. Most of these studies are concerned with time-resolved measurements of the resonance fluorescence. Sections IIIC and IIID deal with a few recent studies of vibrational redistribution in the electronic ground state and in van der Waals complexes.

A. Sequence Congestion

It is extremely difficult to avoid sequence excitations in spectroscopic studies of large molecules in the vapor phase. Usually the experiments are carried out at temperatures between 300 and 500 K because of the low vapor pressure of the substances. At these temperatures many of the low-frequency vibrational modes are excited in the electronic ground state and consequently the absorption spectrum contains many bands which originate from vibrationally excited molecules. Most of these transitions are associated with optically inactive, non-totally symmetric vibrations and therefore in general the vibrational quantum number v_i of the thermally excited mode is preserved in the optical transition: $\Delta v_i = v_i' - v_i'' = 0$. These sequence transitions are usually situated to the red of the parent band in the absorption spectrum since the vibrational frequencies in the upper vibronic state are generally smaller than the frequencies in the ground state.

At higher temperatures more members appear in a sequence, giving rise to an increased sequence broadening mainly in the red wing of the absorption bands. In the absorption spectra of large molecules different sequences overlap, even at moderate temperatures, since the widths of the rotational contours of individual vibronic transitions are in the range of 10–30 cm^{-1}. Hence, any sequence fine structure is completely blurred out. This implies that selective excitation of a single vibronic level is impossible even with extremely narrow excitation band widths. Byrne and Ross (1971) have presented a detailed discussion of sequence congestion in connection with the diffuseness of electronic absorption spectra of large molecules in the vapor phase.

The sequence bands originate from vibrational levels in the electronic ground state with quite different energies. Since so many sequence bands overlap, electronic excitation gives rise to an efficient transfer of thermal energy to the upper electronic state. No matter how the excitation wavenumber is tuned within an absorption band, it is always in resonance with a considerable number of sequence transitions including combined transitions of the type $\Delta v_i = 0$, $\Delta v_j = 0$, etc., in which different vibrational modes are involved simultaneously.

In Fig. 3 this effect is illustrated in a schematic way. The left part of this figure shows the thermal equilibrium distribution of vibrational energies in the electronic ground state. The number of molecules with vibrational energies between E and $E + dE$ is given by

$$N(E)\,dE = \rho(E)\exp(-E/kT)\,dE \Bigg/ \int_0^\infty \rho(E')\exp(-E'/kT)\,dE',$$

where ρ is the density of vibrational states. Since ρ grows rapidly with the vibrational energy, the maximum of the thermal distribution is situated at rather high values of E, especially for large molecules. In the case of anthracene vapor,

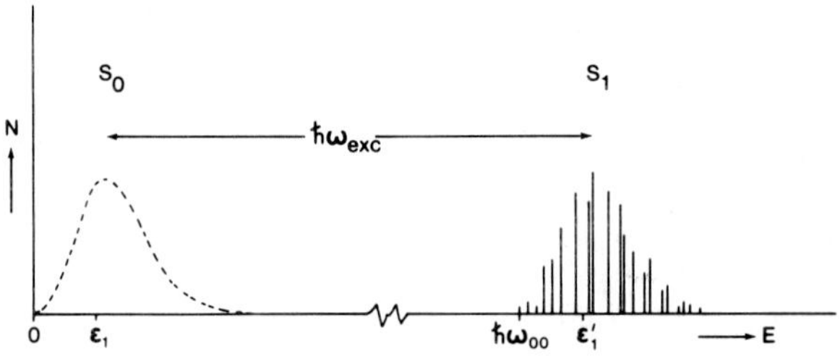

Fig. 3. Transfer of thermal energy from the ground state to the upper electronic state due to optical excitation. The thermal equilibrium distribution of vibrational levels in the electronic ground state is indicated by the dashed line and the population distribution of the optically excited vibronic levels is depicted by vertical lines. The excitation process produces vibronic states on both sides of a threshold energy ε'_1. Fluorescence emission from levels with energies in excess of ε'_1 may be broadened due to anharmonic interactions.

for instance, the maximum of the distribution curve is situated at about 2500 cm^{-1} at a temperature of 380 K. At this temperature, the vapor pressure is 0.1 torr. If we take pyrene vapor as another example, corresponding data are 3300 cm^{-1}, 420 K, and 0.5 torr, respectively. In Fig. 3 the population distribution of the vibrational levels in the upper electronic state is indicated by vertical lines. The occupation numbers $N(\varepsilon)$ of the vibrational levels in the excited electronic state depend upon the population of the vibrational levels in the electronic ground state and the exact position of the excitation wavenumber within the band contours of the individual sequence bands. The energy spread of the vibronic levels in the excited electronic state is similar to the width of the thermal distribution in the electronic ground state (several thousands of wavenumbers). The widths of the absorption bands are much smaller (some hundreds of wavenumbers); these widths are of the order of the frequency changes $\Delta\omega = \omega' - \omega''$, which may be as large as about 100 cm^{-1} for the low-frequency modes. Although the detailed structure of the population distribution of vibronic S_1 levels depends upon the actual value of the excitation wavenumber, it may be expected that in a large molecule the envelope of the population distribution is rather similar to the envelope of the thermal equilibrium distribution of S_0 vibrational levels, even in the case of narrow-band excitation. Broad-band excitation will enhance the similarity of both distributions.

A very efficient transfer of thermal energy under conditions of narrow-band excitation has been found for pyrene (Deinum *et al.*, 1974a) and 3,4-benzpyrene (van den Bogaardt *et al.*, 1976). The resonance fluorescence spectra of these compounds exhibit S_2 emission in addition to the normal S_1 fluorescence. The relative S_2 intensity increases rapidly with the excitation energy and for this reason, the S_2 emission can be used as a sensitive probe of the excess vibrational energy in the upper electronic state. It has been shown that the intensity ratio of both components of the fluorescence spectrum is proportional to ρ_2/ρ_1, the

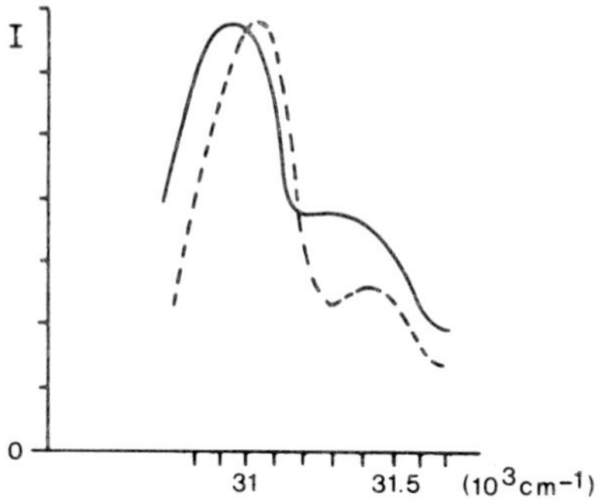

Fig. 4. Part of the fluorescence excitation spectrum of pyrene vapor (150°C and 10^{-2} torr). Solid line: fluorescence detected at 29,800 cm^{-1} (S_2 emission region). Broken line: fluorescence detected at 25,800 cm^{-1} (S_1 emission region). For comparison, both excitation spectra have been normalized to the same maximum intensity in the 0–0 band of the $S_1 \rightarrow S_2$ transition. From Deinum (1974).

ratio of the densities of vibrational states in the S_2 and S_1 manifolds (Deinum *et al.*, 1974b). The observed dependence of the relative S_2 intensity upon both the excitation wavenumber and the temperature can be explained in a satisfactory way if it is assumed that the thermal equilibrium distribution of vibrational energies is transferred completely to the excited electronic state. This assumption may be used only for an approximate description of sequence congestion effects, as can be seen from Fig. 4, where a part of the fluorescence excitation spectrum of pyrene vapor is shown (Deinum *et al.*, 1974b). The solid curve represents the excitation spectrum of the S_2 component of the fluorescence, and the broken line shows the excitation spectrum of the normal S_1 fluorescence. The broken line coincides with the absorption spectrum. This figure clearly shows that the sequence transitions, and especially the higher members of the sequences which give rise to the population of highly excited vibronic levels, are situated in the red wing of the absorption band.

It is a general property of the absorption spectra of large molecules that the pure transition is located near the blue end of the absorption band, whereas most of the accompanying sequence transitions are red shifted with respect to the pure transition. As a consequence, the energy dependence of the fluorescence yield and the fluoresence decay time may display an oscillatory pattern which resembles vibrational structure. This effect has been pointed out by Nitzan and Jortner (1972) in a theoretical study of the influence of temperature changes on the fluorescence decay pattern of an isolated large molecule due to sequence congestion. The congestion of absorption and emission spectra due to sequence transitions in which the vibrational quantum numbers are not preserved, will be discussed in Section IIIB1.

The troublesome effects of sequence congestion in spectroscopic studies of large molecules can be reduced in different ways. By tuning the excitation wavenumber of a narrow-band source to the blue wing of an absorption band, the population of higher vibronic levels is suppressed. This may even give rise to fluorescence spectra showing a sharp vibrational structure for a large molecule like anthracene in the vapor phase (Section IIIB3). On the other hand, it may be profitable to use broad-band excitation, especially in time-resolved experiments. In this case the population of excited vibronic levels will be fairly similar to the thermal equilibrium distribution in the ground state which implies that the energy dependence of quantum yields and decay times are much less sensitive to the exact value of the excitation wavenumber. Furthermore, in such a case the effect of sequence excitation can be taken into account quantitatively if the thermal distribution of vibrational energies in the ground state is known. Obviously, sequence congestion is suppressed most efficiently by cooling down the vibrational degrees of freedom of the molecule in a supersonic beam. Unfortunately, this is not a simple technique, especially if a compound consisting of large molecules has to be mixed with the carrier gas.

B. Vibrational Redistribution in Aromatic Molecules Subsequent to Electronic Excitation

It is easier to demonstrate the absence of vibrational redistribution than to obtain any quantitative information about the redistribution of vibrational energy in an isolated molecule. The simplest experimental proof of the absence of vibrational redistribution in electronically excited molecules is provided by single vibronic level fluorescence spectra in which extra bands and background emission are lacking. For instance, the fluorescence spectra of isolated benzene containing up to about 2300 cm^{-1} of excess vibrational energy are characteristic only of the initially excited levels (Knight *et al.*, 1975). No indication of vibrational redistribution on the time scale of the fluorescence exists in these spectra. In this energy region the vibrational level density is too low to allow a coherent excitation of more than one vibrational eigenstate of S_1. The absorption bands in the benzene spectrum are fairly well separated, also because sequence bands are relatively sparse at normal temperatures. In this section some studies dealing with intramolecular vibrational redistribution in aromatic molecules will be discussed.

1. Benzene Derivatives

Considerable interest has been devoted to the resonance fluorescence spectra of benzene derivatives. The fluorescence spectra of substituted benzenes like toluene, *p*-fluorotoluene, and aniline are not free from diffuseness. The relative intensity of the unresolved background emission increases considerably when higher vibronic levels are excited. This can be ascribed to the simultaneous excitation of different vibronic levels, since the absorption spectrum becomes more crowded at higher frequencies. Because of the lower symmetry of the molecules more vibronic transitions are allowed than in benzene, giving rise to absorption bands with overlapping rotational contours.

It is a general feature of the resonance fluorescence spectra of these molecules that the relative intensity of the underlying background is much less in the hot band region than in the normal fluorescence region. This phenomenon is to be expected if a strong absorption band overlaps the rotational contours of one or more weak absorption bands. For example, the strong transition $v_s'' = 0 \rightarrow v_s' = 2$, represented by arrow **a** in Fig. 1, may be accompanied by a weak transition of some other mode, say $v_i'' = 0 \rightarrow v_i' = n$ (not shown in the figure). The fluorescence originating from level $v_i' = n$ will be extremely weak in the hot band region (covered by arrows **b**). However, the Franck–Condon factors for transitions like $v_i' = n \rightarrow v_i'' = n$ may be more favorable and thus, in spite of its relatively low population, level $v_i' = n$ may give a nonnegligible contribution to the diffuse background in the normal fluorescence region (arrows **d** in Fig. 1).

The congestion in the resonance fluorescence spectra of the benzene derivatives is not only due to spectral crowding, but also to the appearance of more sequence bands than in the case of benzene. If a relatively strong transition $v_j'' = 0 \rightarrow v_j' = n$ does not coincide with the optically excited transition $v_s'' = 0 \rightarrow v_s' = 2$, the hot transition $v_j'' \neq 0 \rightarrow v_j'(=v_j'' + n)$ might be in resonance with the excitation energy due to the nonzero frequency change $\Delta\omega_j$:

$$\hbar\omega_{\text{exc}} = \Delta E_{\text{el}} + 2\hbar\omega_s' + \tfrac{1}{2}\hbar\,\Delta\omega_s \approx \Delta E_{\text{el}} + n\hbar\omega_j' + (v_j'' + \tfrac{1}{2})\hbar\,\Delta\omega_j.$$

The resonance fluorescence from the excited levels v_j' may contribute substantially to the continuous background in the emission spectrum, even in the hot band region.

Blondeau and Stockburger (1971) have measured resonance emission spectra from a number of selectively excited vibronic levels of toluene. It was found that an increase of the excitation wavenumber produces a marked decrease of the relative emission intensity in the hot band region and an increase of the relative intensity of the continuous background, mainly in the frequency region of the normal fluorescence. These effects, which are very distinct even at vibrational energies below 1000 cm^{-1}, where the vibronic level density is less than $1/\text{cm}^{-1}$, have been considered as indications of vibrational redistribution. However, in this energy region any substantial vibrational redistribution can hardly be expected. Therefore it is likely that the strong energy dependence of the relative intensities of the hot bands and the continuous background are mainly due to spectral congestion effects and not to vibrational redistribution, at least at lower energies up to approximately 1000 cm^{-1}. No decisive conclusions about vibrational redistribution can be drawn from the structure of the resonance fluorescence spectra without any additional information from time-resolved measurements.

Jacon *et al.* (1977) and de Vries (1976) have performed time-resolved measurements of the resonance emission of toluene. Their experiments have shown that the decay patterns of the structured and diffuse components of the spectrum are identical. Redistribution of vibrational energy on the time scale of the fluorescence could have given rise to a decay of the structured component in favor of the continuous emission, but no indication of an induction period of the diffuse emission was found in these experiments. Apparently, no detectable vibrational redistribution occurs between about 1 nsec (the time-resolution of the experiments) and about 50 nsec (the fluorescence decay time) at energies up to approximately 2500 cm^{-1}.

It is very unlikely that any significant vibrational redistribution occurs on a time scale shorter than 1 nsec. This can be concluded from the experiments of Jacon *et al.* (1977), who have measured the decay characteristics of isolated toluene as a function of the excitation wavenumber. It was found that the nonradiative decay rate increases gradually with the excitation wavenumber in the lower energy range, but it increases more rapidly when the vibrational energy exceeds 2000 cm^{-1}. This behavior resembles the so-called channel 3 decay of benzene although the effect is somewhat attenuated, which is at least partially owing to sequence congestion effects. In order to explain the experimental results, different thresholds for channel 3 decay had to be assumed for different vibrational progressions. This assumption implies the absence of rapid vibrational redistribution at excess energies up to about 4000 cm^{-1} where the vibronic level density is up to about $10^3/cm^{-1}$.

The spectroscopic properties of benzene derivatives like *p*-fluorotoluene (Rockley and Phillips, 1974) and aniline (Blondeau and Stockburger, 1971) are similar to those of toluene. The structure of the resonance fluorescence spectra of these molecules is very sensitive to the amount of excess vibrational energy. These spectra exhibit an extremely well resolved vibrational structure when the 0^0 level is excited but there is not much structure left when the excitation is about 1000 cm^{-1} above the S_1 electronic origin and the intensity of the hot band fluorescence is relatively weak. These phenomena may be considered as due to intramolecular vibrational relaxation, but no decisive conclusions can be drawn from only the structure of the emission spectra.

As yet, time-resolved measurements of resonance fluorescence spectra of *p*-fluorotoluene have not provided evidence for a rapid redistribution of vibrational energy at excess energies up to about 2000 cm^{-1} (Phillips, 1978).

The fluorescence lifetimes and quantum yields of aniline-h7 and aniline-d5 have been measured by Jacon *et al.* (1977) under collision-free conditions. The experimental results have been explained in terms of different decay rates for closely spaced levels with vibrational energies up to about 6000 cm^{-1}. This interpretation implies that no substantial vibrational redistribution occurs on a nanosecond time scale even when the density of vibronic levels in the S_1 manifold exceeds a value of $10^5/cm^{-1}$.

2. Naphthalene

Indications of an incomplete redistribution of vibrational energy on the timescale of the fluorescence have been found for a number of aromatic compounds by Lim and co-workers (cf. Section III B5). Naphthalene is an interesting example. Hsieh *et al.* (1974) have measured the energy dependence of the fluorescence lifetime and quantum yield of naphthalene under collision-free conditions. The nonradiative decay rate k_{nr}, calculated from these data, is plotted as a function of the excitation energy in Fig. 5. The plot exhibits a remarkable change of slope at about 36,000 cm^{-1}. This excitation energy corresponds with the onset of the $S_0 \rightarrow S_2$ transition. Due to sequence congestion effects the observed nonradiative decay rate may have become less dependent on the excitation energy and hence the actual change of k_{nr} with the excess vibrational energy may be even more pronounced. This change can be interpreted in the following way. Optical excitation above the S_2 threshold produces S_2 vibronic states which are subject to internal conversion into the S_1 manifold. Since the high frequency

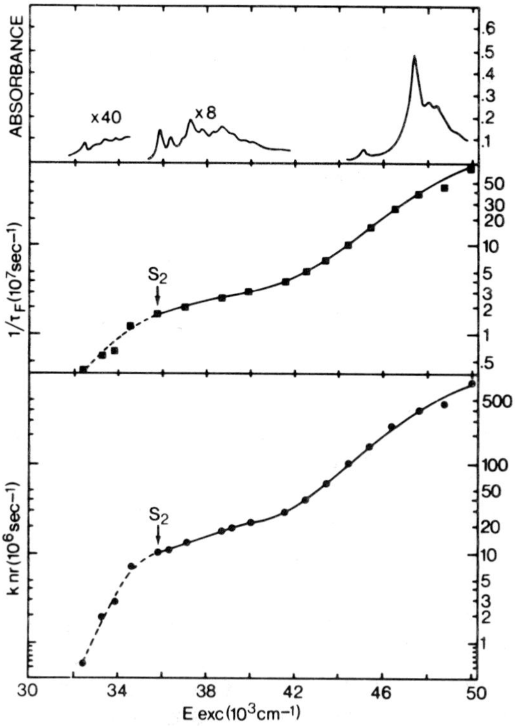

Fig. 5. Top: low resolution absorption spectrum of naphthalene vapor. Center: excitation energy dependence of the total decay rate. Bottom: excitation energy dependence of the nonradiative decay rate. From Hsieh *et al.* (1974).

C–H stretching modes are the dominant accepting modes, these vibrations are primarily excited by the internal conversion. On the other hand, if the excitation is achieved directly into the S_1 manifold, mainly totally symmetric C–C stretching modes are excited. A complete randomization of the excess vibrational energy in the S_1 manifold would imply a gradual change of the nonradiative decay rate with the excitation energy since in that case the distribution of the excess energy over the vibrational modes would be independent of the method of preparation of the emitting states. The abrupt change in the plot of k_{nr} versus the excitation energy as the threshold of the S_2 state is crossed suggests an incomplete vibrational redistribution on the time scale of the experiment (some tens of nanoseconds) at an excess vibrational energy of about 5000 cm^{-1} (including thermal energy).

At vibrational energies below 1400 cm^{-1}, vibrational redistribution appears to be absent in isolated naphthalene molecules. This can be concluded from the discrete vibrational structure of the single-vibronic-level fluorescence observed by Stockburger *et al.* (1975). These resonance emission spectra, which have been obtained with narrow-band excitation, exhibit relatively strong hot bands and the continuous emission is rather weak.

3. Anthracene

Under conventional experimental conditions the absorption and emission spectra of anthracene vapor lack any vibrational fine structure. Due to sequence congestion and band crowding the vibrational structure appears as broad maxima. However, the resonance fluorescence spectrum contains many discrete lines superimposed on a continuous background, when the vapor is excited with a narrow-band source near the origin of the $S_0 \rightarrow S_1$ transition. The sharp vibrational structure in the fluorescence spectrum of anthracene vapor was first reported by Pringsheim (1938) and was subsequently studied by various investigators. Haebig (1968), using a cadmium line at 361.2 nm for the excitation of the vapor, has shown that the positions of the discrete lines correlate with the vibrational structure of the fluorescence from the vibrationless S_1 level of anthracene in a heptane matrix at 4 K. Apparently, the cadmium line coincides with a specific set of sequence transitions in which low vibrational energies prevail, so that the fluorescence spectrum originating from a significant fraction of the excited molecules will display vibrational fine structure.

Klochkov and Smirnova (1967), Mirumyants *et al.* (1974, 1975), and Mirumyants and Demchuk (1976) have extensively studied the so called quasi-line spectrum of isolated anthracene using different excitation sources. They observed 76 quasi-lines, which have been assigned to progression bands and combination bands originating from the vibrationless S_1 level. This assignment was made in terms of ten different fundamental modes which are also active in the low temperature Shpolskii-type spectra of anthracene. In addition, a hot emission band was observed, which was assigned to an excited vibrational level in the S_1 state. The observed widths of the quasi-lines are of the same order of magnitude as the bandpass of the exciting light, which varied between 20 and 60 cm^{-1}. The diffuse background in the fluorescence spectrum was ascribed to vibronically induced transitions as well as to intramolecular relaxation broadening. Mirumyants *et al.* (1975) have shown that the discrete component of the spectrum can still be observed if the excitation wavenumber is varied within a range as large as 250 cm^{-1}, the wavenumbers of the quasi-lines then shifting linearly with the excitation wavenumber. This shift is due to the fact that many sequence transitions $\Delta v_i = 0$ are continuously distributed within the absorption band. A variation of the excitation wavenumber by an amount $\Delta\omega_{exc}$ gives rise to the excitation of a different set of vibronic levels which is characterized by a different value of $\Delta\omega = \omega' - \omega''$. As a consequence, each of the discrete lines will shift by $\Delta\omega_{exc}$. The observed shifts may differ slightly from $\Delta\omega_{exc}$ because of slightly different rotational contours of the emission bands.

The structured spectrum of anthracene vapor under collision-free conditions has been measured very recently by de Vries *et al.* (to be published) with an excitation bandwidth of 0.6 cm^{-1}. The excitation was achieved with a tunable dye laser which was pumped by a nitrogen laser. Figure 6 shows a part of a typical emission spectrum obtained with an excitation wavelength of 361.2 nm and at a temperature of 405 K. A substantial part of the fluorescence appears as well-defined narrow lines. The resonance fluorescence spectrum of anthracene vapor, measured with low spectral resolution, is shown in the inset of Fig. 6.

At a temperature of 400 K the discrete lines can easily be distinguished from the background emission when the excitation wavelength is confined to the range between 361 and 363 nm. This wavelength range is situated in the blue wing of the 0–0 absorption band. The relative intensities of the peaks with respect to the background intensity increase strongly when the excitation wavelength λ_{exc} is varied from 361.0 to 361.3 nm, and they decrease gradually when λ_{exc} is varied from 361.3 to 363 nm. According to the considerations given in Section IIIA, it is likely that the discrete component of the resonance fluorescence originates from the lower S_1 levels, whereas the emission from the higher levels may be broadened due to anharmonic interactions and spectral congestion. If the excitation source is tuned to longer wavelengths, the population distribution of excited S_1 levels is dominated more and more by the higher levels, giving rise to decreasing relative intensities of the discrete lines. A similar effect is observed when the temperature of the vapor is increased. The widths of the sharp lines are in the range of 15–25 cm^{-1}, which is considerably more than the bandpass of the exciting light (0.6 cm^{-1}). This observation shows that the minimum widths of the lines are determined by the rotational contours of the vibronic transitions. Even if an extremely narrow excitation bandpass is used, the widths of separate vibronic transitions in the fluorescence spectrum of isolated large molecules are similar to the bandwidths in a thermally equilibrated emission spectrum. This

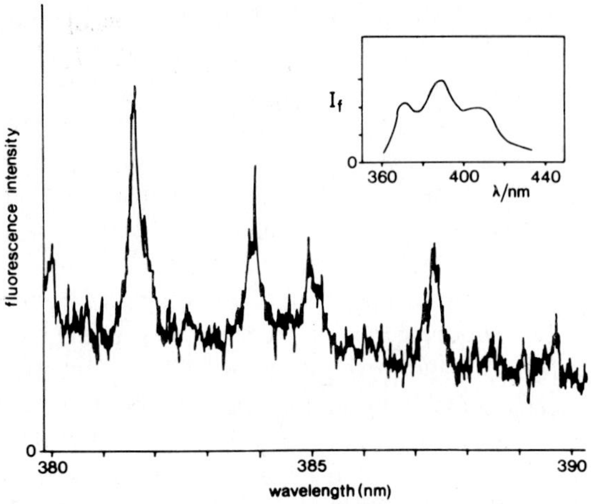

Fig. 6. Part of the resonance fluorescence spectrum of anthracene vapor (405 K and 0.1 torr). Excitation wavelength 361.2 nm, excitation bandwidth 0.6 cm^{-1}, and fluorescence bandpass 4 cm^{-1}. With an excitation wavelength of 361.3 nm the relative intensities of the peaks, with respect to the continuous background, are nearly twice as high as shown in this figure. The inset shows a low resolution fluorescence spectrum of anthracene vapor (430 K and 0.3 torr). Excitation wavelength: 313 nm. The excitation bandwidth and the fluorescence bandpass are approximately 150 cm^{-1}.

is due to the fact that the rotational contours of many sequence bands overlap so that a variety of rovibronic levels is excited with a wide spread of rotational energies.

A variation of the excitation wave number produces a shift of the spectral positions and a change of the intensities of the discrete lines, but no new lines are observed when the wavenumber of the exciting light is changed. These observations can be explained if it is assumed that the sharp emission lines originate from vibronic levels with energies below a certain threshold, for instance ε_1' in Fig. 3, whereas the diffuse component of the spectrum originates from the higher levels in the S_1 manifold. The observation that no alterations in the vibrational fine structure appear when the excitation wavenumber is changed suggests that no vibrational redistribution occurs at energies below ε_1'.

Evidence for a time-dependent redistribution of vibrational energy in molecules with internal energies in excess of the threshold energy could be conclusively obtained only from time-resolved measurements. It would be interesting, though not easy, to measure the time dependence of the fluorescence intensity detected at the positions of the peaks and also in between the lines. A comparison of the decay patterns could give an indication as to what extent the diffuse emission has something to do with vibrational redistribution.

The nonradiative decay rate of anthracene exhibits a steplike change when the excitation energy corresponds to the onset of the second strong electronic transition (Laor *et al.*, 1973; Huang *et al.*, 1974). This phenomenon is comparable with the abrupt variation in the energy dependence of k_{nr} of naphthalene at the origin of the second electronic transition (Section IIIB2). The effect has been interpreted, as in the case of naphthalene, as an indication of an incomplete vibrational redistribution in isolated anthracene following internal conversion (Fisher and Lim, 1974). If the thermal energy is neglected, the steplike variation of k_{nr} occurs at an excess energy of 12,000 cm^{-1} where the total density of vibronic states in the S_1 manifold is of the order of $10^{13}/cm^{-1}$.

The presence of vibrational redistribution is suggested by the absence of hot bands in the resonance fluorescence spectrum of anthracene vapor. The fluorescence intensity in the hot band region is about 1% of the intensity of the normal fluorescence when the excitation wavenumber is tuned into the 0–2 band of the main vibrational progression, i.e., about 3000 cm^{-1} above the origin of the $S_0 \rightarrow S_1$ transition. The measurements were performed under continuous wave conditions at a temperature of 380 K (de Vries *et al.*, to be published). At this temperature, the thermal energy which is transferred from the lower to the upper electronic state is on the average about 2500 cm^{-1}, so that the mean vibrational energy in the excited molecules is about 5500 cm^{-1}. With this excess energy the mean occupation number of a single 1500 cm^{-1} mode is 2×10^{-2} if the total vibrational energy is completely randomized. This number is consistent with the observed relative intensity of the hot band emission, which implies that approximately 1% of the excited molecules is in a vibrational quantum state $v_s' = 1$ of an optically active mode.

Time-resolved measurements of the resonance fluorescence of anthracene at the same temperature and with the same excitation wavenumber have been reported by de Vries *et al.* (1977). They have used excitation pulses with a duration between 5 and 10 psec, obtained by mode locking of a jet stream dye laser which was synchronously pumped with a mode-locked argon ion laser (de Vries *et al.*, 1976). The bandwidth of the frequency doubled pulses varied between 2 and 3 cm^{-1}. The emission signals were measured with a time resolution of 300 psec by using a single-photon counting detection system. The emission was characterized by a monoexponential decay with a decay time of 5 nsec in the frequency region of the hot bands as well as in the normal fluorescence region. The relative intensities in both emission regions are similar to those measured under continuous wave conditions. No trace of a short component in the hot band region was observed.

From these results it is concluded that the harmonic states, situated 4000–6000 cm^{-1} above the vibrationless S_1 state, are deactivated within a period of less than 10 psec. This conclusion is based on the assumption that such a harmonic state would give rise to similar fluorescence intensities in both spectral regions. The rapid decay of the optically excited vibronic state implies that the width of the harmonic states in this energy region measures at least a few wavenumbers.

Anthracene may be a suitable molecule for further studies of intramolecular vibrational relaxation. A combination of frequency-resolved and time-resolved measurements of the fluorescence of this compound, if introduced in a supersonic molecular beam, might provide interesting information about the redistribution of vibrational energy following optical excitation of single vibronic levels.

4. Pyrene

The fluorescence decay characteristics of isolated pyrene-h_{10} and pyrene-d_{10} have been studied by Werkhoven *et al.* (1971, 1973, 1975a, b). They have found that the time-resolved fluorescence can be analyzed in terms of a *biexponential* decay when the excitation proceeds via a vibronically induced absorption band of the $S_0 \rightarrow S_1$ transition. The decay time of the short component was about four times shorter than that of the long component. Such a short component was observed only after excitation into vibronically induced bands. On the other hand, a *monoexponential* decay was observed when the excitation was achieved via the 0–0 band of the weakly allowed $S_0 \rightarrow S_1$ transition or via the $S_0 \rightarrow S_2$ transition. It was concluded from these different decay characteristics that the observed biexponential decay is not due to sequence congestion effects.

The short component has been interpreted as emission from an optically excited vibronic state prior to redistribution of vibrational energy. The long-lived part of the fluorescence was ascribed to emission from the vibrationally redistributed states which have smaller oscillator strengths to the electronic ground state than the initial vibronic state. The different radiative properties of the optically excited and redistributed levels imply the observability of the redistribution process. The absence of a short component in the fluorescence decay of 3-methylpyrene is in accordance with this interpretation, since the radiative rate constants of different vibronic levels of this compound are nearly equal (Werkhoven *et al.*, 1975a, b). The decay rates measured for the long component of the pyrene fluorescence and the decay constants of the exponential decay which is observed after excitation above the S_2 threshold or into the 0–0 band of the $S_0 \rightarrow S_1$ transition increase monotonically with the excess vibrational energy. The dependence of these decay rates on the vibrational energy is in good agreement with calculations based on the assumption that the vibrational energy is completely randomized (Werkhoven *et al.*, 1975b).

The ratio of the rate constants of the short and the long component of the fluorescence decay is at least four times larger than the upper limit that has been calculated for this ratio. This calculation was based on the assumption that the short component corresponds to an initially excited vibrational mode which acts as an inducing mode for the radiative transition and also as a promoting mode for the electronic relaxation, whereas the long component originates from completely randomized isoenergetic vibronic states. The difference between the time constants of both decay components has been ascribed to vibrational redistribution with a rate constant of 10^7–10^8 sec^{-1}.

The excitation of the pyrene vapor into the S_1 manifold was achieved with the second harmonic of a ruby laser (28,800 cm^{-1}) or the third harmonic of a neodymium glass laser (28,300 cm^{-1}), both with a pulse duration of about 3 nsec. A review of this work has been given by van Voorst (1975).

Similar experiments have been carried out recently by Ehrlich and Wilson (1977) who used the third harmonic of a neodymium glass laser for the excitation of the vapor. They observed a growth of the S_1 fluorescence of pyrene-h_{10} with a rise time of 21 nsec subsequent to excitation with a 300 psec pulse and they have ascribed the effect to vibrational redistribution. However, the effects of stray light from the pump pulse and ringing of the detection system cannot be left out of consideration.

The studies of Werkhoven *et al.* have been continued by de Vries *et al.* (to be published) using pulsed tunable dye lasers with narrow bandwidths in combination with a single photon counting system, or a transient digitizer as part of a waveform processing device. This equipment allows time-resolved measurements of the fluorescence decay with a high degree of accuracy. It has been found that the decay characteristics are strongly dependent on the excitation wavenumber even when the wavenumber is varied within a single vibronic absorption band. These effects could not be observed in the preceding experiments since the excitation sources were not tunable. The observed fluorescence decay appears to be multiexponential, both when the excitation wavenumber is tuned to the 0–0 band of the $S_0 \rightarrow S_1$ transition and for energies above the S_2 threshold. The experimental results can be explained on the basis of sequence congestion effects if it is assumed that the fluroescence decay rate depends linearly on the vibrational energy. No indications of any incomplete vibrational redistribution on a nanosecond time scale emerge from these studies.

5. Further Aromatic Molecules

In the following we will focus our attention on some experimental studies dealing with intramolecular vibrational redistribution in a number of different large molecules. Interesting results have been obtained by Lim and co-workers, who studied the decay characteristics of various electronically excited aromatic molecules. They have found discontinuities in the energy dependence of the nonradiative decay rate of the lowest excited singlet state k_{nr}, as well as discontinuities in the magnitude of k_{nr}, when the excitation energy crosses the threshold of the next electronic manifold. The case of naphthalene has been presented as an example in Section IIIB2. Another example is anthracene (Section IIIB3). These results have been considered as an indication of an incomplete redistribution of vibrational energy on the time scale of the fluorescence. If the excess energy does not redistribute randomly among the vibrational modes of the molecule, the distribution of vibrational quanta in the fluorescent electronic state will depend on the method of preparation, i.e., the molecule "remembers" the steps involved in the preparation of the final electronically excited state. This implies that k_{nr} may have different magnitudes for molecules which are directly excited to the S_1 state as compared to molecules which are excited to S_1 via internal conversion from the S_2 state.

Results similar to those for naphthalene and anthracene have been obtained for β-naphthol and β-naphthylamine (Hsieh *et al.*, 1974) and fluorene (Huang *et al.*, 1976). The measurements indicate a remarkable change of k_{nr} when the photon energy crosses the threshold for S_2 excitation, and the experimental data suggest that the excess vibrational energy is not completely randomized within 10 nsec after optical excitation at 4000–6000 cm^{-1} above the vibrationless S_1 state.

In these studies discontinuities in the decay rate have been found only near the excitation energy of the S_2 state. No significant discontinuities were observed when the excitation energy surpasses the threshold of a higher electronic state. This result is not surprising since above the S_2 threshold the final states are produced by internal conversion, so that the method of preparation of the fluorescent states does not change essentially. Furthermore, at high excitation energies the density of vibronic states in the S_1 manifold is so large that an efficient redistribution of vibrational energy due to anharmonic and Coriolis interactions may be expected.

Comparable results have been reported for 3,6-diaminophthalimide and 3,6-tetramethyldioamino phthalimide (Borisevich, 1968). A discontinuous change of the energy dependence of the fluorescence quantum yield, measured under conditions of isolation, has been ascribed to an incomplete randomization of vibrational energy on the time scale of the fluorescence.

In larger molecules an efficient randomization of vibrational energy is to be expected at lower excess energies, since the density of vibrational states increases rapidly with the size of the molecule. Okajima and Lim (1976) have measured the energy dependence of the decay rate and the quantum yield of the resonance fluorescence from tetracene and pentacene vapors. It appears that in tetracene k_{nr} initially increases rapidly with the excitation energy, whereas at higher photon energies the energy dependence of k_{nr} is much less. The incremental growth of k_{nr} is already low at excitation energies below that of the second electronic transition. In this case k_{nr} does not show any remarkable change when the threshold of the second electronic manifold is crossed. The excess vibrational energy in the S_1 manifold at the position of the S_2 threshold is about 10,000 cm^{-1}.

The energy dependence of k_{nr} observed for tetracene can be explained if it is assumed that the optically excited vibrational states of the S_1 manifold can be considered as pure C–C stretching modes at low excess energies ε, whereas at higher energies a substantial mixing of harmonic states occurs. (Okajima and Lim, 1976; Freed, 1976 a, b). The prominent vibrational structure of the $S_0 \rightarrow S_1$ absorption spectrum is due to the excitation of totally symmetric C–C stretching modes. The excitation of these modes leads to a significant growth of k_{nr} with ε (Heller *et al.*, 1972; Fleming *et al.*, 1975). Since the density of vibrational states in the S_1 manifold grows rapidly with ε, the effects of anharmonic mixing are enhanced considerably at higher energies so that the C–C stretching modes contribute only little to the composition of the individual vibrational eigenstates. The major component of these states probably originates from low-frequency bending modes because of their relatively high contribution to the total density of vibrational states. Excitation of the bending modes generally does not produce a substantial increase of k_{nr}, and consequently a strong intrastate mixing of vibronic states at higher excitation energies will give rise to a weak energy dependence of k_{nr}. For the same reason a weak energy dependence of k_{nr} is to be expected when the S_1 vibronic levels are populated by internal conversion at sufficiently high excess energies.

According to Okajima and Lim (1976) the nonradiative decay rate of pentacene initially decreases with ε and then slowly increases at higher photon energies. The decay rates measured at high excitation energies (1.7×10^8 sec^{-1} near 39,000 cm^{-1}) are in accordance with observations of de Vries (1977). However, at low excitation energies (19,000–22,000 cm^{-1}) de Vries has found a constant value of the decay rate (1.0×10^8 sec^{-1}) instead of the decreasing values reported by Okajima and Lim. Anyhow, the observations of both de Vries, and Okajima and Lim can be taken as evidence for an efficient vibrational redistribution in pentacene on a nanosecond time scale, even at low excess energies.

This interpretation is not in agreement with the opinion of Soep and collaborators, who have come to the contradictory conclusion that no complete vibrational redistribution takes place on a microsecond time scale in the electronic ground state of pentacene at excess energies of more than 20,000 cm^{-1}. Soep (1975) has studied the $S_1 \rightarrow S_0^{**}$ internal conversion in pentacene under collision-free conditions by measuring the transient absorption of the vapor after a pulsed excitation of the S_1 state. This transient absorption is ascribed to $S_0^{**} \rightarrow S_1^*$ transitions originating from hot ground state levels produced by $S_1 \rightarrow S_0^{**}$ internal conversion. The transient absorption is strongly red shifted with respect to the normal $S_0 \rightarrow S_1$ absorption and it is quenched by collisions at an almost gas kinetic rate.

Sander *et al.* (1976) have continued this work using an effusive molecular beam which brings about a considerable increase of the collision-free time. Instead of a transient absorption technique to detect the hot S_0^{**} molecules a more sensitive fluorescence technique was applied. The hot ground state molecules were excited into the S_1 manifold with a second pulse from a tunable dye laser and the intensity of the transient fluorescence was measured for various delay times between the pump and the probe pulses (1–10 μsec). The transient excitation spectrum is red shifted with respect to the normal $S_0 \rightarrow S_1$ absorption spectrum by approximately 3000 cm^{-1}.

It is assumed that the initial excitation energy is converted primarily to high frequency accepting modes such as the symmetric C–H stretching vibrations ($\simeq 3000\ \mathrm{cm}^{-1}$), but also to symmetric C–C stretching modes ($\simeq 1400\ \mathrm{cm}^{-1}$). The red shifted transient absorption and excitation spectra are ascribed to absorption originating from excited C–C stretching vibrations, which are assumed to be optically active in the $S_0^{**} \rightarrow S_1^*$ transition. According to Sander *et al.* a complete randomization of the vibrational energy in the hot ground state molecules would have caused much weaker transient spectra. This implies an incomplete redistribution of an excess vibrational energy of approximately $20{,}000\ \mathrm{cm}^{-1}$ on a microsecond time scale.

This interpretation has been challenged by Frank-Kamenetskii *et al.* (1977). They assume a complete vibrational redistribution after the internal conversion to the electronic ground state and they ascribe the observed red shifts entirely to distorted vibrations. This view implies rather high values of $\Delta\omega$, the difference between the vibrational frequency in the upper and lower electronic state.

An alternative explanation of the transient spectra, observed by Soep and collaborators, has been suggested by de Vries (1977). He has calculated ρ, the total density of vibrational states of pentacene, as well as ρ_j, the partial density of states, both as a function of the vibrational energy. The quantity ρ_j is defined as the density of vibrational states in which a $1400\ \mathrm{cm}^{-1}$ mode j is excited. The vibrational frequencies of pentacene are not known, and therefore it has been assumed that the frequency distributions of the normal modes of pentacene and anthracene are similar. The vibrational frequencies of anthracene have been taken from the literature. The calculations show that in an ensemble of pentacene molecules, each containing $30{,}000\ \mathrm{cm}^{-1}$ of vibrational energy ($20{,}000\ \mathrm{cm}^{-1}$ as a result of internal conversion in addition to a thermal energy of $10{,}000\ \mathrm{cm}^{-1}$), more than 10% of the excited molecules will contribute to the transient spectra, if the vibrational energy is completely randomized. From this point of view the observed red shifts of about $3000\ \mathrm{cm}^{-1}$ are due to distorted vibrations and to hot transitions $v_j'' \geq 1 \rightarrow v_j' = 0$. More time-resolved experimental data are required to resolve this question.

C. Vibrational Redistribution in the Electronic Ground State

In recent years numerous experimental investigations have been carried out dealing with unimolecular reactions following a nonrandom excitation of the molecules. Some of these studies provide information about the dynamics of the intramolecular vibrational energy flow. In general it appears that intramolecular energy transfer proceeds extremely rapidly, although some examples have been found in which the energy is not statistically distributed in the molecule before the reaction takes place. Along with such experimental studies, various computer studies of unimolecular decomposition reactions have been performed.

The experimental and theoretical results have been discussed in several review papers (e.g., Rice, 1975; Hase, 1976; Gelbart, 1977). Usually these studies are concerned with highly excited molecules and the results of most of them suggest a picosecond time scale for the vibrational redistribution process.

The discovery of isotopically selective multiphoton dissociation of polyatomic molecules in an intense infrared laser field has led to numerous studies of highly excited molecules in the electronic ground state. The experimental work has been performed primarily on SF_6. An interesting aspect of the multiphoton absorption process is the compensation of the anharmonic shift. The transitions between the higher levels of the pumped vibrational mode are not in resonance with the laser field and therefore the dissociation energy cannot be reached by successive excitations of a single vibrational mode. For example, the anharmonic shift of the ν_3 mode of SF_6 is approximately 5 cm^{-1}/1000 cm^{-1} (Kildal, 1977). Dissociation of the SF_6 molecules requires approximately 30 photons ($\hbar\omega_3 =$ 948 cm^{-1}). The highly excited molecules can be prepared by multiphoton absorption if the energy, which is initially localized within the pumped vibrational mode, is rapidly transferred to the other vibrational degrees of freedom of the molecules, thus compensating the anharmonicity of the ν_3 mode.

Arguments in favor of this picture can be derived from the results of multiphoton dissociation experiments in a molecular beam formed by expansion of pure SF_6 (Coggiola *et al.*, 1977; Brunner and Proch, 1978) and also from measurements of the absorption cross section of SF_6 as a function of the laser fluence (Black *et al.*, 1977). The competition between vibrational redistribution and the absorption process determines to what extent the vibrational energy is redistributed among the 15 vibrational modes of SF_6 prior to its decomposition.

Attempts to measure the vibrational redistribution rate in SF_6 have been made by Frankel (1976) and by Deutsch and Brueck (1978). These authors have carried out infrared–infrared double resonance measurements of the ν_3 absorption band. The pump pulse was obtained from a mode-locked CO_2 laser, and a grating tunable cw CO_2 laser was used to probe the absorption intensity subsequent to the absorption of the pump pulse. The induced transient absorption was probed at different wavenumbers. The results have been ascribed to the energy flow from the pumped ν_3 mode to the other vibrational modes. It has been concluded that the intramolecular vibrational energy transfer in these experiments was taking place primarily in the absence of collisions. The results of Frankel suggest that the energy randomization proceeds on a nanosecond time scale at energies of 1000 cm^{-1} or more above the ground state, whereas according to Deutsch and Brueck the time constant of the collisionless vibrational redistribution is 3 μsec at an energy of approximately 3000 cm^{-1}.

In view of recent experiments carried out by Kwok *et al.* (1978), these data are questionable, the observed effects probably due to collisional interactions. Kwok and collaborators have measured the recovery of the absorption saturation subsequent to the excitation of the ν_3 mode with a powerful picosecond laser pulse from a CO_2 laser. It was assumed that the overall anharmonic shift is approximately 2.6 cm^{-1}/1000 cm^{-1} if the vibrational energy is completely randomized. This number was determined from the temperature shift of the ν_3 absorption band (Nowak *et al.*, 1975). Kwok *et al.* have found an upper limit of 30 psec for the redistribution time. A lower limit of 1 psec follows from the observed linewidth in the absorption spectrum of heated SF_6 (Nowak *et al.*, 1975).

A very interesting experimental study has been reported by Maier *et al.* (1977), who measured the decay of an excited vibrational level of the electronic ground state of coumarin 6 under collision-free conditions. The coumarin vapor ($C_{20}H_{18}O_2N_2S$) was excited by an ultrashort infrared pulse with a frequency of 5950 cm^{-1}, a bandwidth of 60 cm^{-1}, and a duration of 3 psec. The time development of the vibrationally excited molecules was probed by means of a double resonance technique. A second ultrashort pulse of higher frequency (18,910 cm^{-1}) promotes the vibrationally excited molecules to the fluorescent singlet state. The sum frequency of the infrared and the green pulse is situated near the maximum of the diffuse first band of the electronic absorption spectrum of the vapor. The intensity of the fluorescence was measured as a function of the time interval between both pulses. The plot of the fluorescence intensity versus the delay time is characterized by a rise time which is similar to the duration of the infrared pulse and by an exponential decay with a time constant of about 4 psec.

The observed decay must be due to intramolecular processes since the average time between successive collisions is in the order of 100 nsec. The fluorescence is interpreted as due to hot band excitation originating from the primarily excited vibrational state. If the vibrational states which are formed by the intramolecular relaxation do not contribute to the hot band excitation of the fluorescent state, the observed decay rate of 2.5×10^{11} sec^{-1} may be interpreted as a vibrational redistribution rate. The average thermal energy of the molecules must be at least 10,000 cm^{-1} and may be significantly more than that, and therefore the average energy of the vibrationally excited molecules is on the order of 20,000 cm^{-1}.

D. Vibrational Predissociation of van der Waals Complexes

Spectroscopic studies of van der Waals molecules generated in a supersonic free jet expansion are beginning to provide dynamical information about the redistribution of vibrational energy. These complexes seem to be suitable prototype systems for studies of unimolecular dissociation reactions and the intramolecular vibrational redistribution process. In general, vibrational excitation of a molecule which is a part of such a weakly bound complex will give rise to dissociation of the complex as a result of vibrational energy transfer within the van der Waals molecule. Usually the dissociation energy of the van der Waals bond is considerably smaller than a single vibrational quantum of the molecular vibration. Dixon and Herschbach (1977) have found that the van der Waals dimer $Cl_2 \cdot Cl_2$ containing vibrational energy significantly in excess of the dissociation energy of the complex can survive for at least 10^8 vibrational periods of the molecular Cl–Cl mode. They predict long vibrational predissociation lifetimes for several other van der Waals molecules.

Very interesting results have been obtained recently by Smalley and collaborators (Smalley *et al.*, 1976, 1978; Kim *et al.*, 1976). They have studied the vibrational predissociation of electronically excited $He \cdot I_2$. The rate constants for dissociation following the excitation of vibronic levels of the $B^3\Pi$ state were derived from linewidths of the rotational substructure of vibronic bands in the fluorescence excitation spectrum. These rate constants are in the order of 10^{10} sec^{-1} and depend slightly on the vibrational quantum number of the molecular I–I mode. The coupling between the I_2 vibration and the dissociative vibrational mode of the complex is so weak that the complex remains bound for about 10^3 vibrational periods of I_2, in spite of the fact that the energy initially present in the molecular vibration exceeds the dissociation energy of the complex by at least a factor of 10. A theoretical discussion of these results has been presented by Beswick and Jortner (1978a, c).

Up to now, the most detailed information about vibrational predissociation of Van der Waals molecules has been obtained for $He \cdot I_2$. For similar electronically excited complexes only approximate values of the photodissociation lifetimes have been deduced from the linewidths in their fluorescence excitation spectra, that is, approximately 10^{-11} sec for $He \cdot NO_2$ (Smalley *et al.*, 1977b) and a lower limit of 3×10^{-11} sec for the vibronic state $v' = 15$ of $Ar \cdot I_2$ (Kubiak *et al.*, 1978). Recently, Johnson *et al.* (1978) have reported lifetimes of a variety of vibronic levels of $He \cdot I_2$. The predissociation lifetimes change gradually from 221 psec for $v' = 12$ to 38 psec for $v' = 26$. This observed v' dependence is perfectly described by a theory of Beswick *et al.* (1979) for vibrational predissociation of a T-shaped van der Waals complex, $He \cdot I_2$.

IV. CONCLUDING REMARKS

Experimental studies of intramolecular vibrational relaxation are sparse and the conclusions emerging from these studies are sometimes contradictory. The discrepancies between the results of different experiments emphasize strongly the importance of the experimental conditions in spectroscopic studies of large molecules.

The observability of the redistribution process implies different spectroscopic characteristics of the vibronic states which are involved in the redistribution process. These differences are associated with the s and l components of the coherently excited quasi-stationary states. The best evidence for the occurrence of the intramolecular vibrational relaxation process subsequent to optical excitation would be provided by the observation of both the decay of some optically prepared level *s* and the concomitant built up of a population of levels $\{l\}$. Such experiments are difficult for several reasons.

Vibrational redistribution is an intrastate relaxation process and therefore the differences in the spectral properties of the *s* and *l* components are less pronounced than in the case of an electronic relaxation process where *s* and *l* belong to different electronic manifolds. In most of the molecules which have been studied so far, the coupling between the optically active modes and the inactive modes is probably so strong that extremely short light pulses are required to allow the observation of a time dependent vibrational redistribution. This condition is less restrictive in the case of complexes which are bound in part only by van der Waals forces (Section IIID). Another difficulty has been discussed in Section IIIA. Since all experiments involving large molecules have been carried out at conventional temperatures, the available experimental data are obscured by sequence congestion effects.

The cumbersome effects of sequence congestion can be eliminated if the experiments are performed with a seeded supersonic beam. By means of this technique the vibrational and rotational degrees of freedom of a large molecule in the vapor phase can be cooled down to very low effective temperatures. The beam is produced by expanding a carrier gas containing a small fraction of the molecules of interest (the seed) through a nozzle. In this manner a nearly perfect relaxation of the rotational degrees of freedom can be established. For example, Smalley *et al.* (1977a) have obtained a rotational temperature of less than 1 K for *s*-tetrazine in a supersonic beam with helium as the carrier gas. The linewidth of the vibronic bands in the fluorescence excitation spectrum of *s*-tetrazine appeared to be as narrow as 1.7×10^{-2} cm^{-1} which is consistent with lifetimes measured by de Vries *et al.* (1976) at a temperature of 300 K. The spectra observed by Smalley *et al.* exhibit a well-resolved rotational structure.

The vibrational relaxation in the expanding gas proceeds less efficiently than the rotational relaxation. Generally, the effective vibrational temperature of the seed is considerably higher than the temperature of the carrier gas after the expansion. Typical values of the vibrational temperature are 20–40 K. Nevertheless, sequence congestion effects may be eliminated or at least considerably reduced. In this respect also the narrowing of the absorption bands due to the low rotational temperature is of great importance, since this reduces substantially the number of overlapping vibronic bands.

Interesting results are to be expected when supersonic nozzles will be used in combination with sophisticated spectroscopic techniques. The narrow rotational contours of the absorption bands allow the excitation of individual vibronic levels of a large molecule under collision-free conditions. These levels will usually be associated with mixed harmonic states. Evidence of the anharmonic couplings can be obtained from a thorough analysis of the vibrational structure of the fluorescence spectrum and the fluorescence excitation spectrum. Especially the hot band region of the emission spectra would be of interest. It will not be easy to accomplish time-resolved measurements of these spectra, but such experiments are not impossible.

The application of spectroscopic methods based on coherent optical effects might be another prospect for future investigations of intramolecular vibrational relaxation. Recently Jortner and Kommandeur (1978) pointed out the applicability of such techniques to studies of vibrational redistribution in large molecules under collision-free conditions. Measurements of optical-free induction decay and photon echo experiments may provide information about the dynamics of intramolecular dephasing, provided that the excitation process satisfies the condition $\tau_p < \hbar\Delta_s^{-1}$ where Δ_s is the width of the coherently excited initial state of the molecule.

Spectroscopic studies of vibrational redistribution have been devoted primarily to large molecules with moderate excess energies. Some of these studies, discussed in Section IIIB, suggest that the randomization of the vibrational energy following internal conversion is not complete on a nanosecond time scale when the excess energy is relatively low. The density of vibronic states ρ in this energy range is in the order of, say, $10^6/\text{cm}^{-1}$. With this value of ρ, redistribution times of more than 10 nsec could be expected if the effective width of the distribution of anharmonic coupling elements is less than $10^{-5}\ \text{cm}^{-1}$ [Eq. (5)]. In Section II the suggestion was made that in general a few large coupling elements may exist for specific vibrational modes which can give rise to observable Fermi resonances, whereas most of the coupling elements are considerably smaller. The apparently slow vibrational redistribution, for example in the case of naphthalene (Section IIIB2), may be due to a relatively weak coupling between certain C–C and C–H modes. It has been shown by Fischer *et al.* (1974) that the experimental results obtained for naphthalene-h8 and -d8 are not compatible with the communicating states model which assumes a very rapid redistribution of vibrational energy, as compared to electronic relaxation processes, so that k_{nr} should be independent of the specific nature of the excitation process, and therefore k_{nr} should depend only on the excess vibrational energy in the fluorescent state. The calculations of Fischer and collaborators appear to be consistent with the experimental data if it is assumed that the excess energy is only partially redistributed among the accessible vibrational degrees of freedom.

At present, the only examples of spectroscopic studies providing direct information about the dynamics of intramolecular vibrational redistribution seem to be coumarin 6 and a few van der Waals complexes. Coumarin 6 is probably not the most appropriate molecule for such a study because of its complicated structure. An extension of this work to other molecules with simpler spectroscopic features looks promising. These molecules should contain vibrational modes that are infrared active as well as optically active with respect to the $S_0 \rightarrow S_1$ transition. Such investigations will certainly be stimulated by the availability of pulsed tunable infrared sources which presently permit a time-resolution of less than 1 psec.

Vibrational predissociation studies of the van der Waals molecule $He \cdot I_2$ have provided interesting information about the dynamics of the transfer of vibrational energy between the molecular vibration of I_2 and the dissociative mode of the complex. Such studies may yield important information about the mechanism of the redistribution process, especially when polyatomic molecules are involved. Recently, Beswick and Jortner (1978b) have pointed out the interesting implications of vibrational predissociation studies of van der Waals complexes $X \cdot M$, consisting of a rare gas atom X and a polyatomic molecule M, for the study of vibrational redistribution in M. Various weakly bound molecular complexes can be prepared by means of the supersonic expansion technique. It is to be expected that future investigations of such complexes will provide experimental data that will be very useful for the understanding of vibrational relaxation in isolated molecules.

APPENDIX

Recently interesting information about vibrational redistribution in naphthalene has been obtained by Beck *et al.* (1979). These authors have reported measurements of the fluorescence excitation spectrum of isolated naphthalene cooled in a supersonic free jet with helium as the carrier gas.

The excitation spectra have been measured with high spectral resolution: the sum of the spectral bandwidth of the laser and the effective Doppler width was 0.05 cm^{-1}. The width of the narrowest rotational substructure that could be clearly resolved was 0.1 cm^{-1}. Such narrow spectral features have been observed only if the excess vibrational energy is less than 2400 cm^{-1}. At higher energies the vibronic bands are significantly broadened and the width of the bands increases monotonically with the vibrational energy. This broadening has been interpreted by Beck *et al.* as due to intramolecular redistribution of the excess vibrational energy. Other processes than vibrational redistribution within the optically excited singlet manifold can not be considered as responsible for the observed broadening because of the high quantum yield and the long lifetime of the fluorescence.

From an analysis of the band contours, Beck *et al.* have derived vibrational redistribution rates varying from 9×10^{10} to 4×10^{11} sec^{-1} for vibrational energies in the range from about 3100 up to 4300 cm^{-1}.

In Section IIIB2 a study of Hsieh *et al.* (1974) was mentioned which gives rise to the conclusion that in naphthalene molecules with excess vibrational energy of approximately 5000 cm^{-1} vibrational redistribution is not complete within approximately 100 nsec. This conclusion is not in contradiction with the results obtained by Beck *et al.* since a part of the vibrational modes may be very weakly coupled with the vibrations that are involved in the optical excitation process. More detailed information about the redistribution of vibrational energy in naphthalene may be provided by measurements of fluorescence spectra obtained by excitation of single vibronic levels. Beck *et al.* have announced a further study, to include time-resolved measurements.

The author is sincerely indebted to Dr. J. de Vries for commenting on the manuscript and for many helpful discussions. Also some stimulating and valuable discussions with Dr. A. Tramer are greatly acknowledged. The author wishes to acknowledge Dr. T. Deinum and Professor E. C. Lim for agreeing to the use of Figs. 4 and 5.

REFERENCES

Beck, S. M., Monts, D. L., Liverman, M. G., and Smalley, R. E. (1979). *J. Chem. Phys.* **70**, 1062–1063.

Beswick, J. A., and Jortner, J. (1978a). *J. Chem. Phys.* **68**, 2277–2297.

Beswick, J. A., and Jortner, J. (1978b). *J. Chem. Phys.* **68**, 2525.

Beswick, J. A., and Jortner, J. (1978c). *J. Chem. Phys.* **69**, 512–518.

Beswick, J. A., Delgado-Barrio, G., and Jortner, J. (1979). *J. Chem. Phys.* **70**, 3895–3901.

Black, J. G., Yablonovitch, E., Bloembergen, N., and Mukamel, S. (1977). *Phys. Rev. Lett.* **38**, 1131–1134.

Blondeau, J. M., and Stockburger, M. (1971). *Ber. Bunsenges. Phys. Chem.* **75**, 450–455.

Borisevich, N. A. (1968). *In* "Elementary Photoprocesses in Molecules" (B. S. Neporent, ed.), pp. 39–50. Plenum Press, New York.

Brunner, F., and Proch, D. (1978). *J. Chem. Phys.* **68**, 4936–4940.

Byrne, J. P., and Ross, I. G. (1971). *Aust. J. Chem.* **24**, 1107–1141.

Coggiola, M. J., Schulz, P. A., Lee, Y. T., and Shen, Y. R. (1977). *Phys. Rev. Lett.* **38**, 17–20.

Deinum, T. (1974). Thesis, Univ. of Amsterdam.

Deinum, T., Werkhoven, C. J., Langelaar, J., Rettschnick, R. P. H., and van Voorst, J. D. W. (1974a). *Chem. Phys. Lett.* **27**, 206–209.

Deinum, T., Werkhoven, C. J., Langelaar, J., Rettschnick, R. P. H., and van Voorst, J. D. W. (1974b). *Chem. Phys. Lett.* **27**, 210–213.

Deutsch, T. F., and Brueck, S. R. J. (1978). *Chem. Phys. Lett.* **54**, 258–264.

de Vries, J. (1976). Private communication.

de Vries, J., Bebelaar, D., and Langelaar, J. (1976). *Opt. Commun.* **18**, 24–26.

de Vries, J. (1977). Private communication.

de Vries, J., Rettschnick, R. P. H., Langelaar, J., and van Voorst, J. D. W. (1977). *Conf. Digest General Disc. Meeting Radiationless Processes, Breukelen* p. 43.

de Vries, J., Langelaar, J., van Voorst, J. D. W., and Rettschnick, R. P. H. (to be published).

Dixon, D. A., and Herschbach, D. R. (1977). *Ber. Bunsenges. Physik. Chem.* **81**, 145–150.

Ehrlich, D. J., and Wilson, J. (1977). *J. Chem. Phys.* **67**, 5391–5392.

Fischer, S. F., and Lim, E. C. (1974). *Chem. Phys. Letters* **26**, 312–317.

Fischer, S. F., Stanford, A. L., and Lim, E. C. (1974). *J. Chem. Phys.* **61**, 582–593.

Fischer, G., Sharf, B., and Parmenter, C. S. (1975). *Mol. Phys.* **29**, 1063–1072.

Fleming, G. R., Gijzeman, O. L. J., and Lin, S. H. (1974). *J. Chem. Soc. Faraday Trans. II* **70**, 37–44.

Fleming, G. R., Lewis, C., and Porter, G. (1975). *Chem. Phys. Lett.* **31**, 33–36.

Frankel, D. S., Jr., (1976). *J. Chem. Phys.* **65**, 1696–1699.

Frank-Kamenetskii, M. D., Lukashin, A. V., and Puretskii, A. A. (1977). *Chem. Phys. Lett.* **45**, 583–585.

Freed, K. F. (1976a). *In* "Topics in Applied Physics," Vol. 15 (F. K. Fong, ed.), pp. 23–168. Springer-Verlag, Berlin and New York.

Freed, K. F. (1976b). *Chem. Phys. Lett.* **42**, 600–606.

Gelbart, W. M. (1977). *Ann. Rev. Phys. Chem.* **28**, 323–348.

Haebig, J. E. (1968). *J. Mol. Spectrosc.* **25**, 117–120.

Hase, W. L. (1976). *In* "Dynamics of Molecular Collisions" (W. H. Miller, ed.), Part B, pp. 121–169. Plenum Press, New York.

Heller, D. F., Freed, K. F., and Gelbart, W. M. (1972). *J. Chem. Phys.* **56**, 2309–2328.

Hsieh, J. C., Huang, C-S., and Lim, E. C. (1974). *J. Chem. Phys.* **60**, 4345–4353.

Huang, C-S., Hsieh, J. C., and Lim, E. C. (1974). *Chem. Phys. Lett.* **28**, 130–134.

Huang, C-S., Hsieh, J. C., and Lim, E. C. (1976). *Chem. Phys. Lett.* **37**, 349–352.

Jacon, M., Lardeux, C., Lopez-Delgado, R., and Tramer, A. (1977). *Chem. Phys.* **24**, 145–157.

Johnson, K. E., Wharton, L., and Levy, D. H. (1978). *J. Chem. Phys.* **69**, 2719–2724.

Jortner, J., and Kommandeur, J. (1978). *Chem. Phys.* **28**, 273–283.

Jortner, J., and Mukamel, S. (1974). *In* "The World of Quantum Chemistry" (R. Daudel and B. Pullman, eds.), pp. 145–209. Reidel, Dordrecht.

Kay, K. G. (1974). *J. Chem. Phys.* **61**, 5205–5220.

Kildal, H. (1977). *J. Chem. Phys.* **67**, 1287–1288.

Kim, M. S., Smalley, R. E., Wharton, L., and Levy, D. H. (1976). *J. Chem. Phys.* **65**, 1216–1217.

Klochkov, V. P., and Smirnova, T. S. (1967). *Opt. Spectrosc.* **22**, 464–465.

Knight, A. E. W., Parmenter, C. S., and Schuyler, M. W. (1975). *J. Am. Chem. Soc.* **97**, 1993–2005, 2005–2013.

Kubiak, G., Fitch, P. S. H., Wharton, L., and Levy, D. H. (1978). *J. Chem. Phys.* **68**, 4477–4480.

Kwok, H. S., and Yablonovitch, E. (1978). *In* "Picosecond Phenomena" (C. V. Shank, E. P. Ippen, and S. L. Shapiro, eds.), pp. 218–223. Springer-Verlag, Berlin and New York.

Lahmani, F., Tramer, A., and Tric, C. (1974). *J. Chem. Phys.* **60**, 4431–4447.

Laor, U., Hsieh, J. C., and Ludwig, P. K. (1973). *Chem. Phys. Lett.* **22**, 150–153.

Maier, J. P., Seilmeier, A., Laubereau, A., and Kaiser, W. (1977). *Chem. Phys. Lett.* **46**, 527–530.

Mirumyants, S. O., and Demchuk, Yu. S. (1976). *Opt. Spectrosc.* **40**, 23–27.

Mirumyants, S. O., Vandyukov, E. A., Demchuk, Yu. S., and Nagulin, Yu. S. (1974). *Opt. Spectrosc.* **36**, 52–54.

Mirumyants, S. O., Vandyukov, E. A., and Demchuk, Yu. S. (1975). *Opt. Spectrosc.* **38**, 25–27.

Nitzan, A., and Jortner, J. (1972). *Chem. Phys. Lett.* **13**, 466–472.

Nordholm, K. S. J., and Rice, S. A. (1974). *J. Chem. Phys.* **61**, 203–223, 768–779.

Nowak, A. V., and Lyman, J. L. (1975). *J. Quant. Spectrosc. Radiat. Transfer* **15**, 945–961.

Okajima, S., and Lim, E. C. (1976). *Chem. Phys. Lett.* **37**, 403–407.

Parmenter, C. S. (1972). *Adv. Chem. Phys.* **22**, 365–421.

Phillips, D. (1978). Private communication.

Pringsheim, P. (1938). *Ann. Acad. Sci. Tech. Varsovie* **5**, 29–42.

Rhodes, W. (1974). *Chem. Phys.* **4**, 259–268.

Rice, O. K. (1930). *Z. Phys. Chem. B.* **7**, 226–233.

Rice, S. A. (1975). *In* "Excited States" (E. C. Lim, ed.), Vol. II, pp. 111–320. Academic Press, New York.

Rice, O. K., and Ramsperger, H. C. (1928). *J. Am. Chem. Soc.* **50**, 617–620.

Rockley, M. G., and Phillips, D. (1974). *J. Photochem.* **3**, 365–382.

Sander, R. K., Soep, B., and Zare, R. N. (1976). *J. Chem. Phys.* **64**, 1242–1243.

Smalley, R. E., Levy, D. H., and Wharton, L. (1976). *J. Chem. Phys.* **64**, 3266–3276.

Smalley, R. E., Wharton, L., and Levy, D. H. (1977a). *J. Mol. Spectrosc.* **66**, 375–388.

Smalley, R. E., Wharton, L., and Levy, D. H. (1977b). *J. Chem. Phys.* **66**, 2750–2751.

Smalley, R. E., Wharton, L., and Levy, D. H. (1978). *J. Chem. Phys.* **68**, 671–674.

Soep, B. (1975). *Chem. Phys. Lett.* **33**, 108–113.

Stockburger, M., Gattermann, H., and Klusmann, W. (1975). *J. Chem. Phys.* **63**, 4519–4528.

Tric, C. (1976). *Chem. Phys.* **14**, 189–212.

Van den Bogaardt, P. A. M., Rettschnick, R. P. H., and van Voorst, J. D. W. (1976a). *Chem. Phys. Lett.* **41**, 270–273.

Van den Bogaardt, P. A. M., Rettschnick, R. P. H., and van Voorst, J. D. W. (1976b). *Chem. Phys. Lett.* **43**, 194–196.

Van Voorst, J. D. W. (1975). *In* "Lasers in Physical Chemistry and Biophysics" (J. Joussot-Dubien, ed.), pp. 207–209. Elsevier, Amsterdam.

Von der Linde, D. (1972). *IEEE J. Quantum Electron.* **8**, 328–338.

Werkhoven, C. J., Deinum, T., Langelaar, J., Rettschnick, R. P. H., and van Voorst, J. D. W. (1971). *Chem. Phys. Lett.* **11**, 478–481.

Werkhoven, G. J., Deinum, T., Langelaar, J., Rettschnick, R. P. H., and van Voorst, J. D. W. (1973). *Chem. Phys. Lett.* **18**, 171–175.

Werkhoven, G. J., Deinum, T., Langelaar, J., Rettschnick, R. P. H., and van Voorst, J. D. W. (1975a). *Chem. Phys. Lett.* **30**, 504–509.

Werkhoven, G. J., Deinum, T., Langelaar, J., Rettschnick, R. P. H., and van Voorst, J. D. W. (1975b). *Chem. Phys. Lett.* **32**, 328–331.

5

Dynamic Aspects of Molecular Excitation by Light

William Rhodes
Department of Chemistry
and
Institute of Molecular Biophysics
Florida State University
Tallahassee, Florida

I. INTRODUCTION

In most discussions of radiationless transitions in molecules it is tacitly assumed that the absorption of light prepares the molecule in a particular excited Born–Oppenheimer (BO) state. Radiationless transitions thereby result from (1) intramolecular interaction associated with the fact that the prepared BO state is not an eigenstate of the exact molecular Hamiltonian and/or (2) intermolecular interactions of the molecule with its medium. The reason for this assumption is understandable in terms of the history of the development of molecular spectroscopy. It has long been clear that the complexity of molecular electronic spectra is associated with nuclear motions of the molecule. Early

ISBN 0-12-450650-X

researchers attempted to explain spectral structure in terms of electronic energy (potential) surfaces and their associated BO (vibronic) states (Henry and Kasha, 1968; Sponer, 1959). Such a basis has the desirable feature of relating molecular structure (including geometry and symmetry) to spectral structure. In addition to providing a molecular structural framework for understanding radiative transitions, it establishes radiationless transitions in terms of interactions which are related directly to molecular properties, such as vibrational motion. These features have no doubt contributed greatly to the widespread and persistent usage of BO states as the basis of molecular spectroscopy in spite of the fact that such a basis is not unique. For example, there are adiabatic, crude, pure spin, mixed spin, etc. BO bases (Burland and Robinson, 1970; Sharf and Silbey, 1970).

Furthermore, prior to 1965 most spectroscopic experiments were done with conventional chaotic light sources on molecular systems in condensed phase or relatively dense gas phase. In these systems excited state relaxation processes involving the medium play a major role, whereby emission spectra have a different character than absorption spectra, and common behavior patterns, as exemplified by the Kasha (1950) rule, are observed. These characteristics of conventional spectroscopy are easily explicable in terms of electronic energy surfaces and BO bases.

During the past decade there has been a widespread development of alternative approaches to the understanding of radiationless transitions in molecules. Jortner and co-workers (the Chicago–Tel Aviv School), and many others, have made extensive studies of the intramolecular aspects of isolated molecules. (Bixon and Jortner, 1968; Nitzan *et al.*, 1972; Jortner and Berry, 1968; Mukamel and Jortner, 1974; Freed, 1970, 1972; Voltz, 1975; Tric, 1971; van Santen, 1972). Their approach utilizes Green's operator methods for the analysis of excited state coupling schemes involving primarily intramolecular and radiative interactions. A primary purpose has been to distinguish large molecule behavior, in which the internal degrees of freedom act as an internal heat bath producing irreversible radiationless decay, from small molecule behavior, in which radiationless transitions are reversible and may produce oscillations (beats) in radiative decay. In this approach it is assumed, either explicitly or implicitly, that the coupled zeroth-order states are BO states and that light absorption always prepares an excited BO state. In this sense, therefore, the philosophy of the Chicago–Tel Aviv school does not represent a radical departure from that of earlier workers. The difference is one of emphasis on the Green's function analysis of coupling schemes involving BO states and the time-dependent behavior associated with various schemes.

At the time Jortner *et al.* began their studies on radiationless transitions, Bryan Henry and I became involved in the question of describing molecular excitation in terms of molecular stationary states (molecular eigenstates), which are eigenstates of the full molecular Hamiltonian excluding interaction with the radiation field (Rhodes *et al.*, 1969). It appeared that, if the molecular eigenstates remained discrete upon addition of a radiative linewidth to each state, excitation by sufficiently monochromatic light would result in the excitation of a single molecular eigenstate which could decay only radiatively without the existence of radiationless transitions. It quickly became clear that the distribution of molecular eigenstates of most molecules and the properties of the exciting light were such that individual molecular eigenstates are not selectively excited under experimental conditions. Rather, many eigenstates in a region of the spectrum are excited with a definite phase relation among them. *Radiationless transitions are, consequently, a manifestation of the time-dependent dephasing among the excited molecular eigenstates.* This realization establishes the philosophy of our approach to radiationless transitions, which makes use of time-development operator, density operator, as well as Green's operator methods (Rhodes, 1969, 1970, 1971, 1974, 1977).

The cornerstone of our philosophy is that the prepared excited state of the molecule depends on the interplay of the properties of the exciting light and of the absorption spectrum of the molecule. Thus, the prepared excited state may or may not be an allowed BO state, depending on the conditions of the experiment. This represents a very fundamental difference in philosophy from that of the Chicago–Tel Aviv school, and of others, who have assumed that the prepared state is necessarily an allowed BO state. In most other respects concerning methods and principles there is general agreement among workers in the field. However, it is important to realize that the above difference in philosophy is a crucial one, because the assumption that the prepared state is a BO state precludes the very basic theoretical issues that distinguish the modern approach (developed since the late 1960s) from the more traditional approach. In fact, it restricts the problem of radiationless transitions to one of determining the molecular mechanism (vibronic, spin–orbital coupling, etc.) for transitions out of the prepared BO state and to the Green's function analysis of the associated coupling schemes.

In addition to the above theoretical formulations, the rapid development of laser techniques has led to the emergence of a new era of molecular spectroscopy. Use of picosecond, and longer, pulses of near minimum uncertainty width makes it possible to excite molecules in very narrow regions of the spectrum and under essentially isolated molecule conditions (cf. Rentzepis, 1973). Since the work of Williams *et al.* (1974) (and others) on resonance scattering by iodine vapor, there have been numerous papers on the theoretical formulation of, and the conceptual distinctions among, resonance scattering processes (Berg *et al.*, 1974; Friedman and Hochstrasser, 1974; Mukamel and Jortner, 1975; Metiu *et al.*, 1975; Hilborn, 1975). In particular, Hochstrasser and co-workers have been concerned about the interpretations of laser pulse scattering experiments (Novak *et al.*, 1978). Consideration of laser pulse excitation led Robinson *et al.* to formulate molecular excitation in terms of a convolution involving the molecule and field (Robinson and Langhoff, 1974), a method which implicitly allows for the dependence of the prepared excited state on the incident light. More recently, Grigolini and Lami (1978; Grigolini, 1977) have developed a master equation in terms of excited state correlation functions for the molecule which also brings out the role of the light in governing excitation dynamics. In addition, the development of laser techniques has led to the discovery of coherent, non-linear optical phenomena such as photon echo, self-induced transparency, and superradiance (Allen and Eberly, 1975; Sargent *et al.*, 1974).

The rapid experimental and theoretical developments during the past decade have led to a *revolution in molecular spectroscopy*. Prior to this period all of molecular spectroscopy was couched within the Born–Oppenheimer scheme, which provided the language and conceptual framework for all aspects of the subject. Now, a new line of thinking has emerged—one in which absorption and emission of light and all associated processes are regarded in a unified manner. We refer to this as the *dynamic* line of development of molecular spectroscopy (Rhodes, 1977). Of course, the BO line of thinking is still present and is important for considerations of the mechanistic features of excited state processes. We refer to this as the *mechanistic* line of development. It is of basic importance to those who are interested in the intramolecular mechanisms of radiationless processes, as is the case in much of the work of Lin (1966), Siebrand (1967), and many others. Thus, although the BO scheme remains necessary for many facets of molecular spectroscopy, it is no longer sufficient for all aspects, as it was a decade ago.

II. BASIC PREMISES

With the above comments providing a general perspective, let us now consider two sets of premises which establish a basis for further theoretical development consistent with the dynamic line of thinking. Consider first the electronic absorption spectrum of a polyatomic molecule. Such a spectrum typically will consist of highly structured regions and broad, diffuse regions, where the latter are due either to a high density of bound states or to dissociation (or ionization) continua. We are interested here only in the bound state regions of the spectrum for which we make the following premises:

(1) The absorption spectrum consists of *molecular resonances.*

(2) Each molecular resonance is composed of (transitions to) one or more molecular eigenstates. Usually there are many.

(3) The molecular eigenstates are mostly radiatively distinct; i.e., they remain discrete upon the assignment of an appropriate radiative linewidth. This is presumed to be the case even for very large molecules for which the density of states can be extremely high (Rhodes, 1974).

(4) Any given radiatively allowed BO state is distributed among one or more molecular resonances.

The picture which emerges from the above set of premises is that the bound state portion of a molecular spectrum consists of molecular eigenstates which are clustered in bands (resonances) with respect to absorption intensity distribution. Even broad, diffuse absorption bands may consist of sets of such resonances which are not resolved because of heterogeneous structure or lack of experimental resolution. The molecular eigenstates are eigenfunctions of the full molecular Hamiltonian H_m in the absence of radiative coupling. Insofar as the molecular eigenstates are radiatively discrete, this is the natural basis to use for theoretical formulations. Radiative decay can then be accounted for through a damping operator which results in a non-Hermitian total Hamiltonian. However, depending on the structure of the absorption spectrum in the spectral range of interest, it may be desirable to use a zeroth-order molecular basis, consisting of eigenstates of a zeroth-order Hamiltonian, H_0. Obviously, there are many ways of choosing H_0. The individual molecular resonances could provide a basis for choosing H_0, or one could use combinations of molecular resonances which amounts to a BO basis, for example. In any event, the operator $H_m - H_0$ serves as an intramolecular potential for radiationless processes. Judicious choice of H_0 is important for studies directed toward the molecular mechanistic aspects of radiative and radiationless processes.

The other set of premises is concerned with the structure of the excited state prepared by the incident light interacting with a region of the absorption spectrum. They are stated as follows:

(1) The incident light tends to excite those molecular eigenstates which lie within the spectral distribution of the light.

(2) The amplitudes of the excited eigenstates have a phase relation that depends on the properties of the incident light.

The first premise is self-evident and is taken for granted in most studies. It emphasizes the important fact that the incident light tends to be *selective* on the absorption spectrum; i.e., it selects a subspace of the molecular spectrum. The second premise is a little more subtle. It will be shown in the next section that conventional light sources tend to produce a random phase relation among excited states which results in a statistical excited state mixture. On the other hand, minimum-width pulses produce a definite phase relation among excited states, which implies that the prepared excited state is a pure state, as opposed to a statistical mixture.

A special and very important case is one in which the light acts as a *projector on a portion of the absorption spectrum.* This means that every molecular eigenstate in that portion of the spectrum is excited in phase and with a probability which is proportional to the associated absorption intensity. By definition this is the so-called doorway state for radiative transitions into that portion of the spectrum. We refer to such a prepared state as a *projected radiative state.* A projected radiative state subsequently undergoes time-dependent changes by way of dephasing among the molecular eigenstate components.

The result of the latter two premises is that the incident light always tends to be selective on a subspace of the spectrum. It may or may not be projective depending on the properties of both the light and the spectrum. If it is projective, the projected radiative state which is prepared may or may not be an allowed BO state, depending on the circumstances. For many spectra, a prepared BO state would be an exception rather than a rule.

Recently, Zewail *et al.* (1977a; Orlowski *et al.*, 1977) have made an intensive effort to discern excited state selectivity by laser excitation of large molecules, such as pentacene. At first, it appeared that by narrowing the frequency width of the laser light, individual molecular resonances (referred to as stationary states by Zewail *et al.*) could be selectively excited. Later experiments by Orlowski and Zewail (1979) and by Wiersma (De Vries *et al.*, 1977) show, however, that the long lifetime component originally attributed to individual resonances is due to other causes.

The premises stated above serve to reinforce the comments made in the introduction and to firmly establish the philosophy of our approach to the dynamics of molecular excitation with light. The central theme might be stated to be that light acts by *selection* with or without *projection* on the absorption spectrum.

Our purpose is to make use of the absorption spectrum to obtain as much information as possible about the dynamics of all processes associated with light absorption. Since the structure of the spectrum is very important, it is necessary to eliminate artifacts, to consider the effects of instrumental averaging, and to distinguish the role of homogeneous and heterogeneous spectral structure.

In the remainder of this chapter the question of selection with or without projection by the incident light will be discussed in more rigorous, detailed terms. The next section deals with many-molecule and single-molecule features, various kinds of light sources, an exactly soluble model which exhibits selection without projection, and a qualitative model which exhibits selection with projection on the absorption spectrum. In the last section we consider the case of the prepared state being a projected radiative state. Using Green's operator methods we analyze the resulting coupling scheme in a manner which produces *the spectrum for radiationless transitions from the projected radiative state.*

III. SELECTIVE EXCITATION

One of the primary objectives of molecular spectroscopy is to relate observed features of spectroscopic phenomena to the properties of the individual molecules constituting any given system. Once these relationships are known it is possible to design experiments (thought and real) in which the time dependence of the state of a given molecule is followed (at least in principle) during the course of the interaction between the molecular system and the exciting light. The determination of these relationships between spectra and individual molecule properties is complicated by the fact that observations are made on systems of many molecules in which effects of molecular interactions and heterogeneity must be taken into account. Similar remarks can be made about the properties of the exciting light since observed spectroscopic features can be affected by the many-mode, many-photon character of the incident radiation field. Therefore, it is important at the outset to consider the composite nature of both the molecular system and the radiation field.

A. General Formulations

Consider a system of identical, noninteracting molecules. The simplest states are direct product states

$$|\{s\}\rangle = |s_1 s_2 \cdots s_m \cdots\rangle, \tag{1}$$

in which molecule 1 is in state s_1, etc. If the molecules are indistinguishable then $|\{s\}\rangle$ represents the properly symmetrized or antisymmetrized combination of products, depending on whether the molecules are bosons or fermions. At zero temperature the system may be represented by a pure state which may be one of the $|\{s\}\rangle$ states or a superposition of them. However, for $T > 0$ (and often for $T = 0$) the system cannot be represented by a pure state but, instead, must be represented as a statistical mixture,

$$\rho_{\{s\}} = \sum_{\{s\}} |\{s\}\rangle p_{\{s\}} \langle\{s\}|, \tag{2}$$

where $p_{\{s\}}$ is the probability of the corresponding pure state. For thermal equilibrium it is the Boltzmann factor. In the case of indistinguishable molecules the basis $|\{s\}\rangle$ is equivalent to the Fock basis

$$|\{m\}\rangle = |m_1 m_2 \cdots\rangle \tag{3}$$

in which m_1 molecules are in state 1, m_2 are in state 2, etc.

Similarly, the radiation field is a composite system made of different modes, each of which is characterized by its frequency, propagation direction, and polarization. An appropriate basis for description is the set of Fock states

$$|\{n\}\rangle = |n_1 n_2 \cdots\rangle, \tag{4}$$

in which there are n_1 photons in mode 1, n_2 in mode 2, etc. Since photons are bosons, then states are understood to be symmetrized. They are eigenstates of the free field Hamiltonian.

We are interested in several kinds of incident light, corresponding to that produced by both conventional and laser sources.

1. Chaotic Light

An example is light produced by a thermal source. It is described by the field density operator (Glauber, 1970)

$$\rho_r = \sum_{\{n\}} |\{n\}\rangle p_{\{n\}} \langle\{n\}|, \tag{5}$$

where

$$p_{\{n\}} = p(n_1)p(n_2)\cdots p(n_i)\cdots, \qquad \text{with} \quad p(n_i) = \bar{n}_i^{n_i}/(1 + \bar{n}_i)^{n_i}.$$

Here, $\bar{n}_i$ is the average number of photons in mode i. This is the description which we will use for light produced by conventional incandescent or other random sources from which continuous collimated beams may be selected. Note that ρ_r in Eq. (5) is stationary with respect to the free field Hamiltonian, i.e., it is a statistical mixture of stationary states.

2. Incoherent Pulses

These are pure state nonstationary wave packets of the form

$$|\psi_r\rangle = \sum_i C_i |\{n\}_i\rangle, \tag{6}$$

in which $|\{n\}_i\rangle$ is a direct product state with distribution $\{n\}_i$. It is possible for the distribution of the C_i and $\{n\}_i$ to be such that the pulse has minimum uncertainty width. In general, these states are incoherent in the sense that the field correlation functions do not factor as products of space-time field functions (Glauber, 1963a). Also, in general, the average electric field vanishes, meaning that the pulse is a noise pulse of energy density propagating through space.

A special case is a single photon pulse for which each $\{n\}_i$ is a single photon state for mode i. Such pulses are obviously first-order coherent and have been used extensively in recent resonance scattering studies (Williams *et al.*, 1974; Berg *et al.*, 1974; Friedman and Hochstrasser, 1974; Mukamel and Jortner, 1975; Metiu *et al.*, 1975; Novak *et al.*, 1978).

3. Single Coherent Mode

This is the ideal limit of a monochromatic laser and is represented by a pure Glauber state (Glauber, 1963b)

$$|\alpha\rangle = \sum_n C_n |n\rangle \tag{7}$$

for a single mode. The probability distribution $|C_n|^2$ is Poisson. For these states the average value of the electric field, which is proportional to $|\alpha|$, is sinusoidal in space and time, corresponding to a continuous classical monochromatic wavetrain. They provide the basis for the quantum theory of coherent optical phenomena.

4. Coherent Pulses

These are many-mode coherent states

$$|\{\alpha\}\rangle = |\alpha_1\alpha_2\cdots\rangle, \tag{8}$$

which represent a distribution of frequencies, polarizations, etc. For the proper set of α's this corresponds to a minimum uncertainty width pulse, for which the average electric field is a minimum width traveling wave packet. Such states correspond to our mental picture of a classical light pulse for which the electric field $\mathbf{E}(t)$ corresponds to the Fourier transform of the field frequency distribution $\mathbf{E}(\omega)$.

With these brief comments about the model states used for the exciting light and for molecular systems, we are in a position to look at the general features of the excitation process. We assume that the molecular system and light are initially noninteracting and that the molecular system is in a stationary state (pure or statistical). The exciting light may or may not be stationary, depending on the source. We further assume that prior to interaction the density operator for the molecule-field system is uncorrelated; thus,

$$\rho = \rho_{\{s\}}\rho_r \qquad (t \leq 0), \tag{9}$$

where it is understood that, unless otherwise indicated, the interaction begins at $t = 0$.

As usual, the Hamiltonian of the molecule-field system is expressed as

$$H = H_s + H_r + V_r, \tag{10}$$

where H_s represents the N-molecule system, H_r represents the radiation field, and V_r is the interaction between them. Unless otherwise indicated, we consider only electric-dipole radiative coupling, for which

$$V_r = -\sum_m \boldsymbol{\mu}_m \cdot \mathbf{E}_m. \tag{11}$$

The summation is over all molecules, with $\mathbf{E}_m$ being the field at molecule m.

We are interested in the structure of the density operator at time t, which is given formally as

$$\rho(t) = U_s(t, 0)\rho(0)U_s^{\dagger}(t, 0), \tag{12}$$

where U_s is the time development operator in the Schrödinger picture. It is convenient to formulate U_s in terms of the interaction picture, whereby

$$U_s(t, 0) = e^{-i\mathscr{H}_0 t}U(t, 0), \qquad \mathscr{H}_0 \equiv H_s + H_r, \tag{13}$$

and U may be obtained in integral form from the Schrödinger equation (Rhodes, 1970):

$$U(t, 0) = 1 - i\int_0^t dt_1 U(t, t_1)V_r(t_1) \tag{14}$$

Throughout we express energy in units of circular frequency ($\hbar \equiv 1$).

Since we are interested in the changes in the molecular system induced by the radiation field, the integral term in Eq. (14) is the important one for considering the states resulting from photon absorption. The corresponding contribution to the density operator in the interaction picture is

$$\rho_i^{\mathrm{ex}}(t) = \int_0^t dt_1 \int_0^t dt_1' \, U(t, t_1) V_{\mathrm{r}}(t_1) \rho(0) V_{\mathrm{r}}(t_1') U(t_1', t); \tag{15}$$

Eq. (15) contains all terms resulting from transitions in the molecule-field system. On the other hand, for photon scattering processes U may be written explicitly in the more useful form (Rhodes, 1970)

$$U(t, 0) = 1 - i \int_0^t dt_1 \, V_{\mathrm{r}}(t_1) - \int_0^t dt_1 \int_0^{t_1} dt_2 \, V_{\mathrm{r}}(t_1) U(t_1, t_2) V_{\mathrm{r}}(t_2). \tag{16}$$

Scattering is a two-photon process, so it is the last term that is important. That projection of ρ which contains only scattering terms may be written

$$\rho_i^{\mathrm{sc}}(t) = \int_0^t dt_1 \int_0^{t_1} dt_2 \int_0^t dt_1' \int_0^{t_1'} dt_2'$$
$$\times V_{\mathrm{r}}(t_1) U(t_1, t_2) V_{\mathrm{r}}(t_2) \rho(0) V_{\mathrm{r}}(t_2') U(t_2', t_1') V_{\mathrm{r}}(t_1'). \tag{17}$$

At this point a few comments may be useful in delineating the physical features in the above formulas:

(1) The time dependence of $V_{\mathrm{r}}(t)$ is given by

$$V_{\mathrm{r}}(t) = e^{i\mathscr{H}_0 t} V_{\mathrm{r}} e^{-i\mathscr{H}_0 t}. \tag{18}$$

(2) The operator for the electric field in V_{r} [Eq. (11)] is given by

$$\mathbf{E}(\mathbf{r}) = \sum_{k,\lambda} [f_{k\lambda}(\mathbf{r}) a_{k\lambda} + f_{k\lambda}(\mathbf{r})^* a_{k\lambda}^\dagger], \tag{19}$$

where summation is taken over all photon momenta $\mathbf{k}$ and polarizations λ. For each mode there is a photon annihilation operator $a_{k\lambda}$ and a photon creation operator $a_{k\lambda}^\dagger$. The mode functions $f_{k\lambda}$ are generally plane wave functions of the position coordinate in the system. In formulas (15) and (17) each V_{r} acts to produce molecular transitions to states of either higher or lower energy. Frequently, the rotating wave approximation (RWA) is used, in which only the $a_{k\lambda}$ terms of $\mathbf{E}$ are used for excitation transitions and only the $a_{k\lambda}^\dagger$ terms of $\mathbf{E}$ are used for deexcitation transitions. We will use the RWA throughout this discussion.

(3) Because of the linear dependence of V_{r} on the electric-dipole operators, each time V_{r} acts it produces a transition in only one molecule; i.e., multiple excitations require the successive actions of V_{r}.

(4) Frequently, we are interested in the state of the molecular system irrespective of the state of the field, which is obtained by taking the trace of $\rho(t)$ with respect to all field states. This produces the reduced density operator for the molecular system and we denote it as $\bar{\rho}$. In general, we refer to $\bar{\rho}(t)$ as the *prepared state* of the *molecular system.*

(5) All effects of molecule-field interaction are contained in ρ_i^{ex} of Eq. (15). Notice that an essential part of its structure is the initial action of V_r on $\rho(0)$ (from both sides) to produce transitions. The molecular part of this can be expressed as $\boldsymbol{\mu}\rho_{\{s\}}(0)\boldsymbol{\mu}$, which is a density operator (unnormalized) describing the total radiative (electric dipole) strength of the initial state. We refer to this, upon normalization, as the *radiative state* to the initial state of the molecular system. Of particular interest is the case where the initial molecular state is the ground state, corresponding to a pure state direct product. The radiative state is thus a pure state:

$$|\mathrm{R}\rangle = \boldsymbol{\mu}|00\cdots\rangle[\langle 00\cdots|\boldsymbol{\mu}^2|00\cdots\rangle]^{-1/2}. \tag{20}$$

It is understood to be defined for each Cartesian component of $\boldsymbol{\mu}$. The radiative state is an example of a class of states, called *doorway states*, which was discussed briefly in the previous section. In the following discussion we will be interested not only in $|\mathrm{R}\rangle$, but also in the projections of $|\mathrm{R}\rangle$ onto various subspaces of the Hilbert space of the molecule. These are the projected radiative states defined earlier.

Optical processes can be divided into those that are *linear* and those that are *nonlinear* with respect to the incident light intensity. These can further be studied with regards to their multimolecule versus their single molecule features. Consider first the linear regime. The structure of ρ_i^{ex} is simplified by the fact that the U operators in Eq. (15) contain only radiative interactions involving the vacuum state of the field, i.e., only terms involving spontaneous radiative decay. All contributions of the incident field are contained in the portion

$$A(t_1, t_1') = \sum_{m, m'} V_{r,m}(t_1)\rho(0)V_{r,m'}(t_1'), \tag{21}$$

where the summation is taken over molecules. It is clear that this term allows for intermolecular coherence effects, but the propensity for such effects depends on the initial state of the field. Most light sources $[\rho_r(0)]$ are such that m and m' in the above term are not highly correlated, resulting in intermolecular interference effects in such processes as single photon scattering. For the extreme case of a highly localized light pulse (or a statistical mixture of such pulses) localized on single molecules, the above term is diagonal in m (proportional to $\delta_{mm'}$) and there is no intermolecular interference.

We are especially interested in excitation by *stationary light* sources, whose most general form is Eq. (5). For the linear regime, it is easy to show that the use of Eq. (5) in Eq. (15) leads to the same physical results as does the use of the pure field state $|\bar{n}_1\bar{n}_2\cdots\rangle$ specifying the average number of photons $\bar{n}_i$ in each mode (Rhodes, 1969). This can be seen by formulating the probability of any physical quantity, i.e., by evaluating any diagonal element of Eq. (15). The consequence of this equivalence property for linear stimulated processes is that, for the ground state (or any pure state) of the molecular system, the initial state behaves effectively as a pure state,

$$|u(0)\rangle = |00\cdots\rangle|\bar{n}_1\bar{n}_2\cdots\rangle. \tag{22}$$

This greatly simplifies the formulation since $|u(0)\rangle$ evolves as a pure state (albeit correlated) of the molecule-field system.

The absorption spectrum of the molecular system arises largely from the linear response to the radiation field. In this chapter we limit discussion to absorption spectra which can be attributed to the linear regime. One way to generate such an absorption spectrum from the linear limit of Eq. (15) is to use an initial field state $|\bar{n}_1\bar{n}_2\cdots\bar{n}_i\cdots\rangle$ in which the distribution of $\bar{n}$ is such that the incident field intensity is independent of frequency. The absorption spectrum for the ground state of the molecular system, $|0\rangle$, is then given by

$$P(\omega_i) = \lim_{t\to\infty} (\mathrm{tr})_{\mathrm{m}}\langle\bar{n}_1\cdots(\bar{n}_i-1)|\rho_i^{\mathrm{ex}}(t)|\bar{n}_1\cdots(\bar{n}_i-1)\rangle, \tag{23}$$

where ω_i is the frequency of the photon removed from the initial state (i.e., the resulting photon hole state) and $(\mathrm{tr})_{\mathrm{m}}$ indicates the trace with respect to the molecular system.

The absorption spectrum is rich in information which can be used to determine the dynamics of the molecular system prepared under various conditions. However, we must remember that the experimental spectrum contains nonlinear, as well as linear, response components and contains inhomogeneous as well as homogeneous structure. Let us assume, as indicated above, that the nonlinear portion can be removed or neglected. The resulting linear response portion can be used, among other things, to describe the dynamics of the radiative state and its projections. It is important to distinguish dynamical features associated with heterogeneous and homogeneous structure. Fortunately, the formal theoretical expressions in Eqs. (15) and (23) also allow for both. Suppose, for example, that the absorption spectrum contains two neighboring lines which arise from molecules at two kinds of sites. The (projected) radiative state which arises from this portion of the spectrum will oscillate with a frequency equal to the separation of the two lines. This results in a beat in the associated photon scattering, Eq. (17). Since the origin of the beat is in the heterogeneous structure, however, it is an intermolecular interference phenomenon. It arises from the nondiagonal components of the A term, Eq. (21), which is substituted into Eq. (17) for scattering. On the other hand, intramolecular dynamical features (arising from the homogeneous structure) depend on the diagonal elements of Eq. (21) substituted into Eqs. (15) and (17).

Although we are interested primarily in processes associated with linear absorption, a few comments about nonlinear aspects will provide a more complete overall picture. The contribution of nonlinear terms may be obtained by replacing $V_r(t_1)$ and $V_r(t'_1)$ in Eq. (15) by time-ordered products of V_r's and requiring that the terminal U's contain no contributions from the incident light. Each V_r is additive in the molecules and is a superposition of photon creation and annihilation operators. These properties together with the RWA allow us to make several distinctions:

(a) Suppose each V_r in the time-ordered product refers to a different molecule and only photon annihilation operators are included. This is multimolecule absorption in which each molecule responds linearly to the field. Eq. (23) becomes additive in the molecules or, for N identical molecules, it is proportional to N. This is Beers' law and represents linearity on a single molecule basis. Nonlinearity arises only because the N-molecule system is treated as a single quantum system. On the other hand, when photon scattering phenomena are considered [Eq. (17)], multimolecule excitation provides the basis for such phenomena as superradiance (Nussenzveig, 1973), depending on the structure of $\rho_r(0)$.

(b) Suppose each V_r in the product refers to the same molecule and only photon annihilation operators are included. This is single-molecule multiphoton absorption, which is one of the major facets of modern laser spectroscopy.

(c) Suppose each V_r in the product refers to the same molecule with alternating creation and annihilation operators, ordered so that photon annihilation occurs first. With use of the RWA the molecule in question undergoes alternating excitation and deexcitation in correlation with the field operators. If only one molecular excited state is considered, the resulting terms are of the type that contribute to coherent, nonlinear phenomena such as optical nutation, dynamic stark splitting, and self-induced transparency. Of course, the appearance of such phenomena depends on the properties of $\rho(0)$.

In most of the remainder of this chapter we are interested in the dynamic properties of a single molecule interacting with light. In particular, we are interested in the role of both the absorption spectrum and the exciting light. One approach would be to let the experimental absorption spectrum represent the "average molecule." For a system of indistinguishable molecules the corresponding theoretical entity is the reduced density operator obtained by taking the trace of ρ with respect to all molecules but one:

$$\rho_m(t) = (\mathrm{tr})_{m,\,1}\rho(t). \tag{24}$$

This is a single molecule-field density operator which is the same for all molecules in the system. Obviously this approach has the effect of treating heterogeneous structure as though it is homogeneous and in some cases it can lead to false results for intramolecular dynamics.

Our approach will be to treat the spectrum as though it belongs to a single molecule (i.e., is homogeneous) and to develop the theory on a single molecule basis. Heterogeneous and multimolecule aspects will then be considered separately as a modification of this ideal situation.

To this point all of the formulas have been developed in the time domain. An alternative approach would be to use the frequency (energy) domain in which the time-development operator is replaced by the Green's operator $G(\omega)$. For analytical development and for analysis of coupling schemes, the latter is more practical, as we will show later. However, the time domain formulation is more transparent for physical features, so we use it preferably for qualitative and conceptual discussions.

We are interested in the structure and properties of the molecule-field state produced by various kinds of exciting light. For the most part the linear response regime will be considered, whereby the state of the system may be regarded as emerging from Eq. (15) in the following manner: Linearization is introduced into Eq. (15) by first taking the trace of the term $V_r(t_1)\rho(0)V_r(t'_1)$ with respect to all states of the field and then taking the direct product of the density operator so obtained with the density operator for the vacuum state of the field. The result is that the system evolves linearly under the interaction with the incident light until t_1 and t'_1, after which it evolves according to the molecular Hamiltonian and by radiative interaction with the vacuum state. Obviously, for the special case in which $\rho_r(0)$ is a single photon state, the linearization operation has no effect, so no approximation is involved. For all other initial states of the field, however, such a procedure introduces a definite approximation. This results in the following expression for the molecule-field density operator (given as the tensor contraction between two terms):

$$\rho_i^{(1)}(t) = \int_0^t dt_1 \int_0^t dt'_1 \; \mathbf{\Theta}(t, t_1, t'_1) : \mathbf{\Phi}(t_1, t'_1), \tag{25}$$

where the molecule-field operator is

$$\mathbf{\Theta} = U(t, t_1)|\text{vac}\rangle \boldsymbol{\mu}(t_1)\rho_m(0)\boldsymbol{\mu}(t'_1)\langle\text{vac}|U(t'_1, t)$$

and the field correlation tensor (Glauber, 1963a) is

$$\mathbf{\Phi}(t_1, t'_1) = (\text{tr})_r \rho_r(0)\mathbf{E}^{(-)}(t'_1)\mathbf{E}^{(+)}(t_1).$$

The effects of the incident light, described by $\rho_r(0)$, are contained in $\mathbf{\Phi}$, in which $\mathbf{E}^{(+)}$ and $\mathbf{E}^{(-)}$ contain only the photon annihilation and creation operators, respectively. The molecule is assumed to be sufficiently small that spatial variation of the field can be neglected and the coordinate origin is chosen on the molecule.

Equation (25) is our most basic result, in that it contains all information about radiative and radiationless processes associated with the linear response of the molecule to the incident light. Although at first sight it appears to be cumbersome and *ad hoc* in nature, it is really a rigorous first-order formulation with a straight forward physical interpretation. All information about the exciting light is contained in $\boldsymbol{\Phi}$ and all information about the spectrum of the molecule, including spontaneous radiative decay of molecular excited states, is contained in $\boldsymbol{\Theta}$. The molecule and field tensors are contracted in producing $\rho_i^{(1)}$.

The characteristics of the absorption spectrum of the molecule are important. For small molecules such as diatomics and triatomics, it is natural to regard the absorption spectra as being so highly structured that transitions to individual molecular eigenstates are distinguishable, at least in principle. However, for larger molecules, such as aromatic hydrocarbons, the bound state absorption spectrum is usually so dense that it is assumed that individual molecular eigenstates are not discernible, even in principle. Rather, it is assumed that radiative linewidths strongly overlap, making the absorption spectrum effectively continuous. However, this does not appear to be generally the case (cf. Section IIIB). If the existence of a large number of eigenstates is a result of there being a large number of (nonradiative) degrees of freedom of the molecule (e.g., nuclear motion), there is a profound dilution effect (Rhodes, 1974) which increases with the size of the molecule. A simple way of viewing this effect is as follows. First consider the nuclei to be frozen in a given configuration. All electronic transitions would appear as absorption lines, each of which has a radiative width. Now, we let the nuclei be free to move under the molecular Hamiltonian, so that the given electronic linewidth becomes distributed over numerous eigenstates (within the Wigner–Weisskopf approximation). If the resulting molecular absorption band is appreciably broader than the total radiative linewidth, the molecular eigenstates remain discrete for the most part (Rhodes, 1974). Thus, the absorption spectra of large molecules, such as naphthalene and anthracene, can be regarded as being primarily discrete, albeit quite dense.

The discreteness, in principle, of dense spectra of large molecules is quite helpful in the further formulations of Eq. (25). The Wigner–Weisskopf approximation is applied to each excited molecular eigenstate $|s\rangle$, whereby it has a radiative linewidth γ_s. Suppose we are interested in matrix elements of $\rho_i^{(1)}$ for excited states with no photons in the field. For such states it turns out that the matrix elements of U appearing in $\rho_i^{(1)}$ are diagonal, within the WW approximation, and have the simple structure

$$\langle s, \text{vac}|U(t, t_1)|s, \text{vac}\rangle = e^{-(1/2)\gamma_s(t-t_1)} \tag{26}$$

in the interaction picture. For $t \ll \gamma_s^{-1}$ these matrix elements may be replaced by unity in Eq. (25). In the case of large molecules $\gamma_s \leq 10^6\ \text{sec}^{-1}$, so this simplification is valid for long times on the experimental scale of picosecond and nanosecond spectroscopy.

The molecule-field density operator $\rho_i^{(1)}$ contains amplitudes for all molecule-photon states produced by single-photon absorption. It is convenient to separate $\rho_i^{(1)}$ into two major parts: (1) one containing only field vacuum terms, which we refer to as the prepared excited state of the molecule, $\bar{\rho}_{\rm ex}^{(1)}$; and (2) one containing only photon scattering states, $\rho_{\rm sc}^{(1)}$, which is obtained by projecting $\rho_i^{(1)}$ with the projection $1 - |{\rm vac}\rangle\langle{\rm vac}|$. We are particularly interested in the former, which, in the Schrödinger picture, has the relatively simple form

$$\bar{\rho}_{\rm ex}^{(1)}(t) \equiv \langle {\rm vac}|\rho^{(1)}|{\rm vac}\rangle = \int_0^t dt_1 \int_0^t dt'_1\, e^{-iH_{\rm eff}(t-t_1)}\boldsymbol{\mu}\rho(0)\boldsymbol{\mu}e^{iH^{\dagger}_{\rm eff}(t-t'_1)}:\boldsymbol{\Phi}(t_1, t'_1). \tag{27}$$

The bar denotes that the prepared excited state is similar in structure to the reduced density operator used earlier (Rhodes, 1974). Note that, because of the Schrödinger Picture, $\boldsymbol{\mu}$ is independent of t. The term involving the effective Hamiltonian is defined by

$$\langle {\rm vac}|U_s(t, t_1)|{\rm vac}\rangle = e^{-iH_{\rm eff}(t-t_1)} \tag{28}$$

in the Schrödinger picture. The effective molecular Hamiltonian is given by

$$H_{\rm eff} = H_{\rm m} - i\tfrac{1}{2}\Gamma, \tag{29}$$

where Γ is the damping operator for radiative decay. Its features will be discussed later.

It is easy to show that all elements of $\rho_{\rm sc}^{(1)}$ involving single photon scattering are contained in the operator (Schrödinger picture)

$$\bar{\rho}_{\rm sc}^{(1)}(t) = \int_0^t dt_a \int_0^t dt_b e^{-i\mathscr{H}_0(t-t_a)}V_{\rm r}|{\rm vac}\rangle\bar{\rho}_{\rm ex}^{(1)}(t_a, t_b)\langle{\rm vac}|V_{\rm r}e^{i\mathscr{H}_0(t-t_b)}, \tag{30}$$

which is obtained by the use of Eq. (27) in Eq. (17). The double time value of the prepared excited density operator is obtained by replacing t by t_a and t_b in the upper limits of the integrals in Eq. (27). Note that, because of the Schrödinger picture, $V_{\rm r}$ is independent of t.

We are now in a position to use Eqs. (27) and (30) to determine the nature of the state generated from the interaction of a molecule with (1) chaotic light and (2) light pulses of varying structure and degree of coherence. For simplicity, the initial state of the molecule is taken to be the ground state, but this is easily generalized for a statistical state.

B. Linear Response to Chaotic Light*

Consider a molecule, initially in its ground state $|0\rangle$, and a beam of light from a conventional light source. The light is prepared so that it has a definite intensity spectrum $I(\omega)$, and the interaction with the molecule begins at $t = 0$. We are interested in the time development of the structure of the excited state $\bar{\rho}_{\mathrm{ex}}^{(1)}$ as determined by Eq. (27).

In the linear response limit, the dependence of the prepared excited state on the properties of the exciting light is contained entirely in the first-order field correlation tensor. Stationary, chaotic light has the form of Eq. (5), which, when substituted into $\mathbf{\Phi}$, gives a statistical mixture of correlation tensors for various stationary states of the field. As we have indicated earlier, however, Eq. (5) may be replaced in $\mathbf{\Phi}$ by the pure stationary state $|\{\bar{n}\}\rangle$, representing the average photon distribution. Thus,

$$\mathbf{\Phi}(t_1', t_1) = \langle\{\bar{n}\}|\mathbf{E}^{(-)}(t_1')\mathbf{E}^{(+)}(t_1)|\{\bar{n}\}\rangle$$

$$= \int d\omega \langle \mathbf{E}^{(-)}\mathbf{E}^{(+)}\rangle_{\omega} e^{-i\omega(t_1 - t_1')}. \tag{31}$$

This expression has two important features: (1) It is a Fourier expansion in terms of the frequency-dependent correlation tensor $\langle \mathbf{E}^{(-)}\mathbf{E}^{(+)}\rangle_{\omega}$, the trace of which gives the average energy density of the field at the site of the molecule (proportional to $I(\omega)$). (2) It shows that $\mathbf{\Phi}$ depends only on the time difference $\Delta t = t_1 - t_1'$, a property resulting from the field being stationary.

The above properties of $\mathbf{\Phi}$ show that it plays the role of the generating function of the generalized (tensor form) intensity spectrum of the incident light; i.e., the trace of $\mathbf{\Phi}$ is simply the generating function for $I(\omega)$, which is determined experimentally. For typical intensity distributions, $I(\omega)$ is a single band of width $\Delta\omega$, which implies that $\mathbf{\Phi}$ decays monotonically with increasing Δt. The resulting width of $\mathbf{\Phi}$ defines a correlation time τ_c for the stationary light source. There is a straightforward physical interpretation of these results: The action of $\mathbf{E}^{(+)}(t_1)$ is to remove a photon from the field at t_1 at the site of the molecule. This creates a hole wave packet in the field which propagates in time. The operator $\mathbf{E}^{(-)}(t_1')$ creates the same (albeit conjugate) hole wave packet at t_1' by acting to the left on the field state. Thus, $\mathbf{\Phi}$ is a measure of the overlap of the hole which has propagated from t_1 to t_1' with its original form at t.

We wish to determine the structure of $\bar{\rho}_{\mathrm{ex}}^{(1)}$ for various kinds of molecular spectra and for incident light spectra of a typical structure consisting of a single band of width $\Delta\omega$ and centered at ω_0. For example, the molecular spectrum may be broad, with little structure, as in the $S_1 \leftarrow S_0$ band of naphthalene; it may be complex with a lot of structure, as in the $S_2 \leftarrow S_0$ vibronic bands of naphthalene (Fig. 4); or it may have sparse structure as in the absorption spectra of NO_2 or SO_2 (Fig. 5).

* See Rhodes (1971, 1974).

In each case, however, we assume that the molecular eigenstate spectrum is discrete and that the Wigner–Weisshopf approximation is valid. In the molecular eigenstate basis, Eq. (27) becomes

$$\bar{\rho}_{\text{ex}}^{(1)}(t) = \sum_{s,\,s'} e^{-i(\Omega_s - \Omega_{s'}^*)t} \int_0^t dt_1 \int_0^t dt_1' \int d\omega \, e^{-i(\omega - \Omega_s)t_1} |s\rangle$$

$$\times \, \boldsymbol{\mu}_{s0} \cdot \mathbf{F}(\omega) \cdot \boldsymbol{\mu}_{0s'} \langle s'| e^{i(\omega - \Omega_{s'}^*)t_1'}, \tag{32}$$

where

$$\Omega_s = \omega_s - i\tfrac{1}{2}\gamma_s, \qquad \mathbf{F}(\omega) = \langle \mathbf{E}^{(-)}\mathbf{E}^{(+)} \rangle_\omega .$$

The radiative linewidths γ_s allow for the radiative decay of each excited eigenstate as it is being generated by the light. In any quantitative formulation they should be retained; however, we are interested in the propensity for time-dependent molecule field correlations, so the γ_s will be dropped. This amounts to assuming that $t \ll \gamma_s^{-1}$ in the following.

Consider now the time integrals in Eq. (32) for a particular frequency component of the field. It is easily shown that the time-dependent part has the form

$$f_{ss'}(t, \omega) = e^{-i(1/2)(\omega_s - \omega_{s'})t} \, \frac{\sin \frac{1}{2}(\omega_s - \omega)t}{(\omega_s - \omega)} \, \frac{\sin \frac{1}{2}(\omega_{s'} - \omega)t}{(\omega_{s'} - \omega)} . \tag{33}$$

For fixed values of ω and t, each term is a damped sinusoidal function of ω_s (or $\omega_{s'}$) with a large central peak centering on ω and having width $\Delta\omega_c \equiv 2\pi t^{-1}$. We refer to $\Delta\omega_c$ as the *coherence width* (Rhodes, 1974). The important point is that the matrix elements of $\bar{\rho}_{\text{ex}}^{(1)}$ tend to be small unless both ω_s and $\omega_{s'}$ lie within the range $\Delta\omega_c$ of ω, i.e., there is a correlation between the molecular excitation frequency and the photon frequency which increases with t. This results in a correlation between molecular eigenstates which correspondingly increases with t. There is a tendency for $\bar{\rho}_{\text{ex},\,ss'}^{(1)}$ to be large only if $|\omega_s - \omega_{s'}| < \Delta\omega_c$; i.e., those matrix elements tend to be damped out which lie outside of a band of elements along the diagonal of $\bar{\rho}_{\text{ex}}^{(1)}$. This is referred to as the *filtering of the (reduced) density operator* (Rhodes, 1971).

The result of the filtering of the density operator is that only those molecular eigenstates whose energies lie within the coherence width $\Delta\omega_c$ are coherently excited, which means that they are excited with a phase relation that permits interference among these eigenstates in subsequent physical processes such as photon emission. In the large t limit ($t \to \infty$), $\bar{\rho}_{\text{ex}}^{(1)}$ tends to become diagonal, with the structure

$$\bar{\rho}_{\text{ex}}^{(1)}(t) \approx 2\pi t \sum_s |s\rangle \boldsymbol{\mu}_{s0} \cdot \mathbf{F}(\omega_s) \cdot \boldsymbol{\mu}_{0s} \langle s| . \tag{34}$$

This results from use of Eq. (33) in Eq. (32) and integrating over ω. It is entirely consistent with usual first-order time-dependent perturbation theory. The criterion for such complete filtering is that $t \gg \omega_{ss'}^{-1}$, the frequency spacing between every pair of eigenstates.

Obviously, for the opposite limit, $t \ll \omega_{ss'}$ for all adjacent states, whereby $\Delta\omega_c$ brackets many neighboring eigenstates. The extreme limit is $t_{\min} = 2\pi\,\Delta\omega^{-1}$, the minimum-uncertainty width determined by $I(\omega)$, but it would not be correct to regard such light as being stationary. Nevertheless, as t approaches this limit, the light beam begins to act more like a pulse which excites the molecular eigenstates coherently in a manner described in earlier formulations (Bixon and Jortner, 1968; Nitzan *et al.*, 1972; Jortner and Berry, 1968; Mukamel and Jortner, 1974; Freed, 1970, 1972; Tric, 1971; van Santen, 1972; Rhodes *et al.*, 1972; Rhodes, 1969, 1970, 1971, 1974, 1977). This limiting case will be included in Section IIIC, on pulse excitation. For practical purposes the light beam can be regarded as stationary only for t many times the minimum value determined by $I(\omega)$, and it is this limit with which we are concerned here.

It is well known that the Schrödinger equation prescribes that a pure state evolve into a pure state. The question may therefore arise as to how the pure state, $|0\{\bar{n}\}\rangle$, evolves into the statistical state, Eq. (34), for large t. The reason, of course, lies in the fact that we have effectively taken the trace of ρ with respect to the field in defining $\bar{\rho}_{\text{ex}}^{(1)}$, so we are dealing with the reduced density operator for the molecule as a subsystem of a composite molecule–field system. The filtering of the density operator (evolution of a statistical state) is associated with the development in time of correlations between the molecule and field from a state which at $t = 0$ is completely uncorrelated.

The importance of the evolution of a statistical excited state is more than academic. Consider, for example, the probability for photon emission which is governed by Eq. (30). In the molecular eigenstate basis the probability of the photon scattered state $|vk\rangle$ is

$$\langle vk|\rho_{\text{sc}}^{(1)}|vk\rangle = \sum_{s,s'} \int_0^t dt_a \int_0^t dt_b \langle vk|e^{-i\mathscr{H}_0(t-t_a)}V_{\text{r}}|s, \text{vac}\rangle \times \bar{\rho}_{\text{ex},ss'}^{(1)}(t_a t_b)\langle s', \text{vac}|V_{\text{r}}e^{i\mathscr{H}_0(t-t_b)}|vk\rangle. \tag{35}$$

The double time excited state density operator is defined in Eq. (30) and must be considered within the context of Eq. (32). The integration over ω in Eq. (32) imposes a correlation between t_1 and t_1', whereby $|t_1 - t_{1'}| \leq \Delta\omega^{-1}$, the width of the light spectrum. This correlation is carried over to t_a and t_b, implying that $|t_a - t_b| \leq \Delta\omega^{-1}$. On the other hand, for $t \gg \omega_{ss'}^{-1}$ we have the effect that the integrand is significant only if $t_a \approx t_b$ in Eq. (35). Furthermore, $t_a \gg \omega_{ss'}^{-1}$ for most of the integration interval, so the principal contribution to the integral will arise from those values of t_a for which $\bar{\rho}_{\text{ex}}^{(1)}$ is diagonal.

The result is that the photon emission probability arises mainly from the independent excitation and decay of each molecular eigenstate. Fluorescence is a weighted superposition of exponential terms with an average lifetime longer than that calculated from absorption. This dilution effect has long been recognized as a possible explanation of the anomalously long lifetimes (Nitzan *et al.*, 1972; Jortner and Berry, 1968; Mukamel and Jortner, 1974; Rhodes, 1969) of SO_2 and NO_2. In addition, however, the filtering of the density operator due to long-time excitation leads to the absence of beats in the fluorescence (Rhodes, 1971), in contrast to the prediction for minimum-width pulse excitation.

The results obtained here suggest that, in principle, one should find dilution effects in dense, large molecule spectra in which one normally sees only the opposite effects of shortened lifetimes and reduced fluorescence quantum yields associated with coherent excitation of many molecular eigenstates. Because of the great experimental difficulties, little effort has been made to characterize this stationary excitation effect. However, some of the results of Zewail *et al.* (1977) on the effects of the exciting light duration on fluorescence decay characteristics of large molecules may be explained in these terms.

C. Linear Response to Light Pulses

In this section we consider the state prepared by the excitation of a molecule by a light pulse in the linear regime. In contrast to the quasi-stationary pulses discussed in section IIIB, we are interested here in light which is strongly nonstationary, for which the first-order field correlation tensor depends on two time arguments.

The properties of the field correlation for two kinds of nonstationary pulses will first be considered. Then we will determine the structure of the prepared state related to both photon absorption and scattering, as determined for an arbitrary molecule and pulse. These results will be applied to a special case of Lorentzian spectra, in order to distinguish scattering and fluorescence. Finally, the selective excitation of more complex molecular spectra will be discussed.

Consider first a single photon pulse, defined by the pure state

$$|\psi\rangle = \sum_i f_i |k_i\rangle, \tag{36}$$

where $|k_i\rangle$ is the single photon state for mode k_i. It follows that

$$\mathbf{E}^{(+)}(\mathbf{r}, t)|\psi\rangle = \mathscr{E}(\mathbf{r}, t)|\text{vac}\rangle, \tag{37}$$

from which we get for the correlation function

$$\mathbf{\Phi}(x', x) = \mathscr{E}^*(x')\mathscr{E}(x), \tag{38}$$

where $x = (\mathbf{r}, t)$. Thus, the correlation function factors in terms of the complex analytic function $\mathscr{E}(x)$ for the positive frequency part of the field. Such factoring implies that the pulse is first-order coherent (Glauber, 1963a) in spite of the fact that $\langle \mathbf{E}(r, t) \rangle = 0$, contrary to coherent light for which $\langle \mathbf{E} \rangle \neq 0$ and the photon number is indefinite. The reason for property (38), however, lies in the fact that each field component in $\mathbf{E}^{(+)}$ produces $|\text{vac}\rangle$ as the sole intermediate state.

The second kind of pulse corresponds to the coherent Glauber state, Eq. (8). This is an eigenstate of the field operator

$$\mathbf{E}^{(+)}(\mathbf{r}, t)|\{\alpha\}\rangle = \mathscr{E}(\mathbf{r}, t)|\{\alpha\}\rangle. \tag{39}$$

Thus, we have

$$\mathbf{\Phi}(x', x) = \mathscr{E}^*(x')\mathscr{E}(x),$$

which is Eq. (38). The difference in the two cases is that $|\{\alpha\}\rangle$ is coherent to all orders (Glauber, 1963a) and gives $\langle \mathbf{E} \rangle \neq 0$, which describes a classical-like electric field pulse. In contrast, the single-photon pulse is a noise pulse in that the phase is completely undetermined. In both cases, however, $\mathscr{E}(x)$ corresponds to a minimum uncertainty width pulse.

For the single photon case the linear regime, Eq. (25), is exact. But, for $|\{\alpha\}\rangle$ it is clearly an approximation which is obtained by simply inserting the linearization operator $|\text{vac}\rangle\langle\{\alpha\}|$ in Eq. (15) before the first V_r and its adjoint after the second V_r. The severity of the linear approximation depends, of course, on the values of α_i, which for laser pulses may be quite large.

The important result, however, is that the two kinds of pulses lead to identical formulas for $\rho^{(1)}$, as has been stressed by Novak *et al.* (1978). Furthermore, because of the factoring of $\mathbf{\Phi}$, $\rho^{(1)}$ evolves as a pure state,

$$\rho^{(1)}(t) = |u(t)\rangle\langle u(t)|, \tag{40}$$

so it is necessary to consider only the time evolution of $|u(t)\rangle$. The reason for this is that t_1 and t_1' in $\mathbf{\Phi}$ are uncorrelated, in contrast to the case of stationary light.

In the Schrödinger picture the prepared state vector in the linear regime has the simple convolution form

$$|u(t)\rangle = i \int_{-\infty}^{t} dt_1 e^{-iH(t-t_1)}|\text{vac}\rangle\mathscr{E}(t_1)\cdot\boldsymbol{\mu}|0\rangle, \tag{41}$$

where $\mathscr{E}$ is evaluated at the site of the molecule and it is understood that $\mathscr{E}$ vanishes for $t_1 \to -\infty$. Again, it is understood that $|u(t)\rangle$ is that part of the molecule-field state vector which results from interaction. Also, we may again project $|u(t)\rangle$ onto an excited state subspace in which there are no photons,

$$|u_e(t)\rangle = |\text{vac}\rangle\langle\text{vac}|u(t)\rangle, \tag{42}$$

and onto the complementary space of emitted photon states,

$$|u_{sc}(t)\rangle = Q|u(t)\rangle, \tag{43}$$

where $Q \equiv 1 - |\text{vac}\rangle\langle\text{vac}|$. It is easily seen that the single photon scattering part has the simple form

$$|u_{\text{sc},1}(t)\rangle = -i \int_{-\infty}^{t} dt_1\, e^{-i\mathscr{H}_0(t-t_1)} V_r |u_e(t_1)\rangle. \tag{44}$$

Transitions from two-photon states have been dropped.

The total probability that a photon has been absorbed by time t is

$$P_a(t) = P_e(t) + P_{sc}(t), \tag{45}$$

where $P_a(t) = \langle u(t)|u(t)\rangle$, $P_e(t) = \langle u_e(t)|u_e(t)\rangle$, and $P_{sc}(t) = \langle u_{sc}(t)|u_{sc}(t)\rangle$. Here, $P_e(t)$ is the probability that the molecule is excited with no photons in the field and $P_{sc}(t)$ is the probability that there is one or more scattered photons. These relations emphasize the fact that the prepared state is not normalized.

The prepared state, Eq. (41), can be further developed by expanding $\mathscr{E}(t_1)$ in terms of its frequency components $\mathscr{E}(\omega)$. This leads to

$$|u(t)\rangle = -\int_{-\infty}^{\infty} \frac{d\omega}{2\pi} \frac{e^{-i\omega t}}{\omega - H + i\eta} \boldsymbol{\mu}|0, \text{vac}\rangle \cdot \mathscr{E}(\omega), \tag{46}$$

in terms of the Green's operator $G(\omega) = (\omega - H + i\eta)^{-1}$, in which the limit $\eta \to 0$ is understood.

Either form of $|u(t)\rangle$, Eq. (41) or (46), may be used to further develop $|u_e(t)\rangle$ in terms of molecular eigenstates. Again, we assume that the WW approximation, Eq. (26), is valid and in terms of G this amounts to

$$\langle s, \text{vac}|G(\omega)|s, \text{vac}\rangle = (\omega - \omega_s + i\tfrac{1}{2}\gamma_s)^{-1}. \tag{47}$$

This gives

$$|u_e(t)\rangle = -\sum_s |s, \text{vac}\rangle \boldsymbol{\mu}_{s0} \cdot \int_{-\infty}^{\infty} \frac{d\omega}{2\pi} \frac{e^{-i\omega t}\mathscr{E}(\omega)}{\omega - \omega_s + i\tfrac{1}{2}\gamma_s} \tag{48}$$

Also, the WW approximation may be applied directly to Eq. (42) to give

$$|u_e(t)\rangle = i \sum_s |s, \text{vac}\rangle \boldsymbol{\mu}_{s0} \cdot \int_{-\infty}^{t} dt_1 \exp[-i(\omega_s - i\tfrac{1}{2}\gamma_s)(t - t_1)]\mathscr{E}(t_1). \tag{49}$$

Equations (48) and (49) are very useful, alternative forms.

It is instructive at this point to introduce the radiative state $|\text{R}\rangle$, into the expressions for the prepared state. Here, $|\text{R}\rangle$ is the one-molecule version of Eq. (20) and, by definition, is the state which carries all of the dipole strength from (or to) the ground state $|0\rangle$. In terms of $|R\rangle$ Eq. (41) becomes

$$|u(t)\rangle = i \int_{-\infty}^{t} dt_1 e^{-iH(t-t_1)} |\text{R}, \text{vac}\rangle \boldsymbol{\mu}_{\text{R}0} \cdot \mathscr{E}(t_1), \tag{50}$$

with a corresponding structure for Eq. (46). This does not imply that the molecule is physically in the state $|\text{R}\rangle$ during the course of excitation. For any t the physically prepared state is $|u(t)\rangle$! Only for a pulse proportional to $\delta(t_1 - t_0)$ would $|u(t_0)\rangle$ be collinear with $|\mathrm{R}\rangle$.

On the other hand, a valid interpretation of $|u(t)\rangle$ is that it consists of a superposition (integral on t_1) of amplitudes in each of which the molecule (virtually) passes at t_1 through $|\mathrm{R}\rangle$, which then undergoes rotation (during the interval $t_1 \to t$) in Hilbert space under the system Hamiltonian. All dynamic processes, radiative and radiationless, are determined by the structure of $|u(t)\rangle$. In analyzing the structure of $|u(t)\rangle$, various bases may be used, such as the molecular eigenstate basis or the eigenstates of some zeroth-order Hamiltonian H_0. It turns out that frequently a basis which includes $|\mathrm{R}\rangle$ is quite useful. In particular, this is so if the excited state of the molecule undergoes radiative decay to low-lying molecular states for which the radiative state is (almost) $|\mathrm{R}\rangle$. Of course, the radiative state is usually different for different low-lying states of the molecule. However, it turns out that, because of the selective properties of the light, only a projected part of $|\mathrm{R}\rangle$ is effective in Eq. (50). Such projected radiative states are more likely to be common for different low-lying levels. A case in point is a BO model for which the effective (projected) $|\mathrm{R}\rangle$ is one allowed BO state of an excited electronic state and the low-lying decay states are the lower vibrational states of the ground electronic state.

Thus, insofar as $|\mathrm{R}\rangle$ is common to the set of low-lying molecule states, the amplitude $\langle \mathrm{R}, \mathrm{vac}|u(t)\rangle \equiv \langle \mathrm{R}|u_e(t)\rangle$ is a measure of the radiative decay amplitude to this manifold; i.e., the total emission intensity is proportional to

$$P_{\mathrm{R}} = |\langle \mathrm{R}|u_e(t)\rangle|^2, \tag{51}$$

where

$$\langle \mathrm{R}|u_e(t)\rangle = i\int_{-\infty}^{t} dt_1 \langle \mathrm{R}, \mathrm{vac}|e^{-iH(t-t_1)}|\mathrm{R}, \mathrm{vac}\rangle \boldsymbol{\mu}_{\mathrm{R}0}\cdot\mathscr{E}(t_1).$$

For these situations, the single matrix element $U_{\mathrm{RR}} \equiv \langle \mathrm{R}, \mathrm{vac}|e^{-iH(t-t_1)}|\mathrm{R}, \mathrm{vac}\rangle$ and the corresponding element of G in Eq. (46), namely $\langle \mathrm{R}, \mathrm{vac}|(\omega - H + \eta)^{-1} \times |\mathrm{R}, \mathrm{vac}\rangle$, play a vital role. In brief, the single-photon scattering amplitude to any state $|v, k\rangle$ having the same radiative state as $|0\rangle$ is given by [cf. Eq. (44)⟩

$$\langle v, k|u(t)\rangle = \langle v, k|u_{\mathrm{sc}}(t)\rangle = \int_{-\infty}^{t} dt_1 \int_{-\infty}^{t_1} dt_2\, e^{-i\omega_{vk}(t-t_1)} \times \langle v, k|V_{\mathrm{r}}|\mathrm{R}, \mathrm{vac}\rangle U_{\mathrm{RR}}(t_1, t_2)\boldsymbol{\mu}_{\mathrm{R}0}\cdot\mathscr{E}(t_2), \tag{52}$$

where $\omega_{vk} = \omega_v + \omega_k$ and again transitions involving two-photon states in the final step have been ignored.

The derivations thus far in this section have made use of a minimum number of assumptions about the system. These include (1) the nature of the light pulses, (2) the linear response formulation, (3) the WW approximation in Eqs. (48) and (49), and (4) the equivalence of the state $|\mathrm{R}\rangle$ for absorption and single photon emission, Eq. (52). Relations similar to these have been used by a number of groups in recent years to study the nature of resonance Raman scattering and resonance fluorescence (Williams *et al.*, 1974; Berg *et al.*, 1974; Friedman and Hochstrasser, 1974; Mukamel and Jortner, 1975; Metiu *et al.*, 1975; Hilborn, 1975; Novak *et al.*, 1978). We do not wish to repeat these results here. However, there has been expressed a great deal of concern about the interplay among Raman scattering and fluorescence in the near and far from resonance limits. Therefore, we limit our discussion to a simple model which deals with some of these features (Needham and Rhodes, 1980).

D. The Lorentzian Absorption Profile; Negative-Time Light Pulse

Our model (Fig. 1) for the molecule consists of a set of discrete molecular eigenstates $|s\rangle$ having a radiative linewidth γ_s. The profile of the absorption from the ground state is a Lorentzian of width Γ and the distribution of ω_s is such that the constant-coupling (picket fence) model for the radiative state results. The radiative state $|\mathrm{R}\rangle$ has (average) energy ω_{R}, located at the center of the band. In addition there are low-lying excited states $|v\rangle$, for each of which $|\mathrm{R}\rangle$ is the radiative state. The light pulse is taken to be a negative-time Lorentzian

$$\mathscr{E}(t) = \mathscr{E}_0 \theta(-t) \exp[-i(\omega_0 + i\tfrac{1}{2}\gamma)t], \tag{53}$$

which is centered at ω_0 and has width γ:

$$\mathscr{E}(\omega) = -i\mathscr{E}_0/(\omega - \omega_0 - \tfrac{1}{2}i\gamma).$$

The important feature of our model is that we are using only a negative-time pulse, in contrast to the models frequently used which emphasize the role of positive-time decay of the light. We are interested in the behavior of the molecule for both positive and negative values of t.

Consider first the structure of the no-photon prepared state. Substitution of Eq. (53) into Eq. (49) yields

$$|u_{\mathrm{e}}(t)\rangle = \sum_s |s, \mathrm{vac}\rangle \boldsymbol{\mu}_{s0} \cdot \mathscr{E}_0 \exp[-i(\omega_0 + i\tfrac{1}{2}\gamma)t]/\omega_s - \omega_0 - i\tfrac{1}{2}(\gamma + \gamma_s)]. \tag{54}$$

Thus, the excited state amplitudes peak at ω_0. We assume that $\gamma \gg \gamma_s$, so the width of the excitation distribution is the width of the field, γ.

Examination of the probability distribution $P_s(t) = |\langle s|u_{\mathrm{e}}(t)\rangle|^2$ shows that energy from the pulse is deposited in the eigenstate manifold with a distribution given by the product of the light intensity and the absorption profile. For the limit $\gamma \ll \omega_{ss'}$ for all states, we have selective excitation of a single eigenstate, provided $|\omega_s - \omega_0| < \gamma$. We are interested primarily in $\gamma \gg \omega_{ss'}$, so that a large number of eigenstates are covered by the pulse. Obviously, for the limit $\gamma \gg \Gamma$, the entire absorption manifold is excited with a distribution given by the absorption profile.

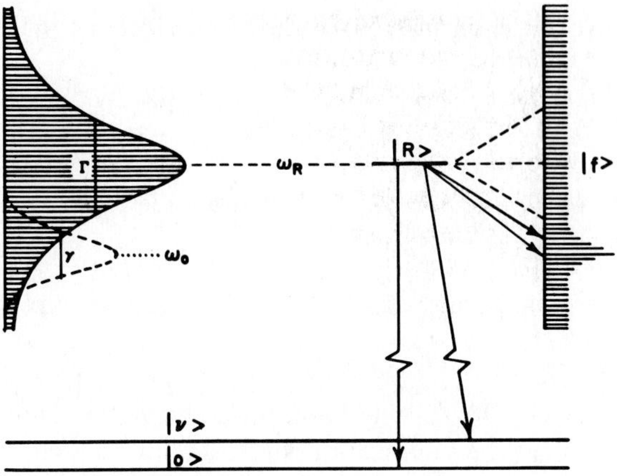

Fig. 1. Lorentzian absorption profile of molecular resonances (overall width Γ) and a negative time Lorentzian light pulse (width γ), with the corresponding coupling scheme in terms of the radiative state.

In order to obtain most of the other quantities of interest we need to know $\langle \mathrm{R, vac} | U_s(t - t_1) | \mathrm{R, vac} \rangle$ in Eq. (51). By expanding this in terms of molecular eigenstates, we obtain in the WW approximation

$$\langle \mathrm{R, vac} | e^{-iH(t-t_1)} | \mathrm{R, vac} \rangle = \sum_s |C_{\mathrm{R}s}|^2 \exp[-i(\omega_s - i\tfrac{1}{2}\gamma_s)(t - t_1)]. \quad (55)$$

The quantity $|C_{\mathrm{R}s}|^2$ is the probability distribution of $|\mathrm{R}\rangle$ among the eigenstate spectrum, which is just the normalized Lorentzian absorption profile

$$|C_{\mathrm{R}s}|^2 = (2\pi)^{-1}\Gamma/[(\omega_s - \omega_{\mathrm{R}})^2 + (\tfrac{1}{2}\Gamma)^2]. \quad (56)$$

Upon substituting Eq. (56) into (55), the summation may be replaced by an integration provided $(t - t_1) \ll \omega_{ss'}^{-1}$. Since Eq. (55) is to be used only in conjunction with $\mathscr{E}(t_1)$, this condition effectively holds provided $\gamma \gg \omega_{ss'}$, which is the range of our interest. Therefore, integration of Eq. (55) and neglecting the contribution of γ_s gives

$$\langle \mathrm{R, vac} | e^{-iH(t-t_1)} | \mathrm{R, vac} \rangle = \exp[-i(\omega_R - i\tfrac{1}{2}\Gamma)(t - t_1)], \quad (57)$$

not a surprising result.

Use of Eqs. (57) and (53) in Eq. (51) yields for $t \leq 0$

$$\langle \mathrm{R} | u_e(t) \rangle = \boldsymbol{\mu}_{\mathrm{R}0} \cdot \mathscr{E}_0 \exp[-i(\omega_0 + i\tfrac{1}{2}\gamma)t]/[\omega_{\mathrm{R}} - \omega_0 - i\tfrac{1}{2}(\Gamma + \gamma)], \quad (58)$$

and for $t > 0$

$$\langle \mathrm{R} | u_e(t) \rangle = \boldsymbol{\mu}_{\mathrm{R}0} \cdot \mathscr{E}_0 \exp[-i(\omega_{\mathrm{R}} - i\tfrac{1}{2}\Gamma)t]/[\omega_{\mathrm{R}} - \omega_0 - i\tfrac{1}{2}(\Gamma + \gamma)], \quad (59)$$

which shows that P_{R} builds up with the light pulse, but decays with rate constant Γ after the light is turned off.

Similarly, use of Eq. (57) and (53) in Eq. (52) yields for $t \leq 0$

$$\langle v, k|u(t)\rangle = \mathbf{e}_k \cdot \boldsymbol{\mu}_{v\mathrm{R}} \frac{\boldsymbol{\mu}_{\mathrm{R}0} \cdot \mathscr{E}_0 \exp[-i(\omega_0 + i\tfrac{1}{2}\gamma)t]}{(\omega_{kv} - \omega_0 - i\tfrac{1}{2}\gamma)[\omega_{\mathrm{R}} - \omega_0 - i\tfrac{1}{2}(\Gamma + \gamma)]}, \tag{60}$$

where $\omega_{kv} = \omega_k + \omega_v$; and for $t > 0$, the amplitude for scattering over the interval 0 to t is given by

$$\langle v, k|u(t, 0)\rangle = \mathbf{e}_k \cdot \boldsymbol{\mu}_{v\mathrm{R}} \frac{\boldsymbol{\mu}_{\mathrm{R}0} \cdot \mathscr{E}_0(\exp[-i(\omega_{\mathrm{R}} - i\tfrac{1}{2}\Gamma)t] - \exp(-i\omega_{kv} t))}{(\omega_{kv} - \omega_{\mathrm{R}} + i\tfrac{1}{2}\Gamma)(\omega_{\mathrm{R}} - \omega_0 - i\tfrac{1}{2}(\Gamma + \gamma))}. \tag{61}$$

This is obtained from Eq. (44) by setting the lower integration limit equal to zero. The vector $\mathbf{e}_k$ specified the polarization and contains the field factors for the emitted photon. Inspection shows that the emission spectra are quite different for the two time ranges. For $t \leq 0$, the emission spectrum is like that of the incident light and the intensity builds up with the light pulse; for $t > 0$, the spectrum is that of the state $|\mathrm{R}\rangle$ decaying into the manifold of states $|f\rangle$ with decay constant Γ.

A reasonable interpretation of these results is that all of the emission for $t \leq 0$ is Raman scattering (or Rayleigh scattering for $v = 0$). The effects of approaching resonance, $\omega_0 \to \omega_{\mathrm{R}}$, are purely quantitative. On the other hand, for $t > 0$, the emission is ordinary fluorescence, but may be termed resonance fluorescence as $\omega_0 \to \omega_{\mathrm{R}}$. The reason we have obtained such a clean separation of the two cases is that the light pulse is turned off suddenly at $t = 0$. If we had included a positive-time Lorentzian pulse with decay constant γ', there would have been an additional scattering component (Rousseau and Williams, 1976) whose dependence on $\omega_{\mathrm{R}} - \omega_0$ would be different from Eq. (61).

The total probability that a photon has been emitted is obtained by integrating the absolute value squared of Eqs. (60) and (61) over all photon modes k and summing over molecule states $|v\rangle$. This gives, for $t \leq 0$,

$$P_{\mathrm{sc}}(t) = \gamma_v' |\boldsymbol{\mu}_{\mathrm{R}0} \cdot \mathscr{E}_0|^2 e^{\gamma t}/\gamma[(\omega_{\mathrm{R}} - \omega_0)^2 + \tfrac{1}{4}(\Gamma + \gamma)^2], \tag{62}$$

with $\gamma_v' \equiv (\omega_0/\omega_{\mathrm{R}})^3 \gamma_v$, where γ_v is the total decay constant for radiative decay from $|R\rangle$ to all $|v\rangle$. For $t > 0$ Eq. (61) gives the total single-photon fluorescence probability. In the large t limit this is

$$P_{\mathrm{fl}}(t \to \infty) = \gamma_v |\boldsymbol{\mu}_{\mathrm{R}0} \cdot \mathscr{E}_0|^2/\Gamma[(\omega_{\mathrm{R}} - \omega_0)^2 + \tfrac{1}{4}(\Gamma + \gamma)^2]. \tag{63}$$

Note that we have tacitly ignored any recurrences that exist due to the fact that the excited state manifold is discrete. These were eliminated by use of integration on ω_s for all values of $t > 0$. Thus, the ratio of total scattering to total fluorescence is Γ/γ, which is independent of $\omega_0 - \omega_{\mathrm{R}}$.

The total probability at $t \leq 0$ that the molecule is in the no-photon excited manifold is obtained by taking the norm of Eq. (54), with use of Eq. (56) and integrating over ω_s. Thus,

$$P_e(t) = (\Gamma + \gamma)|\boldsymbol{\mu}_{R0} \cdot \mathscr{E}_0|^2 e^{\gamma t}/\gamma[(\omega_R - \omega_0)^2 + \tfrac{1}{4}(\Gamma + \gamma)^2]. \tag{64}$$

The total probability that a photon has been absorbed at $t \leq 0$ is obtained by the sum of Eqs. (62) and (64),

$$P_a(t) = [(\Gamma + \gamma + \gamma_v')/\gamma]Be^{\gamma t}, \tag{65}$$

where B is defined by the equations. Corresponding relations for $P_R(t)$ are obtained from Eqs. (58) and (59), which yield $P_R = Be^{\gamma t}$ and $P_R = Be^{-\Gamma t}$ for $t \leq 0$ and $t > 0$, respectively.

The above probabilities produce the quantum yields:

Scattering ($t \leq 0$);

$$Q_{sc} = P_{sc}(t)/P_a(t) = \gamma_v'/(\Gamma + \gamma + \gamma_v'),$$

Fluorescence,

$$Q_{fl} = P_{fl}(t \to \infty)/P_a(0) = \gamma\gamma_v/\Gamma(\Gamma + \gamma + \gamma_v'). \tag{66}$$

It is instructive to consider the physical meaning of these results for various values of the parameters γ (which determines the duration of the pulse), Γ (the rate of radiationless decay of $|R\rangle$), and γ_v (the rate of radiative decay from $|R\rangle$ to all $|v\rangle$). Normally γ_v' [cf. Eq. (62)] is much smaller than Γ and can be neglected in the denominator term of Eq. (66). For $\gamma \gg \Gamma$ (short pulse) Q_{sc} tends to be small while Q_{fl} increases to a plateau value γ_v/Γ. This is physically reasonable since the brief pulse tends to prepare the molecule in state $|R\rangle$, with little time for scattering, and $|R\rangle$ decays ($t > 0$) radiatively and radiationlessly with the standard branching ratio.

For the opposite limit, $\gamma \ll \Gamma$ (long pulse), Q_{sc} increases to the constant branching ratio, $\gamma_v'/(\Gamma + \gamma_v')$, as γ decreases, while Q_{fl} tends toward zero. Thus, Q_{sc} and Q_{fl} complement one another upon variation of γ. It is interesting that as γ decreases the values of $P_e(0)$ and P_{sc} increase, although the spectrum of the light pulse covers less and less of the absorption spectrum. The reason for this is that the duration of the pulse is longer for smaller γ, a property which manifests itself in an increased peak height of the frequency spectrum of the light.

The physical aspects of the excitation and decay can be understood in terms of either the molecular eigenstate basis, $\{|s\rangle\}$, or the zeroth-order basis, $|R\rangle$ and $\{|f\rangle\}$, in Fig. 1. There are three important cases:

(1) $\gamma \gg \Gamma \gg \gamma_v$ and $\omega_0 = \omega_R$. This corresponds to broad band, short pulse excitation in which every molecular eigenstate is excited in phase and with a magnitude dictated by the absorption intensity (we call this a coherent superposition of eigenstates). For $t > 0$, these excited states change in time according to $\exp(-i\omega_s t)$ and the phase of the photon emission amplitude changes, correspondingly, in a manner that produces destructive interference. This dephasing results in loss of radiative decay rate which follows $\exp(-\Gamma t)$. The zeroth-order basis provides the more conventional picture. The initial coherent superposition of $|s\rangle$ is identical to $|R\rangle$, which undergoes competitive radiationless

decay into $\{|f\rangle\}$ and radiative decay into $\{|v, k\rangle\}$. This is an example of selection with projection (cf. Section II).

(2) $\Gamma \gg \gamma \gg \omega_{ss'}$ and $\omega_0 \neq \omega_R$. This corresponds to narrow-band, off-resonance excitation as is shown schematically in Fig. 1. The narrow pulse brackets many states $|s\rangle$. According to Eq. (54) the incident light selectively favors excitation of those eigenstates lying within γ, i.e., there is phase matching between the components of the field and those eigenstates within γ which results in emission amplitudes being favored for this region. At $t = 0$ the light is turned off and the molecular eigenstates in the neighborhood of ω_0 are out of phase. Phase matching at $t = 0$ occurs weakly over the entire spectrum, thus giving weak fluorescence with decay constant Γ for $t > 0$. This is an example of selection without projection.

In terms of the zeroth-order basis the long duration pulse excites the molecule via (virtual) transitions through $|R\rangle$. Because of the frequency distribution of the light the excitation is selectively channeled through $|R\rangle$ to the manifold $|f\rangle$ within the range of γ. However, the phase relations among the $|f\rangle$ change during excitation in such a way that the superposition of $|f\rangle$ will not return to $|R\rangle$ during the time interval $2\pi\varepsilon^{-1}$, where ε is the spacing between the $|f\rangle$. In other words, because of the phase relations, $|R\rangle$ channels irreversibly into $\{|f\rangle\}$ (for the positive t interval $2\pi\varepsilon^{-1}$). This means that only the $|R\rangle$ component of $|u_e(0)\rangle$ contributes to fluorescence and that component decays just as in case 1. For both case 1 and case 2 the $t > 0$ fluorescence is from the state $|R\rangle$. The fluorescence intensity depends on P_R, Eq. (59), but another important quantity is the fraction of the excited (no-photon) state which is $|R\rangle$, given by

$$P_R/P_e = \gamma/(\Gamma + \gamma). \tag{67}$$

For the present case this is a small fraction. The remainder is in the radiatively inactive $|f\rangle$ manifold.

(3) $\Gamma \gg \omega_{ss'} \gg \gamma$ and $\omega \cong \omega_s$. This is very narrow band excitation in which only a single eigenstate $|s\rangle$ becomes excited to an appreciable extent. The result is that only $|s\rangle$ exhibits fluorescence and with time dependence $\exp(-\gamma_s t)$. This is the ultimate limit of selective excitation and is a feature which Zewail (Zewail *et al.*, 1977a, b; Orlowski *et al.*, 1977; DeVries *et al.*, 1977; Orlowski and Zewail, 1979) has attempted to find experimentally. Case 1 and case 3 represent the two extremes to which Zewail alludes; however, because of the extremely high density of states in large molecule spectra, it is very doubtful that the case 3 limit can be reached.

E. Complex Spectral Structure

Let us turn now to some more general considerations of real molecule spectra (Rhodes, 1977). In the electronic transition range there is always a certain amount of band structure which becomes more and more diffuse as ω increases into the continuum region. Sometimes, however, a higher-lying band has more discrete structure than those at lower frequencies, as with the $S_2 \leftarrow S_0$ band of naphthalene (Fig. 4). Our picture of a typical electronic absorption spectrum is one consisting of isolated bands at lower frequency and continuous absorption with various amounts of structure superposed at higher ω.

Consider now the prepared state $|u_e(t)\rangle$ resulting from excitation by a pulse centered at ω_0 and having width γ. We have seen that it is convenient to express $|u_e(t)\rangle$ in terms of $|\mathrm{R}\rangle$ as in Eq. (50). However, $|\mathrm{R}\rangle$ is the radiative state for the entire molecular spectrum, which is generally unknown. Thus, it is impractical to consider using $|\mathrm{R}\rangle$ alone in any quantitative study of $|u_e(t)\rangle$. On the other hand, most spectra consist of bands which are isolated to some extent. These bands or sets of bands are separated by distinct gaps where the absorption is weak. Suppose that the incident light spectrum overlaps totally, or partially, a certain number of bands which are distinct from the rest of the spectrum. We may define a projection operator P_1 which projects all of the molecular eigenstates within this set of isolated bands. If P_0 is its complement then we have

$$|\mathrm{R}\rangle = c_1|\mathrm{R}_1\rangle + c_0|\mathrm{R}_0\rangle, \tag{68}$$

with $c_1|\mathrm{R}_1\rangle = P_1|\mathrm{R}\rangle$ and $c_0|\mathrm{R}_0\rangle = P_0|\mathrm{R}\rangle$. The c's are normalization constants. Substitution of Eq. (68) into Eq. (50) reveals (in the molecular eigenstate basis) that $|\mathrm{R}_0\rangle$ contributes only rapidly oscillating terms which are vanishingly small upon integration. Thus, only $|\mathrm{R}_1\rangle$ makes a significant contribution to $|u_e(t)\rangle$.

The result of the above analysis is that any distinctly isolated region of the absorption spectrum may be used to define a projected radiative state $|\mathrm{R}_1\rangle$, which determines the excitation response of the molecule to light pulses which fall within this region of the spectrum. Of course, the choice of spectral region is usually not unique for a given pulse. Suppose P_1 defines the entire spectral region range considered in a given experiment. If the light pulse falls well within one or more bands making up a small part of the total range, then it is possible to choose a smaller subspace which brackets the pulse. Successively smaller subspaces, each of which contains the bands effectively coupled to the pulse, are projected by $P_2, P_3, \ldots,$ etc.

The smallest possible subspace which contains all of the essential spectral structure for describing the response to a given pulse will be denoted by the projector P_m and the corresponding projected radiative state $c_m|\mathrm{R}_m\rangle = P_m|\mathrm{R}\rangle$. This subspace brackets the frequency range $\omega_{21} = \omega_2 - \omega_1$. Clearly, ω_{21} is determined not only by the light pulse, but by the spectral structure on the edges of the pulse, as well. For example, in Fig. 1 it would be incorrect for ω_{21} to correspond to the pulse, since we have seen that all of the states in the Lorentzian manifold are active in the fluorescence of the molecule. Therefore, ω_{21} must bracket the entire absorption band.

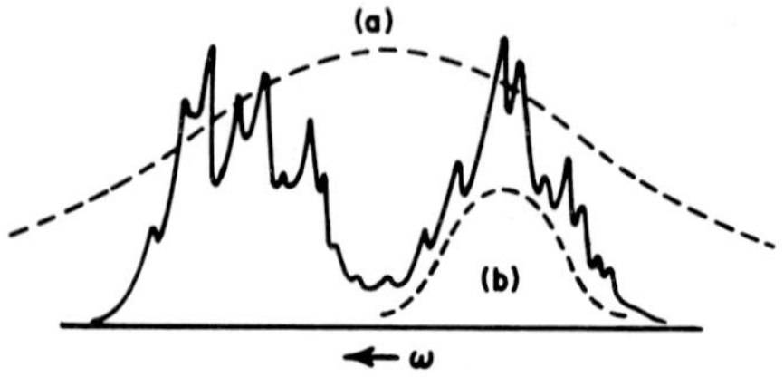

Fig. 2. Model molecular absorption spectrum originating from one allowed Born–Oppenheimer state, with (a) excitation by broad-band light and (b) narrow-band light.

Consider now the model spectrum in Fig. 2. Suppose that the entire spectrum shown is well isolated and that it corresponds to a single vibronic band; i.e., all of the structure arises from a single allowed BO state which is coupled to the vibronic manifolds of lower-lying electronic states. The entire spectrum, as shown, defines the projector P_1 and the radiative state $|\mathrm{R}_1\rangle$.

We wish to compare the results of excitation with a broadband pulse (a) and a narrow band pulse (b). For the broad band case we must use the full spectrum and its radiative state $|\mathrm{R}_1\rangle$, which is identical to the allowed BO state. If γ is much greater than the width of the spectrum, $|u_e(0)\rangle$ is almost identical to $|\mathrm{R}_1\rangle$ for a negative time pulse, as Eq. (51) shows. This is short pulse excitation which prepares the molecule in the allowed BO state and corresponds to selection with projection. That portion of the fluorescence due to $|\mathrm{R}_1\rangle$ is proportional to probability $P_{\mathrm{R}_1}(t)$, which undergoes a complicated oscillating decay due to the interaction of $|\mathrm{R}_1\rangle$ with the radiative continuum and with the manifold of secondary states $|f\rangle$ (other BO states).

Next consider the narrow band case (b). Here, the effective radiative state is $|\mathrm{R}_m\rangle$, the projected radiative state for that portion of the spectrum covered by the major band system on the right. Since γ is nearly equal to ω_{21}, it follows that $|u_e(0)\rangle$, for a negative time pulse, is not $|\mathrm{R}_m\rangle$; rather, it is a superposition of $|\mathrm{R}_m\rangle$ and the manifold $|f\rangle$. However, $|\mathrm{R}_m\rangle$ is the principal component and the fluorescence due to $|\mathrm{R}_m\rangle$ is proportional to $P_{\mathrm{R}_m}(t)$ which undergoes oscillating decay. In this case the oscillations are due to beats among the peak frequencies of the $|\mathrm{R}_m\rangle$ band, so the strong high frequency beat between the two band system, which was present in case (a), is missing here.

It should be emphasized that P_1 and $|\mathrm{R}_1\rangle$ could have been used as the basis of description in case (b). In fact, this would be more accurate for a description of scattering, since the contribution of the off-resonance band system would be included. But for resonance fluorescence ($t > 0$) this would be a more cumbersome description since the prepared state $|u_e(0)\rangle$ is so far removed from $|\mathrm{R}_1\rangle$ and contains so few of its higher frequency components.

The most important physical aspect of case (b) is that the prepared state is not even close to being the allowed BO state; i.e., we have selection without projection. Rather, it is a strong superposition of the allowed BO state $|R_1\rangle$ and the forbidden BO manifold $|f\rangle$. This is a situation for which the conventional spectroscopic description is not valid. The $S_2 \leftarrow S_0$ vibronic bands of naphthalene, Fig. 4, are a good example of a spectrum for which selective excitation may not prepare an excited BO state. There are probably many more cases than is commonly recognized.

Insofar as Fig. 2 represents a molecular spectrum, it is understood that each of the resonances is composed of many molecular eigenstates. For practical reasons it would probably not be possible to excite a single eigenstate. In fact, it would not be possible even to excite a single one of the resonances independently since there is such strong overlap. This situation is probably typical for the homogeneous spectra of intermediate size and some large molecules. It is suggested, therefore, that the selective excitation experiments of the type previously done (Zewail *et al.*, 1977a, b; Orlowski *et al.*, 1977; DeVries *et al.*, 1977; Orlowski and Zewail, 1979) could involve the selection of one or more resonances in the fine structure of a vibronic band, each resonance consisting of many molecular eigenstates and being far removed from a single BO state.

In this section we have considered the linear response properties of a molecule to light pulses which are coherent to at least first order. The general formulations of the prepared excited and photon scattering states, the results for a single Lorentzian profile of discrete transitions, and the case of a single vibronic band having complex structure each point out the following basic fact: The essential portion of the spectrum which gives the simplest description of the dynamics of excitation and decay depends on both the molecular absorption spectrum and the spectral properties of the exciting light pulse.

IV. SPECTRAL ANALYSIS OF COUPLING SCHEMES

In this section we consider the case in which the molecule portion of the prepared state $|u_e(0)\rangle$ is a projected radiative state $|R_m\rangle$. This occurs whenever the light pulse completely brackets an isolated portion of the spectrum, with the pulse width much greater than the width of the isolated band system. The study of absorption and emission (exclusive of scattering) then centers around the formation and decay of $|R_m\rangle$, which is determined solely by the isolated portion of the absorption spectrum. The important quantity is $\langle R_m, \text{vac}|U_s(t)|R_m, \text{vac}\rangle$, which is the amplitude that the molecule field is in the state $|R_m, \text{vac}\rangle$ at time t given that it is in that state at $t = 0$. Note that this kind of matrix element is important in Eqs. (51) and (55), and related formulas.

Our approach will be to consider an isolated portion of the absorption spectrum, as in Fig. 3, as though it is the entire molecular spectrum. Thus, in the following $|R\rangle$ will replace $|R_m\rangle$ and the expression of the field state $|\text{vac}\rangle$ will be suppressed where there is no ambiguity. The matrix element $\langle R|U_s(t)|R\rangle$

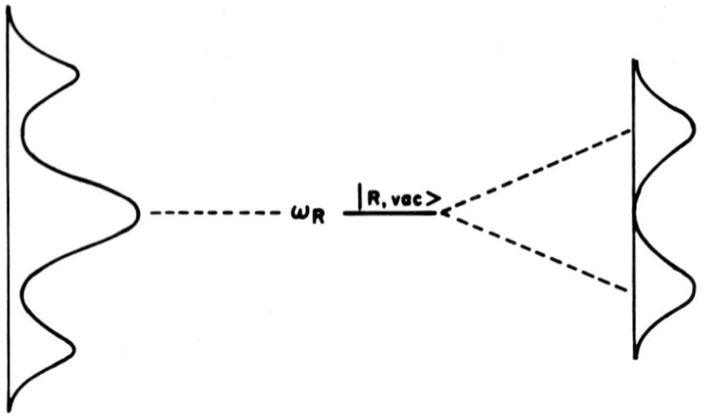

Fig. 3. Model absorption spectrum (left) with the associated radiative state and its interaction density spectrum (right).

is best analyzed formally in terms of the Laplace transform of $U_s(t)$, which is the Green's operator (Messiah, 1966)

$$G(\omega) = (\omega - H)^{-1}. \tag{69}$$

Because of the assumption that $U(t) = 0$ for $t < 0$, $G(\omega)$ is analytic in the upper half plane.

In terms of $G(\omega)$ the absorption spectrum is proportional to

$$\sigma_a(\omega) \sim -\text{Im}\langle 0, \omega | V_r G(\omega) V_r | 0, \omega\rangle, \tag{70}$$

where $|0, \omega\rangle$ is the ground state of the molecule and the single-photon state for mode ω and Im represents the imaginary part. By definition of $|\text{R}\rangle$ this can be rewritten as

$$\sigma_a(\omega) \sim -|V_{\text{R}0}|^2 \text{ Im } \mathscr{G}_{\text{R}}(\omega), \tag{71}$$

where

$$\mathscr{G}_{\text{R}}(\omega) \equiv \langle \text{R, vac}| G(\omega) | \text{R, vac}\rangle.$$

Thus we see that $\mathscr{G}_{\text{R}}(\omega)$ determines both the absorption spectrum and the time dependence of $|\text{R, vac}\rangle$.

By use of standard technique (Messiah, 1966) we may express $\mathscr{G}_{\text{R}}(\omega)$ in the standard form

$$\mathscr{G}_{\text{R}}(\omega) = [\omega - \omega_{\text{R}} - \mathscr{F}_{\text{R}}(\omega)]^{-1}, \tag{72}$$

where ω_{R} is the average frequency (center of gravity) of the absorption band (Fig. 3) and $\mathscr{F}_{\text{R}}(\omega)$ is the matrix element of the level shift operator. The latter is defined by decomposing the total Hamiltonian into a zeroth-order part H_0, of which $|\text{R, vac}\rangle$ is, by definition, an eigenvector with eigenvalue ω_{R}. Thus,

$$H = H_0 + V_0 + V_r = H_0 + V, \tag{73}$$

whereby the level shift operator is

$$F(\omega) = V + VG_1(\omega)V. \tag{74}$$

The reduced Green's operator G_1 is defined solely on the subspace $Q \equiv I - |\mathrm{R, vac}\rangle\langle\mathrm{R, vac}|$; namely, $G_1 = [Q(\omega - H)Q]^{-1}$. By construction the diagonal elements of V vanish, so

$$\mathscr{F}_{\mathrm{R}}(\omega) = \langle \mathrm{R, vac}|VG_1(\omega)V|\mathrm{R, vac}\rangle. \tag{75}$$

This can be separated into two terms: (1) the primary radiative part, due to V_r, which accounts for the direct radiative decay of $|\mathrm{R, vac}\rangle$; and (2) a nonradiative part due to V_0, which accounts for the fact that $|\mathrm{R}\rangle$ is not an eigenstate of H_m. If we assume that the former is independent of ω, then it becomes purely imaginary and we have

$$\mathscr{F}_{\mathrm{R}}(\omega) = -i\tfrac{1}{2}\Gamma_{\mathrm{R,r}} + |V_{1\mathrm{R}}|^2\mathscr{G}_1(\omega), \tag{76}$$

where $V_{1\mathrm{R}} = \langle 1|V_0|\mathrm{R, vac}\rangle$ defines the doorway state for V_0 acting on $|\mathrm{R, vac}\rangle$ and $\mathscr{G}_1(\omega)$ is the diagonal element of G_1 for the state $|1\rangle$.

We are now in a position to discuss the physical meaning and the pattern of the relations emerging above. $\mathscr{G}_1(\omega)$ plays a role analogous to $\mathscr{G}_{\mathrm{R}}$, in that its Laplace transform gives the time dependence of $|1\rangle$ and its imaginary part determines the spectrum of transitions from the state $|\mathrm{R, vac}\rangle$ due to the interaction V_0. The latter is represented by

$$\tfrac{1}{2}\Gamma_{\mathrm{R},n}(\omega) = -|V_{1\mathrm{R}}|^2 \,\mathrm{Im}\, \mathscr{G}_1(\omega), \tag{77}$$

which we refer to as the *interaction density spectrum* (IDS) for the state $|\mathrm{R}\rangle$. Note the parallel relation for the absorption spectrum of the ground state, Eqs. (70) and (71), which implies that the IDS is the spectrum for radiationless transitions from $|\mathrm{R, vac}\rangle$. Figure 3 includes the qualitative features of the IDS derived from the model absorption spectrum.

There exists an interesting parallel between the absorption spectrum and the IDS. The absorption spectrum is the spectrum of radiative interaction of the states $|0, \omega\rangle$ with the excited state manifold, where the state $|0, \omega\rangle$ is the ground molecule state and one photon of an appropriate mode of frequency ω. Since the zero of energy is chosen to be that for the state $|0, \mathrm{vac}\rangle$, the result is that ω is the photon frequency required for energy conserving transitions into the excited state manifold. The IDS has a similar interpretation, except that the starting state is $|\mathrm{R, vac}\rangle$, of energy ω_{R}, and the interaction generating the spectrum is nonradiative. The frequency $\omega - \omega_{\mathrm{R}}$ is the energy required for energy conserving nonradiative transitions from $|\mathrm{R, vac}\rangle$ into the remaining manifold of states. Such frequencies could be provided by a phonon source or sink (depending on the sign of $\omega - \omega_{\mathrm{R}}$) which couples to the molecule. This involves the concept of *phonon assisted radiationless transitions* which could be useful in designing experiments on specific medium effects.

Because of its relation to both the absorption spectrum and the IDS the state $|R, vac\rangle$ is placed between them in Fig. 3. Notice that the normalized absorption spectrum is simply $-\pi^{-1}$ Im $\mathscr{G}_R(\omega)$. Insofar as the absorption is composed of discrete transitions to molecular eigenstates the normalized spectrum is given by $|C_{sR}|^2$, the probability distribution of $|R\rangle$ among the molecular eigenstates [Eq. (55)]. Similarly, the normalized IDS is given by $-\pi^{-1}\mathscr{G}_1(\omega)$, but this is determined uniquely by $|R, vac\rangle$ and the total Hamiltonian.

Berg *et al.* (1974) and Robinson and Langhoff (1974) were first to emphasize the importance of the IDS, which they refer to as the *weighted density function.* They pointed out the peak to valley correlation tendency for the two spectra and developed a trial and error method of computing the IDS from the absorption spectrum which they applied to naphthalene (Langhoff and Robinson, 1974). Shortly afterwards, a rigorous, analytical method for calculating the IDS was developed independently by Berg (1976) and by Ziv and Rhodes (1976). This method has been used by Brand to calculate the IDS for regions of the NO_2 spectrum (Brand and Hoy, 1977).

Figure 4 shows the spectrum of the (0, 0) vibronic band of the $S_2 \leftarrow S_0$ transition of naphthalene, together with the corresponding IDS (Cable and Rhodes, 1980). Naphthalene has an unusually large amount of structure in the

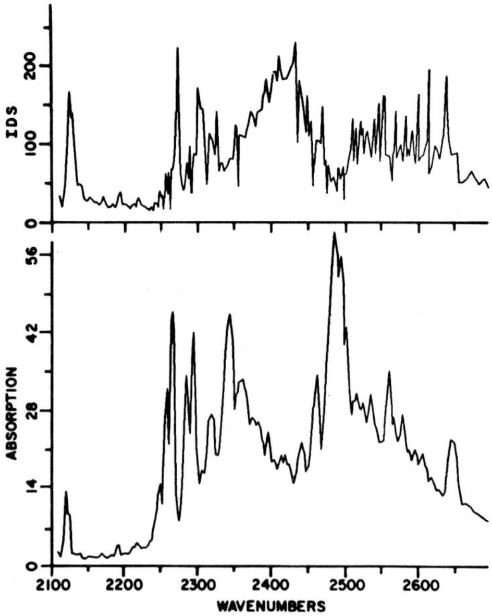

Fig. 4. Portion of the naphthalene absorption spectrum (bottom) and its derived interaction density spectrum (top). The absorption region shown corresponds to the 0-0 vibronic band of the $S_2 \leftarrow S_0$ electronic transition and was taken from Wessel (1970).

region of the spectrum, presumably due to the low density of states in this spectral region. There is a definite peak to valley relation between the two spectra which (as will be shown below) is related to the value of $|V_{1R}|^2$. These results agree well with those of Berg (1976).

In order to make comparison with small molecule spectra having sparse spectral structure we have repeated Brand's NO_2 calculation on a portion of the $^2B_2 \leftarrow {}^2A_1$ band which Brand believes, on the basis of rotational constant analysis, to originate from a single allowed BO state (Brand and Hoy, 1977). The results (Cable and Rhodes, 1979) are given in Fig. 5. All of the resonances in the portion of the absorption spectrum shown presumably arise from coupling of the single allowed BO state, belonging to the $^2B_2$ electronic state, with the highly excited vibrational states of the ground $^2A_1$ electronic state. Using a simple linelike spectrum brings out some of the elementary features, such as the fact that the number of resonances in the IDS should be one less than its parent spectrum.

It is instructive to consider some of the quantitative features of this kind of spectral analysis. The normalized absorption spectrum is the only information needed to obtain ω_R, $|V_{1R}|^2$, and $-\pi \operatorname{Im} \mathcal{G}_1(\omega)$ (the normalized IDS). Let us refer to $|V_{1R}|^2$ as the *interaction strength* of the IDS and to $-\pi^{-1} \operatorname{Im} \mathcal{G}_1(\omega)$ as the *structure* of the IDS. In a reverse sense it is the relation between the interaction strength and the structure, together with ω_R, which determines the normalized absorption spectrum. Consider the example in Fig. 3. We may regard the normalized absorption spectrum as arising from the interaction of $|R\rangle$ and two

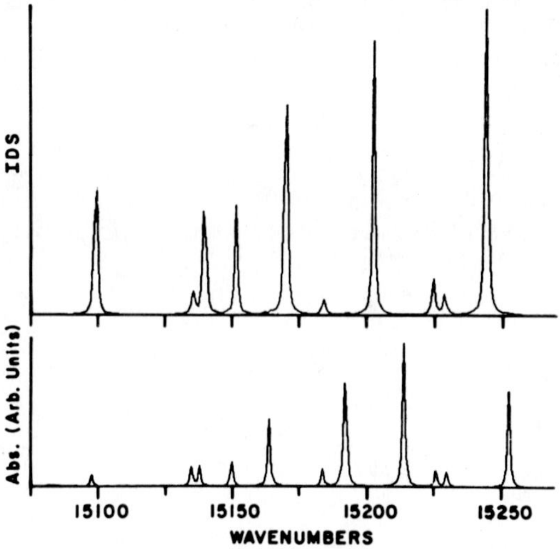

Fig. 5. Portion of the $^2B_2 \leftarrow {}^2A_1$ absorption band of NO_2 (below) presumed to originate from a single allowed vibronic state (cf. Brand and Hoy, 1977). The corresponding interaction density spectrum is shown at the top. Absorption data were taken from Smalley *et al.* (1975).

distinct resonances, each represented by one of the bands of the IDS. This interaction produces three resonances in the absorption spectrum, with the greatest intensity residing in the central resonance, which therefore has the largest component of $|R\rangle$. These features indicate that $|V_{1R}|^2$ is of moderately weak coupling strength, so that the three resonances in the absorption spectrum maintain their identity to a great extent in correlating with $|R\rangle$ and the two resonances of the IDS. If the coupling had been in the strong limit, the two outer resonances of the absorption would have been split farther apart and most of the intensity would have resided there.

The structures of Eqs. (71) and (77) indicate clearly that it is possible to continue the process, where the IDS for $|R, \text{vac}\rangle$ is used to generate the IDS for the state $|1\rangle$. The concept and method for this has been discussed in detail by Ziv and Rhodes (1976). The processes can be repeated until either all resonances are depleted (for the case of a strictly discrete spectrum) or presumably until the IDS is essentially a constant. The latter presumption is based on the idea that all bound state spectra of molecules consist of a finite number of resonances. For a continuous spectrum, the result of iterated IDS formulation is a sequence of doorway states, each coupled to its nearest neighbors, in which the last one is coupled uniformly to a continuum. This basis is termed the spectroscopic channel basis (SCB). The existence of c members of the SCB implies that the absorption spectrum consists of c resonances. Ziv has developed the procedure for formulating the molecular resonances from the SCB (Ziv and Rhodes, 1976). The excited states associated with these resonances are referred to as *effective system states*, since they are eigenfunctions of an effective, non-Hermitian Hamiltonian. Ziv is continuing to study the properties of these states (Ziv, 1978).

Figure 6 shows the (qualitative) results of continuing the SCB formulation for the spectrum shown in Fig. 3. The IDS for $|R, \text{vac}\rangle$ is on the left. The state $|1\rangle$ has an IDS with a single resonance and, insofar as it is Lorentzian, the IDS of the resulting doorway state $|2\rangle$, is a constant.

We may use the results of Fig. 6 to discuss a fundamental feature of the dynamics associated with these coupling schemes. Suppose the system were pre-

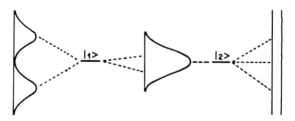

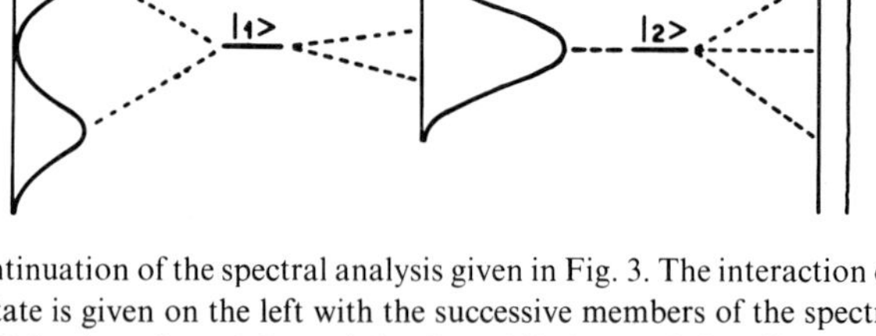

Fig. 6. A continuation of the spectral analysis given in Fig. 3. The interaction density spectrum of the radiative state is given on the left with the successive members of the spectroscopic channel basis and their IDSs given from left to right. State $|2\rangle$ is coupled uniformly to a final quasi-continuum.

pared at $t = 0$ in the state $|2\rangle$ and V_{12} ignored. The constant IDS of $|2\rangle$ specifies clearly that the probability of $|2\rangle$ decays exponentially with decay constant Γ_2. On the other hand, suppose the system were prepared at $t = 0$ in state $|1\rangle$. The shape alone of the IDS for $|1\rangle$ is not sufficient to determine the qualitative dynamics of $|1\rangle$. Knowledge of the interaction strength $|V_{12}|^2$ also is required. For very weak coupling, $|V_{12}| \ll \Gamma_2$, the time dependence of $|1\rangle$ is almost exponential decay, but in this limit the parent spectrum for $|1\rangle$ would also contain only a single resonance. As the case is, the parent spectrum is bimodal, showing that the strong coupling limit exists ($|V_{12}| \gg \Gamma_2$). In this case the time dependence of $|1\rangle$ exhibits oscillatory decay.

Another interesting feature of the spectral analysis of such coupling schemes is that the absorption spectrum itself may be regarded as the IDS for the ground state of the molecule, as was alluded to above. Suppose we have the molecule in its ground state with a single photon of frequency ω_0. The corresponding state $|0, \omega_0\rangle$ has an energy on the spectral scale equivalent to that for some excited state $|s, \text{vac}\rangle$. Thus the absorption spectrum is simply the IDS for the state $|0, \omega_0\rangle$; consequently the time dependence of the probability of $|0, \omega_0\rangle$ is completely determined by the absorption spectrum.

It would be suggestive to continue the process further in the reverse direction and to determine the parent spectrum for the absorption spectrum of the molecule. However, one runs into the question of the physical meaning of the doorway states obtained. One particular case is worth noting. Suppose the absorption spectrum of the ground state is a single Lorentzian of width Γ. Furthermore, instead of a single photon initial state, suppose we have a single-frequency mode with n photons, where n can be arbitrarily large. Thus, the starting state is $|0, n\rangle$ and the variation of n provides a way of varying the coupling strength $|V_{\mathrm{R}0}|^2$ or, equivalently, the Rabi frequency $|V_{\mathrm{R}0}|$. Now, for $|V_{\mathrm{R}0}| \ll \Gamma$ the initial state $|0, n\rangle$ decays monotonically into the excited state. This is the weak coupling saturation limit of ordinary spectroscopy. However, if $|V_{\mathrm{R}0}| \gg \Gamma$, we have the limit of nonlinear optical resonance, whereby the state $|0, n\rangle$ oscillates with the excited state $|\mathrm{R}, n - 1\rangle$. This produces optical nutation and dynamic Stark splitting. The basis of these phenomena can be more easily seen in the parent spectrum for the single Lorentzian molecular absorption spectrum. In Fig. 6, $|0, n\rangle$ corresponds to $|1\rangle$ and $|\mathrm{R}, n - 1\rangle$ corresponds to $|2\rangle$. For strong coupling the parent spectrum is bimodal, as in Fig. 6, and the two resonances correspond to the correlated molecule-photon states (Cohen-Tannoudji, 1977)

$$|\mathrm{r}_{\pm}\rangle = 2^{-1/2}(|0, n\rangle \pm |\mathrm{R}, n - 1\rangle). \tag{78}$$

Dynamic Stark splitting results from the absorption (or emission) by the resonance states [Eq. (78)]. This is a good example of a physical phenomenon being associated with the analysis of a molecular absorption spectrum in the reverse direction.

V. SUMMARY: PERSPECTIVES AND PREDICTIONS

In this chapter we have discussed a new, dynamical line of thinking concerning molecular excitation by light and other radiation sources. This line of thinking has emerged over the past decade in our and other laboratories as a result of the revolutionary impact made by the rapid development of laser technology.

One of the major purposes of this chapter has been to establish basic premises concerning the nature of molecular spectra and of molecular excitation by light. These premises have evolved from our own philosophy of approach regarding the conceptualization and description of radiative and radiationless processes, the basic feature of which is that the time dependence of all such processes is a manifestation of the time dependence of the amplitudes of all states of the system. Dynamics results from dephasing of these amplitudes, as prescribed by the basic tenets of quantum theory. The most elementary aspect of our approach, consequently, is that light absorption prepares the molecule in a state in which the molecular eigenstates (or resonances) have prescribed phase relations determined by the absorption spectrum and the light.

The structure of the absorption spectrum is obviously very important in the dynamics of excitation, and it is important to eliminate artifacts and to distinguish heterogeneous and homogeneous structure. It should be noted, however, that heterogeneous structure may provide information on intermolecular aspects of photon processes. A heterogeneous spectrum has (projected) radiative states which are multimolecule in character. The analysis of their interaction density spectra by the method described in the previous section can be useful for understanding the intermolecular features of coherent excitation for both single and multiphoton processes.

The formulations discussed here have led to several results and predictions:

(1) The excitation of molecules by chaotic light from conventional sources tends to prepare a statistical mixture of molecular resonances among which there is an absence of interference and associated beat phenomena.

(2) The bandwidth of the exciting light (with frequencies within the envelope of a set of molecular resonances) has a profound effect on the decay characteristics of the molecule.

(3) The use of exponentially increasing light pulses (negative-time pulses) can distinguish resonance scattering and resonance fluorescence and help to resolve ambiguities which have arisen (Novak *et al.*, 1978).

(4) The spectrum of nonradiative interaction of a prepared (projected) radiative state with the remaining states of the molecule provides a conceptual framework for (a) the sequential analysis of coupling schemes derivable from an absorption spectrum and the physical implications of such coupling schemes; and (b) the coupling of the molecule in question with other systems to provide phonon assisted radiationless transitions. These systems may possibly be designed in accordance with the structure of this interaction density spectrum (IDS).

(5) The method of spectral analysis in terms of coupling schemes of sequential doorway states and their associated IDSs provides a basis for a further understanding of the distinctions between oscillatory and dissipative limits in both radiative and radiationless processes. This method may help to give a more unified perspective of a broad class of phenomena which includes linear and nonlinear limits, as well as intra- and intermolecular aspects of energy transformations.

This work was supported (in part) by National Science Foundation Grant No. CHE 76-80292 and by Contract No. EY-76-S-05-2690 between the Division of Biomedical and Environmental Research of the Department of Energy and Florida State University.

REFERENCES

Allen, L., and Eberly, J. H. (1975). "Optical Resonance and Two-Level Atoms." Wiley, New York.
Berg, J. O. (1976). *Chem. Phys. Lett.* **41**, 547.
Berg, J. O., Langhoff, C. A., and Robinson, G. W. (1974). *Chem. Phys. Lett.* **29**, 305.
Bixon, M., and Jortner, J. (1968). *J. Chem. Phys.* **48**, 715.
Brand, J. C. D., and Hoy, A. R. (1977). *J. Mol. Spectrosc.* **65**, 75.
Burland, D. M., and Robinson, G. W. (1970). *Proc. Nat. Acad. Sci. U.S.* **66**, 257.
Cable, R., and Rhodes, W. (to be published).
Cohen-Tannoudji, C. (1977). *In* "Frontiers in Laser Spectroscopy" (R. Balian, S. Haroche, and S. Liberman, eds.), Vol. 1. North-Holland Publ., Amsterdam.
DeVries, H., DeBree, P., and Wiersma, D. (1977). *Chem. Phys. Lett.* **52**, 399.
Freed, K. (1970). *J. Chem. Phys.* **52**, 1345.
Freed, K. (1972). *Topics Current Chem.* **31**, 105.
Friedman, J. M., and Hochstrasser, R. M. (1974). *Chem. Phys.* **6**, 155.
Grigolini, P. (1977). *Chem. Phys. Lett.* **47**, 483.
Grigolini, P., and Lami, A. (1978). *Chem. Phys.* **30**, 61.
Glauber, R. (1963a). *Phys. Rev.* **130**, 2529.
Glauber, R. (1963b). *Phys. Rev.* **131**, 2766.
Glauber, R. (1970). *In* "Quantum Optics" (S. M. Kay and A. Maitland, eds.). Academic Press, New York.
Henry, B. R., and Kasha, M. (1968). *Ann. Rev. Phys. Chem.* **19**, 161.
Hilborn, R. C. (1975). *Chem. Phys. Lett.* **32**, 76.
Jortner, J., and Berry, R. S. (1968). *J. Chem. Phys.* **48**, 2757.
Kasha, M. (1950). *Disc. Faraday Soc.* **9**, 14.
Langhoff, C. A., and Robinson, G. W. (1974). *Chem. Phys.* **6**, 34.
Lin, S. H. (1966). *J. Chem. Phys.***4**, 3759.
Messiah, A. (1966). "Quantum Mechanics," Vol. 2. Wiley, New York.
Metiu, H., Ross, J., and Nitzan, A. (1975). *J. Chem. Phys.* **63**, 1289.
Mukamel, S., and Jortner, J. (1974). *In* "The World of Quantum Chemistry" (R. Daudel and B. Pullman, eds.), pp. 145–209. Reidel, Dordecht.
Mukamel, S., and Jortner, J. (1975). *J. Chem. Phys.* **62**, 3609.
Needham, C., and Rhodes, W. (1980, to be published).
Nitzan, A., Jortner, J., and Rentzepis, P. M. (1972). *Proc. R. Soc. London Ser. A* **327**, 367.
Novak, F. A., Friedman, J. M., and Hochstrasser, R. M. (1978). *In* "Laser and Coherence Spectroscopy" (J. Steinfeld, ed.). Plenum Press, New York.
Nussenzveig, H. M. (1973). "Introduction to Quantum Optics." Gordon and Breach, New York.

Orlowski, T. E., and Zewail, A. H. (1979). *J. Chem. Phys.* **70**, 1390.
Orlowski, T. E., Jones, K. E., and Zewail, A. H. (1977). *Chem. Phys. Lett.* **50**, 45.
Rentzepis, P. (1973). *Adv. Chem. Phys.* **23**.
Rhodes, W. (1969). *J. Chem. Phys.* **50**, 2885.
Rhodes, W. (1970). *Symp. Rad. Transitions Mol., J. Chim. Phys. Suppl.* 40, 1970.
Rhodes, W. (1971). *Chem. Phys. Lett.* **11**, 179.
Rhodes, W. (1974). *Chem. Phys.* **4**, 259.
Rhodes, W. (1977). *Chem. Phys.* **22**, 95.
Rhodes, W., Henry, B., and Kasha, M. (1969). *Proc. Nat. Acad. Sci. U.S.* **63**, 31.
Robinson, G. W., and Langhoff, C. A. (1974). *Chem. Phys.* **5**, 1.
Rousseau, D. L., and Williams, P. F. (1976). *J. Chem. Phys.* **64**, 3519.
Sargent, M., III, Scully, M. O., Lamb, W. E., Jr. (1974). "Laser Physics." Addison-Wesley, Reading, Massachusetts.
Sharf, B., and Silbey, R. (1970). *Chem. Phys. Lett.* **4**, 561.
Siebrand, W. (1967). *J. Chem. Phys.* **47**, 2411.
Smalley, R., Wharton, L., and Levy, D. (1975). *J. Chem. Phys.* **63**, 4977.
Sponer, H. (1959). *Radiat. Res. Suppl. No. 1* 558.
Tric, C. (1971). *J. Chem. Phys,* **55**, 4303.
van Santen, R. A. (1972). *Physica* **62**, 51, 84.
Voltz, R. (1975). *In* "Organic Molecular Photophysics" (J. B. Birks, ed.), Vol. 2. Wiley, New York.
Wessel, J. E. (1970). Ph.D. Dissertation, Univ. of Chicago, Chicago, Illinois.
Williams, P. F., Rousseau, D. L., and Dworetsky, S. H. (1974). *Phys. Rev. Lett.* **32**, 196.
Zewail, A. H., Orlowski, T. E., and Jones K. E. (1977a). *Proc. Nat. Acad. Sci. U.S.* **74**, 1310.
Zewail, A. H., Godar, D. E., Jones, K. E., Orlowski, T. E., Shah, P. R., and Nichols, A. (1977b). "Laser Spectroscopy I (*Proc. Soc. Photo-Opt. Instrum. Eng.*), Vol. 113, p. 42.
Ziv, A. R. (1978). *J. Chem. Phys.* **68**, 152.
Ziv, A. R., and Rhodes, W. (1976). *J. Chem. Phys.* **65**, 4895.

6

Spectroscopic and Time Resolved Studies of Small Molecule Relaxation in the Condensed Phase

L. E. Brus and V. E. Bondybey
Bell Laboratories
Murry Hill, New Jersey

I. INTRODUCTION

Over ten years ago the large molecule and small molecule limits of radiationless transition theory were first recognized (Robinson, 1967). In a large isolated molecule (e.g., benzene) the density of internal vibronic states is sufficiently

ISBN 0-12-450650-X

high that internal radiationless transitions are essentially irreversible in real laboratory experiments. In this limit, transition rates are independent of the molecular environment. However, in the small molecule limit, (di- and tri-atomics) the internal vibronic states are well separated in energy, and therefore energy conservation requires that the environment directly accept energy during an irreversible radiationless transition. A small molecule in condensed phase is then an interesting case in radiationless transition theory, in that the observed rates provide direct information about the coupling between the molecule and its environment.

Radiationless transition rates are essentially controlled by the shapes of potential surfaces. A diatomic molecule in a solid rare gas host actually is a polyatomic species with a multidimensional surface. Is there a physically useful way to project this multidimensional surface onto the diatomic axis? How are the observed spectra related to the true multidimensional surface and to the guest–host interaction? Spectra actually provide invaluable information about the dynamical interaction between the guest and its environment. The molecule–environment interaction both modifies the static energy level structure of the guest and provides low frequency external modes that may accept energy. The classical spectroscopy and the time resolved dynamical behavior to a great extent provide complementary information about this interaction.

This review begins with a discussion of spectra, and then proceeds to treat various different types of radiationless transition phenomena. In the latter sections the interplay between time resolved studies and spectra is apparent. Our review is almost entirely confined to physical processes which occur in rare gas matrices. We also note that an excellent review of some aspects of the energy transfer and vibrational relaxation problems has recently appeared (Legay, 1976). Our review was completed in March 1978.

Rare gas solids provide a very weakly interacting solid environment. This weakness provides important advantages in the study of radiationless transitions. First, the rates of fundamental processes such as interelectronic cascading, vibrational relaxation, and exothermic energy transfer may be slowed down into the nanosecond (or longer) time regime. In the $\geq 10^{-9}$ sec range, the mechanisms of these processes can be investigated using the high sensitivity of modern electronics. Secondly, the weakness of the guest–host coupling produces molecular geometries that are very close to those of free isolated molecules. The rare gas environment provides essentially no hindrance to large amplitude molecular motion, and diatomic hydrides are even able to undergo essentially unperturbed free rotation at 4 K! The fact that sharp guest spectra are often observed allows one to spectrally differentiate between radiationless processes that are *vertical* and *adiabatic* with respect to solvent motion, and to observe subpicosecond internal processes via lifetime broadening and/or splittings of initially sharp spectral lines.

II. SPECTRA AND THE PHYSICAL INTERPRETATION OF MOLECULAR LINESHAPES

We give a brief description of the Rebane lineshape theory (Rebane, 1970; DiBartolo and Powell, 1976) emphasizing *solvation* of a molecule by nearby rare gas atoms.

The vibrational quanta of rare gas solids are accoustical lattice phonons with a continuous frequency range up to the Debye cutoff near 65 cm^{-1}. Normally lineshapes of molecules in rare gas solids are determined by coupling between the molecule and these external, low frequency phonon modes. One can propose, to a good approximation, an adiabatic Born–Oppenheimer separation of the high frequency internal molecular vibrations from the low external lattice modes. For each internal vibronic state of the molecule there exists a potential energy curve governing the motion of the rare gas solvent nuclei.

The changes in shape of the phonon surfaces from one vibronic state to another govern the shape of the electromagnetic transition (electronic, infrared, or Raman) between the two molecular vibronic states. This principle is schematically illustrated in Fig. 1, in which we consider a simple case where the degrees of freedom associated with molecular rotation and center of mass translation are ignored. There are $3N$ lattice vibrations, where N is the number of atoms in the crystal. In practice the transition preferentially couples to a smaller number of pseudolocalized lattice vibrations that have high amplitude on the nearby rare gas nuclei. Figure 1 schematically shows, for one lattice vibration, a potential function that has a different equilibrium position in the excited state of the molecule. The excited state is solvated differently than the ground state; i.e., the rare gas nuclei have different positions around the excited state. The Franck–Condon principle when applied to these curves in absorption when only $v'' = 0$ is populated indicates that a progression of transitions (0, 0), (1, 0), (2, 0), (3, 0), etc., will occur.

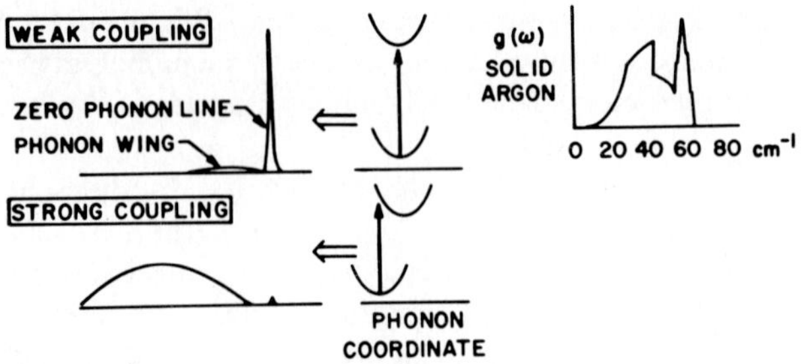

Fig. 1. Schematic diagram of phonon potential curves and molecular lineshapes for a transition between two guest vibronic states in solid hosts. Both weak coupling and strong coupling limits appear. The lattice acoustical phonon density of states function $g(\omega)$ is also shown for solid Ar. Other effects not illustrated are due to local modes (rotation), lifetime broadening, and inhomogeneous broadening (sites).

The experimental absorption band connecting these two molecular states will be a superposition of such progressions—one for each different frequency coupled to the molecular transition. The observed band consists of a sharp feature, which is the (0, 0) transition for all lattice phonons, followed to the blue by an absorption wing which is the superposition of all the (1, 0), (2, 0), etc., lattice phonon transitions. Normally many frequencies are coupled to the transition, and this phonon wing appears continuous.

The sharp (0, 0) feature is labeled the zero phonon line (ZPL), since no lattice phonons are created by absorption at this frequency. ZPLs are especially important in solid state spectroscopy, as they identify the transition energy which simultaneously takes the molecule to its excited state *and* the surrounding solvent to its new equilibrium position. The ZPL is the *adiabatic* (nonvertical) transition with respect to solvent reorganization. The transition occurring at the maximum of the phonon wing is the impulsive (or vertical) solvent transition. After photon absorption at the phonon wing maximum frequency, the solvent is initially at the ground state equilibrium configuration. The energy liberated during subsequent solvent relaxation creates phonons which propagate away from the molecule.

In rare gas solids, guest infrared and Raman transitions invariably correspond to a weak coupling limit: The integrated intensity of the phonon wing is negligible with respect to the ZPL. Electronic transitions occur over the entire range between this weak coupling limit and the opposite strong coupling limit. The ZPL in the weak coupling limit is analogous to a Mossbauer line (Rebane, 1970). The fact that no lattice phonon is created during ZPL absorption necessarily implies that the photon (gamma, visible, or infrared) momentum is absorbed by the lattice as a whole

Two physically distinct procedures could be used to produce effective diatomic potential surfaces from experimental spectra. ZPL positions can be used to calculate an adiabatic surface while the phonon wing maxima can be used to calculate a vertical surface. These surfaces in general differ from each other, and both may be different from the potential surface in vacuum. Only the adiabatic surface has useful predictive value for dynamics in condensed phase (Brus and Bondybey, 1976).

"Adiabatic" is used in two different senses. First, the lineshape theory is adiabatic in the classical sense that the internal vibrations occur fast with respect to lattice vibrations and resolvation. The solvent nuclei move on a potential energy surface created by the "fast" internal motions of the guest molecule. Secondly, the adiabatic guest potential energy surface refers to guest vibronic states where the lattice has resolvated anew about each different vibronic state. The guest nuclei move on an effective potential energy surface that includes the effect of adiabatic solvation. This effective adiabatic solvation surface is theoretically well defined, independently of the fact that a vertical transition (with respect to solvent motion) is more probable than an adiabatic transition. If the observed spectra consist simply of intense ZPLs, then the vertical and adiabatic surfaces are the same. Negligible resolvation accompanies excitation of the molecular guest. However, in chemical situations appreciable resolvation often occurs and the distinction between the vertical and adiabatic surfaces is important.

Any internal radiationless transition which takes place on a time scale slow with respect to solvent reorganization $\sim 10^{-13}$ sec is best described using the adiabatic surface. The formal theory is analogous to weak coupling radiationless transitions within isolated polyatomic molecules. An initial guest–host state, vibrating about an initial equilibrium configuration, crosses over into an excited phonon level of the final guest–host state, where the final state vibrates about new solvent nuclei positions. A diatomic molecule in solid rare gases possesses spectroscopic states corresponding to the level structure of the adiabatic surface, in the sense that lifetime broadening of these levels (due to nonadiabatic effects) is normally negligible with respect to the spacing between such levels. Changes in the guest adiabatic surface, from the molecular potential in vacuum, are a valid measure of the guest–host solvation energy as a function of vibronic level.

On the $\leq 10^{-13}$ sec time scale as the molecule is moving, but the lattice is frozen in its ground state configuration, the vertical potential is of little use. There actually are no states, in which the diatomic nuclei only contain energy, at the energies given by the vertical surface. An impulsive collisional mechanism is best used to describe energy transfer between the guest and host on this short time scale (Brus and Bondybey, 1976).

III. LARGE AMPLITUDE VIBRATIONAL AND ELECTRONIC MOTION

A. Photodissociation Cage Effect

The vibrational relaxation mechanism of bound states is discussed in Section VI. Here we ask, how does the environment respond to large amplitude vibrational motion and the limiting case of dissociation? Correspondingly, do adiabatic potential curves show a distinct change in shape from the gas phase?

The classical experimental method for determination of potential curve shapes is electronic spectroscopy. Spectral analysis for many covalently bound small molecules in general shows a vanishing small change in shape, in the sense that the ω_e and the first one or two anharmonic constants are essentially identical to the values in vacuum. For example, the ground state of O_2 in N_2 exhibits $\omega_a = 1577.4$ cm^{-1}, $\omega_e x_e = 11.79$ cm^{-1}, and $\omega_e y_e = 0.022$ cm^{-1}, while the values in vacuum are $\omega_e = 1580.36$, $\omega_e x_e = 12.07$ cm^{-1}, and $\omega_e y_e = 0.054$ cm^{-1} (Goodman and Brus, 1977b). The ground state $r_e = 1.21$ Å, and the highest observed level ($v = 13$) has an outer turning point at 1.60 Å.

Greater expansions from ground state bond lengths occur when observing an excited electronic state just below its gas phase dissociation limit. Figure 2

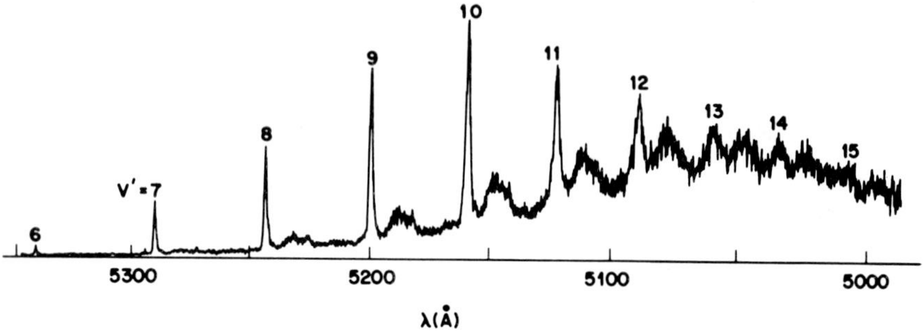

Fig. 2. Excitation spectrum of Cl_2 fluorescence in solid Ar. Vibrational assignments for ZPLs in the $^3\Pi_0$ state are shown. (From Bondybey and Fletcher, 1976.)

shows the $^3\Pi_0$ state excitation spectrum of Cl_2 fluorescence in solid Ar, in the region below the dissociation limit near 4750 Å (Bondybey and Fletcher, 1976). Here the ground state $r_e = 2.0$ Å, and the highest $^3\Pi_0$ level observed ($v' = 15$) has an outer turning point near ~3.5 Å. Despite this large change, there is no perceptible change in the anharmonic spacing of the last few observed ZPL. The effect of the solvent is to transfer intensity from the ZPL to the phonon wing as one goes from low v' to high v'. For high v' the Ar nuclei must undergo greater resolvation around the excited state. However, this resolvation motion, which displaces Ar atoms as the Cl_2 bond expands, experimentally requires little energy since the adiabatic potential curve is not changed in shape from the gas phase.

A direct study of the cage effect which occurs upon excitation into gas phase photodissociation continua was first carried out for the interhalogen ICl (Bondybey and Brus, 1975, 1976a). Figure 3 shows that in the gas phase two excited 0^+ curves exhibit an avoided crossing, such that the lower one exhibits four bound vibrational levels below the barrier to dissociation formed by the avoided crossing. Experimentally in rare gas matrices it is observed that excitation as much as 0.8 eV above this dissociation barrier produces $v' = 0$ fluorescence from the lower 0^+ potential and from the still lower A $^3\Pi_1$ bound state, with subnanosecond risetimes. A complete cage effect occurs with the quantum yield of permanent dissociation being $\leq 10^{-4}$. Nevertheless, the solvent does not induce quantized vibrational levels in the dissociative region, as the fluorescence excitation spectra in the dissociative region are perfectly smooth. In the 0^+ bond region below the dissociation barrier, five quantized levels are observed in the condensed phase experiments. This appears to occur because there is a differential shift in the T_e values of the zero-order bound and repulsive 0^+ states such that the repulsive 0^+ crosses the bound state at a higher energy in the solid than in the gas phase.

Two apparently contradictory facts emerge. (a) There is a strong *kinetic*

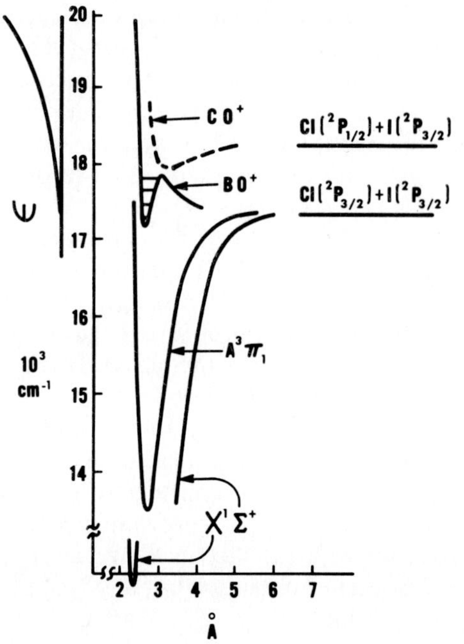

Fig. 3. Gas phase potential energy curves for ICl. (From Bondybey and Brus, 1975a)

cage effect with the yield of permanent dissociation being $\leq 10^{-4}$. (b) There is a negligible *spectroscopic* cage effect, in the sense that there are no bound levels in the dissociative region and the gas phase anharmonicity, just below the dissociation limit, is unchanged. This discrepancy is resolved if we recall the short time scale of the kinetics. The adiabatic surface only describes the position of guest energy levels such that resolvation continuously occurs. However, the kinetic cage effect occurs on the 10^{-14} sec time scale such that the rare gas solvent is locked into its ground state configuration. A Cl atom carrying 0.8 eV kinetic energy collides impulsively with its nearest neighbor Ar atom, and transfer of nuclear kinetic energy is efficient. Apparently after a few such collisions the excess energy is lost, and ICl relaxes into an intramolecularly bound excited state. If ICl were to photodissociate slowly in a *gedanken* experiment, the Cl atom would collide with a coupled group (the lattice) of Ar atoms, whose effective mass would be much higher than that of a single Ar atom. Nuclear energy transfer would be much less efficient.

It is striking that the adiabatic surface shows such little hindrance of the large amplitude vibrational motion. This can be rationalized if we realize that rare gas solids at 4.2 K appear to be as elastic as any known solid material. The macroscopic bulk modulus of solid Ne at 4.2 K is essentially equal to that of room temperature liquids such as ethanol. Let us consider the solvent to be a continuous, elastic material, and ask how much energy is required to open up a cavity of $\approx$4 Å diameter that would accommodate molecular motion. We find that a negligible amount is required if the resulting solvent compression is allowed to propagate to regions far removed from the cavity. In other words, if we wait for solvent relaxation the energy required to open a small cavity is negligible (Brus and Bondybey, 1976a).

This simple picture ignores possible barriers to viscous flow or activation energies for solvent motion. Such effects, if important, could create host-induced negative anharmonicities and barriers to dissociation (Dellinger and Kasha, 1976). To our knowledge such phenomena have not been observed in rare gas solids. Apparently the isotropic nature of the van der Waals forces allows elastic flow without appreciable activation energies.

We may state the following concerning solvent involvement in molecular reactions and isomerizations: Any molecular state, which in the gas phase would isomerize or react, should not be appreciably hindered from doing so in rare gas solids, *if* this state lives for a time long with respect to resolvation times. These ideas are consistent with the fact that diverse organic and inorganic synthetic reactions can be carried out in solid Ar at 4.2 K (Chapman *et al.*, 1973; Davies *et al.*, 1977). For example, the predissociation of s-tetrazine, which is a ring fragmentation producing N_2 + $(HCN)_2$ dimer, occurs apparently unhindered in Ar matrices (Dellinger *et al.*, 1977).

In general one could imagine that an excited vibrational level, which in vacuum would react within a certain lifetime, might vibrationally relax in solid Ar to a nonreactive $v' = 0$ level. This would be an apparent matrix hindrance of reaction. Such fast vibrational relaxation appears to prevent multiphoton IR photodissociation processes from occurring in matrices, despite an early report to the contrary (Ambartsumyam *et al.*, 1976).

The shapes of fluorescence excitation spectra in photodissociation regions, even though they contain no vibrational structure, do yield information about ultrafast radiationless transitions occurring before arrival of the population in the bound fluorescent $v' = 0$ state. In the ICl case, the normalized fluorescence excitation curve dropped sharply when crossing above the gas phase dissociation barrier, and it was concluded that the recombination quantum yield into the fluorescent $v' = 0$ BO^+ dropped from $Q \simeq 1.0$ below the barrier to $Q \simeq 0.01$ above the barrier. Above the barrier, the remaining population explores longer ICl bond lengths and then recombines along other bound potential curves.

In the case of Br_2 in solid Ar (Bondybey *et al.*, 1976), the dissociative region excitation spectrum of $v' = 0$ BO^+ emission showed only the BO^+ dissociation continuum, and not the stronger $^1\Pi_1$ continuum. The repulsive $^1\Pi_1$ curve, which correlates with ground state $^2P_{1/2}$ Br atoms, does not radiationlessly feed the BO^+ state, which correlates with the excited dissociation limit $^2P_{1/2} + {}^2P_{3/2}$.

B. "Bubble" Solvation of Rydberg States

The photodissociation cage effect occurs as the matrix responds to large amplitude vibrational motion. A related phenomenon is resolvation around large diameter excited Rydberg states. The first observation of Rydberg states in solid rare gases occurred via VUV absorption spectra. For some polyatomic guests, progressions of Rydberg-like broad absorptions were observed and interpreted in terms of a modified Wannier model (Gedanken *et al.*, 1973). In a Wannier state, the Rydberg wavefunction physically expands into the solvent, and the binding energy between the electron and ionic core is decreased due to dielectric screening by the host. Resolvation is not explicitly considered.

Diatomic NO presents a fascinating transformation of Rydberg structure between the gas phase and the solid. NO gas phase VUV spectra reveal strong homogeneous perturbations between several different Rydberg and valence excited states. In solid rare gases, sharp ZPL absorption is observed into *de-perturbed* valence excited states (Boursey and Roncin, 1975). The excited Rydberg states are not seen, and it is concluded that Rydberg absorptions have broadened into a continuous grey background.

The broadening mechanism was studied by observing fluorescence from the lowest Rydberg ($3s\sigma$ $A\,^2\Sigma^+$) in solid Ar, Kr, and Xe (Goodman and Brus, 1977a). Despite the extreme broadening, the Rydberg fluorescence quantum yield is $\simeq 1.0$ in all three hosts. Comparison of the fluorescence and absorption spectra shows that (at least for $3s\sigma$) the broadening represents a large phonon wing.

The comparison of the absorption and fluorescence lineshapes, and the values of the radiative lifetimes in each host, allow us to observe resolvation. The data support a "bubble" model in which the solvent cavity expands radially, creating a void in which the largely unperturbed Rydberg wavefunction exists. The lifetimes support this bubble mode, as only $\simeq 10\%$ of the radiative lifetime lengthening expected in the Wannier model limit is observed. The absorption and fluorescence lineshapes are asymmetric, with the phonon wing being a factor of $\simeq 7$ more pronounced (in Ar host) in absorption than in fluorescence. In Fig. 4, this asymmetry is explained by bubble potential energy curves, where the cavity breathing mode creates the phonon wing. The essential idea is that the radial molecule–solvent potentials are *not* harmonic. There is a steep inner limb due to molecule–solvent hard sphere repulsion, and a shallow outer limb representing the small amount of expansion work the molecule must do on the solvent in order to create a bubble. The steep potential is probed in Franck–Condon absorption, and the shallow outer limb is probed in Franck–Condon fluorescence.

Thus the NO $3s\sigma$ Rydberg study supports the conclusion of the adiabatic cage effect studies—the solid rare gas solvent is extremely plastic, and easily

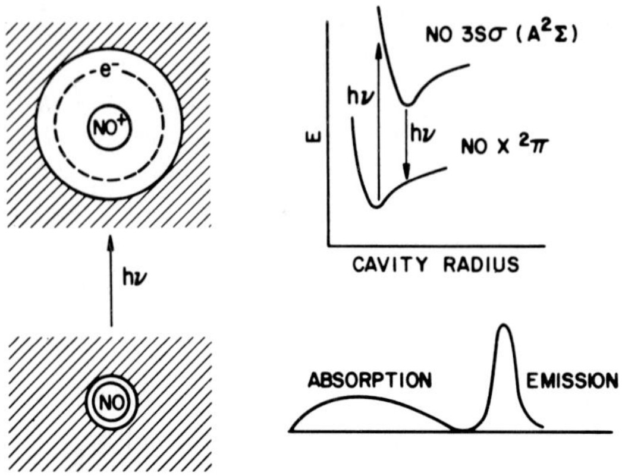

Fig. 4. Schematic diagram for "bubble" Rydberg model. The Franck–Condon principle, when applied to the NO-solvent potential curves, predicts the phonon wing will be much wider in absorption than in fluorescence. (From Goodman and Brus, 1977a.)

accommodates molecular changes in shapes. The fact that the Rydberg line-shape is a phonon wing implies that there is an appreciable difference between *vertical* and *adiabatic* (with respect to resolvation) solid phase ionization. Earlier solid phase ionization limits derived from Wannier absorption studies should be reconsidered. Asymmetry between absorption and fluorescence has also been observed for Rydbergs in room temperature fluids (Hirayama and Lipsky, 1975) and apparently a small bubble forms here also.

IV. ELECTRONIC SOLVATION AND ADIABATIC POTENTIAL CHANGES

Several exceptions are known to our previous conclusion that adiabatic potential energy curves undergo a negligible change in shape in rare gas solids. These exceptions appear to be related to electronic solvation, and not to solvent hindrance of large amplitude vibrational motion. Electronic solvation in this context implies that the molecular wavefunction in the bonding region between guest nuclei is altered.

Alkali halides such as LiF typically have their vibrational frequencies ω_e red shifted by 7–14% in rare gas solids (Linevsky, 1961). This can be understood as an electronstatic interaction between the large permanent dipole and the surrounding polarizable atoms. The ω_e values of the excited charge transfer (e.g., Xe^+F^-) states of diatomic rare gas halides (e.g., XeF) are also lowered by similar amounts (Ault and Andrews, 1976; Goodman and Brus, 1976a), apparently for the same reason given for the chemically similar ground state alkali halides.

Matrix changes in shape are apparent for some very weakly bound molecular states. The ground $X\,^2\Sigma^+$ state of XeF is slightly deeper and more harmonic in solid neon than in the gas phase (Goodman and Brus, 1976a). A similar effect occurs for the similar, nominally covalent $C\,^1\Pi$ state of XeO (Goodman *et al.*, 1977). These observations can be understood by considering these weakly bound states to be Mulliken charge transfer complexes. The rare gas solids preferentially stabilize (i.e., red shift the T_e) the excited charge transfer states which mix with the nominally covalent lower states. Thus, in the solid phase the ionic mixing into the lower state increases, and the potential wells are somewhat deeper.

Ca_2 dimers (Miller and Andrews, 1977; Bondybey and Albiston, 1978) are a remarkable case in which a ground state homonuclear, weakly bound well deepens and becomes more harmonic in heavier rare gas solids. The vibrational relaxation rate also slows in the heavier rare gases where the ΔG values are larger. Even in this case, the observed potentials are well described by a Morse potential without negative anharmonicity or a host induced barrier to dissociation. Some similar species, such as Pb_2 show no sensitivity to the host (Bondybey and English, 1977a).

Free radicals having strongly acidic protons also show electronic solvation effects. The $X\,^2\Pi \rightarrow A\,^2\Sigma^+$ spectra of OH radical in solid Ar do not exhibit free rotation fine structure, as is normally the case for diatomic hydrides, but rather show low frequency vibrational structure that can be interpreted as a progression in the hydrogen bond vibration of a linear complex OHAr (Goodman and Brus, 1977e). The complex binding energy appears to be 675 cm^{-1} in the excited $A\,^2\Sigma$ state, and over 1000 cm^{-1} for OHKr. The OH ΔG_{10} values in both excited and ground states are lowered by $\simeq 100$ cm^{-1} in argon host. Rare gas atoms are weakly basic (electron donating), and appear to hydrogen bond to acidic protons producing these spectral changes.

We also mention a somewhat related example. The out-of-plane vibrations of nitrogen heterocycles are known to be sensitive to environment. For example ν_{10a} in the ^1n–π^* state of pyrazine is 383 cm^{-1} in vacuum and 486 cm^{-1} in the neat crystal (Zalewski *et al.*, 1974). In principle this sensitivity can also be related to electronic solvation. Such out-of-plane modes vibronically couple ^1n–π^* states with nearby higher $^1\pi$–π^* states, and the shape of the ν_{10a} potential is affected by the strength of this coupling. In condensed phases, differential solvent shifts on the T_e values of the two states will change the ΔT_e between them, and thus affect the vibronic coupling strength.

V. RADIATIONLESS TRANSITIONS OBSERVED BY POLARIZATION AND FLUORESCENCE LINE NARROWING

In photoselection (Albrecht, 1961), an aligned excited state population (created by polarized excitation) will emit polarized fluorescence if the excited molecules do not rotate between absorption and fluorescence. For example, in the ICl photodissociation cage effect experiments, the $v' = 0$ BO^+ fluorescence is experimentally polarized following excitation in the bound region of the Fig. 3 BO^+ curve (below the dissociation barrier). The several hundred wavenumbers energy released during the bound state vibrational relaxation to $v' = 0$ does not cause the molecule to permanently reorient in the solid; and in addition the $v' = 0$ BO^+ state does not freely rotate during its $\simeq 2$ μsec lifetime.

Polarized fluorescence is still observed if ICl is excited $\simeq 0.8$ eV above the dissociation barrier. One might have imagined that fast deposition of so much nuclear kinetic energy in the local neighborhood might have produced a transient hot spot (or local melting). If the local Ar atoms transiently melted and refroze in different positions, the ICl molecular axis would have a different final orientation. However, depolarized fluorescence is not observed, and we must conclude the geometry of the local environment is preserved during the ultrafast transient cage effect.

Elegant infrared laser driven photochemistry experiments by Turner and co-workers (Davies *et al.*, 1977) have also demonstrated that orientational specificity is preserved during elementary photochemical events. That is, they also find no evidence for transient local melting.

Fluorescence line narrowing experiments demonstrate that differing local geometries are also preserved during vibronic relaxation of bound excited states. Molecular spectra in rare gas solids exhibit an inhomogeneous linewidth ($\approx$several inverse centimeters) due to differing local environments. If a small subset of the guests are excited with a narrow frequency laser, it is invariably observed that vibronically relaxed fluorescence only occurs from this same small subset (Brus and Bondybey, 1975a). The energy released during relaxation does not convert one local environment into another, even in cases where individual quanta are 2000–3000 cm^{-1}.

These observations are relevant to the role of polarization in detection of transient free rotation. Suppose a diatomic molecule at 4.2 K does not freely rotate in solid rare gases. If vibrational relaxation produces transient free rotation, followed by fast rotational relaxation into the lowest librating level, will fluorescence from the lower level be unpolarized? If the diatomic sits in an environment of low symmetry, such that the lowest librational level has a unique orientation, then the fluorescence remains polarized even if transient free rotation occurs during relaxation. However, if the molecule sits in a high symmetry site (such as O_h) with degenerate equivalent orientations, then the relaxed fluorescence will be substantially depolarized.

VI. DIRECT VIBRATIONAL RELAXATION

Vibrational energy relaxation by small molecules in condensed phases has recently received an increasing amount of both theoretical and experimental attention. Contrary to the common belief that vibrational relaxation in solid phase should be fast, Tinti and Robinson in 1968 concluded that excited vibrational levels of matrix isolated N_2 in the $A\,^3\Sigma_u^+$ state have lifetimes of the order of 1 sec at 4.2 K. Similarly long lifetimes were estimated for vibrationally excited ground electronic state C_2^- molecules in solid argon (Bondybey and Nibler, 1972). More recently it was concluded that lifetimes of CO and CN in solid rare gases are purely radiative with immeasurably slow vibrational relaxation. It is remarkable to note that the lifetime of $v = 1$ N_2 in liquid N_2 is ≈ 1 minute, and is apparently controlled by collision induced infrared emission (Brueck and Osgood, 1976).

Lately, considerable advances have been made in our understanding of the relaxation mechanism. Early theories of vibrational relaxation postulated a simplified model, in which it was generally assumed that the vibrational energy is accepted directly by harmonic, delocalized lattice phonon modes. Furthermore, the lattice spectrum was approximated by a single frequency. The major predictions of these early theories were (Nitzan *et al.*, 1974) (a) an exponential dependence of the relaxation rates on the size of the vibrational quantum to be dissipated (the energy gap law) and (b) a strong temperature dependence of the relaxation rates due to lattice phonon stimulated emission.

In 1975 we studied several diatomic hydrides and proposed a different mechanism (Brus and Bondybey, 1975b; Bondybey and Brus, 1975b). Frequency doubled dye lasers were used to excited single vibronic levels of the $A\,^3\Pi$ state of NH and ND in solid rare gases. By following time resolved emission as in Fig. 5, from both the originally populated level, and levels populated by vibrational relaxation, the relaxation rates could be deduced. It was found that relaxation was considerably faster than the aforementioned cases of N_2 and C_2^-. Contrary to the predictions of the energy gap law, relaxation in NH (which has a larger vibrational quantum) was faster by more than one order of magnitude than in the ND species. Furthermore, the expected strong temperature dependence was not present. A similar result was obtained for relaxation in the excited $A\,^2\Sigma^+$ state of OH and OD in solid Neon.

More recently vibrational relaxation in the ground electronic state of NH was also studied using two separate lasers. (Bondybey, 1976b). The first laser populated the $v = 1$ level of the ground state by an absorption–emission sequence and the second laser probed the ground state $v = 1$ population with variable delay times after the pump laser. The overall relaxation rates were found to be slower by more than a factor of 200 than in the $A\,^3\Pi$ state. Yet the trend remained the same, with NH relaxing a factor of 150 times faster than ND, and no temperature dependence.

The explanation of these anomalous results lies in the approximations made in the simple theories. The assumption that harmonic delocalized lattice phonons directly accept the energy implies that accepting modes are unchanged by isotopic substitution in the guest molecule. However, a diatomic guest introduces local phonon structure into an otherwise perfect lattice. The diatomic has, in

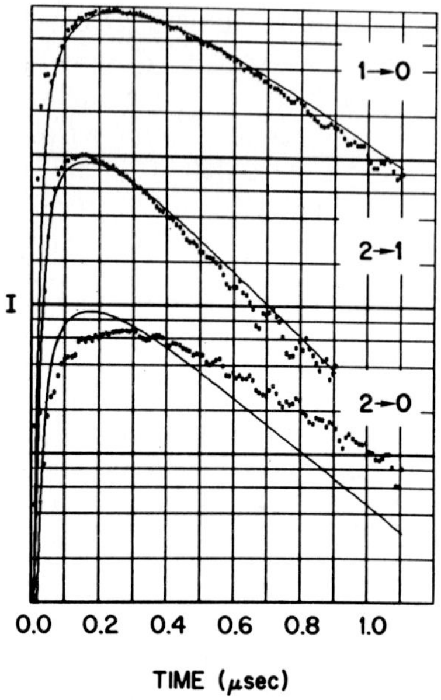

Fig. 5. Time resolved fluorescence from $NH(^3\Pi)$ following excitation of specific vibrational levels. Upper trace: $v = 0$ emission following $v = 1$ excitation. Middle trace: $v = 1$ emission following $v = 2$ excitation. Lower trace: $v = 0$ emission following $v = 2$ excitation. The lower curve is delayed with respect to the middle trace because of the lifetime of the $v = 1$ level. (From Bondybey and Brus, 1975b.)

addition to the internal vibration, five other degrees of freedom. The magnitude of the associated localized modes is also strongly dependent on the isotopic composition of the guest molecule. In particular, in nearly freely rotating hydrides, the two rotational modes will approximately scale with the rotational constant B, i.e., with the reduced mass of the guest diatomic molecule. One can qualitatively explain the observations if one assumes that the rotational local modes are the primary acceptors of the vibrational energy. In a diatomic hydride, deuterium substitution will reduce the rotational constant by approximately a factor of 2. On the other hand, the vibrational energy, which scales as a square root of the reduced mass, will change only by a factor of the square root of 2. Thus, in spite of the larger energy gap, relaxation of the hydride will be a lower order process than that of the deuteride. This assumption also explains the absence of temperature dependence. The expected strong temperature dependence has its origin in the properties of Hermite polynomials and will not be expected if the strongly anharmonic local modes accept the energy.

In his recent review, Legay (1976) has discussed additional evidence in the cases of CH_3F and HBr (studied via infrared laser techniques) that supports this mechanism. He has proposed that the rate is controlled by the minimum number of rotational quanta necessary to dissipate the vibrational energy:

$$J_m = \sqrt{\omega_e/B}. \tag{1}$$

Using some of the experimental data available at the time of his review, he has obtained an approximately linear plot of logarithm of relaxation rate as a function of J_m. It should, however, be noted that the relaxation rate should be extremely sensitive to the extent of guest–host coupling and to the height of the barrier to free rotation. This is evidenced by comparing the relaxation rates in the $X\,^3\Sigma$ and $A\,^3\Pi$ states of NH. Although the vibrational energy gap is almost identical in both states, the relaxation rate in the excited state increased by more than a factor of 200. This behavior can be attributed to the fact that while NH undergoes an almost unperturbed rotational motion in the ground electronic state, the barrier to free rotation is considerably higher in the more strongly interacting upper state. We suggest that Legay's correlation will be semiquantitatively correct if only ground electronic states are included.

This mechanism requires a deviation from spherical symmetry in the local potential energy "seen" by the diatomic guest. The local potential must break the conservation of angular momentum which occurs in the gas phase. A double doping experiment has been performed to test how the relaxation rate depends upon local symmetry for the case of NH($A\,^3\Pi$) (Goodman and Brus, 1976b). The NH guest exists in a O_h symmetry substitutional vacancy (12 nearest neighbors) in both pure Ar and pure Kr. The relaxation rate in Kr is a factor of 15 higher than in pure Ar. The basic idea of the experiment is to compare these rates with those where the NH has 11 Ar and 1 Kr nearest neighbor, 10 Ar and 2 Kr nearest neighbors, etc. First, this experiment established that vibrational relaxation is controlled by short range forces; i.e., the rates are unaffected by lattice defects and different guest molecules that are *not* in the first coordination sphere. Second, this experiment showed that a saturation effect occurs in both vibrational relaxation and spectral shifts. The relaxation rate increase observed between pure Ar and pure Kr is reproduced in the local environment of 11 Argon atoms and 1 Kr atom. Moreover, in this local environment the electronic spectra ZPL shift almost to the position observed in pure Kr. This saturation effect is suggestive of chemical valence. Klemperer and co-workers have found that the structure of gas phase complexes between rare gas atoms and small molecules is predicted by Mulliken donor–acceptor complex schemes (Harris *et al.*, 1975). For NH($A\,^3\Pi$), then, the major interaction with the surrounding matrix may be chemical, i.e., a linear hydrogen bonded NHKr change transfer complex may form. In this situation we have proposed that the diatomic vibrational relaxation might be profitably modeled as intramolecular energy redistribution from the high frequency stretch into the two low frequency triatomic modes (Goodman and Brus, 1976b). Rotation accepting energy corresponds to energy flow into the v_2 low frequency bend.

An even stronger interaction should occur for OHAr because of the increased electronegativity of O with respect to N. $OH(A\,^2\Sigma^+)$ in matrices shows remarkable behavior (Goodman and Brus, 1977e). (a) Unlike all other diatomic hydrides, the OH spectra in Ar do not show perturbed free rotation structure, but rather ν_3 OH–Ar vibrational progressions. (b) The vibrational relaxation rate of $OH(A\,^2\Sigma^+)$ in Neon is increased by a factor of $\geq 10^3$ by introducing one Ar atom nearest neighbor. The hydrogen bonded complex model certainly appears justified for $OH(A\,^2\Sigma^+)$. However, it is not yet clear if this complex model is appropriate for weakly interacting, closed shell cases such as ground electron state HCl in Argon.

Recently formal theories have appeared which show how rotation may accept energy during relaxation processes (Freed *et al.*, 1977; Gerber and Berkowitz, 1977). Gerber and Berkowitz have been able to *quantitatively* (within a factor of 2) calculate the relaxation rate of ground state NH in Ar by invoking the known gas phase potential of the ArHCl complex. The relaxation of $v' = 1$ produces $J' = 13$ of $v' = 0$ and 1 phonon in the NH center of mass local mode. Indeed, it is rare in radiationless transition theory to be able to quantitatively calculate any measured rate.

While hydride molecules and also several other relatively light molecules are known to undergo a relatively free rotation in solid matrices, for heavier species the barriers to free rotation are high and rotational degrees of freedom might not be expected to participate in the relaxation mechanism. As discussed in Section V, in cases of irregularly shaped molecules such as ICl that presumably occupy low symmetry cavities in the solid, fluorescence polarization data does not experimentally bear on the question of whether transient free rotation occurs during relaxation.

A systematic isotopic study of the direct vibrational relaxation mechanism in a nonhydride species was recently carried out for the case of isolated $(O_2)_2$ dimers in solid Neon host (Goodman and Brus, 1977d). An intersystem crossing process produced a vibrational cascade in the dimer electronic state $^3\Sigma(v = 0) + {}^1\Sigma_g^+(v)$. The cascade occurs in the electronically excited $^1\Sigma_g^+(v)$ state; it's neighbor $^3\Sigma(v = 0)$ is in the O_2 ground state. The energy gap law was found to be systematically violated, with the rates for $(^{18}O_2)_2$ being a factor of ≈ 40 slower than for $(^{16}O_2)_2$ which has the larger vibrational spacings. Moreover, the rates for $^{16}O_2(^3\Sigma, v = 0) + {}^{18}O_2(^1\Sigma, v)$ are a factor of 2 higher than those for $^{18}O_2(^3\Sigma, v = 0) + {}^{18}O_2(^1\Sigma, v)$. This is, the relaxation rate also depends upon the isotopic identity of the unexcited nearest neighbor O_2 in the dimer. These results are consistent with (but do not necessarily prove that) energy being accepted by the O_2–O_2 low frequency stretching motion or alternately O_2 rotation accepting energy during the vibrational relaxation. The nearest neighbor involvement comes about as the two rotational manifolds are coupled by strongly anisotropic exchange forces, which are known to control the observed dimer (rotationally relaxed) structure. An O_2 in a $(O_2)_2$ dimer isolated in Neon must see a strongly anisotropic local potential, which creates the necessary rotation–vibration coupling. It is not clear whether, in the more symmetric case of isolated O_2 in Neon, this coupling would be sufficiently strong for the same mechanism to occur.

An interesting heavy, low frequency molecule which shows evidence of 10^{-9} sec time scale relaxation is Ca_2, which has only weak van der Waals bonding in the ground state and stronger chemical bonding in one upper electronic state. This molecule was recently studied by Miller and Andrews (1977) and by Bondybey and Albiston (1978). The upper state in solid Kr has a vibrational frequency of 120 cm^{-1} and a radiative lifetime of 9 nsec. Observation of extensive vibrationally unrelaxed emission of levels up to $v = 7$ permits determination of a vibrational lifetime of ~ 3 nsec for $v = 1$ and of progressively decreasing values for the higher levels. In solid Ar the upper state vibrational frequency decreases to 109 cm^{-1}, and no unrelaxed emission is seen. One can conclude that the $v = 1$ lifetime decreases to less than ~ 500 psec in Ar. The ground state frequency of Ca_2 is only 72 cm^{-1}. The ground state relaxation rates have not been measured directly but can be deduced from the spectral line shapes in the emission spectrum. While the $v = 0$ ZPL is sharp, the higher levels show increasing lifetime broadening reflecting an increase in relaxation states. In this way one can deduce a lifetime of near 2 psec for $v = 2$ and lifetimes of less than 1 psec for levels above $v = 8$.

VII. VIBRATIONAL RELAXATION THROUGH REAL INTERMEDIATE STATES

In the previous section we dealt with the situation of relaxation within a given electronic state. A more complex situation arises when the vibrational manifolds of more than one electronic state overlap in the region of interest. In this circumstance the molecule may, in addition to direct vibrational relaxation, undergo a radiationless transition into a nearby vibronic level of another electronic state. In fact, some time ago Frosch (1971) suggested that the selective appearance of some vibrational levels of the $d^3\Pi_g$ state of C_2 in emission spectra is due to selective feeding by vibrational levels of some other nearby electronic state.

The first time resolved study of interelectronic energy cascading following tunable excitation of specific levels occurred in isolated C_2^- (Bondybey and Brus, 1975c). The time dependence of the $B\,^2\Pi$ state vibrational relaxation proved that the relaxing population went through real intermediate levels of another electronic state. By the systematic use of isotopic substitution, the $v = 0$ level of this other state was located between $v' = 1$ and $v' = 0$ of $B\,^2\Pi$. The number of intermediate levels observed coupled with the $v = 0$ location and the MO calculations of Barsuhn (1974) allowed us to identify this other state as a predicted yet previously unobserved $^4\Sigma$ state. The potential energy curves, which favor

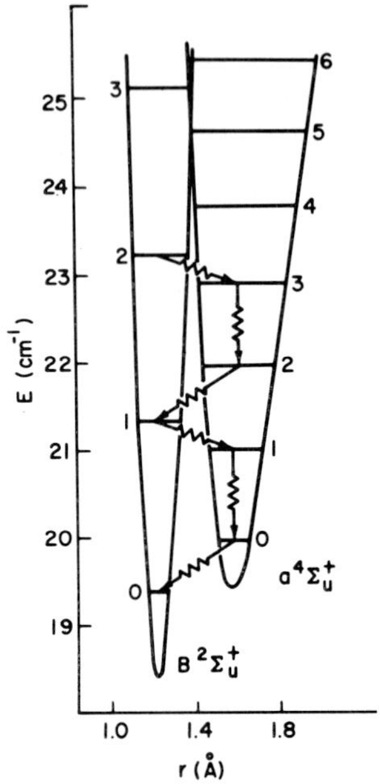

Fig. 6. Potential energy curves and cascading sequence for matrix isolated C_2^-. Broken arrows represent radiationless transitions. (From Bondybey and Brus, 1975a.) The Franck–Condon factors ($\times 10^8$) are as follows:

$B^2\Sigma$–$a^4\Sigma$	12–12	12–13	13–13
2–3	122,380	99,120	79,030
1–2	5086	3927	2980
1–1	1333	1008	748
0–0	7	5	4

interstate cascading due to a curve crossing, appear in Fig. 6. In fact, the matrix data could be used to assign previously reported $B^2\Pi$ rotational perturbations, and thus to determine the spectroscopic parameters of $^4\Sigma$ in the gas phase. This study experimentally established the relative facility of spin forbidden diatomic radiationless transitions in rare gas solids. However, no quantitative data concerning an intersystem crossing energy gap law were obtained.

The principal drawback of the C_2^- study was that the exact relative positions of the $^4\Sigma$ states were not known in the condensed phase. Very recently a detailed study of relaxation in the $A^2\Pi$ state of CN was carried out, in a system where the relative positions of all vibronic levels could be spectroscopically measured (Bondybey, 1977a; Bondybey and Nitzan, 1977). By monitoring the fluorescence occurring following selective excitation of a $A^2\Pi$ vibrational level it was

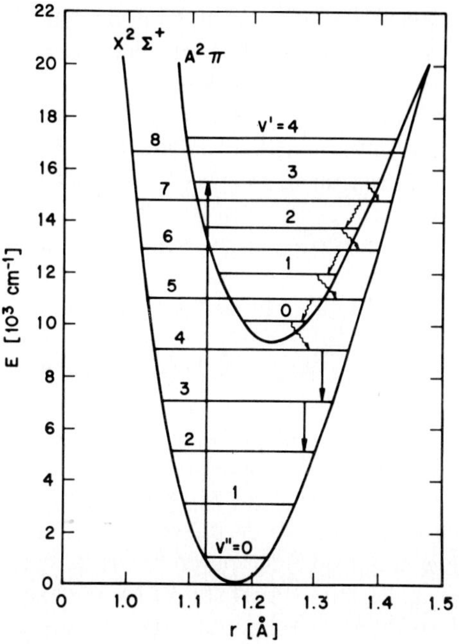

Fig. 7. Potential energy curves and cascading sequence for CN radical in solid neon. The population crosses back and forth from the $A^2\Pi$ state to the $X^2\Sigma^+$ state. The laser initially populates $v = 3$ of $A^2\Pi$. (From Bondybey, 1977a.)

shown that the predominant relaxation channel involves crossing from a given level of the $A\,^2\Pi$ state into the nearest lower level of the ground state, and then back-crossing into the $v - 1$ level of the A state. It was concluded that neither exact energy resonance, nor high molecular Franck–Condon factors, were required for efficient internal conversion to occur. A plot of the observed relaxation rates versus the respective energy gaps gave approximately linear dependence, indicating the existence of an energy gap law.

The cascading process was independently confirmed by directly monitoring the ground state level population following A state excitation, as shown in Fig. 7. This was accomplished by tuning a second probe laser to a wavelength connecting the desired level of the ground state with one of the low levels of the higher lying $B\,^2\Sigma^+$ state, and monitoring $B\,^2\Sigma$ fluorescence as a function of delay between the pump and probe lasers. This permitted not only confirmation of transient population in ground state levels following A state excitation, but also allowed measurement of the lifetimes of ground state levels below $v' = 1$ $A\,^2\Pi$. These lower levels can only relax by infrared emission or direct vibrational relaxation. The lifetimes of $v = 4$ and 3 levels were found to be 2.8 and 5.4 msec, respectively—values that are at least three orders of magnitude longer than the lifetimes of the higher lying levels. Comparison with similar molecules suggests that these values are probably controlled by radiation and that direct vibrational relaxation may be unmeasurably slow.

Bondybey and Nitzan (1977) compared the observed rates of the internal conversion processes with the rates calculated by a simple theoretical model of multiphonon electronic relaxation. It was found that although the observed rates qualitatively satisfy the energy gap law, quantitative agreement could not be established. The theory predicted an even steeper dependence of rates on energy gap than observed. More recently Weismann *et al.* (1978) have shown that better agreement could be reached by considering quadratic coupling effects.

Bondybey (1976c) studied similar spin forbidden processes in C_2 radical which has numerous low lying electronic states. The lowest triplet state, which previously was assumed to be the ground state, was recently shown to lie $\sim 700\ cm^{-1}$ above the $X\,^1\Sigma$ ground state. Intersystem crossing from the lowest triplet into the ground state is forbidden by both spin selection rules and by orbital symmetry. We have measured directly the lifetimes of the low triplet levels in rare gas matrices, using the pump-probe laser technique. The pump laser excited C_2 molecules into a $A\,^1\Pi$ state which relaxed ultimately into low triplet vibrational levels. The triplet population was monitored by a probe laser tuned to one of the Swan transitions which populates a higher fluorescent $d\,^3\Pi_g$ state. The study revealed that the low lying triplet levels all have lifetimes in the low microsecond range. The $v = 0$ lifetime in solid Ar is approximately 65 μsec, apparently controlled by spin and orbitally forbidden radiationless transitions into $v = 0$ $X\,^1\Sigma$. Actually the population of the triplet state levels was so efficient that excitation on the Swan transitions resulted in stimulated emission on some of the $d\,^3\Pi \rightarrow a\,^3\Pi$ vibronic transitions. This probably represents the first observation of stimulated emission in matrices.

The ability of matrices to induce interaction, and radiationless transitions, between states of different orbital symmetries was observed spectroscopically in a recent study of O_2 excited states (Goodman and Brus, 1977b). Near 30,000 cm^{-1} O_2 has $C\,^3\Delta_u$ and $A\,^3\Sigma_u^+$ excited states with nested potential energy curves. In the nonrotating gas phase molecule, there is only a weak spin–spin interaction between them. In solid N_2 host, a progression of strong perturbations between levels v of $C\,^3\Delta$ and $v-2$ of $A\,^3\Sigma_u^+$ is observed in fluorescence excitation spectra. The upper $\Omega = 2$ component of $^3\Delta$ is "pushed blue" of its zero-order position and lifetime broadened by interaction with the lower $^3\Sigma_u^+$ state. Analysis of this host-induced interaction yields a 27 cm^{-1} interaction matrix element. This interaction appears to require a phonon promoting mode, as the experimental spectroscopy is consistent with S_6 electronic symmetry in which $^3\Delta_u \rightarrow F_u$ and $^3\Sigma_u^+ \rightarrow A_u$.

The molecules discussed above have fairly large vibrational frequencies. It was therefore of interest to determine whether in heavier molecules with small vibrational spacing vibrational relaxation will prevail over interstate electronic relaxation processes. We have examined the heaviest member of the group IV molecules, Pb_2 (Bondybey and English, 1977a), Pb_2 spectroscopy is very complex with a large number of low lying and uncharacterized electronic states. We have studied emission from a state we denote F 0^+. Excitation of individual levels of this state results in emission from the level directly excited with no vibrational relaxation in spite of the low (~ 150 cm^{-1}) vibrational spacing. Two intense infrared emission systems are seen from two low lying electronic states populated by radiationless transitions from the F 0^+ levels. It was concluded that internal conversions and electronic radiationless processes prevail over vibrational relaxation. Unlike the other cases discussed above, the Pb_2 population does not return into lower F vibrational levels. This irreversibility appears to be due to the large number of electronic states present. Once the molecule leaves the F state it has statistically a low probability of returning.

All of the above examples indicate that electronic relaxation processes are remarkably efficient. This efficiency can be rationalized by considering the phonon line shapes and electron–phonon coupling. As discussed earlier, the guest–host interaction potential is controlled by the periphery of the electron density distribution in the guest species and is insensitive to motion of the nuclei and changes in vibrational state. Accordingly, pure vibrational transitions in matrix isolated molecules show very little intensity in the phonon wings. The vibrational transition therefore requires no rearrangement in the guest lattice atoms and results in poor Franck–Condon factors for multiphonon relaxation. Electronic transitions on the other hand, result in more drastic changes in electron density distribution and, accordingly, are often accompanied by intense phonon wings and better Franck–Condon factors. This argument has, in fact, been previously advanced (Bondybey and Brus, 1975b) to suggest that delocalized lattice phonons would accept energy during electronic relaxation processes, while local modes (e.g. rotation) would accept energy during vibrational relaxation processes. Atomic relaxation studies, to be discussed in Section VIII, directly demonstrate the ability of lattice phonons to accept energy during electronic relaxation processes.

VIII. ATOMIC RELAXATION

For sometime it has been known that alkali atomic spectra in rare gas solids are broadened, and occur as triplets when initially degenerate ^2P states are involved (Meyer, 1971). Moskovits and Hulse (1977) have recently suggested that this splitting is not a static Jahn–Teller effect due to a local environment distortion, but rather reflects a diatomic-like binding with one (perhaps fluxional) nearest neighbor rare gas atom. In the alkalis, the absence of appreciable fluorescence suggests a high quantum yield for radiationless transitions. In the case of isolated Be atoms, fluorescence from metastable ^3P apparently produced by radiationless transitions from an absorbing ^1P state, has been observed (Brom and Broida, 1975).

Recently tunable laser excitation has been used to prove that population of either the ^1P or ^1D states of matrix isolated Ca atoms leads to intense ^3P emission (Bondybey, 1978). This fast, direct multiphonon electronic relaxation, in the absence of vibrational or local mode degrees of freedom, confirms the strong coupling between electrons and lattice phonons previously postulated in the diatomic studies. In this study ZPLs were resolved on the atomic bands for the first time. This proves that the line broadening is a homogeneous phonon contour. There appears to be no special source of inhomogeneous broadening for isolated atoms when compared with isolated molecules.

IX. VIBRATIONAL RELAXATION IN TRIATOMICS

Within one electronic state of a triatomic molecule, mode to mode conversion of energy may compete with direct vibrational relaxation within one mode. This competition makes relaxation in polyatomics particularly interesting. Energy gap law considerations favor crossing to the closest lower-lying vibrational level, regardless of mode type, while the general weakness of anharmonic intermode coupling elements tends to favor relaxation within one normal mode. A few studies of intermode relaxation pathways have been carried out by Raman techniques (Laubereau *et al.*, 1973, 1974) in room temperature liquids. The lifetimes are in the low picosecond range.

Normally, electronic emission spectra of polyatomic molecules in matrices are completely vibrationally relaxed. Fast relaxation reflects a high density of vibrational states, and slower relaxation would be expected for strongly bound smaller polyatomics having only a few modes of high frequency.

The first clear observation of vibrationally unrelaxed emission was reported in the CF_2 radical (Bondybey, 1977c). Emission from the (0, 1, 0) and (0, 2, 0) states was seen in solid Ar, and lifetimes of 10 and 5 nsec, respectively, were reported for the vibrational relaxation of this level. Additional information about relaxation dynamics was provided by two independent studies of CNN (Wilkerson and Guillory, 1977; Bondybey, 1976d). In this molecule all the modes are active in the electronic spectrum. The observations demonstrate that the stretches relax via intermode conversion into overtones of the bend. Excitation of the (0, 2, 0) state produces emission from both (0, 2, 0) and (0, 1, 0) in addition to the relaxed (0, 0, 0) emission. Excitation of any of the stretching vibrations results in identical relative intensities from these three states, 1:10:90. Thus all higher states relax through these bending states.

However, a study of ClCF (Bondybey, 1977e) leads to different conclusions. In solid Ar a small fraction of the emission intensity originates from the (0, 0, 1)

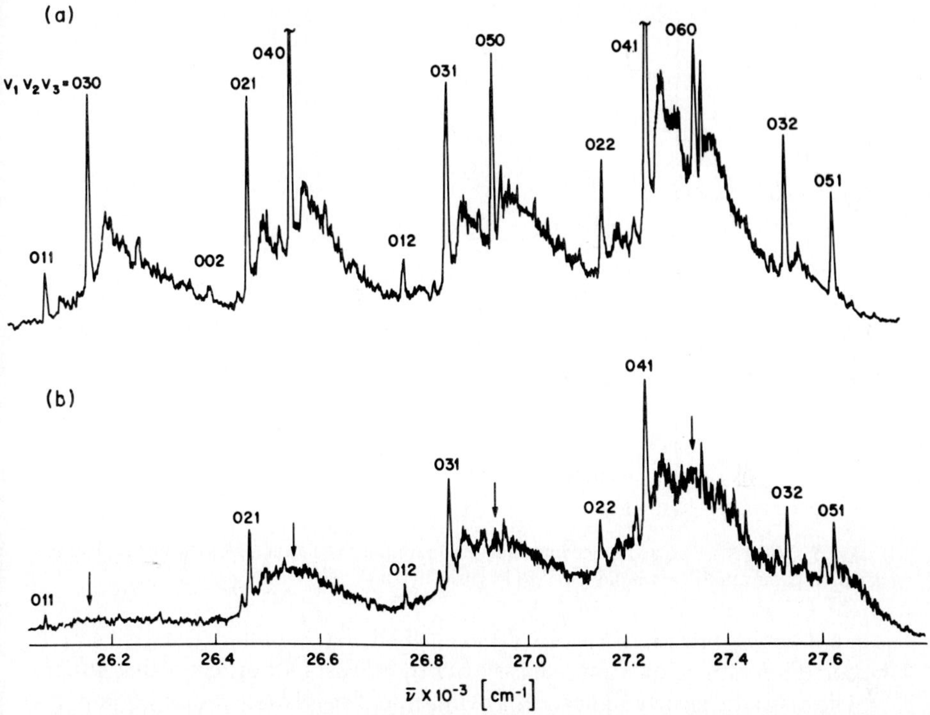

Fig. 8. Excitation spectra of ClCF fluorescence in solid Argon: (a) relaxed (0, 0, 0) fluorescence, exhibiting both bending and stretching model (24,097 cm^{-1}); (b) unrelaxed (0, 0, 1) fluorescence, exhibiting only higher stretching states (24,809 cm^{-1}). (From Bondybey, 1977c.)

state, containing one quantum of the C–Cl stretch. The excitation spectrum of the relaxed fluorescence in Fig. 8 can be analyzed in terms of all three upper state modes. Figure 8 also shows the excitation spectrum of the unrelaxed fluorescence. This contains only states which involve excitation of the ν_3 mode; those states corresponding to the pure bending progression are absent. This implies that states involving overtones of ν_2 undergo intramode relaxation rather than undergo intermode conversion into the nearby states containing ν_3.

NCO presents a particularly interesting example (Bondybey and English, 1977b). NCO has a rather complex spectroscopy with Renner splitting in the ground electronic state and numerous Fermi resonances in the excited electronic state. These resonances occur between the symmetric stretching mode ν_1 and overtones of the bending mode. Selective excitation of individual vibronic levels of the excited electronic state reveals a fast relaxation between groups of levels in Fermi resonance. However, no relaxation involving other levels is observed. For instance, as in Fig. 9 the (0, 0, 2) state relaxes completely into the (0, 4, 0) state 200 cm^{-1} below in less than about 5 nsec. On the other hand, this state

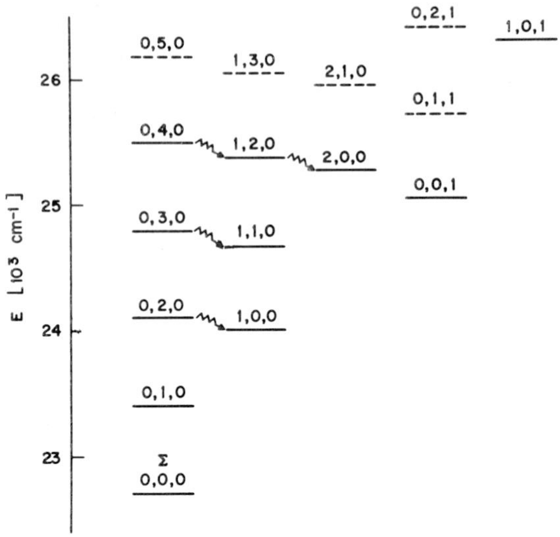

Fig. 9. NCO $A^2\Sigma^+$ excited state vibrational structure. The broken arrows indicate the observed relaxation pathways. (From Bondybey and English, 1977b.)

does not further relax into the (1, 0, 0) state which lies 220 cm^{-1} lower in energy; the corresponding relaxation rate is slower by at least four orders of magnitude.

These results clearly indicate that vibrational relaxation processes in polyatomic molecules cannot be explained by an energy gap law alone. The presently unknown intermode anharmonic coupling elements must also be taken into consideration. Effects which enhance intermode coupling, such as Fermi resonance, enhance intermode transitions in the solid phase just as they do in gas phase intermode collisional transitions (Weitz and Flynn, 1974).

X. VIBRATIONAL ENERGY TRANSFER

A. Long Range Multipole Interaction

We have seen that vibrational relaxation times can be very long. Even in relatively dilute samples, intermolecular vibrational energy transfer competes with relaxation and may dominate the excited state kinetics if relaxation can be neglected. In a classic study, (Dubost and Charneau, 1976; Legay, 1976) extensive transfer of vibrational energy among dilute CO guests in solid Ar has been spectroscopically observed. Two phenomena, both related to the fact that entropy is unimportant in Boltzmann equilibria near 4.2 K, differentiate the energy flow from similar gas phase transfer and equilibration processes at 300 K. First, the bulk of the energy flows into the minor isotopes $^{13}C^{16}O$ and $^{12}C^{10}O$ (present only in natural abundance) when the infrared laser initially excites

$^{12}C^{16}O$. This occurs because the transfer exothermicities ΔH to $^{12}C^{18}O$ and $^{13}C^{16}O$ are large with respect to $kT \simeq 3\ cm^{-1}$. Second, successive exothermic processes of the sort $v = n + v = 1 \rightarrow v = n + 1 + v = 0$ pump considerable energy into the higher vibrational levels $v > 1$. At 4.2 K the CO vibrational anharmonicity is larger than kT. CO($v > 1$) excitons are effectively trapped on the particular CO in which they are formed, as the long range dipole–dipole interaction only weakly transfers such excited states from one CO to another. Population inversions can be formed on $v = 1 \rightarrow v = 0$ of the minor isotopes and between $v > 1$ levels of the $^{12}C^{16}O$.

The first time resolved study of a vibrational energy transfer rate dependence upon transfer exothermicity ΔH involved processes such as (Goodman and Brus, 1976c)

$$\mathrm{ND}^*(v = 1) + \mathrm{CO}(v = 0) \rightarrow \mathrm{ND}^*(v = 0) + \mathrm{CO}(v = 1) + 78\ \mathrm{cm}^{-1}.$$

The asterisk indicates that the donor is electronically (as well as vibrationally) excited. Electronic excitation effectively stopped the competing exactly resonant energy transfer among donor ND molecules, as ND* has a lower vibrational frequency than ND. The observed nonexponential donor decay is well described by the Forster theory of long range R^{-6} dipole–dipole interaction. Figure 10 shows that by using both ^{12}CO and ^{13}CO as acceptors and by using ND*($v = 2$) and ND*($v = 1$) as donors, the rate dependence upon exothermicity ΔH was obtained. An exponential energy gap law of the form $\exp(-\Delta H/28\ \mathrm{cm}^{-1})$ was observed in solid Ar. However, transfer from NH* to CO did not follow this law and was anomalously fast by several orders of magnitude.

Lin and co-workers (1976) have theoretically formulated the relationship between nonresonant energy transfer and vibrational relaxation. Both processes involve a guest's ability to convert internal vibration energy into lattice phonons. The discovery above that NH* is especially effective in nonresonant transfer is consistent with the previously discussed faster vibrational relaxation of NH* as compared with ND*. In this study, the rate of dissipation of 78 cm^{-1} into the lattice could be compared with the vibrational relaxation data by dividing out the transition dipole factors in Lin's formulas. Very recently, Gerber and Berkowitz (1978) theoretically formulated the nonresonant energy transfer process with explicit involvement of the rotational accepting modes. Excellent agreement was obtained for ratios of the rate processes experimentally measured.

B. Exchange Energy Transfer

Little is known about fast *nearest neighbor*, exchange vibrational energy transfer. Energy transfer by highly vibrationally excited N_2 sitting next to an excited N atom in N_2 matrices has been observed (Dressler *et al.*, 1975). A nearest neighbor transfer time of 17 nsec was indirectly obtained for these perturbed N_2 molecules.

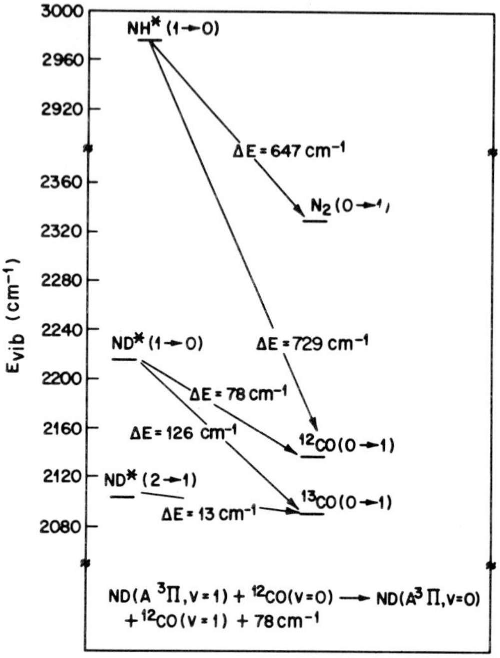

Fig. 10. Schematic diagram of long-range vibrational energy transfer processes in solid argon (4.2K). Donor NH* and ND* levels indicated on the left, and acceptor ^{12}CO, ^{13}CO and N_2 levels indicated on the right. (From Goodman and Brus, 1976c.)

An experiment has been performed upon isolated $(O_2)_2$ dimers in solid neon host (Goodman and Brus, 1977c). The strategy of this latter experiment was to detect the fast transfer as a splitting of vibronic ZPLs in the electronic spectra of the dimer. Oxygen dimers absorb red photons producing an excited electronic state in which both halves of the dimer are in the metastable a $^1\Delta$ state of O_2:

$$O_2(^3\Sigma) + O_2(^3\Sigma^-) + h\nu(6297\text{Å}) \rightarrow O_2(^1\Delta) + O_2(^1\Delta). \qquad (2)$$

An analysis of the electronic fine structure shows the dimer geometry is rectangular D_{2h} and allows one to obtain values for the intermolecular exchange integrals between the two O_2. Figure 11 shows the expected vibrational structure of both upper and lower electronic states of the dimer. The anharmonicity of the monomer vibration produces certain splittings, such as between the levels $(2 + 0)$ and $(1 + 1)$. The notation $(n + m)$ implies that one O_2 contains n quanta, and the other O_2 contains m quanta. Levels such as $(1 + 0)$ are further split by vibrational energy transfer. Such splittings cannot be simply observed in absorption spectra, as the upper component is forbidden under electric dipole selection rules. However, this level is thermally populated following laser excitation of the lower component, and the splitting can be observed in the sub-

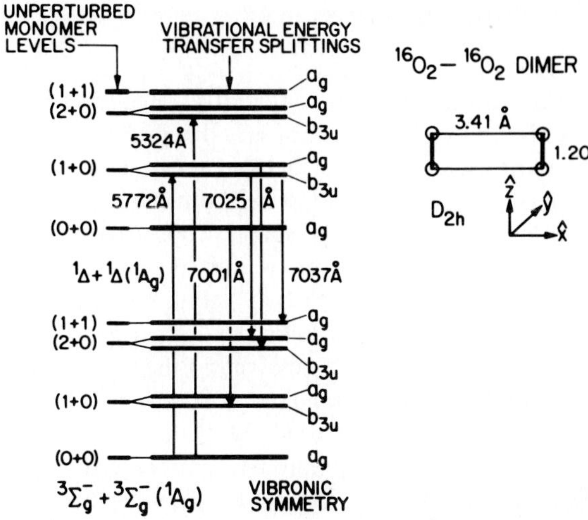

Fig. 11. Schematic diagram indicating $(O_2)_2$ dimer geometry and vibronic level structure. The correlation diagram shows how monomer states split due to vibrational energy transfer. Vertical lines indicate observed transitions. Normal modes: A_g—symmetric O_2 stretch, b_{3u}—antisymmetric O_2 stretch, A_g—dimer bond stretch, A_u—out of plane torsion, b_{2g}—symmetric libration, b_{2u}—antisymmetric libration. Electronic fine structure components are not shown. (From Goodman and Brus, 1977c.)

sequent fluorescence. The exactly resonant transfer time from $^{16}O_2(^1\Delta, v = 1)$ to nearest neighbor $^{16}O_2(^1\Delta, v = 0)$ is 14 psec. An attempt to obtain the isotopic dependence via this technique failed, as isotopic substitution drastically lowers the fluorescence quantum yield due to enhanced intersystem crossing into the $^3\Sigma^- + {}^1\Sigma_g^+$ state. A strong energy gap law, if applicable, would yield a nonresonant transfer time from $^{16}O_2(^1\Delta, v = 1)$ to $^{18}O_2(^1\Delta, v = 0)$ of approximately 10^{-9}–10^{-10} sec.

A quantitative analytical theory of vibrational energy transfer due to weak intermolecular exchange forces does not exist. In the case of a stronger interaction, intermolecular transfer begins to resemble intramolecular vibrational redistribution in a tetra-atomic molecular species. Such intramolecular redistribution is also poorly understood.

Long range energy transfer from ground state C_2^- to C_2H_2 has been studied by Nibler and co-workers using the two laser pump-probe technique. (Allamandola *et al.*, 1977). They report that the transfer rate falls off much more slowly than R^{-10}, which is expected from quadrupole–quadrupole electrostatic coupling. They intriguingly suggest a direct through Ar interaction, with the entire system being considered as a large molecule. Another possibility is that C_2^- has a small permanent dipole in these matrices because of the influence of the unknown counterion.

XI. OBSERVATION OF MISSING STATES

Both the photodissociation cage effect, and the occurrence of multiple cascading processes through various different excited electronic states, allow the observation of molecular states not observable in direct absorption spectroscopy. Valuable information about structure can be obtained in this fashion, especially when the spectra are sharp ZPL without the hot bands and rotational structure which complicate gas phase absorption spectroscopy. We have already described how the $a\,^4\Sigma$ state of C_2^- was found and characterized via the time dependent relaxation of the well known $B\,^2\Sigma^+$ state. In the cases of alkly iodides and perfluoroalkyl iodides, the photodissociation cage effect allowed us to determine that the repulsive 0^+ state involved in the I atom photodissociation laser actually has a shallow bound well of $D_e \leq 2000\ \mathrm{cm}^{-1}$ (Brus and Bondybey, 1976). In the case of Cl_2, excitation of either BO_u^+ or $^1\Pi_u$ produces luminescence from the optically forbidden and previously unobserved $^3\Pi_2$ state, which is determined to lie 650 cm^{-1} below BO_u^+. (Bondybey and Fletcher, 1976). In the heavier halogens the population does not reach $^3\Pi_2$ as larger spin–orbit coupling creates larger energy splittings between the Ω components of the zero-order $^3\Pi$ state. In the cases of TiO and ZrO, Broida and co-workers have observed several new excited states populated by interelectronic cascading sequences (Brom and Broida, 1975; Lauchlan *et al.*, 1976). As previously discussed in the case of O_2 in N_2 host, luminescence was observed from the poorly characterized $C\,^3\Delta_u$ and $c\,^1\Sigma$ states, with the result that the absolute gas phase vibrational numbering could be obtained (Goodman and Brus, 1977b). In the S_2 species, whose gas phase spectra are dominated by the $X\,^3\Sigma_g^- \rightarrow B\,^3\Sigma_u^+$ transition, the matrix fluorescence occurs from a lower lying $^3\Pi_u$ state (Bondybey, 1978).

Similar phenomena occur in polyatomics. In CrO_2Cl_2, the observed emission comes from a state lying 290 cm^{-1} below the state observed in absorption (Spoliti *et al.*, 1974; Bondybey, 1976a). Excitation of the known $^1\Pi$ state of C_3 near 4050 Å produced strong emission from a previously unknown lower lying triplet near 5800 Å (Weltner and McLeod, 1966; Bondybey and English, 1978). In the case of $(O_2)_2$ dimers, the spectra are sufficiently sharp in Neon host that the electronic fine structure splittings and singlet–triplet separation can be observed for the zero-order $^1\Delta + {}^1\Delta$ and $^3\Sigma + {}^3\Sigma$ dimer states (Goodman and Brus, 1977c). The spectra of CF_3NO in Neon show that the n–π^* excited state is staggered, and a reported eclipsed isomer appears to be spurious (Goodman and Brus, 1978).

REFERENCES

Albrecht, A. C. (1961). *J. Mol. Spectrosc.* **6**, 84.

Allamandola, L. S., Rojhantalab, H. M., Nibler, J. W., and Chappell, T. (1977). *J. Chem. Phys.* **67**, 99.

Ambartsumyam, R. V., Gorokhov, Y. A., Markov, G. N., Puretskii, A. A., and Furzikov, N. P. (1976). *JETP Lett.* **24**, 256.
Aulf, B. S., and Andrews, L. (1976). *J. Chem. Phys.* **65**, 4192.
Barsuhn, J. (1974). *J. Phys. B* **7**, 155.
Bondybey, V. E. (1976a). *Chem. Phys.* **18**, 293.
Bondybey, V. E. (1976b). *J. Chem. Phys.* **65**, 5138.
Bondybey, V. E. (1976c). *J. Chem. Phys.* **65**, 2296.
Bondybey, V. E. (1976d). *J. Mol. Spectrosc.* **63**, 164'
Bondybey, V. E. (1977a). *J. Chem. Phys.* **66**, 995.
Bondybey, V. E. (1977b). *J. Chem. Phys.* **67**, 664.
Bondybey, V. E. (1977c). *J. Chem. Phys.* **66**, 4237.
Bondybey, V. E. (1978). *J. Chem. Phys.* **68**, 1308.
Bondybey, V. E., and Albiston, C. (1978). *J. Chem. Phys.* **68**, 3172.
Bondybey, V. E., Bearder, S. S., and Fletcher, C. (1976). *J. Chem. Phys.* **64**, 5243.
Bondybey, V. E. and Brus, L. E. (1975a). *J. Chem. Phys.* **62**, 620.
Bondybey, V. E. and Brus, L. E. (1975b). *J. Chem. Phys.* **63**, 794.
Bondybey, V. E., and Brus, L. E. (1975c). *J. Chem. Phys.* **63**, 2223.
Bondybey, V. E., and Brus, L. E. (1976a). *J. Chem. Phys.* **64**, 3724.
Bondybey, V. E., and English, J. H. (1977a). *J. Chem. Phys.* **67**, 3405.
Bondybey, V. E., and English, J. H. (1977b). *J. Chem. Phys.* **67**, 2868.
Bondybey, V. E., and English, J. H. (1978). *J. Chem. Phys.* **68**, 4641.
Bondybey, V. E., and Fletcher, C. (1976). *J. Chem. Phys.* **64**, 3615.
Bondybey, V. E., and Nibler, J. W. (1972). *J. Chem. Phys.* **56**, 4719.
Bondybey, V. E., and Nitzan, A. (1977). *Phys. Rev. Lett.* **38**, 889.
Boursey, E., and Roncin, J. Y. (1975). *J. Mol. Spectrosc.* **55**, 31.
Brom, J. M., and Broida, H. P. (1975). *J. Chem. Phys.* **63**, 3718.
Brueck, S. R. J., and Osgood, R. M. (1976). *Chem. Phys. Lett.* **39**, 568.
Brus, L. E., and Bondybey, V. E. (1975a). *J. Chem. Phys.* **63**, 3123.
Brus, L. E., and Bondybey, V. E. (1975b). *J. Chem. Phys.* **63**, 786.
Brus, L. E., and Bondybey, V. E. (1976). *J. Chem. Phys.* **65**, 71.
Chapman, O. L., De La Cruz, D., Roth, R., and Pacansky, S. (1973). *J. Am. Chem. Soc.* **95**, 1337.
Davies, B., McNeish, A., Poliakoff, M., Tranquille, M., and Turner, J. J. (1977). *Chem. Phys. Lett.* **52**, 477.
Dellinger, B., and Kasha, M. (1976). *Chem. Phys. Lett.* **38**, 9.
Dellinger, B., King, D. S., Hochstrasser, R. M., and Smith, A. B. (1977). *J. Am. Chem. Soc.* **99**, 7138.
DiBartolo, B., and Powell, R. C. (1976). "Phonons and Resonance in Solids." Wiley, New York.
Dressler, K., Oehler, O., and Smith D. A. (1975). *Phys. Rev. Lett.* **34**, 1364.
Dubost, H., and Charneau, R. (1976). *Chem. Phys.* **12**, 407.
Freed, K. F., Yeager, D. L., and Metiu, H. (1977). *Chem. Phys. Lett.* **49**, 19.
Frosh, R. P. (1971). *J. Chem. Phys.* **54**, 2660.
Gedanken, A., Raz, B., and Jortner, J. (1973). *J. Chem. Phys.* **43**, 2997.
Gerber, R. B., and Berkowitz, M. (1977). *Phys. Rev. Lett.* **39**, 1000.
Gerber, R. B., and Berkowitz, M. (1978). *Chem. Phys. Lett.* **56**, 108.
Goodman, J., and Brus, L. E. (1976a). *J. Chem. Phys.* **65**, 3808.
Goodman, J., and Brus, L. E. (1976b). *J. Chem. Phys.* **65**, 3146.
Goodman, J., and Brus, L. E. (1976c). *J. Chem. Phys.* **65**, 1156.
Goodman, J., and Brus, L. E. (1977a). *J. Chem. Phys.* **67**, 933.
Goodman, J., and Brus, L. E. (1977b). *J. Chem. Phys.* **67**, 1482.
Goodman, J., and Brus, L. E. (1977c). *J. Chem. Phys.* **67**, 4398.
Goodman, J., and Brus, L. E. (1977d). *J. Chem. Phys.* **67**, 4408.
Goodman, J., and Brus, L. E. (1977e). *J. Chem. Phys.* **67**, 4858.

Goodman, J., and Brus, L. E. (1978). *J. Am. Chem. Soc.* **100**, 2971.
Goodman, J., Tully, J. C., Bondybey, V. E., and Brus, L. E. (1977). *J. Chem. Phys.* **66**, 4802.
Harris, S. J., Janda, K. C., Norvick, S. E., and Klemperer, W. (1975). *J. Chem. Phys.* **63**, 881.
Hirayama, F., and Lipsky, S. (1975). *J. Chem. Phys.* **62**, 576.
Laubereau, A., Kirschner, L., and Kaiser, W. (1973). *Opt. Commun.* **9**, 182.
Laubereau, A., Kehl, G., and Kaiser, W. (1974). *Opt. Commun.* **11**, 74.
Lauchlan, L. J., Brom, J. M., and Broida, H. P. (1976). *J. Chem. Phys.* **65**, 2672.
Legay, F. (1976). "Chemical and Biological Applications of Lasers" (C. B. Moore, ed.), Vol. II. Academic Press, New York.
Lin, S. H., Lin, H. P., and Knittel, D. (1976). *J. Chem. Phys.* **64**, 441.
Linevsky, M. J. (1961). *J. Chem. Phys.* **34**, 1956.
Meyer, B. (1971). "Low Temperature Spectroscopy." Elsevier, New York.
Miller, J. C., and Andrews, L. (1977). *Chem. Phys. Lett.* **50**, 315.
Moskovits, M., and Hulse, J. E. (1977). *J. Chem. Phys.* **67**, 4271.
Nitzan, A., Mukamel, S., and Jortner, J. (1974). *J. Chem. Phys.* **60**, 3929.
Robinson, G. W. (1967). *J. Chem. Phys.* **47**, 1967.
Rebane, K. (1970). "Impurity Spectra of Solids." Plenum Press, New York.
Spoliti, M., Thirtle, J. H., and Dunn, T. M. (1974). *J. Mol. Spectrosc.* **52**, 146.
Tinti, D. S., and Robinson, G. W. (1968). *J. Chem. Phys.* **49**, 3229.
Weissman, Y., Jortner, J., and Nitzan, A. (1978). *Chem. Phys.* **26**, 413.
Weitz, E., and Flynn, G. (1974). *Ann. Rev. Phys. Chem.* **25**.
Weltner, W., and McLeod, D. (1966). *J. Chem. Phys.* **45**, 3096.
Wilkerson, J. L., and Guillory, W. A. (1977). *J. Mol. Spectrosc.* **66**, 188.
Zalewski, E. F., McClure, D. S., and Narva, D. L. (1974). *J. Chem. Phys.* **61**, 2964.

7

High Pressure Studies of Luminescence Efficiency

H. G. Drickamer, G. B. Schuster, and D. J. Mitchell

School of Chemical Sciences
and
Materials Research Laboratory
University of Illinois
Urbana, Illinois

I. INTRODUCTION

High pressure has proved to be a very powerful tool for studying luminescence efficiency and energy transfer in condensed systems. In this paper we present examples in five areas. First, we discuss internal conversion involving localized excitations, presenting one qualitative and one quantitative example. Second, we discuss intersystem crossing, again with localized excitations. We demonstrate the effect on fluorescence efficiency and on the quantum yield of a photochemical reaction. The third example involves a more delocalized system—impurity emission in ZnS phosphors. The fourth aspect exhibits a test of the theory of sensitized luminescence by dipole–dipole energy transfer. The final example concerns the effect of viscosity on the luminescence efficiency of dye molecules.

ISBN 0-12-450650-X

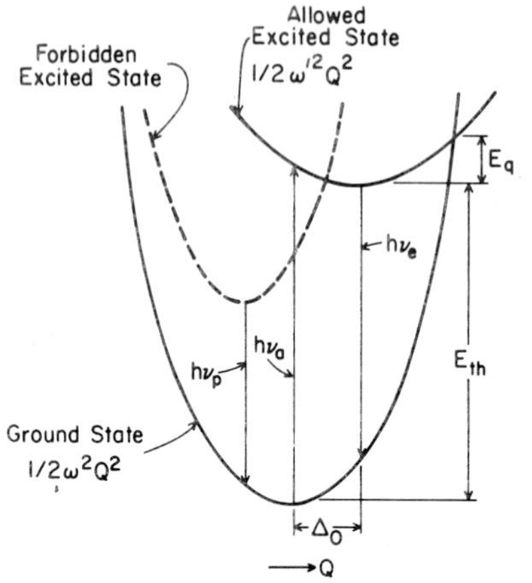

Fig. 1. Schematic configuration coordinate diagram—classical treatment. $R = (\omega'/\omega)^2$; $\Delta = \Delta_0 + P(R - 1)/\omega^2 R$.

For localized excitations the possible processes can conveniently be described in terms of a single configuration coordinate (CC) diagram. Figure 1 exhibits schematically a classical CC diagram from which we can describe qualitatively the various processes: Along the vertical axis is plotted energy, while the horizontal axis is some normal mode of vibration of the system. Pressure couples to the volume, so that the configuration coordinate most affected in these experiments is the breathing mode of the *system*; it is an *inter*-molecular coordinate, i.e., it involves the interaction of an atom or molecule with the environment. This point is discussed in detail elsewhere (Okamoto and Drickamer, 1974; Drotning and Drickamer, 1976). Both optical absorption and emission occur vertically on such a diagram (the Franck–Condon principle) because they are rapid compared with atomic or molecular motions. Each of the electronic potential wells contains a series of vibrational levels of some average spacing $\hbar\eta$ or $\hbar\eta'$, not shown in the figure in order to keep the electronic processes clear.

When the electron absorbs a quantum of light and moves to the upper electronic state it rapidly gives off vibrational energy (phonons) and reaches the bottom of the well. In this classical picture the fluorescence efficiency depends on the relative probability of emitting a phonon or crossing thermally to the ground state at the intersection of the two electronic wells. Later we introduce a quantum mechanical analysis based on the work of Struck and Fonger (1975). In addition to the internal conversion (IC) discussed above, a second factor which limits fluorescence efficiency is also illustrated in Fig. 1. The dotted potential well represents a state nominally not accessible by direct optical excitation (e.g., a triplet state). The excited electron may cross thermally from the allowed state to the triplet state [this process is called intersystem crossing (ISC)] from which it may return either by a thermal or an optical (phosphorescent) path to the ground state. Finally, while the electron is excited, a chemical reaction may occur. Because of the long lifetime of the triplet state, photochemical processes involving triplet states are especially common.

Besides strictly localized excitations, luminescence may occur via band-to-band excitations and subsequent trapping of the excited electron and/or resulting hole. A process of this type involving ZnS is discussed later in the chapter.

II. INTERNAL CONVERSION

A. Qualitative Treatment—Azulene

Let us first take up some qualitative predictions for a localized excitation where intersystem crossing is the only thermal process, and compare them with experiment. In Fig. 2 we illustrate some consequences of the relative displacement of the potential wells of a CC diagram on the rate of the (radiationless) internal conversion process, and thus on the fluorescence efficiency. If the excited state well is displaced (relative to the ground state well) vertically to higher energy—corresponding to increased energy of the emission with no change of shape—the height of the thermal barrier for IC increases, the thermal process becomes less probable, and the fluorescence efficiency increases. Similarly, a

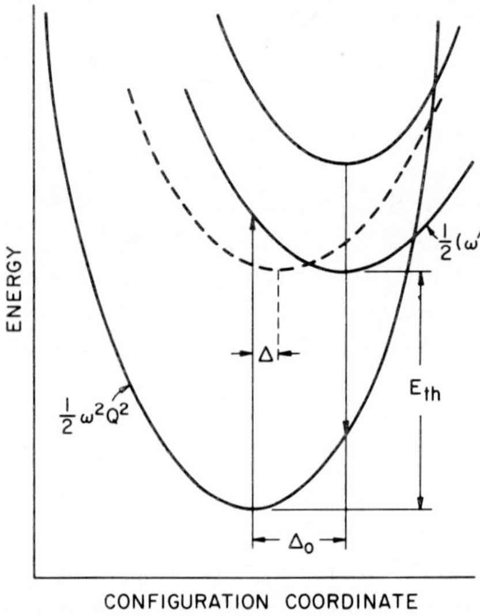

Fig. 2. Schematic configuration coordinate diagram illustrating internal conversion effects. $R = (\omega'/\omega)^2$; $\Delta = \Delta_0 + P(R-1)/\omega^2 R$. (From Okamoto and Drickamer, 1974.)

shift to lower energy would decrease the efficiency of fluorescence. It can also be seen that a decrease in relative displacement along the configuration coordinate ($\Delta_0 \rightarrow \Delta$) would decrease the probability of radiationless relaxation and increase emission intensity.

For organic molecules in particular, the direction and magnitude of shift will depend on the polarizability and/or dipole moment of the excited state relative to the ground state and on the polarizability of the medium. If the excited state is more polarizable, it exerts a greater attraction for the environment and a red shift (shift to lower energy) ensues. If it is less polarizable than the ground state one may expect a blue shift. These directions and magnitudes are modified by the polarizability of the environment.

Here we present a study (Mitchell *et al.*, 1977b) of azulene and an azulene derivative—dicarboethoxychloroazulene (DCECA)—in two solvents: polymethylmethacrylate (PMMA) and polystyrene (PS). Azulene is an isomer of naphthalene with one five- and one seven-membered ring. For most molecules, no matter what the state to which an electron is excited, it returns thermally to the lowest excited state S_1 before it emits, so information is obtained only about S_1. Azulene emits from the second excited state S_2, and some derivatives emit from both S_2 and S_1. It is known that S_2 is significantly more polarizable than the ground state S_0. For DCECA there is reason to believe that S_1 is less polarizable than S_0. PMMA is a significantly less polarizable solvent than PS.

In Figs. 3 and 4 we exhibit the shift of the emission peak and change in integrated intensity with pressure for the $S_2 \rightarrow S_0$ emission of azulene in PMMA and in PS. The peak shifts to lower energy with increasing pressure, as would be expected from the above discussion. It shifts about twice as rapidly in PS as in PMMA, which is consistent with the greater polarizability of the former solvent. By 140 kbar the efficiency in PMMA has dropped to ~2% of its initial value, while in PS at the same pressure the efficiency is less than 1% of the atmospheric value. Thus the shifts and intensity changes correlate with the discussion based on the configuration coordinate diagram. Substituted azulenes give qualitatively similar results for the $S_2 \rightarrow S_0$ emission.

In Fig. 5 we show similar data for the $S_1 \rightarrow S_0$ emission of a substituted azulene in PMMA. For this material the S_1 state is less polarizable than the ground state. As one would predict, the peak shifts to higher energy with increasing pressure. At the same time the intensity increases by a factor of 40 in 130 kbar. This too follows the prediction of our CC diagram. Finally, in Fig. 6 we exhibit data for the $S_1 \rightarrow S_0$ transition in the substituted azulene, in PS. Here there are more complicated interactions involving both a vertical and horizontal displacement of the potential wells. In the low pressure region the peak shifts to higher energy; it reaches a maximum near 80 kbar and then shifts to lower energy. Similarly, the intensity increases at low pressure, maximizes, and then decreases. This is a striking illustration of the relationship between emission energy and luminescent efficiency.

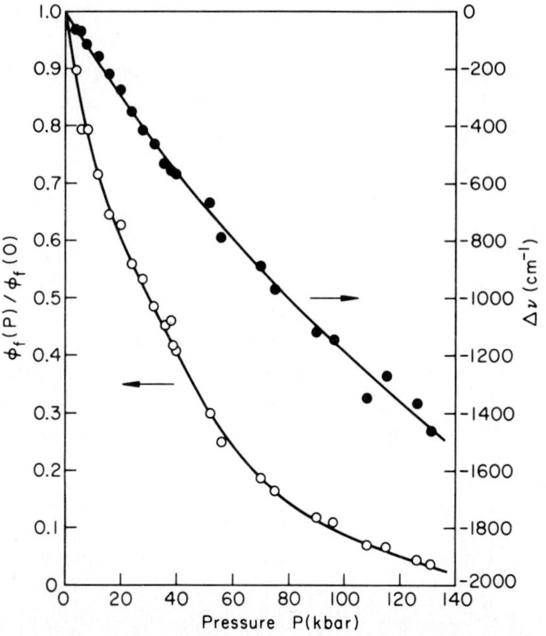

Fig. 3. $S_2 \rightarrow S_0$ emission for azulene in PMMA. (From Mitchell *et al.*, 1977b.)

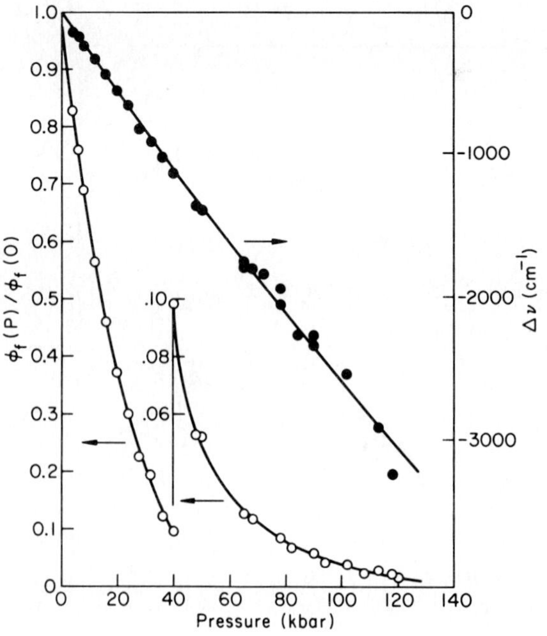

Fig. 4. $S_2 \rightarrow S_0$ emission for azulene in PS. (From Mitchell *et al.*, 1977b.)

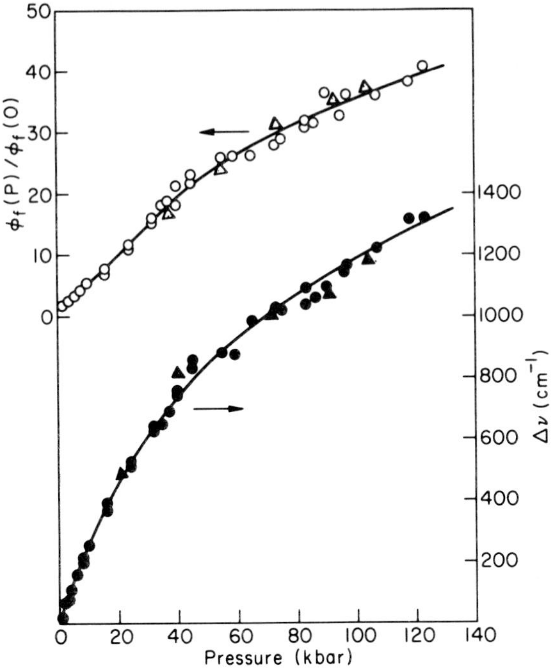

Fig. 5. $S_1 \rightarrow S_0$ emission for dicarboethoxychloroazulene in PMMA. (From Mitchell *et al.*, 1977b.)

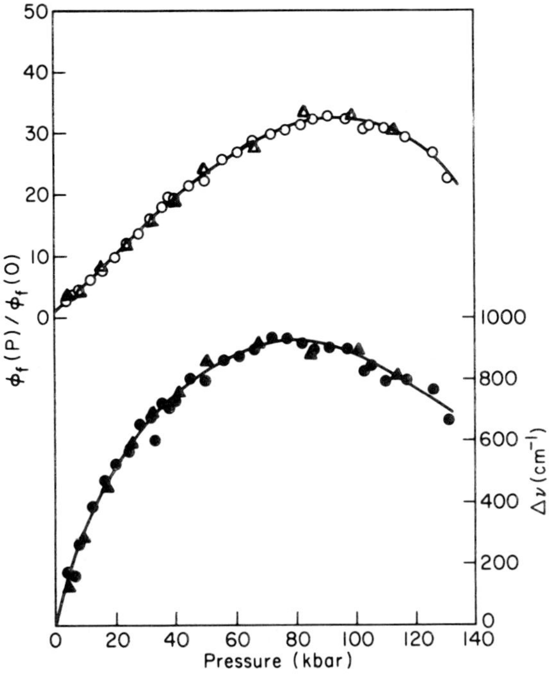

Fig. 6. $S_1 \rightarrow S_0$ emission for dicarboethoxychloroazulene in PS. (From Mitchell *et al.*, 1977b.)

B. Quantitative Treatment—Molybdates

We turn now to a more quantitative treatment of a localized charge transfer, which can be studied in terms of a single configuration coordinate model. We briefly discuss here both classical and quantum mechanical treatments of quenching and apply the latter analysis to $CaMoO_4$ and $HgMoO_4$. Parameters for the classical model are shown in Fig. 1.

ω^2 and $(\omega')^2$ are the harmonic force constants associated with the ground and excited state electronic potential wells. We denote as $h\eta$ and $h\eta'$ the vibrational splittings within these electronic potential wells. Δ is the relative displacement of the excited state potential well with respect to the ground state well along the configuration coordinate of interest. Pressure couples most strongly to the volume, i.e., to the totally symmetric vibrational coordinate. The work done in displacing a well a distance Δ under a pressure p is $pA\Delta$, where A is an (undetermined) cross-sectional area. It is sometimes convenient, for the totally symmetric vibration, to absorb A into Δ. Then the conversion from the units used here, inverse centimeters per kilobar, to, e.g., cubic centimeters per mole is a factor of 0.12.

A study of optical band shapes based on the configuration coordinate model has been published previously (Tyner *et al.*, 1976). The results of this work and an extension of it by Bieg and Drickamer (1977) provide general relationships between two experimentally measurable properties of the emission bandshape (halfwidth and skewness) and parameters of the configuration coordinate model. The resulting expression for the first of these properties, the halfwidth $E_{1/2}$ (full width of the emission band at half its maximum height) is

$$E_{1/2} = N\omega|\Delta|/R^{1/2} \qquad (1)$$

where

$$N = (8kT^* \ln 2)^{1/2}, \qquad T^* = (\hbar\eta'/2k)\coth(\hbar\eta'/2kT), \qquad R = (\omega')^2/\omega^2,$$

and k is the Boltzmann constant. The second property, peak skewness or asymmetry, is measured by the parameter b:

$$b = N(R-1)/2\omega|\Delta|R^{1/2}. \qquad (2)$$

If the rate associated with the radiative transition from the excited state is r_r and that with the nonradiative transition is r_{nr}, the luminescence efficiency ϕ is given by

$$\phi = r_r/(r_r + r_{nr}). \qquad (3)$$

The observed lifetime is given by

$$1/\tau_{obs} = r_r + r_{nr} \qquad \text{or} \qquad \tau_{obs} = \phi/r_r. \qquad (4)$$

The pressure dependence of the radiative rate is generally negligible. The radiative rate of parity-forbidden transitions, however, is temperature dependent since odd-parity vibrations in the crystal tend to relax the selection rules. For this case, known as the phonon-assisted transition, the radiative rate is (Martin and Fowler, 1970)

$$r_{\mathrm{r}} = r_{\mathrm{r}0} \coth(\hbar\eta'/2kT). \tag{5}$$

Classically, the nonradiative transition has been thought of as an activated process with a single activation energy E_{q} (Mott, 1938) (see Fig. 1). The classical model is quantitatively poor for quenching calculations but is quite accurate for peak shape parameters and more convenient to use for evaluating the latter.

Recently Struck and Fonger (1975) developed a quantum mechanical single configuration coordinate (QMSCC) model which quantitatively describes the thermal quenching of luminescence. In this model a nonradiative transition can occur between ground and excited state vibrational levels of approximately equal energy. For the somewhat simple case of $\omega'^2 = \omega^2$ (the linear coupling case), this means that $n = m + p$, where the transition of interest is between the mth excited state level and the nth ground state level; p is the number of vibrational spacings between the ground and excited state minima (integer part of $E_{\mathrm{th}}/\hbar\eta$). See Fig. 7.

If u_n and v_m are the quantum mechanical wavefunctions of the ground and excited state levels n and m, respectively, the rate of the transition is proportional

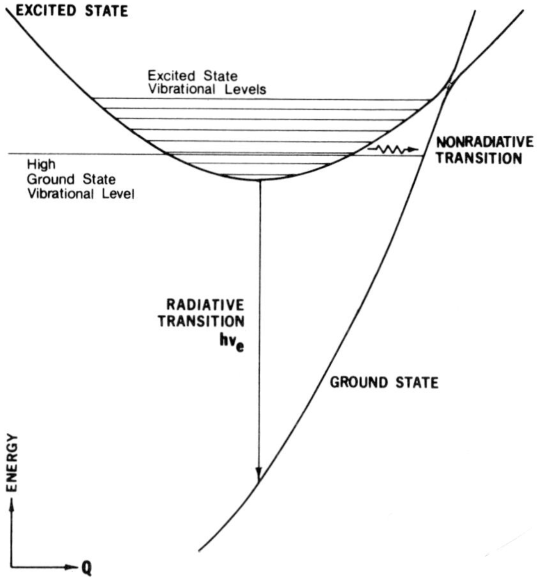

Fig. 7. Schematic configuration coordinate diagram—quantum mechanical treatment. (From Tyner and Drickamer, 1977.)

to the squared wavefunction overlap integral $\langle u_n | v_m \rangle^2$ weighted to account for the thermal probability of the electron being in the mth level of the excited state. The total nonradiative rate r_{nr} is then the sum of these rates from individual levels over all excited state levels:

$$r_{\text{nr}} = r_{\text{n0}} \sum_{m=0}^{\infty} (1 - \alpha)\alpha^m \langle u_{p+m} | v_m \rangle^2, \tag{6}$$

where $\alpha = \exp(-\hbar\eta'/kT)$ is the Boltzman factor between levels and $(1 - \alpha)\alpha^m$ is the thermal weight of the mth excited vibrational level. The nonlinear coupling case, i.e., $\omega'^2 \neq \omega^2$, is similar, though algebraically more complicated. It may be found in Struck and Fonger (1975).

Qualitatively this expression has the same form as the classical single activation energy expression given earlier. However, it is effectively a multiple activation energy model (the α^m factors) weighted for the probability of a transition at that activation energy (the $\langle u_{p+m} | v_m \rangle^2$ factors) and hence provides a much more quantitative picture of the nonradiative process. The overlap integrals $\langle u_n | v_m \rangle^2$ [which depend on the configuration coordinate model parameters R, ω^2, Δ, E_{th} (or $h\nu_{\text{e}}$), and $\hbar\eta'$] can be calculated without great difficulty from Manneback recursion formulas (Mott, 1938). A computer program, QMSCC, has been written to calculate the nonradiative transition rate r_{nr} and the efficiency ϕ as functions of the configuration coordinate model parameters based on this model. All thermal quenching calculations done for this work were made with this program.

Drickamer *et al.* (1972) developed an analysis of the classical configuration coordinate model using pressure as a variable. Their analysis extracts values of Δ, ω^2, and R at any pressure from $h\nu$ and $E_{1/2}$ but does not allow explicitly for pressure dependence of ω^2 and R due to anharmonicity. A differential analysis based on Wilson and Drickamer (1975) gives the following results:

$$d\Delta/dP = (R - 1)/\omega^2 R, \tag{7}$$

$$dh\nu_{\text{e}}/dP = \Delta(P)/R - \Delta^2(P)\omega(P)\, d\omega(P)/dP, \tag{8}$$

with $h\nu_{\text{e}}(0) \equiv h\nu_{\text{e0}}$ *and* $\Delta(0) \equiv \Delta_0$.

These equations may be integrated for various functional dependences of R and ω on pressure. For the simple case of ω and R constant, integration yields the results of Drickamer *et al.* (1972):

$$\Delta(P) = \Delta_0 + P(R - 1)/\omega^2 R, \tag{9}$$

$$h\nu_{\text{e}}(P) = h\nu_{\text{e0}} + P\Delta_0/R + P^2(R - 1)/2\omega^2 R^2. \tag{10}$$

The pressure-dependent halfwidth becomes

$$E_{1/2}(P) = \frac{N\omega}{R^{1/2}} \left| \Delta_0 + \frac{P(R - 1)}{\omega^2 R} \right|. \tag{11}$$

The change in luminescence efficiency with pressure can be calculated using the QMSCC model described previously, substituting the pressure-dependent configuration coordinate model parameters. Because of the complexity of the QMSCC model, an analytical expression of these changes is not enlightening; however, illustrative calculations for specific phosphors are given later.

For qualitative purposes it is useful to note how the single activation energy of the classical model, E_q, varies with pressure. For $R \neq 1$, E_q has the general form

$$E_q(P) = \frac{R}{2(R-1)^2}\{\omega|\Delta| - [R\omega^2\Delta^2 + 2(1-R)E_{th}]^{1/2}\}^2, \qquad (12)$$

where E_{th} and Δ are pressure dependent. The case of $R = 1$ yields the simpler expression

$$E_q(P) = (h\nu_{e0} + P\Delta_0)^2/2\omega^2\Delta_0^2. \qquad (13)$$

The electronic transition responsible for luminescence in molybdate phosphors is of a charge transfer nature within the $MoO_4^=$ ion, with an oxygen 2p electron being transferred to an empty 4d molybdenum orbital on excitation. The reverse occurs during emission. Because they bind much less strongly to surrounding metal ions than they are bound internally, the $MoO_4^=$ ion may be considered as a molecular complex.

T_d symmetry has been assumed for molybdate crystals. Local symmetries are probably somewhat lower. Regardless of deviations from T_d symmetry, however, it can generally be concluded that within the first excited configuration are a number of states, with transitions between some of these states and the ground state being electric-dipole allowed (and thus responsible for the strong absorption). The energetically lowest-lying state is, however, probably a triplet with transitions to the ground state forbidden except via spin–orbit or phonon interaction. Grasser and Scharman (1976) point out that experimental evidence indicates phonon interactions are primarily responsible for making the emission transition partially allowed, while spin–orbit effects are smaller. Thus a temperature-dependent radiative rate, as given in Eq. (5), is required to describe the transition.

1. $CaMoO_4$

The values of Δ_0 and ω_0 extracted from the peak shift and shape change data given in the original paper (Tyner and Drickamer, 1977) are listed in Table 1. The method of obtaining $\hbar\eta'$ is discussed below.

Figure 8 shows experimentally measured values of emission intensity and observed lifetime as functions of pressure at room temperature. The intensity values are relative to a low temperature value of 1.0 and hence are essentially efficiencies. The zero-pressure value is approximately 0.32, in agreement with

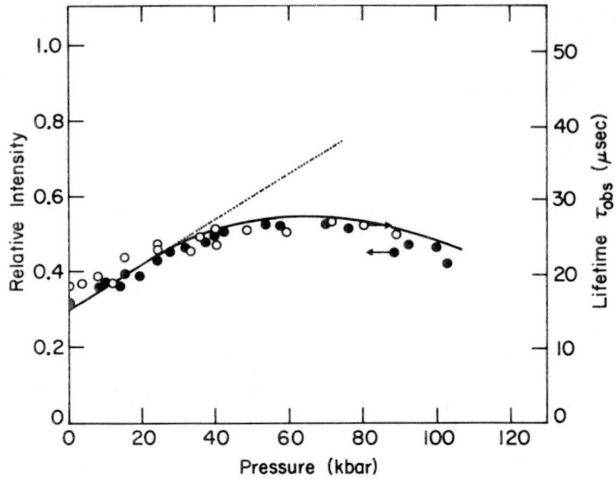

Fig. 8. Effect of pressure on (●) intensities and (○) lifetimes—$CaMoO_4$. See text for meaning of dotted and solid lines. (From Tyner and Drickamer, 1977.)

published values (Kröger, 1948). The observed luminescence decays are exponential in form, with the zero-pressure lifetime approximately 16 μsec. For comparison purposes the lifetimes have been plotted such that they match graphically the intensity value at low pressure. Intensities and lifetimes both increase by around 50% up to 70 kbar, above which they begin to decrease somewhat. At all pressures their ratio remains approximately constant, in agreement with Eq. 4. The dotted and solid lines in Fig. 8 will be discussed shortly.

Figure 9 gives the observed temperature dependence of the intensity and lifetime at each of five pressures. At zero (actually atmospheric) pressure noticeable thermal quenching of the emission intensity begins at about 150 K. The

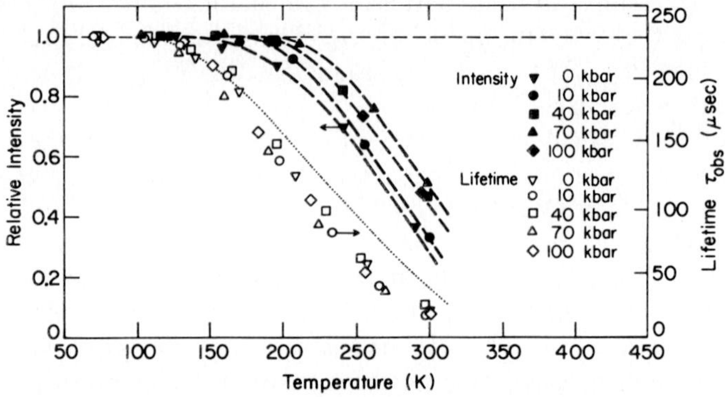

Fig. 9. Relative intensities and lifetimes versus temperature at various pressures for $CaMoO_4$. See text for meaning of dashed and dotted lines. (From Tyner and Drickamer, 1977.)

intensity has dropped to half its low temperature value by 271 K ($T_{1/2}$). The effect of increasing pressure, up to 70 kbar, is to shift the location of the quenching curve to higher temperatures. $T_{1/2}$ is about 300 K at 70 kbar. Above 70 kbar the quenching curve begins to shift back to lower temperatures.

The observed luminescent lifetime of $CaMoO_4$ begins to decrease from its low temperature value of 240 μsec (note this is, from Eqs. (4) and (5), $1/r_{r0}$) at a much lower temperature than the intensity. This decrease is due to the phonon-assisted hyperbolic cotangent dependence of r_r $(=1/\tau)$, not from thermal quenching (which affects r_{nr}). This decrease in τ_{obs} at these temperatures should be independent of pressure, and the data indicate this. In temperature regions where thermal quenching is important, the observed lifetime will change as a result of both effects. The lifetimes will thus be pressure dependent in this range until at room temperature they are as plotted in Fig. 8. (The scale in Fig. 9 is such that low temperature intensities and lifetimes are plotted together, with the result that at room temperature differences in τ_{obs} are difficult to see. They agree, however, with Fig. 8.)

The shape of the quenching curve is an analytically complicated function of $\hbar\eta'$ (as can be seen by differentiation of the QMSCC expression for η). It is possible to obtain a value of $\hbar\eta'$ by trial and error, varying it and comparing the slope of the calculated quenching curve with the experimental data, using the optimum match to determine $\hbar\eta'$. This was the method used to determine the value of $\hbar\eta' = 310 \text{ cm}^{-1}$.

The final variable as yet undetermined in the QMSCC calculation is r_{n0}, the "frequency factor" involved in the nonradiative rate r_{nr}. Its value should be near 10^{13} sec^{-1} (Struck and Fonger, 1975). The effect of changing r_{n0} is to shift the quenching curve, without changing its shape substantially, along the temperature axis. Since there is no way in this work to determine this parameter directly by experiment, r_{n0} has been chosen to match the calculated and observed quenching curves at atmospheric pressure. This value is then assumed independent of pressure and held constant for calculations at other pressures. The magnitude of the *shift* of the quenching curve with pressure is negligibly affected by the value of r_{n0}. The value of r_{n0} found for $CaMoO_4$ is $1.5 \times 10^{13} \text{ sec}^{-1}$, close to the typical expected value.

The QMSCC calculations of intensity versus pressure using the parameters as determined above (and summarized in Table 1) yield the results shown by dotted lines in Fig. 8. Over the first 40 kbar, agreement between calculated and observed values is good.

Above 40 kbar, the intensity at room temperature begins to fall as the quenching curve shifts back to lower temperatures. The simple model, i.e., $R = 1$ and ω and $\hbar\eta'$ constant, does not predict this behavior. While there is no evidence of R not being equal to unity, the peak shift at high pressure is nonlinear [see Tyner and Drickamer (1977)], which indicates ω may be changing slightly. In addition, $\hbar\eta'$ may be changing, as small differences in the slope of the quench-

TABLE 1

Configuration Coordinate Parameters for Molybdates

	$CaMoO_4$	$HgMoO_4$
$h\nu_{e0}$ (cm^{-1})	17300	13880
$E_{(1/2)0}$ (cm^{-1})	6300	4800
b	0	0.1
ϕ_0	0.32	0.25
$\tau_{abs,0}$ (μsec)	16	12
$T_{(1/2)0}$ (K)	271	256
R	1	1
ω_0 (kbar/(cm^{-1})$^{1/2}$)	12.7	8.11
Δ_0 (cm^{-1}/kbar)	13.6	15.9
(cm^3/mole)	1.63	1.91
$\hbar\eta'_0$ (cm^{-1})	310	300
r_{n0} (10^{13} sec^{-1})	1.5	6

See text for definitions of quantities. The subscript 0 refers to the values of these quantities at 1 atm.

ing curve indicate. (This will be discussed in more detail in connection with $HgMoO_4$ in Section IIB2, since the changes in slope are much more dramatic for that compound.)

As indicated above the parameters R, ω, and $\hbar\eta'$ can be determined in principle directly from the peak shift, halfwidth, and quenching curve slope data. However, the experimentally determined quantities are not sufficiently accurate or sensitive to determine small changes in these parameters with pressure. As discussed in the original paper, it is necessary to assume a 10% increase in ω and 15% in $\hbar\eta'$ in 100 kbar to fit the intensity data at high pressure. These changes are well within the scatter of the measured quantities. Using these pressure-dependent values of ω and $\hbar\eta'$, calculations yield the results shown by solid curves in Fig. 8. While this procedure gives closer agreement with the observed data, it is rather speculative since evidence indicating changes in ω and $\hbar\eta'$ is inconclusive.

Finally, the temperature dependence of the observed lifetime τ_{obs} has been calculated. The dotted line on Fig. 9 shows the calculated values at 10 kbar. While the calculated and observed values differ slightly, the agreement is quite satisfactory.

2. $HgMoO_4$

The parameters extracted from the peak shift and shape data appear in Table 1. Figure 10 shows the pressure dependence of intensity and observed lifetime for $HgMoO_4$. The intensity (and hence efficiency) increase from a

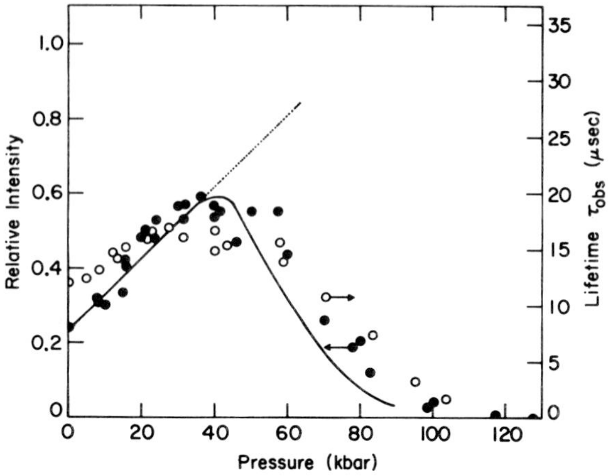

Fig. 10. Effect of pressure on (●) intensities and (○) lifetimes for $HgMoO_4$. See text for meaning of dotted and solid lines. (From Tyner and Drickamer, 1977.)

zero pressure value of 0.25 (in close agreement with the published value) (Blasse and van den Heuvel, 1974) to almost 0.6 at 50 kbar. Above this pressure the intensity decreases rapidly, reaching a value of less than 0.01 at 128 kbar. The decrease is much more pronounced than in $CaMoO_4$. Lifetimes run from 12 μsec at low pressure through a maximum of 17 μsec at 50 kbar and then decrease to 2 μsec at 100 kbar. As with $CaMoO_4$ the ratio of intensity of lifetime is approximately constant with pressure.

The results of QMSCC calculations based on the CC model parameters found above are shown in Fig. 10. The dotted lines indicate calculations done with the simple model ($R = 1$; ω and $\hbar\eta'$ constant at their atmospheric pressure values). Agreement with the observed data is very good up to 40 kbar. Beyond this the simple model, as before, cannot describe the increased quenching.

Proceeding as in the $CaMoO_4$ case, ω and $\hbar\eta'$ were varied in a manner consistent with the nonlinear dependence of the peak shift and the change in quenching curve slope with pressure. Values ranging from 8.11–8.45 kbar/$(cm^{-1})^{1/2}$ for ω and 300–430 cm^{-1} for $\hbar\eta'$ were used [see Tyner and Drickamer (1977)]. The results of this calculation are shown as solid lines in Fig. 10. Agreement in this case is good over the entire pressure range.

It is not clear, however, why $\hbar\eta'$, in particular, should show such a large pressure dependence. One possibility is that this variation could stem from the somewhat unusual crystal structure of $HgMoO_4$. Unlike the scheelite structure of other molybdates where the $MoO_4^{=}$ ion is in a slightly distorted T_d symmetry, the octahedral coordination of the $MoO_4^{=}$ ion in $HgMoO_4$ is very irregular, with two long Mo—O bond distances (Blasse and van den Heuvel, 1974). In addition, two of the four Hg—O bond distances are much shorter than the others. It does not seem unreasonable that high pressure might affect this structure, uncharacteristic of molybdate compounds, in an anisotropic fashion. This would result in significant pressure coupling to modes other than the single configuration coordinate Q considered here. This possibility is only conjecture, however, since no x-ray studies of the crystal structure at high pressure are available.

In summary, one sees that with reasonable assumptions one can characterize quantitatively the quenching and lifetime data as a function of pressure from parameters extracted from measurements of peak location and shape versus pressure.

III. INTERSYSTEM CROSSING

As we mentioned in the introduction, another method of dissipating excitation is via intersystem crossing to a (forbidden) triplet state. Two consequences of intersystem crossing are quenching of fluorescence and enhancement of some photochemical reactions.

For many aromatic compounds with π–π^* excitations the effect of pressure is to decrease substantially the S_1–S_0 energy separation. This occurs because the excited (S_1) state provides an electron configuration with a larger dipole moment and polarizability than the ground state, so that attractive interactions with the surroundings are more significant for the molecule in the excited state than in the ground state. Because of the repulsion involved for electrons with unpaired spins the excited triplet states are generally less polarizable than the excited singlet states of the same symmetry. Typically the red shift (shift to lower energy) with pressure of a triplet π–π^* state is one-fourth to one-sixth as great as that of the corresponding singlet. This results in a displacement to higher energy of the triplet vis-à-vis the singlet state and a decrease in the rate of intersystem crossing. The situation is illustrated in Fig. 11.

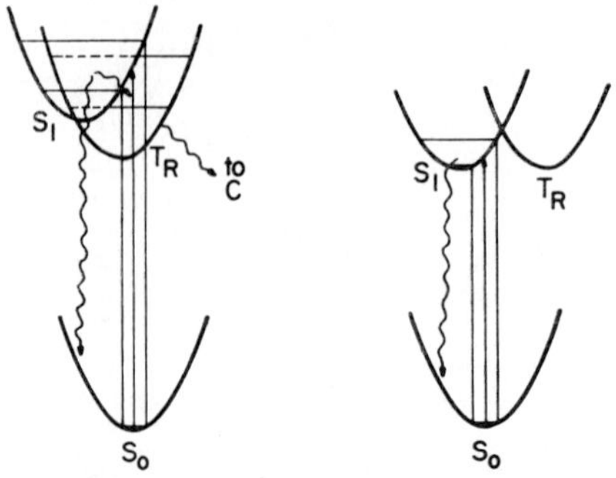

Fig. 11. Schematic configuration coordinate diagram illustrating effect of pressure on the relative locations of triplet and excited singlet states for π–π^* excitation: left—low pressure; right—high pressure.

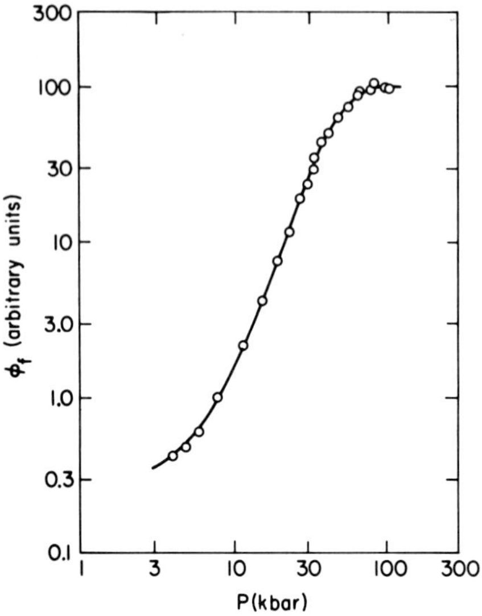

Fig. 12. Relative fluorescence efficiency versus pressure for 9 Anthraldehyde. (From Mitchell *et al.*, 1977a.)

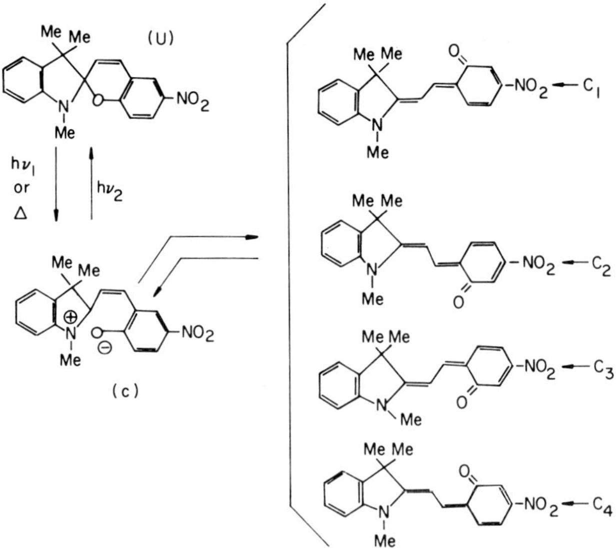

Fig. 13. Photochromic reactions of spiropyrans. (From Wilson and Drickamer, 1975.)

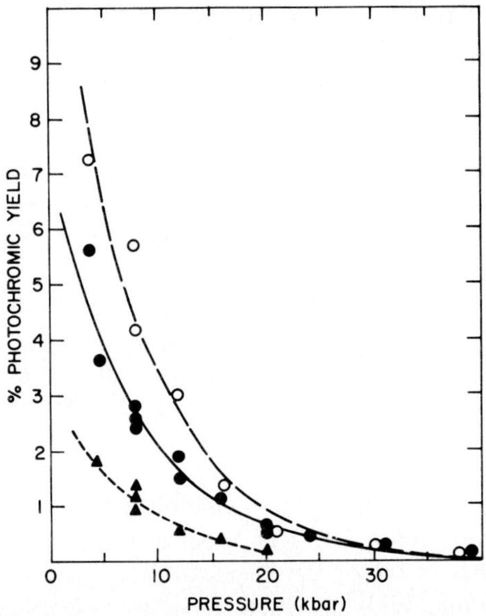

Fig. 14. Photochromic yield versus pressure for substituted spiropyrans: (●) 6-N-BIPS (PS); (○) 6-N-BIPS (PMMA); (▲) BMN-BIPS (PS). (From Wilson and Drickamer, 1975.)

One consequence of this relative shift should be an increase in fluorescence efficiency for molecules where intersystem crossing is the major quenching mechanism. This phenomenon is illustrated by a study of 9-carbonyl substituted anthracenes (Mitchell *et al.*, 1977a). These compounds exhibit very low fluorescence efficiency at 1 atm, and the quenching is attributed to ISC. In Fig. 12 we illustrate the change in fluorescence efficiency with pressure for one compound. The increase is by several orders of magnitude, and, at the highest pressure the quantum yield at room temperature approaches 0.3–0.5.

A second consequence of the decreased rate of intersystem crossing with pressure is the quenching of reactions which occur via the triplet state. This is illustrated in a study of the spiropyrans (Wilson and Drickamer, 1975). There are photochromic compounds which, when indicated in the ultraviolet, undergo a heterolytic cleavage and rearrangement into a quinodal form which has an absorption peak in the visible. The chemistry is illustrated in Fig. 13. The cleavage step is assumed, on the basis of chemical quenching studies, to occur via a triplet state. In Fig. 14 we exhibit the photochromic yield versus pressure for two spiropyran derivatives in polystyrene (PS) and one in polymethylmethacrylate (PMMA). While there is an effect of medium and of substituents on the molecule on the yield, the pressure quenches all the photochemistry by 40 kbar.

These two studies illustrate quite clearly the effect of pressure on the availability of triplet π–π^* states via intersystem crossing.

IV. DELOCALIZED EXCITATION—DOPED ZnS

A somewhat different example of the use of pressure to test a theory of phosphor efficiency involves ZnS doped with Cu^+ or Ag^+ and a coactivator such as Cl^- or Al^{+3} (House and Drickamer, 1977). These materials find application in cathode ray tubes, are model compounds for various semiconducting phosphors such as light emitting diodes, and for photoelectric devices used in solar energy conversion. The situation is represented schematically in Fig. 15. The Cu^+ and compensating ion generate levels in the gap between the top of the valence band and the bottom of the conduction band. An electron is excited from the valence to the conduction band. The Cu^+ forms a deep trap for the holes generated in the valence band while the coactivator acts as a shallow electron trap. There is evidence that changes in the energy of the hole trap E_A are relatively unimportant while the coactivator (electron) trap energy E_D is much more significant for phosphor efficiency.

ZnS involves a complex combination of ionic and covalent binding and there are various theories of the behavior of dopants which are useful in describing one aspect or another of the observed phenomena (Schön, 1942; Lambe and Klick, 1955; Prener and Williams, 1956; Kläsens, 1946). The theory we apply here is perhaps the most generally useful one. It pictures the luminescence as an electron from the coactivator as donor to the Cu^+ (with trapped hole) as acceptor. The donor wavefunctions are spread out over a number of lattice parameters with the effective radius of the orbit a strong function of the trap depth E_D. It, in fact, decreases rapidly with increasing E_D. The acceptor wavefunctions are much more localized. The transfer efficiency depends directly on the overlap of donor and acceptor wavefunctions, and thus on E_D.

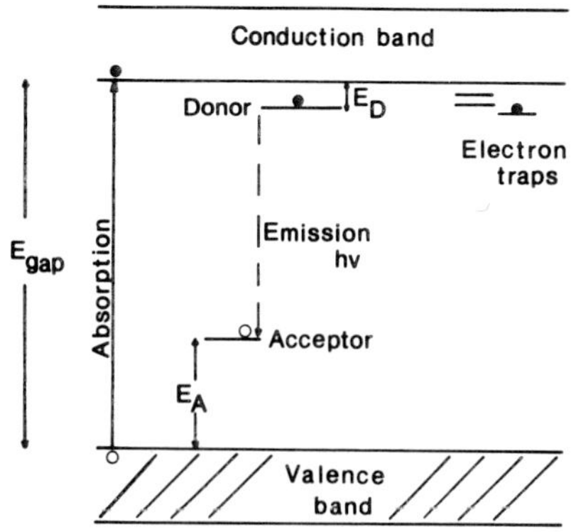

Fig. 15. Schematic band scheme for ZnS with donor–acceptor pair emission.

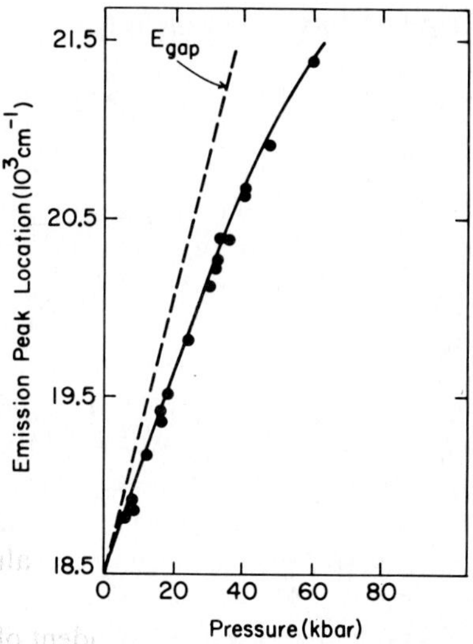

Fig. 16. Shift of emission peak and absorption edge with pressure—ZnS:Cu:Al. (From House and Drickamer, 1977.)

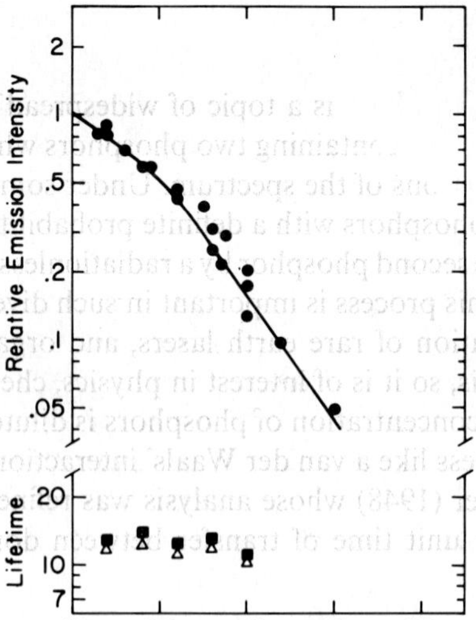

Fig. 17. Measured and calculated luminescence effectiveness versus pressure for ZnS:Cu:Al. ———, theory. Also lifetimes of the two observed components of the luminescence decay versus pressure: (■) 10^{-6} sec, (△) 10^{-4} sec. (From House and Drickamer, 1977.)

The emission intensity can be expressed in the form

$$I(E_{\mathrm{D}}, r) = \text{const} \times \int \tau W(E_{\mathrm{D}}, r) G(r)\, dr, \tag{14}$$

where $W(E_{\mathrm{D}}, r)$, the transition probability, depends exponentially on E_{D}, and $G(r)$ is a distribution function for impurity ions. τ is the emission lifetime.

In this example (House and Drickamer, 1977) we use Al^{+3} as the donor and Cu^{+} as the acceptor, although a similar analysis would apply to other combinations. The coactivator (donor) depth E_{D} is well established at 1 atm. In Fig. 16 we exhibit the shift of the emission peak with pressure, compared with the shift of the absorption edge of the ZnS crystal is given by the dashed curve. To first order the difference represents the change in E_{D} with pressure. This can be used in Eq. (14) to calculate the change in luminescent efficiency with pressure. In Fig. 17 we compare this calculation with the measured values. The agreement is remarkably good.

The model demands that the lifetime be independent of pressure. The actual decay is complex but is fit by two exponentials with lifetimes of the order 10^{-6} and 10^{-4} sec. As can be seen from Fig. 17, these lifetimes are approximately independent of pressure.

V. ENERGY TRANSFER

Energy transfer in phosphors is a topic of widespread interest. Consider a medium (crystal or solution) containing two phosphors which absorb and emit in somewhat different regions of the spectrum. Under some circumstances one can excite one of these phosphors with a definite probability that the excitation can be transferred to the second phosphor by a radiationless process so that both phosphors may emit. This process is important in such diverse areas as fluorescent lighting, the excitation of rare earth lasers, and organic photochemistry including photosynthesis, so it is of interest in physics, chemistry, biology, and materials design. If the concentration of phosphors is dilute the transfer will be by a dipole–dipole process like a van der Waals' interaction. This situation was first discussed by Förster (1948) whose analysis was refined by Dexter (1953).

The probability per unit time of transfer between donor and acceptor is given by

$$P_{\mathrm{DA}} = \frac{\alpha}{n^4}\left(\frac{1}{\tau_{\mathrm{D}}}\right)\frac{1}{R_{\mathrm{DA}}^6}\int \frac{f_{\mathrm{D}}(E) F_{\mathrm{A}}(E)}{E^4}\, dE \tag{15}$$

$$= \frac{R_{\mathrm{DD}}^6}{R_{\mathrm{DA}}^6}\left(\frac{1}{\tau_{\mathrm{D}}}\right). \tag{15a}$$

P_{DA} is then averaged over values of R_{DA} (Dexter, 1953). Here α contains only constants, and the integral is an overlap integral between donor emission and acceptor absorption peaks; τ_D is the donor emission lifetime and n is the refractive index of the medium. R_{DD} is the effective donor–acceptor distance such that there is an equal probability that the donor will fluoresce or transfer its energy to the acceptor. The transfer efficiency is given by

$$\psi = P_{DA}\tau_D/(1 + P_{AD}\tau_D). \tag{16}$$

This can be determined experimentally by measuring the relative areas under the donor and acceptor emission peaks in the mixed crystal. On the other hand, all of the quantities on the right hand side of Eq. (15) can be determined on the pure medium or on the medium singly doped with donor or acceptor ions or molecules. A comparison between the relative areas under the donor and acceptor peaks and the calculated efficiency is one test of the theory.

A second test depends on measuring the intensity of emission as a function of time. The theory gives this intensity for sensitized luminescence by the relation

$$I_D(t) = I_0[\exp(-t/\tau_D)] \exp[(-\beta R_{DD}^3)(t/\tau_D)^{1/2}]. \tag{17}$$

The first bracket gives the lifetime of the donor as a single dopant. The second bracket represents the intensity lost by dipole–dipole transfer to the acceptor. β represents a combination of coefficients independent of pressure. From a comparison of intensity versus time measurements on the singly and doubly doped medium R_{DD} can be calculated and thus P_{DA} and the efficiency ψ established.

This theory has been widely used, but most of the tests of the theory have been based on varying R_{DA} by changing the concentration. The use of pressure permits a variation of R_{DA} through the compressibility of the medium, and of the refractive index. The largest potential effect of pressure is, however in changing the overlap integral. We present results for a system (Bieg and Drickamer, 1977) KCl:Ag:Tl where there are no important secondary effects and where there is a large change in the overlap integral with pressure. As can be seen schematically in Fig. 18, the Ag^+ ion absorbs at high energy ($\sim$220 nm) and, at low pressure, the overlap with the Tl^+ absorption, and the consequent Tl^+ emission, is small. By 18 kbars the overlap and the Tl^+ emission have increased considerably. The calculated efficiency of transfer [Eq. (16)] increases from 7–28% in 18 kb.

In Fig. 19 we compare the calculated efficiency (the solid line) with that extracted from measuring the time-dependent emission intensity of the donor (Ag^+) in KCl:Ag:Tl compared with that in KCl:Ag, using Eq. (17) to obtain R_{DD} and then calculating η from Eq. (16). The agreement between theory and experiment is excellent.

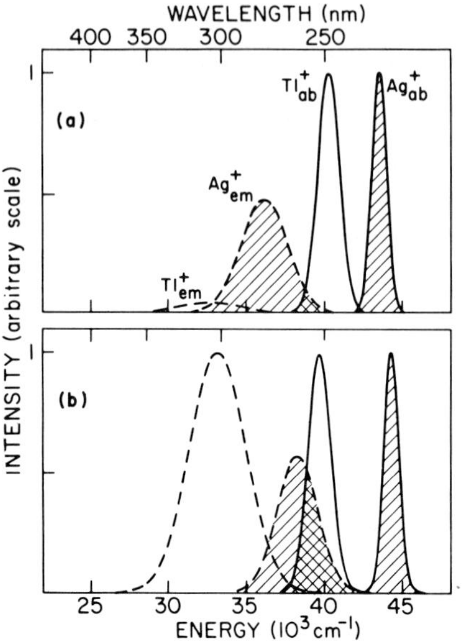

Fig. 18. Schematic representation of spectra for KCl:Ag:Tl at (a) low pressure (0 kbar) and (b) high pressure (18 kbar). (From Bieg and Drickamer, 1977.)

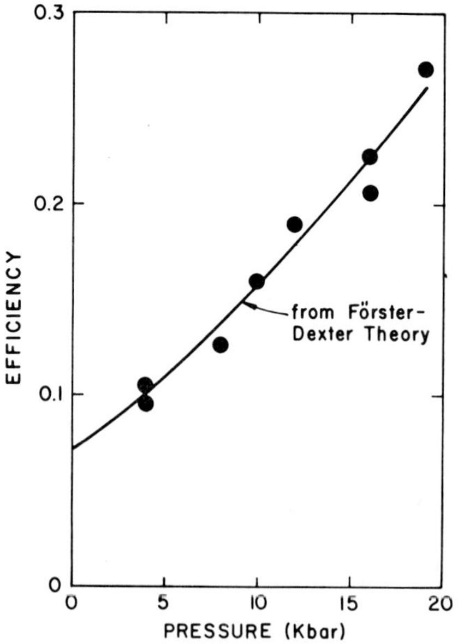

Fig. 19. Comparison of calculated energy transfer efficiency versus pressure with that measured from donor emission decay for KCl:Ag:Tl. (From Bieg and Drickamer, 1977.)

As discussed above, it is also possible to test the theory by measuring the relative intensity of donor and acceptor emission compared with values calculated from Eq. (15). In taking this path it is necessary to make a correction for absorption by the acceptor of photons emitted by the donor and consequent emission from the acceptor (the cascade effect). This correction has been worked out by Dexter (1953), and depends on the crystal thickness. [See Bieg and Drickamer (1977) for detailed application.] Experiments were performed with crystals 0.75 mm thick and with "thin" crystals 0.20 mm thick to be sure the correction was done adequately. As we see in Fig. 20, although there is some scatter due to the number of corrections, the agreement with theory is again excellent.

The efficiency can also be varied by changing temperature at constant pressure. The primary effect of lowering the temperature is to decrease the peak width and thus modify the overlap. This effect is most important in regions of small overlap. In Fig. 21 we compare the measured and calculated efficiencies as a function of temperature at two pressures. The agreement is very good.

These results constitute an excellent illustration of the use of pressure to perform a critical test of a theory important in a wide variety of atmospheric pressure applications.

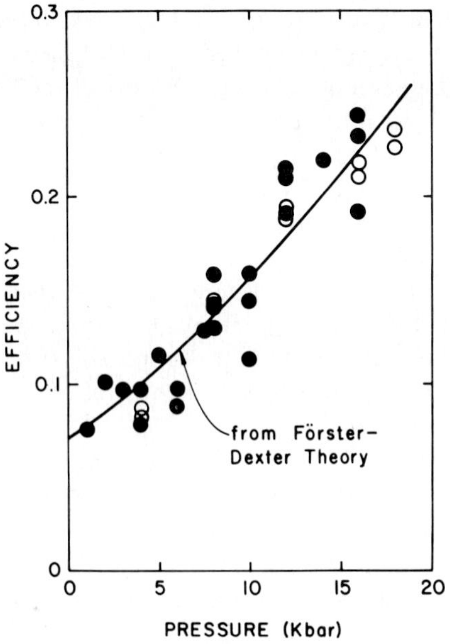

Fig. 20. Comparison of calculated energy transfer efficiency versus pressure with that measured from ratio of emission peak intensities for KCl:Ag:Tl, for (●) thick and (○) thin samples. (From Bieg and Drickamer, 1977.)

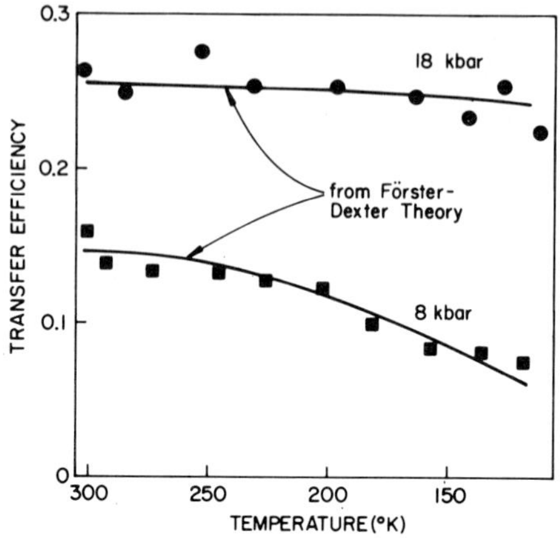

Fig. 21. Calculated and measure energy transfer efficiency versus temperature at 8 and 18 kbar for KCl:Ag:Tl. (From Bieg and Drickamer, 1977.)

VI. VISCOSITY EFFECTS ON LUMINESCENCE EFFICIENCY

A final example of the use of high pressure concerns the viscosity-dependent quenching of the fluorescence of the triphenylmethane dye crystal violet (CV) and the diphenylmethane dye auramine O(AO) in alcoholic solution.

Studies of viscosity-dependent processes in fluid media are usually carried out by varying the composition or the temperature of the solvent. The problem then arises of separating the viscosity dependence from purely temperature and/or solvent effects. The use of pressure allows a significant range of viscosities to be attained in a single solvent at a single temperature. If a wider range of viscosities is desired, chemically similar solvents for which the attainable viscosities overlap can be utilized. In this way a very large viscosity range can be investigated at one temperature with only a few solvents.

In this example (Brey *et al.*, 1977) pressures to 11 kbar were used on methanol, isopropanol, isobutanol, and glycerol to obtain solvent viscosities from less than 10^{-2} P to more than 10^{+3} P at room temperature. The solute concentrations were in the range $(2–4) \times 10^{-6}$ mole.

Förster and Hoffman (1971) (FH) have investigated the fluorescence efficiency of several triphenylmethane dyes including CV in a variety of solvents at different temperatures. They found that the only relevant solvent property was the viscosity μ and that for each dye there was a range of low viscosities for which the quantum efficiency ϕ was given by $\phi = C\mu^{2/3}$ where C is a constant for a particular dye. For CV, $C = 2.75 \times 10^{-3}\ \mathrm{P}^{-2/3}$ and the equation holds for $\mu < 200$ P. At higher viscosities ϕ approaches a limiting value (0.35). FH propose a model to explain the observed dependence of ϕ on μ in which the dye molecule is excited to a Franck–Condon vertical state with the phenyl rings still at the ground state equilibrium angle θ_0. The rings then rotate toward a new equilibrium angle θ_0' at a rate controlled by Stokes-like viscous damping. $\theta - \theta_0'$ therefore decreases exponentially with time with a relaxation time proportional to μ. The nonradiative viscosity dependent deactivation rate of the excited singlet is taken to be proportional to $(\theta - \theta_0)^2$. (For stilbenes, Sharafy and Muszkat (1971) attribute this deactivation to the effect of μ on the twisting and out-of-plane bending motions of the double bond in the excited state. These vibrations are thought to have a strong effect on the Franck–Condon factor and hence on the rate of the IC transition $S_1 \rightarrow S_0$.) There is also a nonradiative viscosity independent deactivation which accounts for the limiting value of ϕ in highly viscous media. The radiative emission rate is independent of θ so ϕ is proportional to the excited state lifetime.

In solvents of very low viscosity, the model predicts a viscosity independent minimum value for ϕ. FH calculate an approximate minimum ϕ for these dyes of about 2×10^{-6}. This limiting case was not observed. In more viscous solvents in which the excited singlet is deactivated nonradiatively before ring rotation proceeds very far, there are two limiting cases. The most interesting is where deactivation by θ dependent nonradiative processes predominates (intermediate μ). This gives $\phi = C\mu^{2/3}$. The other limiting case is where radiative deactivation is dominant (high μ). Here $\phi = \tau_0/\tau_s$ where τ_0 is the characteristic time for deactivation of the excited singlet by radiative and θ independent nonradiative processes and τ_s is the natural radiative lifetime of the excited singlet.

The data for CV are shown in log–log form in Fig. 22. The AO data has been displaced along the arbitrary intensity axis to show that for $\log \mu > -0.5$

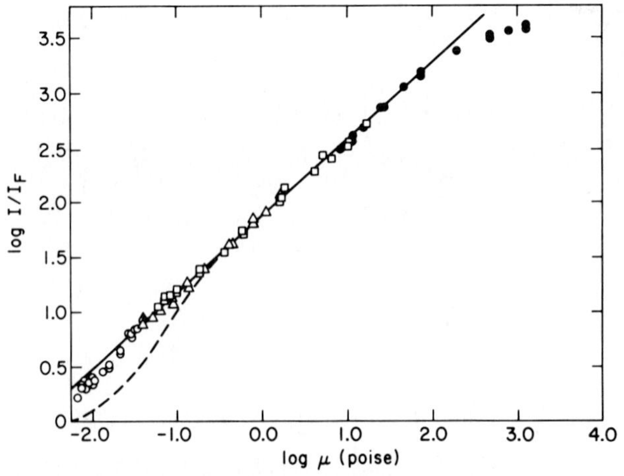

Fig. 22. Log of relative luminescence efficiency versus log viscosity for crystal violet (CV) in a series of alcohols and glycerol: (○) MeOH, (△) iso-prop OH, (□) iso-But OH, (●) glycerol. (From Brey *et al.*, 1977.)

the two dyes exhibit identical dependence of fluorescence efficiency on solvent viscosity. The dashed line in Fig. 22 indicates the deviation of the AO intensities at low μ. The line of slope 0.7 drawn in Fig. 22 shows excellent agreement with the CV data over 3.5 orders of magnitude in viscosity. At very low μ there is a small but definite deviation from linearity and at high η the intensities begin to level off. The AO data is also linear with a slope of 0.7 over a smaller range of 2.5 orders of magnitude in viscosity. It exhibits identical limiting behavior at high η, but much larger deviations from linearity at low η. It should be noted that the deviation at low viscosity is towards a *larger* viscosity dependence of the fluorescence intensity. The FH theory predicts a leveling, i.e., a lower viscosity dependence at low viscosities. We do not at present have an explanation for this discrepancy. Oster and Nishijima (1956) have measured ϕ for AO in glycerol as a function of temperature from about 5–35°C. Their experimentally determined viscosities varied from about 2–25 P in this temperature range. Analysis of their data also shows that ϕ is proportional to $\mu^{2/3}$. The FH model is thus shown to adequately describe the dependence of ϕ on μ for AO and CV over a large viscosity range with deviations occurring only at low viscosities.

The theory of Förster and Hoffman does not bear directly on the mechanism of the energy loss other than giving the quadratic interaction discussed above. Clearly, because of the viscosity dependence it is an *inter*molecular process. One can gain some insight on the mechanism using the peak shift and halfwidth data presented in the original paper, together with Eqs. (10) and (11), which relate these data to the parameters of the single configuration coordinate model. Let us consider the crystal violet data. In methanol one observes an essentially linear red shift and a halfwidth independent of pressure. This would imply a negative Δ_0 and $R \cong 1$. As the solvents become more viscous, there is an increasing tendency for the rate of shift to decrease with increasing pressure and

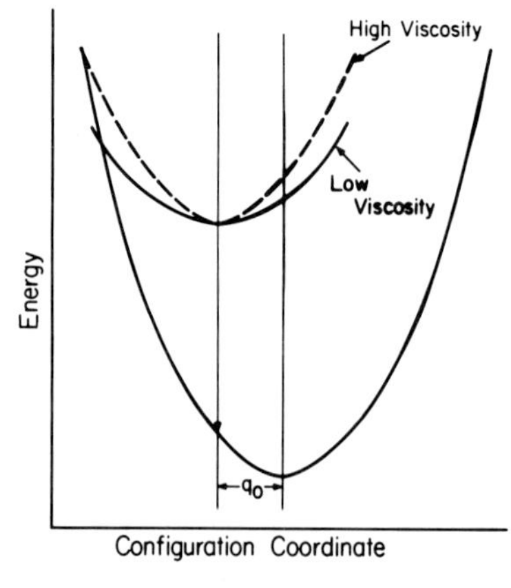

Fig. 23. Schematic representation of the effect of increased viscosity on the force constant of the excited state potential well, and thus on the luminescence efficiency. (From Brey *et al.*, 1977.)

for the half width to decrease also as P increases. Both these observations are consistent with Δ_0 being negative and with an R which is increasingly greater than 1; i.e., the excited state potential well becomes increasingly stiffer with increasing solvent viscosity. The same trends are noted for auramine O. Since R and ω^2 are very probably pressure dependent, a quantitative discussion is difficult.

In Fig. 23 we show the effect of stiffening the wells, everything else being constant. One must keep in mind that both vertical and horizontal shifts of the excited state well with respect to the ground state well must occur, although they are not shown. Nevertheless, one can see from the picture that an increase in R could profoundly effect both the probability of internal conversion and the reversibility of the process by raising the energy of the intersection of the ground and excited state surfaces.

These examples are representative of the wide variety of ways in which high pressure studies can improve our understanding of luminescence properties, especially efficiency.

This work was supported in part by the Department of Energy under Contract DOE-76-C-02-1198.

REFERENCES

Bieg, K. W., and Drickamer, H. G. (1977). *J. Appl. Phys.* **48**, 426.
Bieg, K. W., and Drickamer, H. G. (1977). *J. Chem. Phys.* **66**, 1437.
Blasse, G., and van den Heuvel, G. P. M. (1974). *L. Lumin.* **9**, 74.
Brey, L., Schuster, G. B., and Drickamer, H. G. (1977). *J. Chem. Phys.* **67**, 2648.
Dexter, D. L. (1953). *J. Chem. Phys.* **21**, 836.
Drickamer, H. G., Frank, C. W., and Slichter, C. P. (1972). *Proc. Nat. Acad. Sci. U.S.* **69**, 933.
Drotning, W. D., and Drickamer, H. G. (1976). *Phys. Rev. B* **13**, 4568.
Förster, Th. (1948). *Ann. Phys.* **2**, 55.
Förster, Th., and Hoffman, G. (1971). *Z. Phys. Chem.* **NF75,** 63.
Grasser, R., and Scharman, A. (1976). *J. Lumin.* **12**, 473.
House, G. L., and Drickamer, H. G. (1977). *J. Chem. Phys.* **67**, 3221.
Kläsens, H. A. (1946). *Nature (London)* **158**, 306.
Kröger, F. A. (1948). "Some Aspects of the Luminescence of Solids." Elsevier, Amsterdam.
Lambe, J., and Klick, C. G., (1955). *Phys. Rev.* **98**, 909.
Manneback, C. (1951). *Physica (Utrecht)* **17**, 1001.
Martin, T. P., and Fowler, W. B. (1970). *Phys. Rev. B* **2**, 4221.
Mitchell, D. J., Schuster, G. B., and Drickamer, H. G. (1977a). *J. Am. Chem. Soc.* **99**, 1145.
Mitchell, D. J., Schuster, G. B., and Drickamer, H. G. (1977b). *J. Am. Chem. Soc.* **99**, 7489.
Mott, N. F. (1938). *Proc. R. Soc. London Ser. A* **167**, 384.
Okamoto, B. Y., and Drickamer, H. G. (1974). *J. Chem. Phys.* **61**, 2870.
Oster, G., and Nishijima, K. (1956). *J. Am. Chem. Soc.* **78**, 1581.
Prener, J. S., and Williams, F. E. (1956). *J. Chem. Phys.* **25**, 361.

Scharafy, S., and Muszkat, K. A. (1971). *J. Am. Chem. Soc.* **93**, 4119.
Schön, M. (1942). *Z. Phys.* **119**, 463.
Struck, C. W., and Fonger, W. H. (1975). *J. Lumin.* **1**, 10
Tyner, C. E., and Drickamer, H. G. (1977). *J. Chem. Phys.* **67**, 4103.
Tyner, C. E., Drotning, W. D., and Drickamer, H. G. (1976). *J. Appl. Phys.* **47**, 1044.
Wilson, D. G., and Drickamer, H. G. (1975). *J. Chem. Phys.* **63**, 3649.

8

Relaxation of Electronically Excited Molecular States in Condensed Media

*Robin M. Hochstrasser and R. Bruce Weisman**
Department of Chemistry
and
Laboratory for Research on the Structure of Matter
University of Pennsylvania
Philadelphia, Pennsylvania

I. INTRODUCTION

A large part of chemistry has always been concerned with processes occurring in solutions, and physical chemists have traditionally dealt with the problem of understanding the nature of solutions through studies on the structure of liquids

* Present address: Department of Chemistry, Rice University, Houston, Texas.

ISBN 0-12-450650-X

and on the kinetics of solution phase processes. Diffusion of molecules in solutions is sufficiently slowed by the liquid dynamics that it can be studied relatively simply by means of direct time measurements in the microsecond regime. However, properties of individual solute molecules such as energy and rotational relaxation or internal rearrangements of structure or energy are generally associated with the nanosecond to picosecond time scale. The study of many of these processes is therefore very dependent on technological developments that extend the range and flexibility of ultrashort time scale experiments.

This article will be concerned mainly with the energy transfer processes that occur after a molecule in solution is excited by light, so that both electronic and vibrational states of the molecule are to be considered. We will deal with solid solutions in which the large amplitude molecular motion is largely quenched, as well as liquid solutions in which the molecules can translate and rotate under the influence of fluctuating forces from the liquid.

In solid solutions such as rare gas matrices or mixed crystals it is possible to study relaxation processes at extremely low temperatures with the environment limited essentially to zero-point motion. The solute molecules are then restricted to small oscillations about equilibrium configurations. Under these circumstances one normally observes in optical spectra sharp lines corresponding to transitions between vibrational electronic levels of the solute accompanied by sidebands indicating the different degrees of coupling to the solvent of the two states involved in the transition (Hochstrasser and Prasad, 1974). The relaxation times for such levels appear to vary from seconds for vibrational levels of some diatomics (Legay, 1977) to the subpicosecond regime for the removal of excess lattice energy from the site of the solute. Accordingly, there exist a wide variety of techniques that may be employed to study such systems, ranging from pulsed laser experiments to the use of continuous (cw) lasers and the measurement of spectral lineshapes and quantum yields. Solid solutions have the advantage that meaningful temperature-dependent studies are easily performed on them. As the temperature is raised the sharp spectral structure gradually becomes less distinct because of the widening frequency distribution and increasing amplitude of environmental fluctuations. The system thereby becomes increasingly homogeneous and begins to approach the liquid in that regard.

In liquid solutions near ambient temperatures the vibrational relaxation of polyatomic molecules occurs exclusively in the subnanosecond regime. In cold liquids of some very simple molecules, such as nitrogen, vibrational relaxation proceeds much more slowly than this, but we are considering here mainly polyatomic molecules dissolved in liquids consisting of other polyatomics. Under these circumstances there are numerous relaxation pathways to be considered, including vibrational state to vibrational state energy transfer from solute to solvent and the conversion of vibrational energy into other degrees of freedom in the solution. Transitions between different electronic surfaces of the solute molecule may require the transformation of a significant amount of electronic into vibrational energy; such processes involving ground and excited electronic surfaces frequently occur in the microsecond to nanosecond regime. Prohibitions on electronic coupling matrix elements (e.g., spin) can result in even slower electronic relaxation times. Phenomena of this sort fall in the category of conventional ronradiative processes and have been studied for many years, especially through their effects on the radiative characteristics of molecules. In particular, the kinetics of internal conversion and intersystem crossing between pairs of states taken from the lowest excited singlet and triplet and the ground state have been explored for a vast number of different types of molecules (Birks, 1970). However, direct studies of relaxations between other excited states have been rare. Modern picosecond pulse techniques may be used to learn about these relaxations, and in addition lasers make possible many indirect techniques that utilize the radiative properties of highly excited states as internal clocks to time the relaxation processes.

In a discussion of relaxation of molecules in solutions, the contrasts and similarities between dissolved and isolated molecules must be considered. When a sufficiently large number of molecular states is involved in an electronic relaxation the rate constants are often qualitatively independent of the surrounding medium. However, the effect of the environment on molecules with low densities of states remains as an interesting problem, especially for situations in which the statistical limit is nearly reached. Although the experimental observations that focus on these issues have been made mainly on the microsecond to nanosecond scale, modern spectroscopic techniques allow such studies to be extended into the subnanosecond regime.

Obviously there are qualitative differences between the relaxation behavior of small molecules in rare gas matrices and polyatomics in polyatomic environments that are brought about by the existence of the complex structure of vibrational energy levels in the medium. The extent to which the vibrational relaxation of polyatomics involves these states of the medium is not yet known from direct experiments. In order to accomplish this goal it would be necessary to explore the excitation of the vibrational states of the solvent following an initial population of a solute state. Nevertheless, considerable indirect evidence on this point can be obtained from studies of the radiative efficiencies of sets of relaxing levels using cw laser techniques.

Closely allied to vibrational and electronic relaxation are chemical relaxation processes. In particular, intramolecular rearrangements might be viewed as nonradiative relaxation between states having substantially different equilibrium geometries. In such cases the nuclei of the molecule acquire an excess of kinetic energy in the isomerization coordinate which must be dissipated into other modes in order for the rearrangement to be completed. In solutions this excess kinetic energy is transformed into heat during the period in which frictional forces damp the nuclear motion. This implies that such molecular rearrangements will depend on macroscopic parameters of the liquid such as temperature and viscosity. In dilute gases, on the other hand, the total molecular energy is fixed and a rearrangement is equivalent to a redistribution of vibrational and electronic energy. Some of the best known photoisomerizations of moderate sized molecules seem to occur very rapidly in solutions and their study requires fast pulse techniques. In isolated molecules, though, very little is known about the time scales for energy reorganization. This is an intriguing topic for which modern laser experiments are expected to provide new insights.

The present article is focused on events occurring on the subnanosecond scale and the methods used to study them. Because of the immense variety of molecules that can be designed and studied in the laboratory there is reason to expect that some systems could be found for which the relevant processes of relaxation and energy transfer occur on very long time scales. And while this may also be the case for some isolated molecules, there are many condensed phase relaxation phenomena that simply cannot be slowed down sufficiently to be studied by any but the most advanced techniques.

II. DYNAMICAL EFFECTS IN CONDENSED PHASES

The immediate result of optical excitation of molecules in solution is qualitatively different for low temperature solid solutions and liquid solutions at normal temperatures. Since a vital feature of the ultimate relaxation is the nature of this initial optically prepared state, it seems worthwhile to provide a general description of optical spectra and relaxation in various condensed media.

A. Rare Gas Matrix Spectra and Relaxation

There have been numerous theoretical studies of vibrational relaxation in simple solids (Legay, 1977). A realistic treatment of the coupling of the intramolecular vibrational excitation to the phonons of a bath presents formidable theoretical difficulties even though the physics of the fundamental process is understood. Vibrational energy can be transferred into delocalized states of the lattice formed from phonons, librons, or vibrons. There are in addition highly localized modes into which the vibrational energy may be transferred that involve nuclear motions, and thus structural details, in the neighborhood of the impurity molecule. Indeed recent studies of small molecules in rare gas matrices have shown that the localized modes associated with hindered rotation of the guest are important energy sinks for the vibrational energy of the guest (Legay, 1977; Bondybey and Brus, 1979).

The simplest condensed phase situation that can be contemplated for a molecule is one for which the spectrum of the heat bath is least complex. A rare gas matrix environment fulfils this condition because the densities of one-phonon states cut off in the range of 45 cm^{-1} (Xe) to 65 cm^{-1} (Ar). Thus in order to transfer energy from molecular vibrational or electronic states directly into the lattice it is necessary to involve multiphonon transitions whenever the energy exceeds ~60 cm^{-1}. If the differences in lattice equilibrium coordinates between the initial and final molecular states are relatively small fractions of the equilibrium coordinates themselves, then direct processes involving many phonons are extremely improbable compared with those involving just one or two. This is simply a reflection of the Franck–Condon factors that govern phonon generation probabilities and it has the effect of severely disallowing relaxation processes involving large energy gaps when only small changes of the coordinate are involved. For the case of many small molecules such as diatomics and triatomics, experiments have recently shown that at energy gaps corresponding to the vibrational fundamental region the relaxation is dominated not by these slow direct multiphonon processes but instead by the excitation of hindered rotor states of the molecule in the matrix. The extensive research reported for small molecules has not yet been duplicated for moderate sized systems, so in these cases there is little definitive knowledge of relaxation pathways. However, qualitative differences can be anticipated between small systems where the intramolecular energy gaps are large and moderate sized systems for which the mean level spacing is less than the Debye frequency, even at relatively small total vibrational energy contents. For the larger molecular systems and sufficiently large total energies, there would appear to be no necessary involvement of multiphonon processes in the energy transfer from guest to host. Of course the lowest energy fundamentals corresponding to any particular electronic surface will not normally be able to relax by generating single phonons, so these levels might be expected to have longer lifetimes. It is also possible that they are mainly relaxed by mechanisms other than energy transfer to host lattice states.

Vibrational relaxation between levels of a guest polyatomic molecule in an inert gas matrix probably involves factors other than the level structures of the molecule and the unperturbed lattice. Moderate sized molecules clearly replace host atoms and thereby introduce strain in the lattice. The specifics of the deformation potential are very likely to differ for different molecules. Perhaps information on the coupling of inert gas atoms to guest molecules in condensed phases will ultimately become available from structural studies of gaseous van der Waals complexes.

The foregoing discussion about vibrational relaxation refers in principle to any particular electronic surface. Experimental studies of vibrational relaxation have been carried out on small molecules for both the ground and electronically excited states. A major possible difference between these two cases may arise from relaxation of the excited state through intervention of ground state levels. Such an effect has been demonstrated for the system CN in neon (Bondybey, 1977). For larger polyatomics, on the other hand, the density of ground state levels at the energy of electronic excitation is sufficiently large that there is little likelihood of excited state repopulation from them.

One would therefore expect similar sorts of vibrational relaxation behavior in different states, apart from variations related to the specifics of the nuclear potential surfaces. Furthermore, the higher barriers to rotation of matrix-isolated large molecules make it appear less likely that at low temperatures the relaxation can occur by coupling vibrations to rotational motions, as apparently occurs with smaller systems (Brus and Bondybey, 1975; Bondybey and Brus, 1975). It is to be hoped that experimental data bearing on these issues will soon become available.

Although electronic spectra of molecules in rare gas matrices have been studied for many years, there is still very little information on the effects of such matrices on electronic relaxation except for systems that are likely to be in the statistical limit. The cases of interest are those for which the isolated molecule level structure is sufficiently sparse to limit the relaxation, or where mixed states are actually excited by the light. Situations of this sort arise either in small molecules or in large molecules having small energy gaps (Wannier *et al.*, 1971).

The optical spectra of smaller molecules in rare gas matrices are generally typical of impurity spectra in solids (Rebane, 1970). At low temperatures the lowest energy spectral feature is known as the zero-phonon line. This corresponds to a transition of the guest molecule between two Born–Oppenheimer states which involves no excitation of lattice phonons. It is often the case that the equilibrium configuration of the guest molecule and nearby host atoms differs for the two guest vibronic states involved in the transition. Then, in a manner analogous to the Franck–Condon patterns of free molecules, there will be a tendency for lattice phonon excitation to accompany the guest transition. This effect is manifested in the absorption spectrum as a phonon sideband, or wing, to the high frequency side of the zero-phonon position. The shape of the phonon sideband depends on the density of phonon states: it forms a broad envelope which vanishes near the zero-phonon line and may peak at $\sim$30–100 cm^{-1}. The intensity of the sideband relative to the zero-phonon line meaures the magnitude of displacement of the excited impurity site from its equilibrium ground state configuration. If the electronic transition involves a large enough lattice deformation, it is possible for the zero-phonon line to be undetectably weak. Such is apparently the case for the $A \leftarrow X$ transition of NO in argon. Goodman and Brus (1977) explain its broad spectra in terms of the Rydberg nature of the excited states, which results in a significant increase in the effective size of the guest molecule.

The electronic spectrum of an ensemble of polyatomic molecules in a matrix usually has a width which reflects the configurational distribution of transition energies in the sample. Even in a case such as *s*-tetrazine in argon, for which the visible zero-phonon lines are only about 1 cm^{-1} wide, their inhomogeneous aspect can be clearly demonstrated in spectral hole-burning experiments (Dellinger *et al.*, 1977).

B. Molecular Solid Solutions

There has been a large amount of spectroscopic research on optically induced processes in organic mixed crystals. At low temperatures these materials are generally rigid, and in the normal case of substitutional solutions there is little likelihood of free molecular rotation at an impurity site. The crystals themselves have optical lattice modes, some of which correspond to librations of molecules about the equilibrium angles of orientation defined by the crystal structure. The impurity molecules may also be involved in these motions which, along with small translations of the center of mass of the impurity and a distortion of the host structure, form the most facile way for the system to adjust to optically induced changes in electronic or nuclear structure of the impurity. The low temperature optical spectra of such mixed crystals generally consist of a zero-phonon line for each vibrational level reached and a phonon sideband that shows peaks generally in the range of 30–100 cm^{-1}. These phonon sideband peaks usually fall in the region of the optical phonon bands that are observed in Raman spectra. The integrated absorption strength of the phonon sidebands typically exceeds or is at least comparable with that of the zero-phonon line. These facts indicate that on excitation the impurity region of the lattice is generally displaced from the equilibrium position of the initial state by about the width of the distribution of configurations appropriate to the zero-point librational amplitudes.

The majority of well-studied mixed crystal systems consist of guest and host molecules that are structurally alike, such as naphthalene in durene, azulene in naphthalene, and pentacene in terphenyl. In each case the guest–host and host–host potentials are similar and are determined by the same types of atom–atom interactions that explain the host crystal structures quite well (Kitaigorodsky, 1973). Generally the guest and host molecules adopt closely similar orientations in the lattice and, since the principal moments of inertia are also usually similar, the lattice spectrum of the host is not strongly perturbed by the presence of a low concentration of such guest molecules. The ground state vibrational levels of the host form into bands (vibron bands) that usually have very slight dispersion: bandwidths may range from 1 to 20 cm^{-1}.

An important difference between mixed molecular crystals and rare gas matrix systems is the existence of these host crystal vibrational states, which often serve to accept energy from excited guest molecules. There are three energy regions to consider in any particular spectrum:

(i) The region incorporating the lowest energy guest vibrational levels where the host vibrational structure is also sparse.

(ii) The region up to ~ 1000 cm^{-1}, where the level spacing in the host crystal might be comparable with or greater than the one-phonon edge of the host lattice spectrum.

(iii) The region in excess of 1000 cm^{-1} or so where the level spacing of the host vibron bands becomes uniformly much less than the Debye frequency.

The guest levels in region (iii) are expected to be efficiently relaxed into host states. It is difficult to generalize about the intermediate case (ii), but the expectations are that energy transfer to both the host vibron bands and the host lattice modes will occur. Each of these processes can result in cascading to lower energy guest levels located in region (i). Under normal circumstances energy is transferred to the host bands irreversibly, although there may be special cases (of low dimensionality) where recurrences might occur even in an infinitely dilute mixed crystal. There does not appear to have been any experimental work reported to date on this point.

Case (i) is that most accessible to experimental study. Single vibrational levels of the guest can be excited directly in optical experiments and either their decay characteristics or the time evolution of other states of the guest or host may be studied directly. This is the situation to which the theories of vibrational relaxation may be most applicable. However, it is important to remember that the host may have one or two very low frequency modes (100–200 cm^{-1} for aromatics larger than benzene) that are significantly mixed with the lattice phonons. Under these circumstances it might transpire that the foregoing separation into vibron and libron–phonon mechanisms becomes artificial. A number of different approaches have been used to learn about the relaxation times for levels in cases (i) and (ii), ranging from linewidth studies to measurements of resonance fluorescence that use the radiative rate as an internal clock. Some of these results will be discussed in later sections.

It is well known that the distinct (~ 1 cm^{-1}) inhomogeneously broadened optical spectral lines broaden as the temperature is raised above a few degrees kelvin. Gradually the vibronic transitions become homogeneously broadened until at ambient temperatures the linewidths are often great enough that the vibrational structure of the electronic transition is essentially washed out (Rebane, 1970). In contrast, the Raman spectra of molecular crystals remain moderately sharp (a few inverse centimeters) at ambient temperatures even though some broadening and shifting of the transitions occurs on raising the temperature. It is not difficult to understand this observation that the effect of crystal temperature on the homogeneous linewidths of vibrational transitions is much more pronounced for vibronic transitions than for those occurring within the ground state. The linewidths are caused by fluctuations of the energy levels due to coupling to nuclear motions in the heat bath, and the magnitudes of the energy shifts caused by guest–host interactions are very much larger for transitions between different surfaces than for those within a given surface. Clearly, this effect renders it difficult to use linewidth studies to determine excited state depopulation times once the temperature is sufficiently high that fluctuation broadening becomes an important contribution. In this regime it is necessary either to measure directly the population lifetime or to measure the quantum yield of a process such as radiative emission while having knowledge of the radiative rate of decay of the vibronic state.

Electronic relaxation processes in mixed molecular crystals have been widely studied. The pathways of intersystem crossing between the lowest excited singlet and triplet states, and between the lowest triplet and the ground state have been worked out in detail from optical–microwave double resonance experiments on mixed crystals (El Sayed, 1974). Apparently there is always statistical limit behavior observed in these media. The radiationless processes observed were irreversible and exponential even for cases in which gaseous samples display more interesting behavior [e.g., pyrazine (Frad *et al.*, 1974), quinoxaline (McDonald and Brus, 1973), and benzophenone (Busch *et al.*, 1972; Hochstrasser and Wessel, 1973)].

Internal conversion between excited electronic states in mixed crystals has hardly been explored in time resolved experiments. In regions where vibronic levels from different electronic states overlap, their coupling appear as spectral perturbations. Such effects are well known for small molecules in the gas phase, but they are observed by various forms of spectral diffuseness in gaseous spectra of larger molecules. On the other hand, molecular spectra in low temperature solid solutions show these perturbations very clearly indeed and much can be learned about the nature of the optically excited states and their relaxation from spectral analysis (Hochstrasser and Prasad, 1974; Wessel, 1970; Hochstrasser, 1968; Langhoff and Robinson, 1974). It is apparent from the detailed studies of naphthalene in various mixed crystals that excitation in the region of the second excited state (S_2) generates vibronic levels that are dominantly those of the S_1 surface (Wessel, 1970). The electronic relaxation process is therefore expected to depend on the nature of the excitation pulse in this case. The state S_2 cannot be excited to the exclusion of S_1 unless an extremely short laser pulse is employed, since the many discrete levels of S_1 that are perturbed span a range of $\sim 50\ \mathrm{cm}^{-1}$, corresponding to a decay time of 0.1 psec. When the density of coupled S_1 states in the region of S_2 (or other higher states) exceeds the inverse linewidth of the levels, the optical spectrum becomes intrinsically diffuse and the linewidths signal the electronic relaxation times for delta function pulsed excitation. As the temperature is raised the fluctuation width convolutes into each vibronic transition and the resulting spectral bands usually become somewhat broader. The broadening of these already diffuse higher transitions is not normally so significant as that for the discrete vibronic transitions discussed above.

Shpolskii "matrices" form a class of mixed molecular crystals having paraffinic hydrocarbon hosts (Shpolskii, 1963). In these lattices many types of organic molecules display electronic spectra with sharp zero-phonon lines and relatively weak phonon sidebands. The low intensity of the phonon sidebands suggests that case (i) and (ii) vibrational relaxation might be slowed down compared with other organic mixed crystals, but there is no direct evidence on whether or not this occurs. The low temperature spectral transitions in such media often display discrete as well as the usual continuous inhomogeneous effects, giving rise typically to multiplets in place of the single vibronic transitions observed for the same guests in other mixed crystals or as gases.

In summary, the following are the characteristic spectral and dynamical features of organic solutions as compared with inert gas matrices:

(1) Organic solid solutions generally show smaller inhomogeneous widths.
(2) The vibron bands of the organic host lattice must be considered in understanding guest relaxation processes.
(3) The lattice spectrum of the host is less strongly perturbed by the guest in many of the organic solutions.

The discussion here has been concerned with guest molecules of moderate size without regard to special chemical properties of the guest or host. Obviously if the guest is highly polar or photochemically reactive towards the host there arise a wide range of possible spectral and dynamical effects that would require individual consideration.

C. Spectra of Liquid Solutions

It is well known from statistical physics that the motion of particles surrounded by a fluid medium is subject to irreversible frictional processes. The friction stops, or damps, the motion, and the relaxation process is the dissipation of kinetic energy into heat. In contrast to the case of ordered solutions, an exact quantum mechanical treatment of this thermalization is not feasible. The dissipated kinetic energy is transferred into modes of the medium as well as into those of the particle (which in our case would be a solute molecule). In order to generate a detailed picture of such a process it would be required to solve the equations of motion for the entire system of solute plus solvent. Such a task is impossible to accomplish even for a model system of classical bodies; it is therefore a problem for statistical mechanics.

The exact mechanical solution cannot be achieved because of the dependence of the state of motion of the system at a given time on its previous history. It is thus often assumed that the dynamics depend only on the instantaneous coordinates and momenta of the bodies. In this stochastic case one can derive equations of motion that can be solved for certain situations.

Some aspects of the spectra of molecules in liquid solutions also may have to be understood in terms of statistical models. An important factor in optical spectra is the mean shift of the transition from the gas phase to solution. These shifts are on the order of 10^3 cm^{-1} for the lowest energy electronic excitations of aromatic type molecules. Thus at any instant one can expect a wide range of transition energies corresponding to a distribution of molecule–solvent configurations. This distribution will evolve into others over a time scale dictated by the frictional forces. In contrast, the rapidly varying force will tend to average out over the characteristic period of the molecular motion but will contribute to the width of the spectral line in the form of conventional fluctuation broadening. The time scale of ~ 10 psec is likely to fall in the intermediate regime, so it is not appropriate to assume that the whole linewidth of a liquid solution spectrum is homogeneous on the scale of picosecond pulse excitation.

The optical excitation of individual vibronic transitions is clearly very unlikely for moderate sized molecules in solutions at ambient temperatures. Monochromatic light causing electronic transitions will generally excite all molecules in the sample with uniform probability. Pulsed excitation with a picosecond source may intercept only part of the inhomogeneous distribution, but very rapid fluctuation broadening will still result in the overlapping of many vibronic transitions. In cases where the solute molecules themselves have very low frequency modes (e.g., small barriers to internal rotation) the evolution of the inhomogeneous distribution will be determined also by the intramolecular potential function.

III. EXPERIMENTAL METHODS

A. Introduction

In this section we will review the principles and some specifics of experimental techniques suitable for the spectroscopic study of rapid relaxation processes in excited electronic states. These can be classified as either indirect methods, which are normally made in the frequency domain, or direct, time-resolved techniques. We will briefly discuss the frequency domain category and then deal in greater detail with various aspects of direct measurements. One time-resolved technique, designed for accurate transient absorption spectrometry, is described thoroughly as an illustrative example. Finally, we present an assessment of the general experimental prospects for the future in this area.

B. Indirect Techniques

Information about vibrational population lifetimes in excited electronic states is embedded in the spectral lineshapes of optical transitions to those levels. However, in order to extract this information it is necessary to remove or account for the other components of the lineshape. A distinction can usually be made between homogeneous and inhomogeneous contributions to spectral widths, since the latter result from a distribution of apparent transition frequencies that remains static during the characteristic measurement period and often assumes a Gaussian form. The Lorentzian homogeneous linewidth component reflects the mean rate of dephasing of the optical transition for individual molecules, as given by the relation $\Delta\nu_{1/2} = (2\pi\tau)^{-1}$ in which $\Delta\nu$ is the half frequency width at half maximum of the transition and τ is the phase relaxation time. Processes contributing to the homogeneous width include not only population relaxation, which we wish to determine, but also pure dephasing events such as result from phonon-induced fluctuations in the transition frequencies of molecules in condensed media. The homogeneous dephasing time therefore yields only an upper limit on the rate of population decay.

The characteristic times T_1 and T_2 that are frequently used to describe the dynamics of optically pumped systems are defined by the optical Bloch equations for the state vector of a two level system driven by a light field. The off-diagonal density matrix elements decay with a time constant T_2, and the population difference decays with a time constant T_1, in the simplest model. Molecular systems are generally not well described by such a set of equations with two relaxation parameters because the populations in the two states do not usually decay with the same time constant. Thus in general three or more phenomenological decay parameters for the populations are needed. Nevertheless, the off-diagonal elements of the part of the density matrix corresponding to the optically coupled levels will decrease with time constant T_2. It follows that the macroscopic polarization also decays at the same rate and that the absorption spectrum for weak light sources is therefore proportional to $[\Delta\omega^2 + (1/T_2)^2]^{-1}$, where $\Delta\omega$ is the frequency mismatch. The value $1/T_2$ is identical to the half width at half maximum of the absorption spectrum in a homogeneous system. For samples where there is no pure dephasing, as in very low temperature solids or sufficiently dilute gases, the spontaneous processes representing the population decays of the two levels fully determine the spectral width. In this coherent limit $1/T_2 = \frac{1}{2}(1/\tau_i + 1/\tau_f)$, where τ_i and τ_f are the population relaxation times of the coupled levels. This relation forms the basis for deducing population decay rates from spectral lineshapes.

For an inhomogeneous sample, the absorption spectrum corresponds to a convolution of the homogeneous lineshape with the inhomogeneous distribution function. It is necessary to experimentally distinguish these two components for condensed phase systems of interest. In some cases mixed crystals with very small host inhomogeneities may be prepared; the Lorentzian lineshape contribution then dominates the Gaussian part. But for most solid solutions and liquid samples, the minimum inhomogeneous width of guest electronic transitions is 0.5–1.0 cm^{-1}, so more sophisticated methods must be employed to determine homogeneous dephasing times longer than a few picoseconds. One of these techniques is photochemical hole burning (Kharlamov *et al.*, 1974), in which a configurational subset of guests is selectively decomposed and the linewidth of this homogeneous subset is then measured from the resulting spectral "hole."

Nonlinear optical methods are employed in another class of experiments designed to extract homogeneous widths in the presence of large inhomogeneous broadenings. For example, it was recently shown (Yajima and Souma, 1978; Yajima *et al.*, 1978; Hänsch and Toschek, 1970) that three-wave mixing spectroscopy is readily adapted to this purpose. When beams at frequencies ω_1 and ω_2 are incident on the sample, emission at $\omega_3 = 2\omega_1 - \omega_2$ is generated through the sample's third-order susceptibility $\chi^{(3)}(\omega_3, -\omega_1, -\omega_1, \omega_2)$. As one applied frequency is tuned through the resonance at $\omega_1 = \omega_2$, the resulting variation in intensity at ω_3 provides information about sample dephasing unaffected by the inhomogeneous linewidth. In a variant of this experiment, Song *et al.* (1978) used counterpropagating ω_1 and ω_2 beams with the ω_1 light elliptically polarized and the ω_2 probing light plane polarized at some angle to the principal axes of the ellipse. A crossed polarizer transmits ω_2 only in the presence of the ω_1 light at the sample, and again the nonlinear transmission lineshape contains the homogeneous linewidth information of interest. Three-wave mixing also provides a spectroscopic method for measuring lineshapes of electronic transitions with a dynamic range in intensity of 6–8 decades (Trommsdorff *et al.*, 1979). The technique exploits two-photon resonances in $\chi^{(3)}$ and is advantageous compared with conventional attenuation measurements because it allows observation of lineshapes far into the spectral wings where inhomogeneous contributions normally are negligible compared with the Lorentzian tails of the homogeneous components. These various methods of lineshape analysis are most readily applied to the study of very rapid (i.e., picosecond or subpicosecond) relaxation processes, corresponding to spectral widths of many wavenumbers. They thus nicely complement direct time-resolved measurements, which become very difficult in this regime.

There exists another indirect experimental approach to condensed phase vibrational relaxation dynamics in electronically excited states. During continuous excitation of a vibronic state of the sample, weak fluorescence is emitted from various vibrational levels, some of which are not thermally populated. This hot luminescence can be measured and the intensities of the bands analyzed to derive information about the lifetimes and decay pathways for vibrational levels involved in the relaxation. Normally some kinetic assumptions must be invoked in studies of this sort, and knowledge of absolute quantum yields is very helpful.

C. Direct Techniques

1. Pulsed Laser Characteristics

Direct methods of highly time-resolved spectroscopy are based almost exclusively on the use of ultrashort light pulses generated by one of a number of types of mode-locked laser systems. The operating characteristics of these lasers are crucially important in determining the degree of feasibility of specific experimental schemes, so we will begin by briefly describing the most significant performance parameters and their relevance to spectroscopic applications.

Clearly, the optical pulse duration τ_p must be short enough to resolve the dynamical effects under study. A given pulse has a spectral width $\Delta\nu$ which cannot be smaller than the bandwidth limit corresponding to its temporal pulse shape. The product $\tau_p \Delta\nu$ of the spectral and temporal full widths at half maximum (expressed in cm^{-1} psec) has the value 14.7 for a pure Gaussian pulse, 10.7 for a $sech^2$ form, 3.7 for an exponential, and 29.6 for a rectangular pulse shape (Ippen and Shank, 1977). Pulses with temporal substructure will show excessive spectral widths and will therefore have limited utility in some applications.

The magnitude of a light pulse may be described in terms of the number of photons it contains. This number is proportional to the total pulse energy and varies inversely with wavelength for a given energy content. The instantaneous peak power, typically measured in watts, is approximately given by the pulse energy divided by its characteristic duration. Associated with the light pulse is an optical electric field whose squared magnitude is proportional to the instantaneous power density or intensity (watts per square centimeter, in the commonly used mixed units). Intensity is linear in photon flux for a fixed wavelength. Indispensable in picosecond spectroscopy are several nonlinear optical methods used to generate new frequencies. The efficiencies of these shifting processes increase at least linearly with the instantaneous intensities of the original beams. For this reason, it is advantageous to use pulses that can readily be focused to intensity levels well in excess of 10^6 W/cm^2. This implies that high peak power is a desirable characteristic for a pulsed laser. Another consideration which favors powerful ultrashort pulsed sources relates to the number of photons contained in a pulse. For example, in order to excite 10^{-7} moles of sample molecules in single photon visible absorption, at least 20 mJ of light is required, which, if delivered within 5 psec, corresponds to a very high pulse power of $\sim$4 GW.

The pulse repetition frequency must normally be low enough that full relaxation of the sample (or replacement of the illuminated volume) can occur between pulses. Otherwise the buildup of metastable states or species will interfere with the intended measurement or lead to misinterpretation. At the other extreme, excessively low repetition rates can make optical alignment, signal optimization, and data collection extremely tedious and difficult.

Also experimentally relevant is the cross-sectional intensity profile of the laser beam. This is described by the transverse mode structure of the oscillator, which can range from TEM_{00}, a smooth and radially symmetric Gaussian distribution, to admixtures of very high-order modes, characterized by variable and closely spaced patterns of nodes and antinodes. Because of the high local intensities at the antinodes, multimode beams are more likely to cause damage in optical materials. In addition, such beams have larger divergences and, often, reduced nonlinear generation efficiencies compared with TEM_{00} lasers.

Reproducibility of the pulsed laser output is a highly desirable characteristic. For example, if pulse-to-pulse intensity fluctuations are low, then it may be possible to obtain optimally efficient nonlinear generation by employing power densities very close to the material damage threshold while still avoiding optical damage problems. Good reproducibility is also a great help in reducing the noise level of measurements obtained by summing over a number of pulses. This is particularly true when highly nonlinear processes are involved, as these tend to magnify the fluctuations in the original laser output.

The final laser characteristic to be considered is a central one for spectroscopic applications—tunability. Whether used for specific sample excitation or for selective probing, the experimental flexibility associated with tunable wavelength generation is of great and obvious scientific value. Tunability may be achieved either in the primary laser oscillator or alternatively in a secondary device such as a dye laser (Glenn *et al.*, 1968), parametric oscillator (Weisman and Rice, 1976), parametric generator (Laubereau *et al.*, 1974), or continuum generator (Alfano and Shapiro, 1970) driven by a fixed wavelength primary source. Properly designed secondary devices can provide tunability in the near to midinfrared, visible, or ultraviolet spectral regions. Many of the criteria outlined above will determine the suitability of a fixed wavelength laser for driving a tunable light generator. Also useful in extending the set of available optical frequencies are the nonlinear methods of stimulated Raman shifting and harmonic generation (Yariv, 1975).

2. Properties of Specific Lasers

There are a number of mode-locked laser sources currently available for picosecond spectroscopic applications. Of these, the passively mode-locked Nd:glass system has probably been the most widely used. It emits trains of intense pulses at 1.06 μm with durations near 6 psec. After amplification, peak powers of the order of 10 GW may be obtained, but repetition rates are restricted to the 1 min^{-1} range. When care is taken to operate the oscillator in its TEM_{00} mode and to extract single pulses from very early in the pulsetrain, the resulting $\tau_p \Delta\nu$ product is reported to be nearly ideal (von der Linde, 1972). Lasers of this sort may also be operated in a less controlled manner, characterized by multimode transverse structure and excessive spectral broadening (resulting from severe self-phase modulation). Generally, the reproducibility and stability of mode-locked Nd:glass lasers are quite poor, and this feature, combined with the low repetition rates, presents formidable experimental difficulties. Although the output wavelength is fixed, it may be rather efficiently converted to the second, third, or fourth harmonics at 530, 354, and 265 nm, respectively. In addition, the outputs of mode-locked Nd:glass lasers have been successfully used to pump synchronous tunable dye lasers (Royt *et al.*, 1974) and parametric oscillators (Weisman and Rice, 1976) as well as ultrashort cavity dye lasers (Fan and Gustafson, 1976; Cox *et al.*, 1977) and traveling wave parametric generators (Seilmeier *et al.*, 1978). In this way a considerable range of wavelengths can be obtained at the expense of increased experimental complexity.

The mode-locked ruby laser is another traditional source of ultrashort light pulses. It is similar to Nd:glass in its power outputs, repetition rates, transverse mode behavior, and poor stability, but differs in its 694.3 nm fundamental output wavelength and ~25 psec pulse duration. Only the first and second harmonics are commonly utilized.

Higher repetition rates of several hertz may be achieved with the passively mode-locked Nd:YAG laser, whose output power and wavelength are very similar to those of Nd:glass. Although the pulse durations are normally no shorter than 25 psec, systems of this sort may be operated to give stable TEM_{00} and bandwidth limited outputs which are suitable for very efficient harmonic generation (Reintjes and Eckardt, 1977) as well as for driving synchronously mode-locked tunable devices (Goldberg and Moore, 1975).

Primary tunability is available from flashlamp-pumped passively mode-locked dye lasers. These have good pulse-to-pulse stability with repetition rates intermediate between those of Nd:YAG and Nd:glass lasers. However, their pulse energies are normally at least an order of magnitude lower than those of the solid state sources, and multimode transverse mode structure is common. Pulses may be as short as 1 psec and show bandwidth-limited spectral character (Bradley, 1977).

Another class of mode-locked lasers generates continuous streams of pulses rather than short, infrequent bursts. The most useful of these have been based on Ar^+ lasers. In one configuration a cw Ar^+ laser drives a passively mode-locked dye oscillator, allowing ultrashort pulses as short as 0.2 psec to be produced (Ippen *et al.*, 1972; Diels *et al.*, 1978). The basic pulse repetition frequency of ~100 MHz may be reduced to any convenient value by the technique of cavity dumping. Wavelength tunability tends to be considerably more restricted in such systems than in comparable cw dye lasers, but good stability and transverse mode structure is available and peak powers may reach several kilowatts. When these pulses are subjected to multistage amplification in dye cells pumped by a *Q*-switched Nd:YAG laser, subpicosecond pulses containing several millijoules of energy can be obtained (Ippen and Shank, 1978).

The other major type of continuously mode-locked source employs an actively mode-locked Ar^+ laser, with ~100 psec pulses, to synchronously pump a cw-type dye laser. Here, broadly tunable dye output pulses with widths in the 2 psec range may be stably generated at kilowatt peak power levels. A single mode-locked Ar^+ laser can also be used to pump two dye lasers, thereby providing very useful pairs of synchronized and independently tunable ultrashort pulses (Jain and Heritage, 1978). Another effective technique for synchronous generation involves a cascaded three laser approach in which the output of one mode-locked Ar^+-pumped dye laser is used to pump another matched dye laser (Heritage and Jain, 1978). The same methods of cavity dumping and amplification described above for the cw-pumped dye laser can be applied also to synchronously pumped systems to lower their output repetition frequency and increase the pulse energy. Finally, it is possible to combine active and passive mode-locking in a cw dye laser to obtain still another set of operating characteristics (Kurobori *et al.*, 1978).

3. Measurement Methods

A wide variety of experimental schemes have been used to study the relaxation dynamics of electronically excited states. These schemes can be generally categorized by the number of ultrashort pulses incident on the sample: single-pulse experiments are concerned with measuring the temporal and spectral content of induced sample emission, whereas multiple-pulse methods usually probe double resonance properties. By the term double resonance we denote both direct measurements of induced absorption at one or a variety of wavelengths as well as indirect, laser induced fluorescence detection of excited state absorption.

Single-pulse emission experiments are initiated by electronic excitation of the sample with a light pulse, labeled **a** in Fig. 1. The intensity of emitted light **b** is then measured as a function of the time delay from excitation, $t_b - t_a$. This

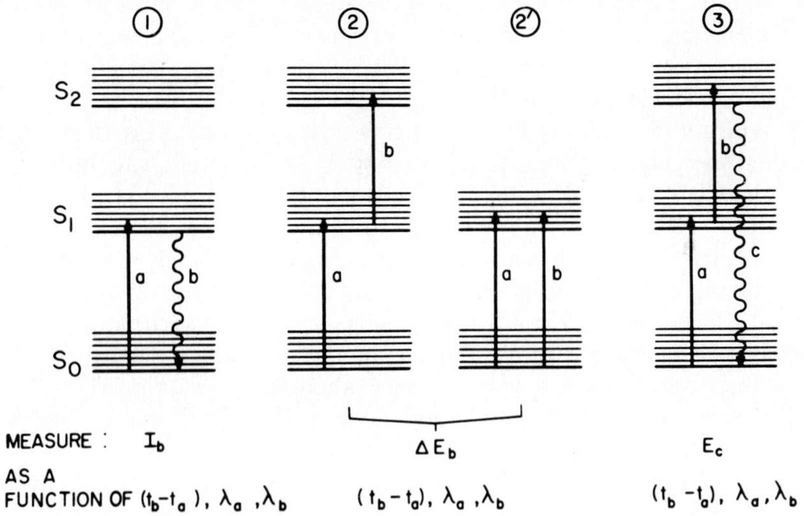

Fig. 1. Classification of schemes for time-resolved spectroscopic measurements. Straight arrows represent laser pulses and wavy arrows represent spontaneous emission. (1) Time-resolved emission: The intensity of sample emission is measured as a function of delay after excitation, and as a function of exciting and emitted wavelengths. (2), (2′) Time-resolved absorption: ΔE_b, the attenuation of the energy of probing pulse **b**, is measured as a function of delay time from excitation, exciting wavelength, and probe wavelength. (3) Time-resolved absorption by induced fluorescence: E_c, the time-integrated amount of sample emission, is measured as a function of delay between excitation pulse **a** and probing pulse **b**, and also as a function of the wavelengths of pulses **a** and **b**.

information may be obtained either as a function of emission wavelength or instead for the spectrally integrated luminescence. Ideally, the excitation wavelength may also be varied.

Several methods are available for time-resolving emitted light. With the optical Kerr shutter technique (Duguay and Hansen, 1969a, b), sample emission is detected through a carbon disulfide cell placed between crossed polarizers. Normally the shutter is closed and transmits no light. But when an intense gating light pulse is incident on the carbon disulfide, a transient orientational birefringence is induced which allows a temporal slice of the sample emission to pass through the second polarizer and be detected. By using many laser shots and varying the arrival time of the gating pulse relative to the excitation pulse, the full time dependence of the luminescence may be determined.

Another method for making measurements of this sort employs gating by nonlinear mixing (Duguay and Hansen, 1968; Mahr and Hirsch, 1975). Here the sample luminescence and an ultrashort gating pulse are both directed into a suitably oriented nonlinear optical crystal. Light at the phase-matched sum or difference frequency of the two beams is generated only when both are simultaneously present, so measuring the total energy at the mixed frequency as a function of delay between the excitation and gating pulses allows one to trace out the time profile of the emission. This method has the advantage of shifting the frequency of the gated light to a different spectral region, allowing improved detection sensitivity and contrast ratio.

The most direct approach to measuring time resolved luminescence makes use of a streak camera. This is an electron optical device which spatially disperses an incident light signal according to arrival time at the photocathode. The spatial intensity profile on the output phosphor of a streak camera therefore directly represents the temporal intensity profile of the light under study. Current streak camera designs achieve time resolutions of $\sim$2–3 psec with dynamic range adequate for many applications. A principal intrinsic advantage of streak cameras in the study of time-resolved emission is that the complete time profile is available from each laser shot; sampling methods are not required. This feature is particularly beneficial when the laser source is of low repetition rate and poor reproducibility.

Another important application of streak cameras is the accurate characterization of widths and shapes of laser pulses. It is well known that indirect autocorrelation methods such as two-photon fluorescence analysis are not useful unless care is taken to determine the contrast ratio accurately and even then are insensitive to the pulse shape or asymmetry (von der Linde, 1972). By contrast, a direct streak camera observation is unambiguous and of considerable value in interpreting experimental results. Figure 2 shows such a measurement of the temporal intensity profile of a single pulse from a mode-locked Nd:glass laser as observed on a high performance streak camera with electronic readout (Liang and Chernoff, 1979).

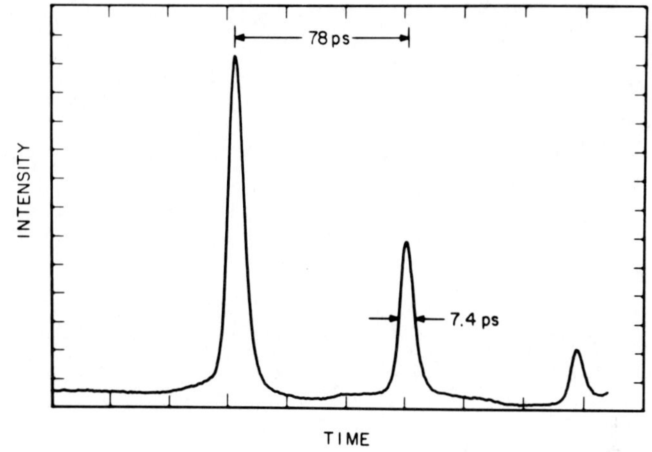

Fig. 2. Streak camera display of a single 530 nm pulse from a mode-locked Nd:glass laser system, after passing through an etalon with a period of 78 psec. The camera is a G.E.A.R. Pico-V and the electronic detector is a P.A.R.C. OMA-2 system.

Transient absorption measurements form an important class of double resonance methods. Here the first laser pulse creates a nonequilibrium population distribution among the electronic states of the system. The resulting changes in the sample's optical absorption properties are then probed by a second pulse which may have either narrow or broad spectral content, depending on the experiment. Frame 2 in Fig. 1 schematically illustrates the nature of such measurements, which include induced absorption, bleaching, and gain effects. When the excitation and probe beams have the same wavelength, certain special types of relaxation measurements are possible, so this case is represented separately as frame 2′.

Ideally one would like to be able to measure induced absorption changes as a function of at least three spectroscopic variables: excitation wavelength, delay time, and probe wavelength. In practice this is prevented by various limitations in the performance of available laser sources. Lack of tunability, for example, severely restricts the set of available excitation wavelengths. An intense fixed wavelength laser may be used to create a broad-band probing pulse by means of continuum generation in various liquid media, but the highly nonlinear nature of this generation process tends to greatly magnify any pulse-to-pulse variations in the original laser output. This means that continuum-based schemes for measuring broad-band induced absorption spectra are subject to relatively large uncertainties and noise levels unless great care is taken in experimental design. The most common optical configurations used with low repetition rate lasers are multiplexed either in delay time or in probe wavelength. That is, an echelon-type device may be used to split a single wavelength probe beam into a number of differently delayed segments (Topp *et al.*, 1971), or alternatively a spectrally broad probe pulse is passed through the sample at a single delay time and dispersed to give a multiwavelength transient spectrum (Magde and Windsor, 1974). Both methods can be effective if employed in double beam arrangements.

Another approach to studying absorption processes in excited electronic states is sketched in frame 3 of Fig. 1. Here the idea is that rather than measuring the decrease in intensity of the transmitted probing laser pulse, one instead detects the small amount of short wavelength fluorescence that is induced by its absorption. Although fluorescence quantum yields are very low for these doubly excited final states, the high detection sensitivity and selectivity that can be achieved help to make this method very feasible. As in the case of direct transient absorption measurements, it is desirable to be able to vary the wavelengths of the excitation and probing pulses as well as the time interval between them. However work reported to date with this technique has not incorporated independent tunability of the two beams. Two general types of experimental configurations can be employed, corresponding to either copropagating or counterpropagating pulses. In the copropagating case one measures the total fluorescence (in the appropriate spectral region) for various delay times (Lin and Topp, 1979). This arrangement is similar to the most common scheme for probe attenuation studies. If the two beams are counterpropagating, however, the entire time dependence of the transient absorption is displayed in the spatial distribution of fluorescence in the sample cell (Rentzepis, 1968). For a system in which the emission is strong enough that its spatial structure can be well resolved, the counterpropagating pulse geometry is advantageous. This is especially so when low repetition rate, irreproducible lasers are involved.

So far we have discussed experimental methods for the measurement of populations, but transient properties of the coherence are also of interest and many experimental techniques for their study have been developed. The time decay of the coherent polarization in a homogeneous medium is the Fourier transform of the spectrum and so is another way of measuring T_2 and, if pure dephasing is absent, the population decay. For inhomogeneous systems the temporal decay of the polarization again transforms to give the lineshape; thus for a Gaussian line the coherence decays in a Gaussian manner. However, in such cases it is still possible to measure T_2, and hence the homogeneous spectrum, by utilizing echo techniques. In the photon echo method the polarization generated by a light pulse is allowed to partially decay before a second pulse arrives to initiate its rephasing. The system emits an echo pulse when the rephasing occurs and variation of the echo amplitude with delay time yields the value of T_2. Hesselink and Wiersma (1978) have employed cw synchronously pumped dye lasers to extend molecular photon echo measurements into the subnanosecond regime.

4. Example of a Direct Double Resonance Technique

Broad-band transient absorption spectroscopy is one of the most generally useful of the double resonance methods. It is considerably more powerful than techniques that use a single probe wavelength because kinetic models for even a simple sequence of light-initiated reactions often cannot be adequately tested on the basis of time resolved absorbance at only one or two optical frequencies. Moreover, the actual molecular relaxation processes which follow electronic excitation may be quite complex and give rise to transient spectral features which change in correspondingly complicated and subtle ways. To study such features fully it is necessary to acquire transient absorption data that not only span a considerable frequency range but also are of high photometric accuracy.

The experimental principles of picosecond absorption spectroscopy were first presented by Alfano and Shapiro (1970), who found that an intense continuum was produced when picosecond pulses were focused into certain glasses. These authors later suggested that the continuum could be employed as a source for obtaining spectra on the picosecond time scale in view of the fact that the light was produced in a coherent process and suffered no time expansion relative to the pumping pulse (Alfano and Shapiro, 1971). There have been many applications of that idea. Busch *et al.* (1973) used the continuum and a vidicon detector to probe the photoprocesses resulting from excitation of the laser dye DODCI. Magde and Windsor (1974) used a photographic technique along with the continuum to study DODCI kinetic spectroscopy. Absorption spectra of porphyrin molecules were investigated by Magde *et al.* (1974), again with the photographic technique. A similar method was employed by Hochstrasser and Nelson (1975) to determine transient spectra of aromatics, *N*-heterocyclics, and ketones and by Anderson *et al.* (1976) to study the spectrum of the second excited state of xanthione.

There were a number of problems with these picosecond spectroscopic methods. Their spectral accuracy was not sufficiently well calibrated that the number of distinct species present in the sample at each time could be properly identified or evaluated, so in the usual case a simple model was constructed incorporating just one or two absorbing species. It is now documented that the picosecond continuum has a transverse spectral distribution that is sensitive to the nature of the pumping pulse (Penzkofer *et al.*, 1975; Smith *et al.*, 1977), so accurate spectroscopic data can be expected only from techniques that compensate for this effect. Another problem was that the laser pulse transverse intensity profile was often not well controlled. No less than with traditional kinetic studies, it is vital in picosecond work to have analytical methods that accurately characterize the species represented in the models; therefore it is important to know the details of the spatial distribution of light and excited molecules in the sample. Recent improvements in the techniques of transient absorption spectroscopy have significantly raised the level of accuracy and reliability that is attainable. We describe below the design and operation of a refined apparatus of this sort (Greene *et al.*, 1979a) which is based on intense single pulses of 1.06 μm light from a mode-locked TEM_{00} Nd:glass laser system.

The transient absorption configuration is illustrated schematically in Fig. 3. It begins with a dichroic beam splitter which transmits the first and second laser harmonics while reflecting the ultraviolet third harmonic, used here as the excitation pulse. The UV beam is filtered to adjust its intensity and to remove residual visible components and is then sent through a fixed optical delay line and focused into the thin sample cuvette through an aperture of 320 μm diameter placed in contact with the cell's front window. Typically, 250 μJ of ultraviolet light is incident on the sample.

The first harmonic pulse transmitted through the dichroic beamsplitter forms the source of the picosecond continuum used to obtain absorption spectra. Light at 1.06 μm is separated from the second harmonic with a color filter and focused with a 10 cm focal length lens into the center of a 5 cm cell of phosphoric acid used as a continuum generator. Emerging light is spatially filtered by transmission through ground glass scatter plates, spectrally filtered by an infrared-blocking color filter, and recollimated by a lens. This continuum

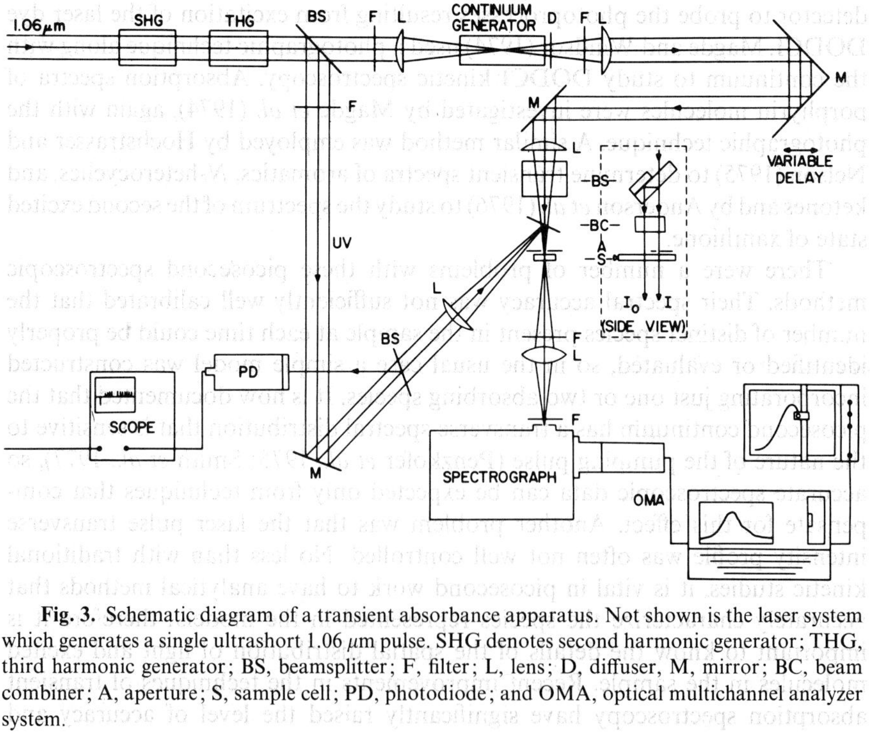

Fig. 3. Schematic diagram of a transient absorbance apparatus. Not shown is the laser system which generates a single ultrashort 1.06 μm pulse. SHG denotes second harmonic generator; THG, third harmonic generator; BS, beamsplitter; F, filter; L, lens; D, diffuser, M, mirror; BC, beam combiner; A, aperture; S, sample cell; PD, photodiode; and OMA, optical multichannel analyzer system.

beam then passes through a reflective optical delay line arranged to permit synchronization with the excitation pulse. Next it is focused through a beam splitting device which produces a displaced but parallel replica beam. Both of these continuum pulses pass through a dichroic beam combiner which serves to reflect the 354 nm excitation pulse and direct it, collinearly with the principal continuum beam, through the apertured sample volume. Meanwhile the replica continuum traverses an unexcited portion of the sample. Following the sample cell, the continuum beams are filtered to remove unabsorbed ultraviolet light and focused with a spherical and a cylindrical lens onto different positions along the height of the entrance slit of a low dispersion spectrograph. This results in two parallel dispersed spectra at the focal plane: one represents the continuum's spectral distribution transmitted through the unexcited sample and the other, through the excited volume.

Both of these spectra are detected, recorded, and digitally processed for each laser shot by a two-dimensional optical multichannel analyzer system (OMA). Each track normally comprises 500 channels of 0.6 nm spectral width and provides a slitwidth-limited resolution of approximately 6 nm. Processing of the raw data begins with the channel-by-channel subtraction of appropriate dark current and flashlamp background spectra. Next the resulting net I_0 spectrum is divided by the net I spectrum and the logarithm of this ratio spectrum calculated and stored. For the next laser shot, the excitation path is blocked and the same processing sequence performed on the resulting data. The difference in the two logarithmic spectra is then calculated, giving a fully double-beam transient absorbance spectrum at the time delay determined by the setting of the variable delay line. To enhance the signal-to-noise ratio and accuracy of the spectrum while guarding against the effects of long term drift, the above data collection and reduction cycle is repeated several times and the resulting spectra are averaged together (after normalization to the relative excitation energy for each shot). Mild seven-point data smoothing of the Savitzky–Golay type is also applied. An important point is that for every shot, the laser pulsetrain and excitation pulse energy are monitored with high bandwidth photodiodes and a storage oscilloscope. Stringent acceptance criteria are applied to each pulsetrain; if one deviates significantly from the standard, the data from that shot are discarded. By using the above procedures it is possible to achieve good suppression of the effects of detector irregularities, laser fluctuations, and optical alignment drifts.

High photometric accuracy is a major design goal for an instrument of this type. A potential limitation, rarely considered in picosecond timescale studies, results from the transverse intensity distributions of the excitation and probe beams. When multimode laser beams are employed for such studies the spatial distribution of excited molecules in the sample becomes highly nonuniform and results in spectral distortions of unpredictable severity. Qualitatively, the effect is to attenuate the peaks of induced absorption bands and to magnify bleaching

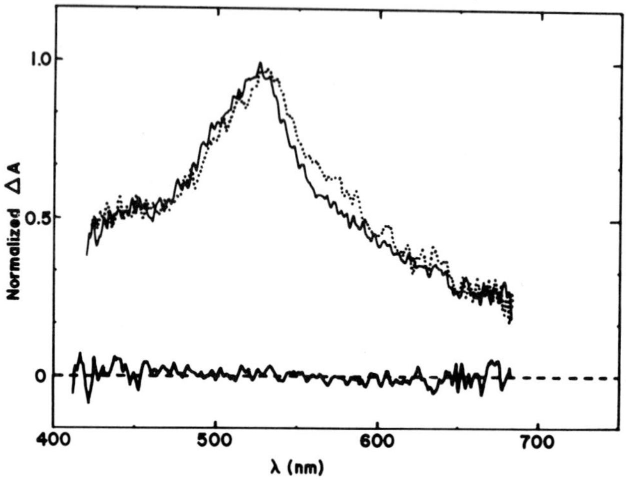

Fig. 4. Transient absorbance spectra of benzophenone in ethanol at room temperature for delays of 10 psec (· · ·) and 22 psec (———) after 354 nm excitation. The solid lower line is the experimental baseline spectrum taken with no excitation, indicating the accuracy and noise levels.

features. This also leads to inaccuracies in single wavelength kinetic data. However, when better characterized beam profiles are used it is possible to reliably estimate the extent of geometrically induced spectrometric errors and also to choose an experimental configuration for which these errors are negligible.

A stringent test of the accuracy, noise level, and stability of a transient spectrometer is provided by a baseline spectrum. Here the data acquisition and processing proceed exactly as usual, except that the excitation beam is kept blocked during both halves of the data cycles. The lowest trace in Fig. 4 shows a typical five-cycle baseline spectrum for this system. The systematic deviation from zero is everywhere less than 0.01 absorbance units and the RMS noise level varies from a low of 0.007 near the center of the spectrum up to 0.03 near the edges. This noise variation follows from the single-beam spectral intensity distribution, which falls at the blue side because of the continuum intensity and the transmission characteristics of the beam combiner, and at the red edge principally because of the response of the detector's intensifier photocathode. It is necessary to adjust conditions to minimize the effects of OMA target lag and local saturation so as to attain the highest linearity in light intensity. The apparatus is then suitable even for studies of bleaching effects in samples whose strong ground state absorption bands make the single beam I_0 spectrum quite irregular.

When either the I or the I_0 continuum path is blocked by an opaque object at the sample cell, less than 1% leakage occurs from either beam into the other, a level which is quite negligible for most purposes. Normally the polarization of the probing continuum is predominantly parallel to that of the excitation beam. For those studies in which the relative polarizations are crucial, a half-wave plate in the 1.06 μm beam allows rotation of its polarization and an added prism polarizer in the continuum path then rejects the unwanted component.

Because of group velocity dispersion in transmissive optics, various frequency components of the continuum have different arrival times at the sample position. It is estimated that this chirp effect delays the blue from the red edge of the continuum by approximately 6 psec for the configuration shown. The spectral distortion resulting from chirp will be greatest when the transient spectrum is changing very rapidly, as is often the case near $t = 0$. If specific corrections are not made, spectra should be obtained at delays of no less than approximately 10 psec so that chirp distortions will be minimized.

This apparatus is readily adaptable to the use of fourth-harmonic excitation (at 265 nm) and to the study of gas phase samples. Because of the laser's low repetition rate, relatively high excitation intensities are unavoidably required. The range of probe wavelengths can be extended somewhat, both to the ultraviolet and to the red, by alterations in optics and detectors.

5. Future Prospects

It is largely true that the methods of picosecond spectroscopy are constrained and shaped by the peculiar characteristics of those lasers which can be made to generate ultrashort pulses. The classic example of this sort consists of the many experiments performed with the first type of mode-locked laser, Nd:glass. The unfortunate properties of this source make multishot signal averaging techniques difficult and have therefore led to the development of a variety of single-shot multiplexed data collection methods. These schemes make use of tools such as echelons, optical multichannel analyzers, and streak cameras, frequently requiring large financial investments in instrumentation. Although some experimental difficulties are overcome in this way, others, such as lack of tunability, persist. As will be shown in the following sections, the performance restrictions of a particular laser lead directly to limitations on the scientific value of experiments carried out with it. Thus, the field of picosecond spectroscopy may currently be described as source limited.

It follows that the most dramatic progress in this field will result from development or implementation of improved sources of ultrashort light pulses. Among the characteristics most needed for scientific flexibility are broad tunability and the option of synchronization with other independently tunable sources. It is possible that a currently existing device such as the synchronously pumped cw dye laser will adequately satisfy these requirements and prove highly versatile, or it may be that the future system of choice will instead be based on a laser that does not yet exist or is still in a developmental stage.

Improved laser performance will likely lead to the redesign of picosecond experiments along more conventional lines, with modulation methods and scanning approaches replacing multiplexed data acquisition. Measurements will be made at fairly high pulse repetition rates and with excitation intensities low enough to avoid serious perturbation of the sample. We anticipate, then, that as progress is made in ultrashort pulse laser technology, picosecond scale experimental work will grow progressively more sophisticated and more productive while its methods become less exotic. Extension of these techniques to the truly subpicosecond scale on a routine basis appears more difficult because of considerations such as pulse distortion in passage through dispersive media.

IV. VIBRATIONAL RELAXATION IN OPTICALLY EXCITED STATES

A. Overview

The dominant factors determining the vibrational decay dynamics of molecules in condensed media seem to be the vibrational structures of the guest species and their environment. At one extreme, diatomics in rare gas matrices frequently have a very limited number of states close in energy to the prepared level. This sparse structure of energy accepting modes and the low frequencies of lattice phonons lead to vibrational relaxation behavior slow enough to be well studied with rather conventional time resolved techniques. A variety of decay mechanisms have been found for different samples; in some cases local rotational modes play a key role, while for others interelectronic state relaxation pathways dominate. In general the theoretical and experimental approaches currently applied to such systems appear adequate. At the next higher level of complexity, triatomics in rare gas matrices, relaxation of the lowest frequency guest mode resembles the behavior of diatomics. But the general understanding of triatomic dynamics is far poorer. The present experimental and theoretical situation for small molecules in rare gas matrices has been well summarized in a recent article by Bondybey and Brus (1979), so we will concentrate our remarks here on more complex molecular systems.

Elucidation of the vibrational dynamics of medium-sized polyatomics at moderate levels of excitation remains an important and unsolved problem even in the case of isolated molecules. Obviously, the situation for polyatomics in condensed media will be still more complex and therefore more poorly understood in certain respects. Experimental difficulties can result partly from higher rates of relaxation associated with the larger densities of vibrational states in the host molecules. An environment of rare gas atoms at low temperature will generally be expected to be the most inert environment; that is, to give the slowest relaxation. Polyatomic hosts contribute their own complex vibrational and librational state densities to the system and can greatly accelerate relaxation by opening many new decay channels, particularly at temperatures high enough that host phonon populations are significant. When the characteristic time scale for vibrational relaxation becomes very short, experimentalists must resort to the difficult and relatively inflexible methods of picosecond spectroscopy. Moreover, in order to characterize vibrational relaxation processes, it is necessary to know more than just the depopulation time of the optically prepared level. One must also identify the major pathways for deexcitation and their individual rates by a means such as observing nonthermal populations in accepting modes of the guest or host. This implies that another obstacle to spectroscopic study of complex decay processes is the dilution of nonthermal populations among a number of accepting modes, an effect which may reduce signals from individual states to undetectable levels.

On the basis of environmental complexity and temperature, it is expected that electronically excited polyatomics dissolved in polyatomic liquids will exhibit rapid and complex relaxation behavior. Laubereau, Kaiser, and co-workers have made impressive progress in determining relatively detailed vibrational dynamics of such systems in the ground electronic state by means of direct time regime experiments (Laubereau and Kaiser, 1978). However no comparable results yet exist for the excited surfaces. Rather, the work reported to date consists only of single lifetimes for each system, representing, for example, the decay of the optically prepared state. One careful study of this sort is the work of Penzkofer *et al.* (1976) on rhodamine dye solutions, the results of which are interesting but incomplete in that the information obtained is far too limited to permit even a general understanding of the relaxation pathways and mechanisms for such large molecules.

Organic solid solutions provide a sample medium which is intermediate between rare gas matrices and fluid polyatomic solutions. These solid solutions may be crystalline or glassy in structure and can be studied at low temperatures in order to retard thermally activated dephasing and depopulation processes. Then the electronic spectra of moderate-sized guest molecules are sharp enough to allow informative vibrational relaxation experiments using direct or indirect spectroscopic methods. In the following section we will describe several studies performed on such systems and, for comparison, one made on a solute molecule of comparable complexity in room temperature fluid solution. It will be seen that there is a considerable disparity in present experimental capabilities for dealing with the two types of systems.

B. Results for Specific Systems

When a molecular system is excited into an absorption band above the origin of its first allowed electronic transition, vibrational population relaxation competes with radiative and nonradiative electronic decay processes. The quantum yield of fluorescence from vibrationally unrelaxed states is normally only 10^{-3} or 10^{-4} even for an allowed radiative transition. Nevertheless, such hot luminescence can often be detected and used to infer rates and pathways for vibrational relaxation. This is true even for the case of continuous excitation and detection.

Rebane, Saari, and co-workers have applied this method to a number of systems in low temperature matrices and crystalline hosts (Saari, 1979; Rebane and Saari, 1976). By measuring the relative intensities of hot luminescence bands they determined kinetic parameters which give the population relaxation times of various vibrational levels relative to the zero-point S_1 lifetime. Perylene was studied in matrices of *n*-alkanes and neon. The lowest Franck–Condon mode (ν_1) at $\sim 350\ cm^{-1}$ excess energy showed a lifetime of 35 psec in heptane and 55 psec in neon. For higher excitations, decay times of $\sim$10–20 psec were deduced, with little evidence uncovered for dominant cascade pathways within the perylene vibrational manifolds. The vibrational relaxation behavior of anthracene showed some qualitative differences. Here the $400\ cm^{-1}$ ν_1 mode, with an 18 psec lifetime in a neon matrix, became significantly populated in the decay of higher vibrational levels. Also, for the $1400\ cm^{-1}$ ν_6 mode, no variation of the lifetime with host composition was found for matrices as different as neon and the polyatomic fluorene. It was concluded that the relaxation of vibrational energy in the S_1 state of matrix-isolated anthracene has a stronger intramolecular character than for perylene.

Recently Hochstrasser and Nyi (1979) applied similar techniques to the study of vibrational relaxation processes in the S_1 state of azulene. Their sample was a dilute mixed crystal of azulene in naphthalene at 2 K; excitation was accomplished with a cw dye laser having a bandwidth of less than $1\ cm^{-1}$. One experimental difficulty in detecting hot luminescence, that of interference from intense relaxed emission, is absent for the first excited singlet of azulene because of rapid nonradiative internal conversion from the vibrationless level. The steady-state populations of all Franck–Condon levels within $\sim 1600\ cm^{-1}$ of the S_1 origin were measured relative to the zero-point population for selective excitation into each of the modes. Vibrational relaxation times deduced from these ratios range from approximately 1–10 psec, values which are longer than or comparable with the mean lifetimes for the levels populated by the vibrational decay process. Importantly, there was no evidence of significant cascade relaxation through the vibrational levels of the azulene guest, and the $384\ cm^{-1}$ lowest Franck–Condon mode was not populated in the decay of higher lying levels. Thus it was concluded that V–V energy transfer to the naphthalene host modes is the dominant pathway for dissipation of vibrational energy in the S_1 state of azulene.

The method of photochemical hole burning provides another indirect probe of vibrational relaxation in polyatomic systems at low temperatures. Voelker and Macfarlane (1979) recently studied free base porphin in a host crystal of *n*-octane at 4.2 K. Sufficient strain was induced in the crystals to give an inhomogeneous broadening greater than the homogeneous linewidths for many vibronic levels of interest. Selective photochemistry within the inhomogeneous line was induced by narrow bandwidth cw dye laser irradiation, after which the width of the resulting spectral hole, which corresponds to the homogeneous broadening, was measured by high resolution absorption spectroscopy. Normally the inhomogeneous contribution to a spectral line shape tends to obscure the dynamically significant homogeneous component, but in this hole-burning method the inhomogeneity is instead essential in uncovering it. The homogeneous widths determined in this way for various levels of vibrational excitation up to 1600 cm^{-1} above the S_1 origin of porphin were reported to range from 4 to 180 GHz. These widths are identified with vibrational depopulation processes, since the corresponding linewidth of the origin transition is only 15 MHz; that is, the low phonon population at this temperature insures that pure dephasing is quite slow. The vibrational decay times for these levels then vary from ~1–44 psec in a way which is not simply related to their excess energies. Unfortunately, hole-burning methods provide no information as to the pathways of energy flow, but it may be possible to identify the extent of host involvement by comparing these results with those obtained on porphin in other low temperature media.

Direct measurements of vibrational relaxation dynamics were attempted by Hochstrasser and Wessel (1974) for a mixed crystal of 10^{-6} mole fraction anthracene in naphthalene at 2 K. Sample excitation was achieved by single photon absorption of the 347 nm second harmonic pulsetrain of a mode-locked ruby laser, or of the 378 nm pulses generated by stimulated Raman scattering in liquid nitrogen. These wavelengths corresponded to vibrational energies of 2950 and 614 cm^{-1} above the origin of the anthracene $^1B_{2u}$ state. Fluorescence emitted from the sample was time resolved with an optical Kerr gate and then dispersed in a spectrograph and recorded. In this way the time evolution of the emission spectra could be observed with ~30 psec temporal resolution and 20 cm^{-1} spectral resolution. Although several experimental difficulties made the results somewhat tentative, it was found that no emission characteristic of vibrationally unrelaxed states was detectable following excitation at 2950 cm^{-1} above the origin. However, the lower energy excitation, which lies in the phonon wing of the 400 cm^{-1} Franck–Condon mode, led to emission spectra with features that changed for a period of ~70 psec after excitation. This evolution was suggested to reflect lattice relaxation processes. At the higher excitation level, thermalization appeared to be more rapid. Refined versions of this experimental technique should be capable of providing more specific information about vibrational relaxation pathways than is obtained from continuous hot luminescence measurements.

There are few informative and reliable experimental results on excited state vibrational relaxation in medium sized molecules dissolved in room temperature polyatomic liquids. Typical of the level of detail currently attainable in this area is the work by Greene *et al.* (1979a) on benzophenone. These investigators used the third harmonic of a single pulse mode-locked Nd:glass laser to excite the $n\pi^*$ singlet state of benzophenone in ethanol solution. Transient absorption measurements in the region of $T_n \leftarrow T_1$ transitions revealed a spectral narrowing and shift on the time scale of ~20 psec which was interpreted as resulting from relaxation of the triplet state C=O stretching mode that had been excited in the rapid intersystem crossing process. Some of the time-resolved absorption spectra for this system are shown in Fig. 4. The specific interpretation of these results remains somewhat uncertain because of possible interference from S_1 absorption or from the effects of solvent reorientation.

In a more general sense, however, detailed pathways of excited state vibrational relaxation are not easily probed by spectroscopic methods involving electronic transitions of complex molecules in high temperature structured environments. Inhomogeneous broadening, rapid dephasing processes, and spectral congestion all tend to obscure the detailed dynamical content that is sought. By contrast, transient vibrational spectroscopies of excited states present greater experimental difficulties but offer more potential for generating highly informative relaxation data.

V. ELECTRONIC RELAXATION

A. General Considerations

Electronic relaxations in isolated molecules are defined as those changes of internal state whereby the difference in energy between two electronic surfaces is converted into nuclear motional excitation. In the classic process of internal conversion from excited states into the ground state, the electronic excitation energy becomes vibrational energy of the ground surface. Similarly, intersystem crossing processes transform the energy difference between the singlet and triplet electronic states into vibrational excitation on the triplet surface. The situation in the condensed phase is a little less clear in that account must be taken of the amount of energy that is directly converted into nuclear motional states of the medium. Particularly those cases where the relaxation is medium induced must be given special consideration. It is also important to decide whether the vibrational relaxation is faster or slower than electronic relaxation in the condensed phase, since the observable relaxation pathways might be quite different for these two cases. A useful categorization of electronic relaxation processes in condensed phases is as follows:

(1) Electronic relaxation rates are comparable to natural radiative rates.

(2) Electronic relaxation rates are slower than vibrational relaxation but still much faster than natural radiative rates.

(3) Electronic relaxation rates are faster than or comparable to vibrational relaxation.

(4) The electronic relaxation is induced by the medium.

Traditional studies of radiationless processes were concerned mainly with the category (1) situation in which measurements of fluorescence or phosphorescence lifetimes, or of moderate emission quantum yields, were utilized to form a framework for understanding nonradiative decay channels. Since the shortest natural radiative lifetimes for molecules fall in the range of 1–10 nsec, the continuing investigation of such systems is well suited to standard nanosecond laser techniques; we will not discuss these studies here.

The category (2) criterion corresponds to the time regime 50–500 psec for the electronic relaxation. As is also true for category (1), this implies that the electronic relaxation processes occur with the system in quasi-thermal equilibrium at all stages. The excesses of vibrational energy on any of the combining surfaces are dissipated into the nearly infinite number of degrees of freedom of the medium rapidly compared with all other transitions or energy transfers. Such cases are now readily studied by means of various subnanosecond laser techniques utilizing decapicosecond sources described in Section III. Moreover, many of the results of time-resolved experiments in this regime can be inferred from measurements of quantum yields for the radiative and radiationless decay channels.

If vibrational and electronic relaxations are occurring on the same time scale or if the vibrational relaxation occurs more slowly, we are dealing with relatively unknown phenomena. Much is currently being learned here through applications of modern laser techniques. One case that is likely to fall into category (3) is that of relaxation from highly excited electronic states of moderately large molecules, such as those processes being explored by Topp and coworkers (Lin and Topp, 1979). In experiments involving multiphoton excitation these investigators excited solution phase samples at room temperature into the 40,000–50,000 cm^{-1} region and studied the emission characteristics of the highly excited levels. The lifetimes of the states explored in this work are apparently in the range of 10^{-13} sec; the processes under study thus often occur before the molecules can thermalize their excess vibrational energy. Other systems in category (3) are those of the azulene type, in which a relatively low energy electronic state is understood to undergo very rapid relaxation into ground state levels. Because of azulene's significance as a prototype photophysical system, we will discuss a number of experimental investigations of its properties in Section VB1.

Systems belonging in category (4) are also of great interest. Whenever the vibrational manifold of the acceptor surface has relatively few levels in the vicinity of the electronic energy gap there arises the possibility of medium-induced relaxation. Obvious examples are to be found with atoms, diatomics and small molecules in crystals or inert matrices. For example gaseous SO_2 under collision free conditions shows only fluorescence for many optical excitation frequencies (Shaw *et al.*, 1976a, b). On the other hand, a crystal of SO_2 at 4.2 K emits mostly phosphorescence for all frequencies of excitation (Hochstrasser and Marchetti, 1970). Obviously in this case the intersystem crossing is medium induced, for it can also be brought about collisionally by addition of foreign gases to the SO_2 vapor (Strickler and Howell, 1968). There are many similar examples among triatomics, tetratomics, and other systems with relatively few vibrational modes and low intramolecular state densities even at rather large energy gaps.

Such dynamical effects also occur for moderate sized molecules when the energy gaps are sufficiently small and when the density of states is low for the motions that are effective in coupling the two electronic states. For example, pyrazine, quinoxaline, and benzophenone show quite different relaxation behavior in dilute gases (Frad *et al.*, 1974; McDonald and Brus, 1973; Busch *et al.*, 1972) than in the condensed phase (Birks, 1970). These systems in the vapor phase also exhibit changes in their relaxation properties on the occurrence of collisions. The medium-induced processes are generally found to occur in the picosecond time regime, suggesting that the electronic interaction is large and that the factor limiting the radiationless transition rate in the isolated molecule is the paucity of levels in the final state. Thus a number of these cases fall into category (3) also and it is therefore not out of the question to experimentally explore the vibrational energy distributions resulting from electronic relaxation. Benzophenone has been the object of such a large number of photophysical investigations that it has become prototypical of the so-called intermediate strong coupling case being discussed here. Accordingly, we provide in section VB2 a detailed review of the subnanosecond studies on this system.

The numerous techniques used to probe the ultrafast processes in categories (2)–(4) include studies of lineshapes and widths in low temperature solids by means of conventional spectroscopy, studies of linewidths in liquid solutions using nonlinear optical techniques, studies of relative quantum yields of radiative processes such as Raman scattering or fluorescence, and measurements of transient effects using pulsed lasers. Many different approaches and schemes have been devised to extract specific information about picosecond time scale events in complex molecular situations. In the following section we review the experimental results relevant to rapid radiationless transitions in three important systems: azulene, benzophenone, and stilbene. These discussions are meant to illustrate the capabilities and limitations of various experimental methods as well as to summarize the current photophysical understanding of these molecules.

B. Prototype Systems

1. Azulene

Azulene is an aromatic hydrocarbon with anomalous photophysical properties. Although fluorescence from its first excited singlet state is too weak to be observed by conventional methods, the $S_2 \rightarrow S_0$ emission is reasonably efficient. Perhaps related to these effects are the unusual energy spacings of its electronic excitations: the S_1 origin lies near 14,000 cm^{-1} while S_2 is at $\sim$28,000 cm^{-1}. This combination of experimentally accessible optical transitions plus the indication of very rapid radiationless decay in S_1 has made azulene an attractive system for the application of picosecond spectroscopic techniques. It will be worthwhile to review these studies here, both to summarize current knowledge of its photophysics and to illustrate the power and shortcomings of different experimental approaches.

One of the major pitfalls in direct picosecond spectroscopic measurements results from inadequately characterized laser pulses. Until recently it has been necessary to rely on indirect methods such as two-photon fluorescence to determine the duration and quality of ultrashort pulses. However, unless such determinations are very carefully performed and include accurate values for the contrast ratios, the results may be misleading. This incomplete knowledge of pulse characteristics has particularly serious consequences when the characteristic time of the physical process under study is shorter than or approximately equal to the pulse duration. For the solid state mode-locked systems that have been the most widely used, the characteristics of pulses vary depending on their location within the pulsetrain. Generally, then, experiments conducted with single pulses extracted from early in the trains may give more reliable results than those employing entire pulsetrains.

It is natural that early picosecond investigations of azulene may have suffered more than recent studies from incomplete knowledge of the pulse characteristics. Nevertheless, the first investigations introduced important experimental methods and ideas. Rentzepis (1968) reported a relaxation time of 7 psec for azulene in solution, a value interpreted as the vibrational decay time in S_1. The experiment represents the first application of the counterpropagating double resonance method. Entire pulsetrains from a Nd:glass laser were used, with the 530 nm second harmonic preparing azulene in vibrationally excited levels of S_1 and the 1.06 μm first harmonic promoting those remaining unrelaxed molecules to the fluorescent S_2 state. The decay time was inferred from comparison of the fluorescence spot shape with that of a two photon absorber. Actually, this time represents not just vibrational relaxation but rather the total decay of the S_1 levels, including electronic as well as vibrational relaxation processes. Another very early study by Drent *et al.* (1968) obtained a relaxation time of 4 psec, but from the present perspective it seems certain that the laser employed for this work was inadequately mode locked.

Wirth *et al.* (1976) used a passively mode-locked dye laser in their study of azulene in benzene and ethanol solutions. They varied the laser wavelength in the range of 590–620 nm, used full trains in a single wavelength counterpropagating geometry, and measured spatial fluorescence patterns with a multichannel photoelectric detector. With pulses in the range of 5–20 psec, these workers reported no detectable broadening in the azulene patterns and therefore deduced S_1 lifetimes of less than 1 psec.

Later that year, a very careful study of the relaxation of azulene in cyclohexane solution was reported by Heritage and Penzkofer (1976). These workers used single, well-characterized pulses from a mode-locked Nd:glass laser to perform three different measurements. In the first, the 530 nm second harmonic was used in a counterpropagating pulse geometry with multichannel photoelectric image recording. Detailed comparison of the fluorescence profiles from azulene and a two photon absorber indicated a sample S_1 relaxation time of less than 1 psec. A second measurement was made with copropagating pulses at 530 and 625 nm, a variable delay line, and spatially integrated fluorescence detection. This checked the first result and provided additional information describing the lifetime of states excited to only 16,000 cm^{-1} above the ground state. Within the noise level of the data (reflecting mainly shot-to-shot pulse fluctuations) both lifetimes were found to be equal and less than 2 psec. The third measurement of Heritage and Penzkofer was a thorough but unsuccessful attempt to verify the early results of Rentzepis using 530 nm excitation and 1.06 μm probing with a variable delay line. It was found that the relevant S_1 absorption cross section was very small.

Shortly thereafter, Ippen *et al.* (1977) reported a similar study of azulene dissolved in cyclohexane, benzene, and other solvents. In contrast to earlier work, however, their experiment employed subpicosecond pulses from a passively mode-locked cw dye laser, which was cavity dumped at 10^5 Hz and operated at 615 nm. The high repetition rate and stability of their laser pulses allowed the use of a copropagating configuration with variable delay line and mechanical chopping. This important modulation method brings to the picosecond regime the signal-to-noise ratio enhancement of phase sensitive detection, a technique long available in conventional experiments. Of course, a high pulse repetition rate is essential for this method. With their extremely short pulses and quiet data, these workers were able to convincingly resolve a solvent-independent 1.9 $\pm$ 0.2 psec decay time for azulene excited at 16,260 cm^{-1}. The relative contributions of vibrational and electronic relaxation could not be separated, but it was suggested that full deactivation of the S_1 state occurs at the measured rate.

Huppert *et al.* (1977) described further measurements on azulene in dichloroethane and methyl cyclohexane solutions and also in the gas phase. Symmetrical counterpropagating configurations were employed using entire trains of either the 530 nm second harmonic or the 625 nm Stokes shifted output derived from a Nd:glass laser. The deduced S_1 lifetimes were 4 $\pm$ 3 psec for 530 nm excitation in solution, 3 $\pm$ 2 psec at 625 nm in solution, and 4 $\pm$ 3 psec at 625 nm in a collision-free gas sample. It was concluded that internal conversion in the S_1 manifold is weakly dependent on vibrational excitation up to 4800 cm^{-1}.

In reviewing all of these studies on azulene made over a period of 10 years, it is clear that the relaxation times involved are very short—shorter in fact than all of the laser pulses except those from Ippen, Shank, and Woerner's subpicosecond dye system. Also evident is the great difficulty of making reliable measurements within the pulse duration using low repetition rate sources, particularly when whole pulsetrains are employed. We conclude, therefore, that the only quantitatively accurate time-resolved value for the azulene relaxation time is 1.9 ± 0.2 psec. Unfortunately, these data are available only for a single level of excitation because of the restricted tunability of the laser system used for the measurement. It would be very interesting to compare this value with the decay time from the S_1 vibrationless level in order to elucidate the role of vibrational effects in the relaxation.

Although a decay time of 1 or 2 psec is quite difficult to resolve directly, it falls in a convenient range for frequency domain measurements. Hochstrasser and Li (1972) reported a high resolution absorption study of the S_1 0–0 transition of azulene dilutely substituted in naphthalene. Although the linewidths obtained in this way reflect the rate of total dephasing, which is generally not the same as depopulation, the use of very low temperatures suppresses pure dephasing processes and insures that the coherent limit is reached. The mixed crystal lineshape at 1.6 K therefore measures the Fourier transform of the same population decay studied by direct time-resolved methods. A recent careful reevaluation of the Hochstrasser and Li spectrum, including deconvolution of the inhomogeneous width, gave a decay time of 3.1 ± 0.3 psec for the low temperature decay of the vibrationless S_1 state of azulene in a naphthalene host. There seem, therefore, to be two reliable values for lifetimes in S_1: this plus the directly measured result of 1.9 psec for $\sim 2000 \text{ cm}^{-1}$ of excess energy in room temperature solution. Since it is likely that the 1.9 psec value represents an electronic rather than a vibrational relaxation time, one must account for the substantial difference between the two measured rates on the basis of a temperature effect, a vibrational enhancement of internal conversion, or a difference in the distributions of states excited in the two media. We suspect that the last two of these are the dominant effects, but a clear understanding of the azulene S_1 dynamics awaits further, more flexible time-resolved measurements.

2. Benzophenone

Another class of compounds that have frequently been studied by subnanosecond methods are the aromatic ketones, typified by benzophenone. These systems were known to undergo very rapid electronic relaxation processes long before subnanosecond laser methods became available for their study (Kasha, 1950). The lowest electronic excitations are of $n\pi^*$ type largely localized on the carbonyl group. The orbital configuration of the excited states is such that the large spin–orbit coupling for an electron on the oxygen atom contributes to the intersystem crossing matrix elements. As a result it is found that in the condensed phase, where there are no level density problems connected with the quasi degeneracy of singlet and triplet levels, the aromatic ketones tend to have very small quantum yields of fluorescence and large yields for intersystem crossing. A similar situation normally prevails for such molecules in gases if the frequency of collisions is sufficiently large. On the other hand, under collision-free conditions the radiative properties of benzophenone are completely different (Naaman *et al.*, 1978). Thus it is safe to conclude that the relaxation processes occurring in this molecule are medium induced, and that it is only in the condensed phase that these systems fall into the statistical limit case. This result is understandable when it realized that the lowest excited singlet and lowest energy triplet states in carbonyl molecules are separated by only $\sim 2000\ cm^{-1}$, which is less than the C—H stretching mode frequencies and nearly equal to two C=O vibrational intervals.

In a low temperature crystal the linewidth of the $S_0 \rightarrow S_1$ transition of benzophenone was measured to be $1.1\ cm^{-1}$, corresponding to a population lifetime for S_1 of at least 5 psec (Dym and Hochstrasser, 1969). Since the linewidth includes a relatively small inhomogeneous component (less than 30%), this deduced lower limit for the intersystem crossing time is expected to be close to the actual value.

Picosecond transients in benzophenone were first reported by Rentzepis (1969). On the basis of the known one-photon absorption spectrum, it was argued that the Stokes Raman-shifted light from the second harmonic of a ruby laser would excite benzophenone into a state that, if unrelaxed into triplets, could absorb the Stokes Raman-shifted fundamental at $10{,}245\ cm^{-1}$. Ruby laser mode-locked pulses having a reported width of 5 psec were employed in this work and lifetimes of 5 and 15 psec were obtained for the S_1 state decay, the latter value arising for excitation at 347 nm. In the light of current knowledge of mode-locked ruby systems, there is sound reason to believe that the pulsewidth in these early experiments was close to 25 psec. Hochstrasser *et al.* (1974) probed the strong triplet–triplet absorption region at 530 nm after exciting the singlet state at 353 nm and found risetimes for the absorption that were definitely longer than the pulsewidth and could be fitted within experimental error to a single time constant of about 10 psec. This value was found to be solvent sensitive (Hochstrasser and Nelson, 1975; Anderson *et al.*, 1975).

These experiments alone cannot provide a detailed picture of the nonradiative processes occurring in benzophenone. The actual situation is complex because the conversion of the optically prepared state (S_1 vibronic levels) into triplet levels and into other vibronic levels of S_1 and of the solvent apparently occurs on the time scale of the Nd:glass pulsewidths being used in these experiments. Other types of experiments that can contribute to sorting out these details include time-resolved spectral measurements and specific probe approaches typified by the method of Lin and Topp (1979) that measures only singlet state amplitude.

Recent absorption spectra of benzophenone after picosecond pulse excitation (see Fig. 4) showed changes indicative of the presence of unrelaxed molecules for times up to ~30 psec after excitation. There was a significant spectral shift and narrowing of the spectrum over this time interval. While these spectra of benzophenone excited by picosecond pulses expose quite new aspects of the dynamics, they do not in themselves solve the problem. Rather, they dramatize the complexity of the processes by indicating that vibrational relaxation is indeed occurring sufficiently slowly that the singlet to triplet conversion cannot be treated as involving only thermalized levels. An additional complication is that the manner in which the solvent is organized around a benzophenone molecule is expected to change after optical excitation and also after the non-radiative relaxation has occurred. These effects might lead to additional spectral shape changes.

Further information can be obtained using the double resonance technique of Lin and Topp (1979). The signal in this case derives from the dipole radiative coupling of highly excited electronic states with the ground state, so population in triplet states is not observed to a first approximation. The S_1 decay time of 13 psec obtained by this technique is in the same range as the S_1 lifetimes measured previously by means of pump and probe experiments (Hochstrasser *et al.*, 1974). Unfortunately, measurements of this sort do not provide direct information about the relative populations of singlet and triplet states.

The aromatic ketones typify the complexities involved in solution phase molecular relaxation processes. A next step in unraveling the pathways of relaxation may be to measure time resolved vibrational spectra in order that the molecular structures and excitations involved with each stage can be understood. Refined methods of transient Raman spectroscopy may soon provide the tools required for further study of vibrational relaxation in excited electronic states.

3. Stilbene Photoisomerization

Photochemical reactions are another class of phenomena that may occur on the subnanosecond time scale in condensed phases. Picosecond pulse methods have been employed to study a variety of light induced reactions such as the photodissociation of I_2 (Chuang *et al.*, 1974), tetraphenylhydrazine (Anderson and Hochstrasser, 1976), and dioxetanes (Smith *et al.*, 1978). In such systems, as well as for many photobiological processes, the changes that occur are complex since they may involve either bimolecular recombination effects or intermediates of unknown structure. Intramolecular rearrangement is conceptually simpler; the process of cis–trans isomerization of ethylenes is perhaps one of the best known examples. Stilbenes (1,2-diphenylethylenes) are probably the most widely studied photoisomerizing systems. The electronic coupling of the phenyl and ethylenic parts of the molecule results in large cross-section absorptions in the near ultraviolet that are readily excited with conventional lamps.

These excitations lead to isomerization about the double bond and in certain cases to other types of photochemical reactions. Therefore there exists an extensive literature on stilbene photochemistry (Saltiel *et al.*, 1973) that can form a solid base for modern laser experiments.

Experimental studies of *trans*-stilbene in solution have led to a picture of the isomerization in which ultraviolet light populates the first electronically excited state. The *trans*-stilbene is subsequently converted into a twisted form (or phantom singlet state), which is also thought to be reached by irradiation of *cis*-stilbene. For each starting species the quantum yield of isomerization is determined by the primary photoprocesses in the absorber and by the nonradiative partitioning occurring from the phantom state. As pointed out by Saltiel *et al.* (1973), this qualitative picture is consistent with many experimental studies such as: photostationary state analysis, the viscosity and temperature dependence of trans fluorescence, the yields of isomerization, the low temperature optical spectra, and triplet sensitized isomerization effects.

Such a concept for isomerization about ethylenic bonds had an early theoretical basis in the prediction that the twisted triplet form of excited ethylenes should be nearly degenerate with the ground state surface. More recent theoretical studies on stilbene have associated the photochemically active "phantom" state with a singlet surface arising from doubly excited configurations (Orlandi and Siebrand, 1975). This view is closely related to earlier ideas concerning the photophysics of polyenes (Hudson and Kohler, 1972).

Transient absorption spectra of stilbenes measured on the picosecond time scale in the manner described in Section IIIC4 have helped clarify a number of features of the isomerization (Greene *et al.*, 1979b). In hexane solution at 296 K *trans*-stilbene exhibits a strong induced absorption band which appears immediately following excitation with a 265 nm, 6 psec pulse and shows a peak at 585 nm. In Fig. 5 the transient spectra taken at 12, 28, and 53 psec delays are shown normalized to the same peak height. The full width at half maximum decays with an exponential time constant of ~ 25 psec to an asymptotic width that is virtually the same as in the 53 psec spectrum.

In addition to the rapid (25 psec) change in shape, the induced *trans*-stilbene spectrum was found to decrease in magnitude on a slower time scale and could be fit within experimental error to an exponential rate constant of 1.1×10^{10} sec^{-1}, corresponding to a characteristic time of 90 ± 5 psec. The induced absorption at 585 nm for long delay times was less than $\sim 2\%$ of the maximum value. When similar measurements were performed on a sample of *cis*-stilbene in hexane, no perceptible induced absorption signal could be detected within the covered wavelength range at any delay time.

It is known from measurements of the trans ^1B fluorescence quantum yield and integrated extinction coefficient, as well as from direct fluorescence lifetime measurements (Sumitani *et al.*, 1977; Heisel *et al.*, 1979), that this first excited singlet state has a lifetime in the vicinity of 100 psec. Any other state which is

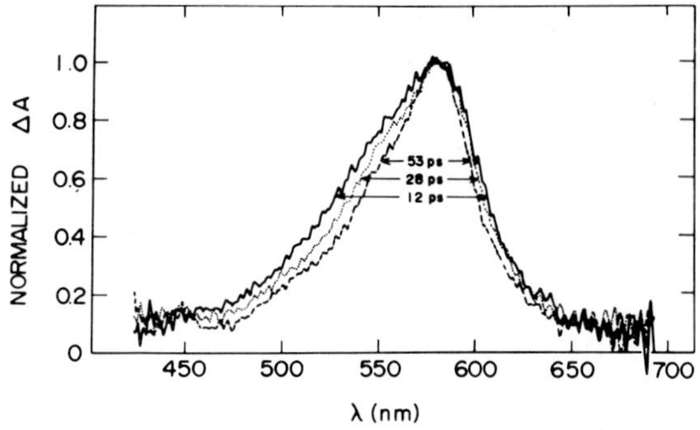

Fig. 5. Transient absorbance spectra of 10^{-3} M *trans*-stilbene in *n*-hexane taken at delays of 12, 28, and 53 psec following 265 nm excitation. The spectra have been drawn normalized to the same peak height; the integrated absorption coefficient actually decays exponentially with a time constant of 90 psec at 296 K.

populated through decay of ^1B would necessarily show a risetime of this order. The observations of a subpulsewidth risetime and a 90 psec decay for the visible transient feature clearly indicate that it represents the absorption of the ^1B excited singlet of *trans*-stilbene. The rapid decay process which this 90 psec reflects cannot be fluorescence or intersystem crossing, for which the quantum yields are known to be low. In addition, the viscosity dependence of the fluorescence quantum yield suggests that direct internal conversion to S_0 is also of secondary importance. Thus, the decay channel through which the ^1B population flows with a rate constant of nearly 1.1×10^{10} sec^{-1} involves twisting about the central double bond—the first step in the process of isomerization.

While the population of the ^1B state decays with a characteristic time of 90 psec the shape of its transient absorption spectrum is observed to evolve on a faster time scale. The 265 nm excitation beam introduces 7000 cm^{-1} of energy excess on the ^1B surface. From the work of Dyck and McClure (1962) it is known that the relevant Franck–Condon modes in *trans*-stilbene are a C=C stretch of $\sim$1600 cm^{-1} and a 200 cm^{-1} in-plane bending vibration. Possible causes for the spectral relaxation would then include thermalization of the excess vibrational energy, reorganization of solvent molecules around the excited solute, rotation of the phenyl groups about their single bonds, and twisting along the isomerization coordinate. The absence of detectable spectral shifts in the transient spectra favors the interpretation that changes in the spectrum signal relaxation of excess vibrational excitation from optically coupled modes into the solvent or into other modes of the stilbene molecule.

The torsional vibrations of stilbene, which correspond to motion along the isomerization coordinate, are not excited by the initial absorption process. Neither is it likely that they are populated significantly in excess of thermal levels in the subsequent relaxation of the Franck–Condon modes. Isomerization would thus be expected to occur with a single rate constant essentially unrelated to the amount of vibrational excitation accompanying the electronic transition. Unfortunately, the kinetic data at very early delay times are not adequate to detect small changes in the ^1B decay rate. It will therefore be of considerable interest to test this speculation by using methods with higher time resolution to determine the early isomerization rate constants with various excitation wavelengths.

Some other features of the picosecond experiments deserve comment. As the magnitude of the 585 nm band decreases, no other absorption is seen to appear within the covered spectral range. This indicates that the state into which the trans ^1B population decays, which would be the twisted phantom ^1A state in the Orlandi and Siebrand (1975) model, has no strong visible absorption or alternatively has a lifetime much shorter than the Nd:glass pulse duration. It is surprising that *cis*-stilbene shows no picosecond tranient S_1 absorption comparable with that seen from trans, because in a first approximation the π-electron states of planar *cis*- and *trans*-stilbene are the same. A probable explanation is that cis undergoes a transition to a twisted form within a time that is short compared with the 10 psec resolution of these experiments. A transient spectrum like that seen for trans having a lifetime of 1 psec would have been observable.

The absence of a spectral shift as the isomerization proceeds is consistent with a thermally activated process in which only a small fraction of electronically excited molecules are able at any given time to surmount the 3–4 kcal mole^{-1} potential barrier associated with twisting into the perpendicular ^1A minimum. The claim that an equilibrium is established between excited populations in the trans and perpendicular configurations finds no support in the observation of near complete exponential decay of trans ^1B (Greene *et al.*, 1979b).

The picosecond transient absorption study of stilbene clarified some of the problems of cis–trans isomerization in solution. Light absorption by *trans*-stilbene in hexane solution yields a thermalized distribution in about 25 psec. These S_1 states disappear (twist) with a rate constant of 1.1×10^{10} sec^{-1}. No evidence of twisted or twisting forms can be identified in the spectrum, so the lifetime of partially twisted forms cannot be long compared with the 6 psec pulsewidth; otherwise a reasonable concentration of these would build up during the evolution of S_1 and then be detected by their visible absorption.

The isomerization of molecules such as stilbene presents some interesting theoretical challenges in solution phase dynamics. The motion of the molecule during the transition from trans to cis, or vice versa, must be controlled by the interactions with the solvent. One can readily imagine that if the solvent molecules were sufficiently small and isotropic, the isomerization would occur by a Brownian motion. Such would very likely be the case for the relative rotation of two phenyl groups if their mutual potential energy were negligible. In the case of stilbene the excited state surface (see Fig. 6) most likely has a shallow

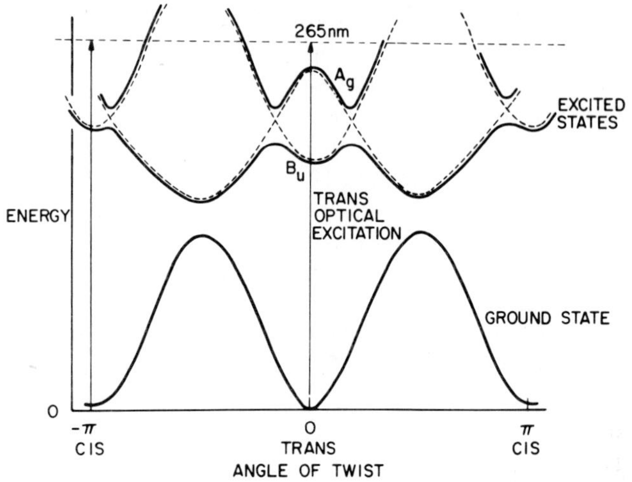

Fig. 6. Sketch of a conceivable potential energy surface for the isomerization coordinate of stilbene.

minimum at trans and a much deeper minimum at the twisted conformation. In order to account for the kinetics of the disappearance of trans and the appearance of the twisted form, or ultimately cis, it would be necessary to include the molecular potential energy function for twisting in the statistical equations of motion. In this case, as well as in the case where the molecular motion occurs without influence from external forces, it would be expected that the dynamical features will depend on the macroscopic parameters of the solvent. In the simplest approach the dynamics would be described by a Langevin equation of motion incorporating an external force due to the twisting potential. A similar treatment (Gilbert and McCaskill, 1979) has been used to explain the visocity dependence of the spectral changes occurring after the picosecond pulse excitation of 1,1′-binaphthyl (Shank *et al.*, 1977). In this case there is presumed to be an optically induced twisting about the single bond connecting the naphthyl groups (Hochstrasser, 1961).

The situation in solutions is expected to differ markedly from that for isolated molecules. In solutions any vibrational energy excess introduced by the optical excitation is very likely to be dissipated before it can be utilized in a molecular process. Obvious exceptions to this will occur when dissociative states are directly excited, but normally configurational changes in solution will occur while the molecular states are maintained in a Boltzmann distribution corresponding to the temperature of the solvent. For the case of *trans*-stilbene vapor the lifetime of the levels excited by 265 nm light is 17 psec under collision-free conditions (Greene *et al.*, 1979c), whereas it is 90 psec in hexane solution at 296 K. This remarkable result dramatizes the differences between gas phase and solution processes. In the solution the lifetime corresponds to a solvent related thermally assisted process; but in the gas phase it reflects the intramolecular redistribution of the large amount of excess vibrational energy introduced by the light pulse. The vapor phase fluorescence lifetime is strongly dependent on this energy excess, whereas in solution the lifetime is apparently insensitive to excitation wavelength. It might appear unlikely that the vibrational redistribution necessary for isomerization of the isolated molecule can occur more rapidly than solution phase vibrational relaxation, as inferred from spectral narrowing. However, the time measured by the solution spectral changes is probably that for full vibrational and orientational thermalization. In addition, the nature of the prepared state is different in solution and vapor because of the large fluctuation broadening occurring in the solutions, and because some energy is directly coupled into solvent states by the light absorption process.

VI. SUMMARY AND PROSPECTS

Dynamical processes in electronically excited polyatomic molecules are strongly influenced by the nature of the surrounding medium. The most important factors in this regard are the density of states and reorientational rigidity of the host. Experimental measurements of excited state vibrational relaxation are few in number and mostly indirect. They point to vibrational depopulation times in the range of a few picoseconds to a few tens of picoseconds for excitation fairly near the origin, values which are similar to those determined in the ground electronic state of fluid media. For low temperature solid solutions and crystals, both direct and indirect experimental methods may be usefully applied to gain relatively specific information on decay pathways for some systems. Although the patterns found have not yet been fully explained, it is expected that they will reflect the effects of energy level structures of the host media. Higher temperature liquid solutions present much greater experimental difficulties and are likely to show more complex behavior because of the additional effects of solvent reorientation.

Radiationless transitions in condensed phases are known to occur on the picosecond time scale for some polyatomics. This implies that vibrational and electronic relaxation rates can be comparable and that one cannot always treat the electronic processes as involving only thermalized vibrational distributions on the various potential surfaces. Since radiationless transitions necessarily entail the conversion of some electronic excitation into vibrational energy, the properties of the medium in accepting and dissipating the vibrational energy play an important role in the overall dynamics. Thus, a full understanding of condensed phase electronic relaxation processes must be based on knowledge of the medium's vibrational characteristics.

In considering the various relaxation possibilities of matrix isolated species and organic mixed crystals it becomes evident that there are large gaps in our knowledge. Virtually no detailed information is available on the pathways of vibrational or electronic relaxation in these condensed phases, since almost all studies have been concerned only with decay of the prepared state. On the other hand, theories of vibrational and electronic relaxation automatically provide detailed predictions of relaxation pathways through their specification of the final state of the system. This is an area in which existing experimental techniques should be able to make important contributions. Especially valuable will be methods such as time-resolved Raman, infrared, and laser-induced fluorescence spectroscopies, which have the potential to measure the excitation of accepting modes.

For liquids, too, there remain many experimental and theoretical difficulties in understanding basic relaxation phenomena. As an example, the conventional optical spectra of solutions present great mysteries even in cases where the isolated molecule spectroscopy is well understood. The various contributions to the observed solution spectral bandwidths have not been elucidated; i.e., the physical relaxation processes that damp the polarization induced in a medium by a resonant light field remain to be fully analyzed. Also, although experiments have succeeded in exposing particular aspects of solution dynamics such as rotational correlation functions, very little is yet known about the mechanisms or time scales of energy exchange between solute and solvent molecules, or about how these energy transfers relate to nuclear configurational changes. For the specific cases in which molecular configurational changes can be measured in the time domain, contact with theory should be possible using statistical methods based on Brownian motion treatments of the solution dynamics. Another liquid phase area in which short pulse optical experiments will be useful is assessing the time scales of inhomogeneity and cross relaxation in simple solutions.

After more than a decade of experimental studies of ultrafast processes in molecular systems, the possibilities still seem unlimited. Present restrictions on the flexibility of direct methods are chiefly due to shortcomings of the pulsed laser systems that are employed. It is anticipated that technological improvements will occur in the near future which will significantly extend the range of highly time resolved spectroscopic techniques. The continuing development of all types of lasers presents exciting prospects for chemistry in general, and particularly for those many areas concerned with phenomena that are intrinsically fast.

REFERENCES

Alfano, R. R., and Shapiro, S. L. (1970). *Phys. Rev. Lett.* **24**, 584.

Alfano, R. R., and Shapiro, S. L. (1971). *Chem. Phys. Lett.* **8**, 631.

Anderson, R. W., and Hochstrasser, R. M. (1976). *J. Phys. Chem.* **80**, 2155.

Anderson, R. W., Hochstrasser, R. M Lutz, H., and Scott, G. W. (1975). *Chem. Phys. Lett.* **32**, 204.

Anderson, R. W., Hochstrasser, R. M., and Pownall, H. J. (1976). *Chem. Phys. Lett.* **43**, 224.
Birks, J. B. (1970). "Photophysics of Aromatic Molecules." Wiley, New York.
Bondybey, V. E. (1977). *J. Chem. Phys.* **66**, 995.
Bondybey, V. E., and Brus, L. E. (1975). *J. Chem. Phys.* **63**, 794.
Bondybey, V. E., and Brus, L. E. (1979). *Adv. Chem. Phys.* (in press).
Bradley, D. J. (1977). *In* "Ultrashort Light Pulses" (S. L. Shapiro, ed.), p. 17. Springer-Verlag, Berlin and New York.
Brus, L. E., and Bondybey, V. E. (1975). *J. Chem. Phys.* **63**, 786.
Busch, G. E., Rentzepis, P. M., and Jortner, J. (1972). *J. Chem. Phys.* **56**, 631.
Busch, G. E., Jones, R. P., and Rentzepis, P. M. (1973). *Chem. Phys. Lett.* **18**, 178.
Chuang, T. J., Hoffman, G. W., and Eisenthal, K. B. (1974). *Chem. Phys. Lett.* **25**, 201.
Cox, A. J., Scott, G. W., and Talley, L. D. (1977). *Appl. Phys. Lett.* **31**, 389.
Dellinger, B., King, D. S., Hochstrasser, R. M., and Smith, A. B. (1977). *J. Am. Chem. Soc.* **99**, 7138.
Diels, J.-C., van Stryland, E., and Benedict, G. (1978). *Opt. Commun.* **25**, 93.
Drent, E., van der Deijl, G. M., and Zandstra, P. J. (1968). *Chem. Phys. Lett.* **2**, 526.
Duguay, M. A., and Hansen, J. W. (1968). *Appl. Phys. Lett.* **13**, 178.
Duguay, M. A., and Hansen, J. W. (1969a). *Opt. Commun.* **1**, 254.
Duguay, M. A., and Hansen, J. W. (1969b). *Appl. Phys. Lett.* **15**, 192.
Dyck, R. H., and McClure, D. S. (1962). *J. Chem. Phys.* **36**, 2326.
Dym, S., and Hochstrasser, R. M. (1969). *J. Chem. Phys.* **51**, 2458.
El-Sayed, M. A. (1974). *In* "Excited States" (E. C. Lim, ed.), Vol. 1, p. 35. Academic Press, New York.
Fan, B., and Gustafson, T. K. (1976). *Appl. Phys. Lett.* **28**, 202.
Frad, A., Lahmani, F., Tramer, A., and Tric, C. (1974). *J. Chem. Phys.* **60**, 4419.
Gilbert, R. G., and McCaskill, J. S. (1979). *Chem. Phys.* (to be published).
Glenn, W. H., Brienza, M. J., and De Maria, A. J. (1968). *Appl. Phys. Lett.* **12**, 54.
Goldberg, L. S., and Moore, C. A. (1975). *Appl. Phy. Lett.* **27**, 217.
Goodman, J., and Brus, L. E. (1977). *J. Chem. Phys.* **67**, 933.
Greene, B. I., Hochstrasser, R. M., and Weisman, R. B. (1979a). *J. Chem. Phys.* **70**, 1247.
Greene, B. I., Hochstrasser, R. M., and Weisman, R. B. (1979b). *Chem. Phys. Lett.* **62**, 427.
Greene, B. I., Hochstrasser, R. M., and Weisman, R. B. (1979c). *J. Chem. Phys.* **71**, 544.
Hansch, T., and Toschek, P. (1970). *Z. Phys.* **236**, 213.
Heisel, F., Miehe, J. A., and Sipp, B. (1979). *Chem. Phys. Lett.* **61**, 115.
Heritage, J. P., and Jain, R. K. (1978). *Appl. Phys. Lett.* **32**, 101.
Heritage, J. P., and Penzkofer, A. (1976). *Chem. Phys. Lett.* **44**, 76.
Hesselink, W. H., and Wiersma, D. A. (1978). *Chem. Phys. Lett.* **56**, 227.
Hochstrasser, R. M. (1961). *Can. J. Chem.* **39**, 459.
Hochstrasser, R. M., (1968). *Accounts Chem. Res.* **1**, 266.
Hochstrasser, R. M., and Li, T. Y. (1972). *J. Mol. Spectrosc.* **41**, 297.
Hochstrasser, R. M., and Marchetti, A. P. (1970). *J. Mol. Spectrosc.* **35**, 335.
Hochstrasser, R. M., and Nelson, A. C. (1975). *In* "Lasers in Physical Chemistry and Biophysics" (J. Joussot-Dubien, ed.), p. 305. Elsevier, Amsterdam.
Hochstrasser, R. M., and Nyi, C. A. (1979). *J. Chem. Phys.* **70**, 1112.
Hochstrasser, R. M., and Prasad, P. (1974). *In* "Excited States" (E. C. Lim, ed.), Vol. 1, p. 79. Academic Press, New York.
Hochstrasser, R. M., and Wessel, J. E. (1973). *Chem. Phys. Lett.* **19**, 156.
Hochstrasser, R. M., and Wessel, J. E. (1974). *Chem. Phys.* **6**, 19.
Hochstrasser, R. M., Lutz, H., and Scott, G. W. (1974). *Chem. Phys. Lett.* **24**, 162.
Hudson, B. S., and Kohler, B. E. (1972). *Chem. Phys. Lett.* **14**, 299.
Huppert, D., Jortner, J., and Rentzepis, P. M. (1977). *Israel J. Chem.* **16**, 277.

Ippen, E. P., and Shank, C. V. (1977). *In* "Ultrashort Light Pulses" (S. L. Shapiro, ed.), p. 83. Springer-Verlag, Berlin and New York.
Ippen, E. P., and Shank, C. V. (1978). *In* "Picosecond Phenomena" (C. V. Shank *et al.*, eds.), p. 103. Springer-Verlag, Berlin and New York.
Ippen, E. P., Shank, C. V., and Dienes, D. (1972). *Appl. Phys. Lett.* **21**, 348.
Ippen, E. P., Shank, C. V., and Woerner, R. L. (1977). *Chem. Phys. Lett.* **46**, 20.
Jain, R. K., and Heritage, J. P. (1978). *Appl. Phys. Lett.* **32**, 41.
Kasha, M. (1950). *Disc. Faraday Soc.* **9**, 14.
Kharlamov, B. M., Personov, R. I., and Bykovskaya, L. A. (1974). *Opt. Commun.* **12**, 191.
Kitaigorodsky, A. I. (1973). "Molecular Crystals and Molecules." Academic Press, New York.
Kurobori, T., Cho, Y., and Matsuo, Y. (1978). *Opt. Commun.* **24**, 41.
Langhoff, C. A., and Robinson, G. W. (1974). *Chem. Phys.* **6**, 34.
Laubereau, A., and Kaiser, W. (1978). *Rev. Mod. Phys.* **50**, 607.
Laubereau, A., Greiter, L., and Kaiser, W. (1974). *Appl. Phys. Lett.* **25**, 87.
Legay, F. (1977). *In* "Chemical and Biochemical Applications of Lasers" (C. B. Moore, ed.), Vol. II, p. 43. Academic Press, New York.
Liang, Y.-L., and Chernoff, D. A. (1979). unpublished results.
Lin, H. B., and Topp, M. R. (1979). *Chem. Phys.* **36**, 365.
Magde, D., and Windsor, M. W. (1974). *Chem. Phys. Lett.* **27**, 31.
Magde, D., Windsor, M. W., Holton, D., and Gouterman, M. (1974). *Chem. Phys. Lett.* **29**, 183.
Mahr, H., and Hirsch, M. D. (1975). *Opt. Commun.* **13**, 96.
McDonald, J. R., and Brus, L. E. (1973). *Chem. Phys. Lett.* **23**, 87.
Naaman, R., Lubman, D. M., and Zare, R. N. (1978). *Chem. Phys.* **32**, 17.
Orlandi, G., and Siebrand, W. (1975). *Chem. Phys. Lett.* **30**, 352.
Penzkofer, A., Seilmeier, A., and Kaiser, W. (1975). *Opt. Commun.* **14**, 363.
Penzkofer, A., Falkenstein, W., and Kaiser, W. (1976). *Chem. Phys. Lett.* **44**, 82.
Rebane, K. K. (1970). "Impurity Spectra of Solids." Plenum, New York.
Rebane, K. K., and Saari, P. (1976). *J. Lumin.* **12–13**, 23.
Reintjes, J., and Eckardt, R. C. (1977). *Appl. Phys. Lett.* **30**, 91.
Rentzepis, P. M. (1968). *Chem. Phys. Lett.* **2**, 117.
Rentzepis, P. M. (1969). *Science* **169**, 239.
Royt, T. R., Faust, W. L., Goldberg, L. S., and Lee, C. H. (1974). *Appl. Phys. Lett.* **25**, 514.
Saari, P. (1979). *In* "Ultrafast Relaxation and Secondary Emission," p. 142. Academy of Sciences of the Estonian S.S.R., Tallinn.
Saltiel, J. *et al.* (1973). *Org. Photochem.* **3**, 1.
Seilmeier, A., Spanner, K., Laubereau, A., and Kaiser, W. (1978). *Opt. Commun.* **24**, 237.
Shank, C. V., Ippen, E. P., Teschke, O., and Eisenthal, K. B. (1977). *J. Chem. Phys.* **67**, 5547.
Shaw, R. J., Kent, J. E., and O'Dwyer, M. F. (1976a). *Chem. Phys.* **18**, 155.
Shaw, R. J., Kent, J. E., and O'Dwyer, M. F. (1976b). *Chem. Phys.* **18**, 165.
Shpolskii, E. V. (1963). *Sov. Phys.-Usp.* **6**, 41.
Smith, K. K., Koo, J.-Y., Schuster, G. B., and Kaufmann, K. J. (1978). *J. Phys. Chem.* **82**, 2291.
Smith, W. L., Liu, P., and Bloembergen, N. (1977). *Phys. Rev. A* **15**, 2396.
Song, J. J., Lee, J. H., and Levenson, M. D. (1978). *Phys. Rev. A* **17**, 1439.
Strickler, S. J., and Howell, D. B. (1968). *J. Chem. Phys.* **49**, 1947.
Sumitani, M., Nakashima, N., Yoshihara, K., and Nagakura, S. (1977). *Chem. Phys. Lett.* **51**, 183.
Topp, M. R., Rentzepis, P. M., and Jones, R. P. (1971). *Chem. Phys. Lett.* **9**, 1.
Trommsdorff, H. P., Hochstrasser, R. M., and Meredith, G. R. (1979). *J. Lumin.* **18**, 687.
Voelker, S., and Macfarlane, R. M. (1979). *Chem. Phys. Lett.* **61**, 421.
von der Linde, D. (1972). *IEEE J. Quantum Electron.* **8**, 328.
Wannier, P., Rentzepis, P. M., and Jortner, J. (1971). *Chem. Phys. Lett.* **10**, 102.

Weisman, R. B., and Rice, S. A. (1976). *Opt. Commun.* **19**, 28.
Wessel, J. E. (1970). Ph.D. Thesis, Univ. of Chicago, Chicago, Illinois.
Wirth, P., Schneider, S., and Dörr, F. (1976). *Chem. Phys. Lett.* **42**, 482.
Yajima, T., and Souma, H. (1978). *Phys. Rev. A* **17**, 309.
Yajima, T., Souma, H., and Ishida, Y. (1978). *Phys. Rev. A* **17**, 324.
Yariv, A. (1975). "Quantum Electronics," 2nd ed. Wiley, New York.

9

Some Considerations of Theory and Experiment in Ultrafast Processes

S. H. Lin

Department of Chemistry
Arizona State University
Tempe, Arizona

I. INTRODUCTION

In the last few years there has been very rapid growth in experimental research on ultrafast processes (see, for example, Govindjee, 1978; Eisenthal, 1977; Laubereau and Kaiser, 1975; Lin and Eyring, 1974; Hochstrasser and Weisman, this volume). The purpose of this paper is to outline a unified treatment of rate processes (or time-dependent phenomena) over a wide range of time scales from the viewpoint of the density matrix method (the master equation approach) and to present some relevant recent experimental results to support the theoretical discussion. It will be shown that this theoretical approach provides not only a description of the population relaxation of the system but also a method for calculating the relaxation rate constant and dephasing rate constant.

ISBN 0-12-450650-X

Since the density matrix method can also be applied to optical processes, the relation between rate processes and the line shape function of an optical property will then be well established. It will also be shown that the density matrix method applies not only to time scales for which the Markoff approximation holds (rate constants well defined in this case) but also to those for which it does not hold (rate constants not defined in this case).

II. TIME EVOLUTION OF A RELAXING SYSTEM

For a typical relaxation experiment, one is concerned with a description of the population of the system versus time and a determination of the rate constants of the relaxation and/or dephasing process involved. In this section, we shall present the master equation approach for this purpose.

The idea is to consider an isolated system which can be divided into a *reservoir* (or *heat bath*) and a *subsystem* (or *system* of interest) and to eliminate the irrelevant part of the density matrix to obtain the equation of motion for the reduced density matrix of the subsystem.

The time dependence of the density matrix of an isolated system $\hat{\rho}(t)$ is determined by the Liouville equation of motion,

$$\partial\hat{\rho}/\partial t = -(i/\hbar)(\hat{H}\hat{\rho} - \hat{\rho}\hat{H}) = -i\hat{L}\hat{\rho}, \tag{2.1}$$

in which $\hat{L}$ represents the Liouville operator. The total system consists of a part called the system of interest, with Hamiltonian $\hat{H}_s$, and of a heat bath with Hamiltonian $\hat{H}_b$. If we let $\hat{H}'$ represent the interaction between the two parts, then the Hamiltonian of the total system can be written

$$\hat{H} = \hat{H}_s + \hat{H}_b + \hat{H}' = \hat{H}_0 + \hat{H}'. \tag{2.2}$$

In accordance with Eq. (2.1), the corresponding Liouville operator takes the form

$$\hat{L} = \hat{L}_s + \hat{L}_b + \hat{L}' = \hat{L}_0 + \hat{L}'. \tag{2.3}$$

Notice that the density matrix of the system of interest at time t can be found from

$$\hat{\rho}^{(s)}(t) = \mathrm{Tr}_b[\hat{\rho}(t)], \tag{2.4}$$

in which Tr_b represents the operator that carries out a trace over the quantum states of the heat bath. To eliminate the reservoir variables, it is commonly assumed that at $t = 0$

$$\hat{\rho}(0) = \hat{\rho}^{(s)}(0)\hat{\rho}^{(b)}(0). \tag{2.5}$$

Using Eq. (2.5), the complete equation of motion for the reduced density matrix $\hat{\rho}^{(s)}(t)$ has been obtained (Lin and Eyring, 1977b; Fano, 1964; Appendix I).

The singular perturbation method has been applied to obtain the approximate equation of motion for $\hat{\rho}^{(s)}(t)$; to second-order approximation, the solution of the Liouville equation (2.1) by the singular perturbation method yields (Cuckier and Deutch, 1969; Lin and Eyring, 1977b)

$$\partial/\partial t\, \hat{\rho}^{(s)} = -i\hat{L}_s \hat{\rho}^{(s)} - \hat{\Gamma}\hat{\rho}^{(s)}, \tag{2.6}$$

where $\hat{\Gamma}$ represents the damping operator. The master equation given by (2.6) can be explicitly written (Appendix I) as

$$\frac{\partial}{\partial t}\rho^{(s)}_{n_s n_s}(t) = -\sum_{n'_s}{}' \Gamma_{n_s n_s : n'_s n'_s}\rho^{(s)}_{n'_s n'_s}(t) - \Gamma_{n_s n_s : n_s n_s}\rho^{(s)}_{n_s n_s}(t) \tag{2.7}$$

and

$$(\partial/\partial t)\rho^{(s)}_{n_s n_s}(t) = -(i\omega_{m_s n_s} + \Gamma_{m_s n_s : m_s n_s})\rho^{(s)}_{m_s n_s}(t), \tag{2.8}$$

where

$$\Gamma_{n_s n_s : n_s n_s} = -\sum_{n'_s}{}'\Gamma_{n'_s n'_s : n_s n_s} = \sum_{n'_s}{}' k_{n'_s} \tag{2.9}$$

$$\Gamma_{n_s n_s : n'_s n'_s} = -k_{n'_s n_s}, \tag{2.10}$$

$$k_{n_s n'_s} = \frac{2\pi}{\hbar}\sum_{n_b}\sum_{n'_b}\rho^{(b)}_{n_b n_b}(0)|H'_{n_s n_b : n'_s n'_b}|^2 \delta(E_{n_s n_b} - E_{n'_s n'_b}), \tag{2.11}$$

$$\Gamma_{m_s n_s : m_s n_s} = \tfrac{1}{2}(\Gamma_{n_s n_s : n_s n_s} + \Gamma_{m_s m_s : m_s m_s}) + \Gamma^{(d)}_{m_s n_s : m_s n_s}, \tag{2.12}$$

and

$$\Gamma^{(d)}_{m_s n_s : m_s n_s} = \frac{\pi}{\hbar}\sum_{n_b, n'_b}\sum \rho^{(b)}_{n'_b n'_b}(0)(|H'_{m_s n_b : m_s n'_b}|^2 + |H'_{n_s n_b : n_s n'_b}|^2)\delta(E_{n_b} - E_{n'_b}). \tag{2.13}$$

Notice that $\omega_{m_s n_s}$ includes the zeroth-order energy resulting from $\hat{H}_s$ and the first- and second-order energy resulting from $\hat{H}'$, i.e.,

$$\omega_{m_s n_s} = (1/\hbar)(E_{m_s} + E^{(1)}_{m_s} + E^{(2)}_{m_s}) - (1/\hbar)(E_{n_s} + E^{(1)}_{n_s} + E^{(2)}_{n_s}), \tag{2.14}$$

where, for example,

$$E^{(1)}_{n_s} = \langle \hat{H}' \rangle_{n_s n_s}, \tag{2.15}$$

$$E^{(2)}_{n_s} = \sum_{n'_s}{}' \frac{|\langle \hat{H}' \rangle_{n_s n'_s}|^2}{E_{n_s} - E_{n'_s}}, \tag{2.16}$$

and

$$\langle \hat{H}' \rangle_{n_s n'_s} = \sum_{n_b} \rho^{(b)}_{n_b n_b}(0) H'_{n_b n_s : n_b n'_s}. \tag{2.17}$$

From Eqs. (2.7)–(2.13) we can see that $\Gamma_{n_s n_s : n_s n_s}$ and $\Gamma_{n_s n_s : n'_s n'_s}$ are related to the rate constants $k_{n'_s n_s}$ in the golden rule expression, $\rho^{(s)}_{n_s n_s}(t)$ describes the population of the system as a function of time, and $\rho^{(s)}_{m_s n_s}(t)$ describes the dephasing of the system and is related to the line shape function of optical spectroscopy (Lin and Eyring, 1977a). Notice that $\Gamma_{m_s n_s : m_s n_s}$, which is important in finding $\rho^{(s)}_{m_s n_s}(t)$, is related to the lifetimes of the n_s and m_s states through $\Gamma_{n_s n_s : n_s n_s}$ and $\Gamma_{m_s m_s : m_s m_s}$ and the rate constant of elastic processes through $\Gamma^{(d)}_{m_s n_s : m_s n_s}$.

To illustrate the application of Eqs. (2.7) and (2.8) we consider a two-level system. In this case, we have

$$(\partial/\partial t)\rho^{(s)}_{n_s n_s}(t) = -\Gamma_{n_s n_s : n'_s n'_s}\rho^{(s)}_{n'_s n'_s}(t) - \Gamma_{n_s n_s : n_s n_s}\rho^{(s)}_{n_s n_s}(t), \tag{2.18}$$

or

$$(\partial/\partial t)\rho^{(s)}_{n_s n_s}(t) = k_{n'_s n_s}\rho^{(s)}_{n'_s n'_s}(t) - k_{n_s n'_s}\rho^{(s)}_{n_s n_s}(t), \tag{2.19}$$

where, for example,

$$k_{n_s n'_s} = \frac{2\pi}{\hbar}\sum_{n_b}\sum_{n'_b}\rho^{(b)}_{n_b n_b}(0)|H'_{n_s n_b : n'_s n'_b}|^2\delta(E_{n_s n_b} - E_{n'_s n'_b}), \tag{2.20}$$

and

$$(\partial/\partial t)\rho^{(s)}_{n_s n'_s}(t) = -(i\omega_{n_s n'_s} + \Gamma_{n_s n'_s : n_s n'_s})\rho^{(s)}_{n_s n'_s}(t), \tag{2.21}$$

where

$$\begin{aligned}\Gamma_{n_s n'_s : n_s n'_s} &= \tfrac{1}{2}(\Gamma_{n_s n_s : n_s n_s} + \Gamma_{n'_s n'_s : n'_s n'_s}) + \tfrac{1}{2}(\Gamma^{(d)}_{n_s n_s : n_s n_s} + \Gamma^{(d)}_{n'_s n_s : n'_s n_s}) \\ &= \tfrac{1}{2}(k_{n_s n'_s} + k_{n'_s n_s}) + \tfrac{1}{2}(\Gamma^{(d)}_{n_s n_s : n_s n_s} + \Gamma^{(d)}_{n'_s n'_s : n'_s n'_s}).\end{aligned} \tag{2.22}$$

The master equation derived in this section is completely general and can be applied to any relaxation process depending on the perturbation $\hat{H}'$, for example, electronic relaxation, vibrational relaxation, electron transfer, and other thermal activated processes (see Section V).

There is now a tendency in the literature to classify coupling terms and relaxation phenomena as dephasing processes (T_2 type), and relaxation of populations (T_1 type). In other words, $\Gamma_{m_s n_s : m_s n_s}$ given by Eq. (2.12) is rewritten as

$$1/T_2 = 1/2T_1 + 1/T'_2 = \Gamma_{m_s n_s : m_s n_s}, \tag{2.23}$$

where

$$1/T_1 = \Gamma_{n_s n_s : n_s n_s} + \Gamma_{m_s m_s : m_s m_s} \tag{2.24}$$

and

$$1/T'_2 = \Gamma^{(d)}_{m_s n_s : m_s n_s}. \tag{2.25}$$

That is, T_2 is the total dephasing decay time constant due to all processes which contribute to the diminution of the phase coherence, T_1 is the population decay time constant and T'_2 is the pure dephasing decay time constant resulting from the fluctuations in the environment to which the system is coupled. Such a classification is based on an analogy with such fields as that of magnetic resonance in condensed phases (Abragam, 1961), pressure broadening in dilute gases (Ben-Reuven, 1975), etc. A dephasing process is any process which causes the decay of a phase coherence (i.e., off-diagonal density matrix element discussed above) and is amenable to direct observation by various experimental techniques such as line shape and relaxation measurements in magnetic resonance (Abragam, 1961), absorption (Ben-Reuven, 1975; Gordon, 1968), etc. Other new experimental techniques that have recently been applied to dephasing phenomena are nonlinear, time-resolved spectroscopy (for example, induced Raman scattering) in liquid and solid phases (Laubereau and Kaiser, 1975, 1978); overtone vibrational spectroscopy of large molecules in gas, liquid, and solid phases (Swofford *et al.*, 1977; Bray and Berry, 1978; Albrecht, 1978; Henry, 1977); IR studies of multiphoton molecular processes (multiphoton absorption, laser chemistry, etc.) (Lee *et al.*, 1979); coherent transient experiments performed both in the gas phase and in low temperature solids (Genack *et al.*, 1976; de Vries *et al.*, 1977; Zewail, 1977; Orlowski and Zewail, 1979).

From the theoretical viewpoint discussed above, we can see that the true dephasing $\Gamma^{(d)}_{m_s n_s : m_s n_s}$ in T_2 is caused by diagonal couplings in the Hamiltonian which do not affect the populations in which we are interested, whereas T_1 originates from off-diagonal coupling terms in the Hamiltonian which affect both the populations and coherences. Thus the occurrence of the dephasing process depends on the way we partition the Hamiltonian and hence our choice of a basis set (Mukamel, 1978). In other words, the definition of T_2 is often not unique but rather depends to a large extent on our level of theoretical description, which in turn is determined by the type and quality of the available experimental information. The choice of the partition of the total system and the kind of basis set we adopt are often dictated by the experimental observables and they will inevitably influence our classification of the interaction Hamiltonians.

Recently, the nature of intramolecular dephasing processes in isolated polyatomic molecules has been studied by Mukamel (1978). He considers several experimental and theoretical approaches for studying intramolecular dynamics and discusses under what conditions it is necessary and useful to introduce explicitly dephasing interactions.

III. SPECTRAL LINE SHAPES

In the previous section we discussed the nature and mechanism of the relaxation and dephasing processes. In this section, we shall show the relation between the dephasing process and spectral line shape. One way to accomplish this will be to calculate the matrix elements of electric and/or magnetic moments. In the following we present the master equation approach to finding the optical properties. The starting point in this case is again the Liouville equation for an isolated system,

$$-\partial\hat{\rho}/\partial t = i\hat{L}\hat{\rho}. \tag{3.1}$$

For a system consisting of the molecular system, the heat bath, and the radiation field, $\hat{L}$ can be written as

$$\hat{L} = \hat{L}_S + \hat{L}_R + \hat{L}_L + \hat{L}'_R + \hat{L}'_{SL} \tag{3.2}$$

where $\hat{L}_S$, $\hat{L}_R$, and $\hat{L}_L$ represent the Liouville operators of the system, the radiation field, and the heat bath, respectively, and $\hat{L}'_R$ and $\hat{L}'_{SL}$ represent the interaction Liouville operator between the radiation field and the system plus the heat bath, and the interaction Liouville operator between the system and the heat bath. Notice that $\hat{L}'_{SL}$ here may be different from that in Section II depending on the nonradiative processes involved.

If at $t = 0$, $\hat{\rho}(0) = \hat{\rho}^{(L)}(0)\hat{\rho}^{(SR)}(0)$, then it has been shown that by tracing out the heat bath variables, we obtain (Lin and Eyring, 1977a)

$$\frac{\partial}{\partial t}\hat{\rho}^{(SR)}(t) = -i(\hat{L}_S + \hat{L}_R)\hat{\rho}^{(SR)}(t) + \int_0^t dz\,[\langle\hat{M}_L(z)\rangle + \langle\hat{M}_{SR}(z)\rangle]e^{-i(t-z)(\hat{L}_S+\hat{L}_R)}\hat{\rho}^{(SR)}(0). \tag{3.3}$$

If $\hat{M}_L(P)$ and $\hat{M}_{SR}(P)$ represent the Laplace transforms of $\hat{M}_L(t)$ and $\hat{M}_{SR}(t)$, respectively, then they are defined by

$$\hat{M}_L(P) = -i\hat{L}'_{SL} + (-i\hat{L}'_{SL})[1/P + i(\hat{L}_S + \hat{L}_R + \hat{L}_L + \hat{L}'_{SL})](-i\hat{L}'_{SL}) \tag{3.4}$$

and

$$\hat{M}_{SR}(P) = -i\hat{L}'_R + (-i\hat{L}'_R)[1/(P + i\hat{L})](-i\hat{L}'_R), \tag{3.5}$$

respectively. Notice that $\langle\hat{M}_L(t)\rangle = \mathrm{Tr}_L\,[\hat{M}_L(t)\hat{\rho}^{(L)}(0)]$ and $\langle\hat{M}_{SR}(t)\rangle = \mathrm{Tr}_L\,[\hat{M}_{SR}(t)\hat{\rho}^{(L)}(0)]$.

Using the Markoff approximation, from Eq. (3.3) we obtain the master equation for the system as

$$\frac{d}{dt}\rho^{(S)}_{n_s n_s}(t) = \sum_{a_s} K_{a_s n_s}\rho^{(S)}_{a_s a_s}(t) - \sum_{a_s} K_{n_s a_s}\rho^{(S)}_{n_s n_s}(t), \tag{3.6}$$

where

$$K_{a_s n_s} = K^{(0)}_{a_s n_s} + K^{(1)}_{a_s n_s} + K^{(2)}_{a_s n_s}, \tag{3.7}$$

with

$$K^{(0)}_{a_s n_s} = \frac{2}{\hbar_2} \sum_{n_b} \sum_{n'_b}{}' \rho^{(L)}_{n'_b n'_b}(0) | H^{(SL)}_{a_s n'_b n_s n_b} |^2 \frac{\Gamma_{a_s n_s}}{\Gamma^2_{a_s n_s} + \omega^2_{a_s n'_b : n_s n_b}}, \tag{3.8}$$

$$K^{(1)}_{a_s n_s} = \frac{2}{\hbar^2} \sum_{n_r} \sum_{a_r} \sum_{n_b} \sum_{n'_b}{}' \rho^{(R)}_{a_r a_r}(0) | H^{(R)}_{a_r a_s n'_b : n_r n_s n_b} |^2 \times \rho^{(L)}_{n'_b n'_b}(0) \Gamma_{a_s n_s} / (\Gamma^2_{a_s n_s} + \omega_{a_r a_s n'_b : n_r n_s n_b}), \tag{3.9}$$

and

$$K^{(2)}_{a_s n_s} = \frac{2}{\hbar^4} \sum_{a_r} \sum_{n_r} \sum_{n_b} \sum_{n'_b}{}' \rho^{(L)}_{n'_b n'_b}(0) \rho^{(R)}_{a_r a_r}(0) \frac{\Gamma_{n_s a_s}}{\Gamma^2_{n_s a_s} + \omega^2_{n_s n_r n_b : a_s a_r n'_b}} \times \left| \sum_{m_r} \sum_{m_s} \sum_{m_b} \frac{H^{(R)'}_{n_s n_r n_b : m_s m_r m_b} H^{(R)'}_{m_s m_r m_b : a_s a_r n'_b}}{i\omega_{m_s m_r m_b : n_s n_r n_b} + \Delta\Gamma_{m_s n_s}} \right|^2. \tag{3.10}$$

Notice that $\Delta\Gamma_{m_s n_s} = \Gamma_{m_s m_s} - \Gamma_{n_s n_s}$ and $\Gamma_{a_s n_s} = \Gamma_{a_s n_s : a_s n_s}$. A similar expression can be obtained for $K_{n_s a_s}$. In Eq. (3.7), $K^{(0)}_{a_s n_s}$ represents the zero-photon process (or the nonradiative relaxation process), $K^{(1)}_{a_s n_s}$ represents the one-photon process, and $K^{(2)}_{a_s n_s}$ represents the two-photon process. Using the relation (Heitler, 1954; Louisell, 1972) for absorption

$$H^{(R)}_{n_r n_s n_b : a_r a_s n'_b} = -(e/m)\sqrt{2\pi\hbar a_r/\omega_r}(\hat{\mathbf{e}}_r \cdot \mathbf{P}_{n_s n_b, a_s n'_b})\delta_{n_r, a_r - 1}, \tag{3.11}$$

where $\hat{\mathbf{e}}_r$ represents the unit vector of polarization of light and frequency ω_r, Eqs. (3.9) and (3.10) can be written as

$$K^{(1)}_{a_s n_s} = \sum_{n_b} \sum_{n'_b} \sum_{a_r} \rho^{(R)}_{a_r a_r}(0) \rho^{(L)}_{n'_b n'_b}(0) \frac{4\pi e^2}{m^2 \hbar \omega_r} | \hat{\mathbf{e}}_r \cdot \mathbf{P}_{a_s n'_b : n_s n_b} |^2 \times \left[\frac{a_r \Gamma_{a_s n_s}}{\Gamma^2_{a_s n_s} + (\omega_r + \omega_{a_s n'_b : n_s n_b})^2} + \frac{(a_r + 1)\Gamma_{a_s n_s}}{\Gamma^2_{a_s n_s} + (\omega_{a_s n'_b : n_s n_b} - \omega_r)^2} \right], \tag{3.12}$$

and

$$K^{(2)}_{a_s n_s} = \sum_{a_r} \sum_{n_b} \sum_{n'_b}{}' \rho^{(R)}_{a_r a_r}(0) \rho^{(L)}_{n'_b n'_b} \frac{8\pi^2 e^4 a_{r_1}(a_{r_2} + 1)}{\hbar^2 \omega_{r_1} \omega_{r_2} m^4} \times \frac{\Gamma_{a_s n_s}}{\Gamma^2_{a_s n_s} + (\omega_{n_s n_b : a_s n'_b} - \omega_{r_1} + \omega_{r_2})^2} \times \left| \sum_{m_s} \sum_{m_b} \left[\frac{(\hat{\mathbf{e}}_{r_1} \cdot \mathbf{P}_{n_s n_b, m_s m_b})(\hat{\mathbf{e}}_{r_2} \cdot \mathbf{P}_{m_s m_b, a_s n'_b})}{i(\omega_{m_s m_b : n_s n_b} + \omega_{r_1}) + \Delta\Gamma_{m_s n_s}} + \frac{(\hat{\mathbf{e}}_{r_2} \cdot \mathbf{P}_{n_s n_b, m_s m_b})(\hat{\mathbf{e}}_{r_1} \cdot \mathbf{P}_{m_s m_b, a_s n'_b})}{i(\omega_{m_s m_b : n_s n_b} - \omega_{r_2}) + \Delta\Gamma_{m_s n_s}} \right] \right|^2. \tag{3.13}$$

Here in the calculation of $K^{(2)}_{a_s n_s}$ we have assumed that we are dealing only with the scattering process $(a_s, a_{r_1}, a_{r_2}) \to (n_s, a_{r_1} - 1, a_{r_2} + 1)$. Similarly, we have

$$K^{(2)}_{n_s a_s} = \sum_{n_r} \sum_{n_b} \sum_{n_b'}{}' \rho^{(R)}_{n_r n_r}(0) \rho^{(L)}_{n_b n_b}(0) \frac{8\pi^2 e^4 (n_{r_1} + 1) n_{r_2}}{\hbar^2 \omega_{r_1} \omega_{r_2} m^4}$$

$$\times \frac{\Gamma_{a_s n_s}}{\Gamma^2_{a_s n_s} + (\omega_{n_s n_b : a_s n_b'} - \omega_{r_1} + \omega_{r_2})^2} \left| \sum_{m_s} \sum_{m_b} \left[\frac{(\hat{\mathbf{e}}_{r_1} \cdot \mathbf{P}_{n_s n_b, m_s m_b})(\hat{\mathbf{e}}_{r_2} \cdot \mathbf{P}_{m_s m_b, a_s n_b'})}{i(\omega_{m_s m_b : n_s n_b} + \omega_{r_1}) + \Delta\Gamma_{m_s n_s}} \right.\right.$$

$$\left.\left. + \frac{(\hat{\mathbf{e}}_{r_2} \cdot \mathbf{P}_{n_s n_b, m_s m_b})(\hat{\mathbf{e}}_{r_1} \cdot \mathbf{P}_{m_s m_b, a_s n_b'})}{i(\omega_{m_s m_b : n_s n_b} - \omega_{r_2}) + \Delta\Gamma_{m_s n_s}} \right] \right|^2. \tag{3.14}$$

The physical meaning of each term in Eqs. (3.6)–(3.14) is clear. For example, let us consider $K_{a_s n_s}$ in Eq. (3.7); it represents the rate constant for the transition $a_s \to n_s$ and consists of three terms $K_{a_s n_s} = K^{(0)}_{a_s n_s} + K^{(1)}_{a_s n_s} + K^{(2)}_{a_s n_s}$ in this case. $K^{(0)}_{a_s n_s}$ denotes the rate constant for the nonradiative transition $a_s \to n_s$, $K^{(1)}_{a_s n_s}$ denotes the rate constant for the one-photon radiative transition $a_s \to n_s$; and $K^{(2)}_{a_s n_s}$ denotes the rate constant for the two-photon radiative transition $a_s \to n_s$. Thus for the Raman scattering measurement, $K^{(1)}_{a_s n_s}$ can be neglected and $K^{(2)}_{a_s n_s}$ will give us the Raman scattering cross section. Similarly, we can see that $K_{n_s a_s} = K^{(0)}_{n_s a_s} + K^{(1)}_{n_s a_s} + K^{(2)}_{n_s a_s}$ and $K^{(2)}_{n_s a_s}$ will give us the inverse Raman absorption cross section.

The main features of the master equation approach presented above can be summarized as follows. It can describe the time evolution of the system undergoing the optical processes for the time range in which the memory effect is negligible. The damping constants are clearly defined; not only the relaxation rate constants but also the elastic rate constants have to be included. Because of the fact that the damping effect is taken into account, the resonance optical phenomena can easily be treated and the temperature effect on the optical phenomena can be studied. Due to the fact that the master equation satisfies the principle of detailed balance, the normal effect and the inverse effect of an optical phenomenon can be obtained simultaneously; for example, as discussed above the ordinary Raman effect and the inverse Raman effect can be obtained directly from the master equation.

It should be noted that although we have derived only the master equation for zero-photon processes, one-photon processes, and the Raman effect, the same approach can be employed to derive the master equation for other optical phenomena.

Let us derive the expression for the one-photon absorption coefficient. For this purpose, we use Eq. (3.12). Expressing a_r in terms of the light intensity (Heitler, 1954), the absorption coefficient for the transition $a_s \to n_s$ is given by

$$\alpha^{(1)}_{a_s n_s} = \frac{4\pi\omega_r}{\hbar c} \sum_{n_b} \sum_{n_b'} \rho^{(L)}_{n_b' n_b'}(0) |\hat{\mathbf{e}}_r \cdot \mathbf{R}_{n_s n_b, a_s n_b'}|^2$$

$$\times \left[\frac{\Gamma_{a_s n_s}}{\Gamma^2_{a_s n_s} + (\omega_r - \omega_{n_s n_b, a_s n_b'})^2} + \frac{\Gamma_{a_s n_s}}{\Gamma^2_{a_s n_s} + (\omega_r + \omega_{n_s n_b : a_s n_b'})^2} \right]. \tag{3.15}$$

Here the following relation has been used:

$$\mathbf{P}_{n_s n_b, a_s n_b'} = (\text{Im } \omega_{n_s n_b, a_s n_b'}/e)\mathbf{R}_{n_s n_b, a_s n_b'}. \tag{3.16}$$

The second term inside the square bracket of Eq. (3.15) is usually negligible compared with the first term. In the absorption spectrum measurement, one does not usually measure the absorption coefficient from one particular level of the system to another. In other words, the distribution of molecules in the initial states should be taken into account, i.e.,

$$\alpha^{(1)} = \sum_{a_s} \sum_{n_s} \rho_{a_s a_s}^{(S)}(0) \alpha_{a_s n_s}^{(1)}. \tag{3.17}$$

In a condensed medium, the solvent effect due to the effective field and effective light velocity should be considered (Basu, 1964). For an isotropic system, one has to carry out the spatial average of $\alpha_{a_s n_s}^{(1)}$ and $\alpha^{(1)}$; in this case, we replace $|\hat{\mathbf{e}}_r \cdot \mathbf{R}_{n_s n_b, a_s n_b'}|^2$ by $\frac{1}{3}|\mathbf{R}_{n_s n_b, a_s n_b'}|^2$.

Next we consider the emission process. In this case, we start with

$$K_{n_s a_s}^{(1)} = \sum_{n_b} \sum_{n_b'} \sum_{n_r} \rho_{n_r n_r}^{(R)}(0) \rho_{n_b n_b}^{(L)}(0) \frac{4\pi e^2}{m^2 \hbar \omega_r} \times |\hat{\mathbf{e}}_r \cdot \mathbf{P}_{a_s n_b', n_r n_b}|^2 \frac{\Gamma_{a_s n_s}}{\Gamma_{a_s n_s}^2 + (\omega_{n_s n_b : a_s n_b'} - \omega_r)^2} \tag{3.18}$$

Using Eq. (3.16) and the density of photon modes $\omega_r^2/\pi^2 c^3$ and carrying out the integration with respect to ω_r yields

$$K_{n_s a_s}^{(1)} = \frac{4}{\hbar c^3} \sum_{n_b} \sum_{n_b'} \rho_{n_b n_b}^{(L)}(0) |\hat{\mathbf{e}}_r \cdot \mathbf{R}_{a_s n_b', n_s n_b}|^2 \omega_{n_s n_b : a_s n_b'}^3. \tag{3.19}$$

Here only the spontaneous contribution to the emission process is retained. Again, if the molecules are randomly oriented, then a spatial average is required. The above expression $K_{n_s a_s}^{(1)}$ is the radiative rate constant for the transition $n_s \to a_s$. If the distribution of the initial states of the system is described by $\rho_{n_s n_s}^{(S)}(0)$, then the observed radiative rate constant is given by

$$K^{(1)} = \sum_{a_s} \sum_{n_s} \rho_{n_s n_s}^{(S)}(0) K_{n_s a_s}^{(1)}. \tag{3.20a}$$

The single vibronic level radiative rate constant is given by (Fleming *et al.*, 1973)

$$K_{n_s}^{(1)} = \sum_{a_s} K_{n_s a_s}^{(1)}. \tag{3.20b}$$

The intensity of the emission spectra for the transition $n_s \to a_s$ is proportional to

$$I_{n_s a_s}^{(1)}(\omega_r) = \frac{4\omega_r^3}{\pi \hbar c^3} \sum_{n_b} \sum_{n_b'} \rho_{n_b n_b}^{(L)}(0) |\hat{\mathbf{e}}_r \cdot \mathbf{R}_{a_s n_b', n_s n_b}|^2 \frac{\Gamma_{a_s n_s}}{\Gamma_{a_s n_s}^2 + (\omega_{n_s n_b, a_s n_b'} - \omega_r)^2}. \tag{3.21}$$

Taking into account the initial distribution $\rho^{(S)}_{n_s n_s}(0)$, we find

$$I^{(1)}(\omega_r) = \sum_{n_s} \sum_{a_s} I^{(1)}_{n_s a_s}(\omega_r) \rho^{(s)}_{n_s n_s}(0). \tag{3.22}$$

Now we consider the two-photon processes. Here we shall discuss only the Raman effect and the inverse Raman effect. Two-photon absorption and emission can be treated in a similar manner as that for one-photon absorption and emission presented above and will not be discussed here.

As mentioned before, $K^{(2)}_{a_s n_s}$ will provide us the transition probability of $a_s \to n_s$ of the Raman effect. In the ordinary Raman scattering measurement, the distribution of molecules in the initial states should be taken into account. Thus, the observed transition probability of the Raman scattering can be expressed as

$$K^{(2)} = \sum_{a_s} \sum_{a_r} \sum_{n_s} \sum_{n_b} \sum_{n_b'}{}' \rho^{(S)}_{a_s a_s}(0) \rho^{(R)}_{a_r a_r}(0) \rho^{(L)}_{n_b' n_b'}(0) \frac{8\pi^2 e^4 a_{r_1}(a_{r_2} + 1)}{\hbar^2 \omega_{r_1} \omega_{r_2} m^4}$$
$$\times \frac{\Gamma_{a_s n_s}}{\Gamma^2_{a_s n_s} + (\omega_{n_s n_b : a_s n_b'} - \omega_{r_1} + \omega_{r_2})^2} |M_{n_s n_b, a_s n_b'}|^2, \tag{3.23}$$

where

$$M_{n_s n_b, a_s n_b'} = \sum_{m_s} \sum_{m_b} \left[\frac{(\hat{\mathbf{e}}_{r_1} \cdot \mathbf{P}_{n_s n_b, m_s m_b})(\hat{\mathbf{e}}_{r_2} \cdot \mathbf{P}_{m_s m_b, a_s n_b'})}{i(\omega_{m_s m_b : n_s n_b} + \omega_{r_1}) + \Delta\Gamma_{m_s n_s}} \right.$$
$$\left. + \frac{(\hat{\mathbf{e}}_{r_2} \cdot \mathbf{P}_{n_s n_b, m_s m_b})(\hat{\mathbf{e}}_{r_1} \cdot \mathbf{P}_{m_s m_b, a_s n_b'})}{i(\omega_{m_s m_b : n_s n_b} - \omega_{r_2}) + \Delta\Gamma_{m_s n_s}} \right]. \tag{3.24}$$

$K^{(2)}$ can be converted into the expression for the scattering cross section,

$$\frac{d^2\sigma}{d\Omega d\omega_2} = \frac{e^4 \omega_2 (a_{r_2} + 1)}{\pi \hbar^2 m^4 c^4 \omega_1} \sum_{a_s} \sum_{n_s} \sum_{n_b} \sum_{n_b'} \rho^{(S)}_{a_s a_s}(0) \rho^{(L)}_{n_b' n_b'}(0)$$
$$\times \frac{\Gamma_{a_s n_s}}{\Gamma^2_{a_s n_s} + (\omega_{n_s n_b : a_s n_b'} - \omega_{r_1} + \omega_{r_2})^2} |M_{n_s n_b, a_s n_b'}|^2. \tag{3.25}$$

For the spontaneous Raman scattering, $a_{r_2} = 0$. For the resonance Raman effect, the first term in Eq. (3.24) can be neglected. Equation (3.25) has been employed by Fujimura and Lin (1979a, b) to study the temperature effect, the high pressure effect and the multimode effect on the resonance Raman scattering.

Notice that $K^{(2)}_{n_s a_s}$ provides the transition probability of the inverse Raman effect for $n_s \to a_s$, which can be converted into the absorption coefficient as

$$\alpha^{(2)}_{n_s a_s} = \sum_{n_{r_1}} \sum_{n_b} \sum_{n_b'} \rho^{(R)}_{n_{r_1} n_{r_1}}(0) \rho^{(L)}_{n_b n_b}(0) \frac{\Gamma_{a_s n_s}}{\Gamma^2_{a_s n_s} + (\omega_{n_s n_b : a_s n_b'} - \omega_{r_1} + \omega_{r_2})^2}$$
$$\times \frac{8\pi^2 e^4 (n_{r_1} + 1)}{\hbar^2 c \omega_{r_1} \omega_{r_2} m^4} |M_{n_s n_b, a_s n_b'}|^2. \tag{3.26}$$

Taking into account the effect of the initial distribution of the system, the observed absorption coefficient is given by

$$\alpha^{(2)} = \sum_{a_s}\sum_{n_s} \rho^{(S)}_{n_s n_s}(0)\alpha^{(2)}_{n_s a_s} \tag{3.27}$$

or

$$\alpha^{(2)} = \frac{8\pi^2 e^4 I_1(\omega_{r_1})}{\hbar^3 c^2 \omega_{r_1}^2 \omega_{r_2} m^4} \sum_{a_s}\sum_{n_s}\sum_{n_b}\sum_{n'_b} \rho^{(S)}_{n_s n_s}(0)\rho^{(L)}_{n_b n_b}(0)$$
$$\times \frac{\Gamma_{a_s n_s}}{\Gamma^2_{a_s n_s} + (\omega_{n_s n_b : a_s n'_b} - \omega_{r_1} + \omega_{r_2})^2} |M_{n_s n_b, a_s n'_b}|^2. \tag{3.28}$$

In particular, if $\Gamma_{a_s n_s} \to 0$, then Eq. (3.28) reduces to

$$\alpha^{(2)} = \frac{8\pi^3 e^4 I_1(\omega_{r_1})}{\hbar^3 c^2 \omega_{r_1}^2 \omega_{r_2} m^4} \sum_{a_s}\sum_{n_s}\sum_{n_b}\sum_{n'_b} \rho^{(S)}_{n_s n_s}(0)\rho^{(L)}_{n_b n_b}(0)$$
$$\times |M_{n_s n_b, a_s n'_b}|^2 \delta(\omega_{n_s n_b : a_s n'_b} - \omega_{r_1} + \omega_{r_2}). \tag{3.29}$$

Notice the similarity between Eq. (3.29) and the expression for one-phonon absorption coefficient,

$$\alpha^{(1)} = \frac{4\pi^2\omega_r}{\hbar c} \sum_{a_s}\sum_{n_s}\sum_{n_b}\sum_{n'_b} \rho^{(S)}_{a_s a_s}(0)\rho^{(L)}_{n'_b n'_b}(0) \, |\hat{\mathbf{e}}_r \cdot \mathbf{R}_{n_s n_b, a_s n'_b}|^2 \delta(\omega_r - \omega_{n_s n_b : a_s n'_b}), \tag{3.30}$$

which has been obtained from Eq. (3.17) by neglecting the term

$$\Gamma_{a_s n_s}/[\Gamma^2_{a_s n_s} + (\omega_r + \omega_{n_s n_b : a_s n'_b})^2].$$

In the above discussion, we have shown the role played by the relaxation process and the dephasing process in determining the line shape functions of optical phenomena. In the following we shall briefly review the recent experimental activity which involves the determination of the rate constants from the spectral linewidths.

From the experimental viewpoint, the linewidths of optical transitions in condensed phases at low temperatures are often inhomogeneously broadened by microscopic and macroscopic strain effects and so usually can be used to determine only lower limits for relaxation times. At higher temperatures the lines may become homogeneous but the linewidth need no longer signify the population relaxation rate. Thus the temperature dependence of the optical linewidth and lineshift need not directly yield information on the variations of the population decay (Hochstrasser and Nyi, 1979). At higher temperatures the main contribution to the linewidth of optical transitions is expected to be the fluctuations in the nuclear coordinates. This is the case for absorption spectra but not necessarily for fluorescence spectra observed in the presence of the exciting field (Hochstrasser and Novak, 1978). Thus a cw spectroscopic investigation of vibrational relaxation by single vibronic level fluorescence requires the unraveling of various coherent and incoherent relaxation effects.

In the last few years, the spectra of vibrational overtones (at ~15,000–20,000 cm^{-1}) in large molecules have received considerable attention. The focus is on three problems dealing with the origin of relaxation at such high energies, the association of spectral band positions with the local modes in molecules, and the relevance of these spectra to possible selectivity in laser-induced chemistry. In liquid benzene and in several other aromatic liquids, typically the fifth overtone is around 6000 Å and the width (FWHM) is ~300 cm^{-1}. A similar transition has been found in gaseous benzene by Reddy *et al.* (1978) but with width of 99 cm^{-1} ascribed to broadening by intramolecular relaxation.

Recently Perry and Zewail (1979) have reported on the observation of CH stretching overtones ($v = 2$–6) in solid naphthalene. The experiments offer the following: (a) the molecules can be frozen to low temperatures (1.3 K) so that the contribution of rotational states to the width of the resonance may be eliminated, (b) at 1.3 K the excitation is from $v = 0$ and hence spectral congestion can be avoided, (c) unlike gases and liquids polarization spectroscopy in solids assigns the bands, and finally, (d) the inhomogeneous broadening may be realized by comparing the width of the overtone to that of nearby crystal electronic states. The line shape at 1.3 K can be fit near perfectly to Lorentzians and not to Gaussians. The width of the Lorentzians compares well with the benzene data. They have made preliminary analysis on the effect of crystal modes on these line shapes and find that although the line will narrow at 1.3 K relative to 300 K, the asymmetry of the line is a measure of the homogeneous width $(\pi T_2)^{-1}$, where T_2 is the total dephasing time. If T_2 is long (~10 psec) the line is asymmetric, while if T_2 is short (~0.3 psec), the line is essentially symmetric. With this in mind they infer that $T_2 \doteqdot 0.1$ psec and that the overtone transition in the solid (no rotation) is a limiting intrinsic one, resembling the equivalent CH modes of benzene in the gas phase. What remains to be determined is the contribution of T_1 to T_2.

Using newly developed techniques with extremely cold supersonic molecular beams, Beck *et al.* (1979) claim that it is possible to measure these vibrational redistribution processes quite clearly in a variety of polyatomic systems. As evidence they offer initial results from the following experiment with naphthalene.

For the spectra obtained by Beck *et al.*, naphthalene at a concentration of ~100 ppm in the He was expanded from a pressure of ~16 atm through a simple 0.1 cm diam orifice into a large chamber where the background pressure was held below 5×10^{-6} torr. The resultant freely expanding supersonic jet was crossed by a probe laser beam 20 cm downstream from the nozzle orifice. Total undispersed fluorescence intensity is recorded as a function of laser frequency.

Band contour analysis of the partially resolved rotational structure for the $\bar{8}_0^1$ band of the $^1Ag \rightarrow {}^1B_{3u}$ ($\pi\pi^*$) transition shows that the rotational temperature is ~0.2 K and vibrational cooling is complete. The experimental spectral bandwidth (laser width plus effective Doppler width) was 0.05 cm^{-1} FWHM throughout these experiments.

The wave packet produced by coherent excitation of any such band will dephase in time at a rate given by

$$k_{vr} = 2\pi c\Delta,$$

where Δ represents the FWHM broadening. Approximate values for such a calculation are

$$\begin{array}{llll} E_v\,(cm^{-1}) = & 4296, & 3765, & 3069; \\ \Delta\,(cm^{-1}) = & 2.10, & 0.96, & 0.47; \\ k_{vr}\,(sec^{-1}) = & 4 \times 10^{11}, & 2 \times 10^{11}, & 9 \times 10^{10}; \end{array}$$

where E_v is the measured vibrational energy from the 0–0 transition at 32018.5 cm^{-1}, and is accurate to ± 5 cm^{-1}. Since the initial wave packet is a well-defined localized vibrational excitation, its dephasing may correspond to a vibrational redistribution and k_{vr} is a vibrational redistribution rate.

Recently, Smalley *et al.* (1976) have reported the fluorescence excitation spectrum of the HeI_2 van der Waals complex and the determination of its vibrational predissociation rates from the spectral linewidth. The HeI_2 van der Waals complex was prepared from a dilute mixture of I_2 in He at a pressure of 100 atm by supersonic expansion through a nozzle into a vacuum. Laser-induced fluorescence excitation spectra were recorded for the $\tilde{X} \rightarrow \tilde{B}$ transition of HeI_2 as well as corresponding spectra for the He_2I_2 and I_2 in the expanding gas. I_2 was found to be cooled by the expansion to a rotational temperature of 0.4 K and a vibrational temperature of 50 K. Similarly, cold internal temperatures were attained by the van der Waals complexes.

Evidence was found for vibrational predissociation of the HeI_2 complex in both the $\tilde{X}$ and $\tilde{B}$ electronic states. The vibrational predissociation rate was found to depend weakly upon the degree of excitation of the I–I stretching mode v_1. For $v_1 = 1$ in the $\tilde{X}$ state the predissociation rate was found to be greater than 5×10^6 sec^{-1}. In the $\tilde{B}$ state the vibrational predissociation rate is $\sim 5 \times 10^{11}$ sec^{-1} for $v_1 = 27$, decreasing to $< 5 \times 10^9$ sec^{-1} for $v_1 \leq 7$. The small (3.4–4.0 cm^{-1}) blue shifts of the vibronic bands of the HeI_2 spectrum relative to the corresponding bands of I_2 indicate (1) the van der Waals complex is slightly more strongly bound in the $\tilde{X}$ state than it is in the $\tilde{B}$ state and (2) the I–I bonding in both the $\tilde{X}$ and $\tilde{B}$ states of I_2 is largely unaffected by the formation of the van der Waals bond with He.

All of the optical transitions that they have been able to examine for the HeI_2 molecule involve an upper state which may predissociate either (1) electronically to form two ground state iodine $^2P_{1/2}$ atoms and a $^1S_0$ He or (2) vibrationally to form a $\tilde{B}$ state I_2 and He. The first possibility would cause a quenching of fluorescence with a consequent loss of intensity in fluorescence excitation spectrum. The second possibility could be noticeable in the spectrum only when the (vibrational) predissociation rate becomes large enough to cause a measurable broadening of the spectral lines.

Smalley *et al.* have obtained the high resolution (± 100 MHz) fluorescence excitation spectrum of the 7–0, 16–0, and 27–0 bands of HeI_2. These bands display a clear v-dependent spectral broadening. This broadening increases to ~ 2 GHz for the 16–0 band, ~ 3 GHz for the 21–0 band, and ~ 10 GHz for the 27–0 band, where the rotational structure is almost completely observed. This broadening of the 27–0 band corresponds to a predissociation rate of $\sim 5 \times 10^{10}$ sec^{-1}, which is roughly four orders of magnitude faster than the radiative decay rate of the $\tilde{B}$ state of I_2. Evidence for vibrational predissociation of the ground electronic state of HeI_2 comes from the fact that they have observed no features attributable to HeI_2 complex associated with hot bands in the I_2 spectrum. The complex is formed as a result of the three-body collisions which occur in substantial number only in the very early stages of the expansion. Since the stream velocity is $\sim 2 \times 10^5$ cm/sec, the inability to observe, say, the 17–1 band of HeI_2 0.5 mm downstream of the nozzle indicates that any $\tilde{X}$ state $v = 1$ complexes that are formed do not live longer than $\sim 2 \times 10^{-7}$ sec before they vibrationally predissociate.

IV. VIBRATIONAL RELAXATION

The purpose of this section is to review experimental and theoretical investigations of vibrational relaxation and dephasing of polyatomic molecules in dense media and in the isolated molecule condition. Progress in the theoretical study of vibrational relaxation and vibrational dephasing of diatomic molecules and small molecules in dense media has been thoroughly reviewed recently by Diestler (1976, 1979) and Oxtoby (1979). Experimental investigations of vibrational relaxation and vibrational energy transfer of small molecules in dense media have also been extensively reviewed lately (Legay, 1978; Blumen *et al.*, 1978; Brus and Bondybey, this volume).

A. Master Equations

This chapter has stressed that to describe the population relaxation of a system it is convenient to employ the density matrix method, i.e., to find the master equation for the system. Here we shall present the derivation of the master equations of vibrational relaxation which are of general interest. The theoretical approach is general and other special cases can be treated in a similar manner. For convenience, we shall suppress the radiative process in the derivation (see Section III; Blumen *et al.*, 1978).

We shall assume that the total density matrix $\hat{\rho}(0)$ can be written as a product of the density matrices of the system and heat bath $\hat{\rho}(0) = \hat{\rho}^{(S)}(0)\hat{\rho}^{(b)}(0)$ with $\hat{\rho}^{(b)}(0)$ being expressed in the equilibrium Boltzmann distribution. It follows that the master equation of the system can be expressed as (Lin, 1974)

$$\frac{d\rho_{m_s m_s}}{dt} = \sum_{n_s} K_{m_s n_s} \exp(\beta E_{m_s})[\exp(-\beta E_{m_s})\rho_{n_s n_s} - \rho_{m_s m_s} \exp(-\beta E_{n_s})], \quad (4.1)$$

where $\beta = 1/kT$ and

$$K_{m_s n_s} = \frac{2\pi}{\hbar} \sum_{n_d} \sum_{m_b} \rho_{n_b n_b} |\langle m_s m_b | \hat{H}' | n_s n_b \rangle|^2 \, \delta(E_{m_s m_b} - E_{n_s n_b}) \quad (4.2)$$

with $\rho_{n_b n_b} = \exp(-\beta E_{n_b})/Q_b(\beta)$ and $Q_b(\beta) = \sum_{n_b} \exp(-\beta E_{n_b})$, the partition function of the heat bath.

First we consider the case in which the molecular oscillator of the system interacts linearly with the heat bath (or medium):

$$\hat{H}' = QF_b, \quad (4.3)$$

where Q represents the normal coordinate of the molecular oscillator and F_b plays the role of a force acting on the oscillator through the medium. In this case, Eq. (4.1) becomes (Fleming *et al.*, 1974)

$$\begin{aligned} d\rho_{m_s}/dt = K\{&(m_s + 1)\rho_{m_s+1} - [m_s + (m_s + 1)\exp(-\beta\hbar\omega_s)]\rho_{m_s} \\ &+ m_s \exp(-\beta\hbar\omega_s)\rho_{m_s-1}\}, \end{aligned} \quad (4.4)$$

where $\rho_{m_s m_s} \equiv \rho_{m_s}$ and

$$K = \frac{\pi}{\omega_s} \sum_{n_b} \sum_{m_b} \rho_{n_b} |\langle m_b | F_b | n_b \rangle|^2 \delta(E_{m_b} - E_{n_b} - \hbar\omega_s). \quad (4.5)$$

Equation (4.4) is known as the master equation of the step-ladder model (Montroll and Shuler, 1957). The solution of this type or master equation has been reached by Montroll and Shuler for various initial conditions.

Notice that for the temperature range $\hbar\omega_s/kT \gg 1$, the master equation given by Eq. (4.4) reduces to

$$d\rho_{m_s}/dt = K[(m_s + 1)\rho_{m_s+1} - m_s \rho_{m_s}]. \quad (4.6)$$

This equation can easily be solved. For example, if the system is initially prepared at the u_s state, then the solution of the above equation is given by

$$\rho_{m_s}(t) = \rho_{u_s}(0) \frac{u_s!}{m_s!(u_s - m_s)!} [1 - \exp(-Kt)]^{u_s - m_s} \exp(-m_s Kt), \quad (4.7)$$

where $\rho_{u_s}(0) = 1$, the initial population at the u_s state. This type of solution has been employed by Fujimura *et al* (1979) to analyze the time resolved absorption spectra of naphthalene reported by Schröder *et al.* (1978) to obtain the information of the time evolution of the triplet T_1 state after the intersystem crossing $S_1 \rightarrow T_1$.

So far we have considered the case in which the molecular oscillator can lose one quantum to the heat bath or gain one quantum from the heat bath during vibrational relaxation. Next we consider another possibility in which the vibrational energy of one molecular oscillator is allowed to flow to another during vibrational relaxation. In this case, in addition to the interaction Hamiltonian given by Eq. (4.3) for each oscillator to describe its relaxation to equilibrium, we need the interaction Hamiltonian to describe the coupling between the two oscillators,

$$\hat{H}' = Q_1 Q_2 F_{12\mathrm{b}}. \tag{4.8}$$

Here we have assumed that only one quantum can be transferred between the two oscillators.

Substituting the interaction Hamiltonians given by Eqs. (4.3) and (4.8) into Eq. (4.1), we obtain (Lin, 1974)

$$\begin{aligned} \frac{d\rho_{m_1 m_2}}{dt} = {} & \sum_{ij=1}^{2} K_i\{(m_i + 1)\rho_{m_i+1,m_j} - [m_i + (1 + m_i)\exp(-\beta\hbar\omega_i)]\rho_{m_i m_j} \\ & + m_i \exp(-\beta\hbar\omega_i)\rho_{m_i-1,m_j}\} \\ & + K_{12}((m_1 + 1)m_2\{\rho_{m_1+1,m_2-1} - \rho_{m_1 m_2}\exp[\beta\hbar(\omega_2 - \omega_1)]\} \\ & + m_1(m_2 + 1)\{\rho_{m_1-1,m_2+1}\exp[\beta\hbar(\omega_2 - \omega_1)] - \rho_{m_1 m_2}\}), \end{aligned} \tag{4.9}$$

$$K_{12} = \frac{\pi\hbar}{2\omega_1\omega_2}\sum_{n_\mathrm{b}}\sum_{m_\mathrm{b}} \rho_{n_\mathrm{b}} |\langle m_\mathrm{b}|F_{12_\mathrm{b}}|n_\mathrm{b}\rangle|^2 \delta(\hbar\omega_2 - \hbar\omega_1 + E_{m_\mathrm{b}} - E_{n_\mathrm{b}}). \tag{4.10}$$

In other words, the vibrational excitation can directly give up its excess energy to the heat bath; it can also redistribute among the molecular oscillators before dissipating its excess energy to the heat bath through low frequency molecular oscillators. The multimode relaxation described by Eq. (4.9) can be applied to both the vibrational relaxation of polyatomic molecules in condensed phases and the collisional vibrational relaxation of polyatomic molecules in gases.

For $\beta\hbar(\omega_1 - \omega_2) \gg 1$, Eq. (4.9) reduces to

$$\begin{aligned} \frac{d\rho_{m_1 m_2}}{dt} = {} & \sum_{i,j=1}^{2} K_i[(m_i + 1)\rho_{m_i+1,m_j} - m_i\rho_{m_i m_j}] + K_{12}[(m_1 + 1)m_2\rho_{m_1+1,m_2-1} \\ & - m_1(m_2 + 1)\rho_{m_1 m_2}]. \end{aligned} \tag{4.11}$$

Here we have assumed that $\omega_1 > \omega_2$. To demonstrate the behavior of the solution of this master equation, let us consider the case that at $t = 0$ only $\rho_{10} \neq 0$. In this case, we have

$$d\rho_{10}/dt = -(K_1 + K_{12})\rho_{10}, \tag{4.12}$$

$$d\rho_{01}/dt = -K_2\rho_{01} + K_{12}\rho_{10}, \tag{4.13}$$

$$d\rho_{00}/dt = K_1\rho_{10} + K_2\rho_{01}, \tag{4.14}$$

which can easily be solved,

$$\rho_{10}(t) = \rho_{10}(0)\exp[-(K_1 + K_{12})t], \tag{4.15}$$

and

$$\rho_{01}(t) = [K_{12}/(K_1 + K_{12} - K_2)]\rho_{10}(O) \times \{\exp(-K_2 t) - \exp[-t(K_1 + K_{12})]\}, \tag{4.16}$$

where $\rho_{10}(0) = 1$. Other cases can be treated similarly.

The types of master equation for vibrational relaxation given by Eqs. (4.4) and (4.9) are general and can be applied to describe the vibrational relaxation in three phases; the difference lies in the rate constants K_i and K_{ij}, which in turn are determined by the choice of the interaction Hamiltonian.

As can be seen from the above derivation, to derive a master equation for the vibrational relaxation it is necessary to decide the interaction Hamiltonian $\hat{H}'$ first; and then the corresponding master equation can be derived by substituting this $\hat{H}'$ into Eq. (4.1).

B. Vibrational Relaxation in Dense Media

In Eqs. (4.1)–(4.7) and Eqs. (4.8)–(4.16), we have demonstrated how to obtain the rate constant of vibrational relaxation once we know the coupling (F_b or F_{12b}) between the relaxing mode and accepting modes. In the following, we shall discuss two general models for calculating the rate constant of vibrational relaxation in condensed media.

The first model is based on the assumption that the interatomic interaction in $\hat{H}'$ is of the Morse-type potential (Berkowitz and Gerber, 1977; Lin, 1977; Lin *et al.*, 1979). For convenience of discussion, we shall retain only the stort-range repulsive portion of the intermolecular force (Sun and Rice, 1965). In this case, we have (Lin, 1974; Nitzan *et al.*, 1975)

$$F_b = C\exp\left(-a\sum_i \gamma_i Q_i\right). \tag{4.17}$$

Here only the dominating term is retained and the instantaneous displacement has been expressed in terms of normal coordinates of the accepting modes. In Appendix II, the case in which the molecular rotation is an important accepting mode is discussed.

Substituting Eq. (4.17) into Eq. (4.5) yields

$$K = \frac{\pi C^2}{\omega_s} \sum_{m_b} \sum_{n_b} \rho_{n_b} \prod_i |\langle m_i| \exp(-a\gamma_i Q_i)|n_i\rangle|^2 \delta(E_{m_b} - E_{n_b} - \hbar\omega_s), \quad (4.18)$$

or

$$K = \frac{C^2}{2\hbar\omega_s} \int_{-\infty}^{\infty} dt\, e^{-it\omega_s} \prod_i H_i(t), \quad (4.19)$$

where

$$H_i(t) = \sum_{n_i} \sum_{m_i} \rho_{n_i} |\langle m_i| \exp(-a\gamma_i Q_i)|n_i\rangle|^2 \exp[it\omega_i(m_i - n_i)]. \quad (4.20)$$

Using the Slater sum (Markham, 1959), Eq. (4.20) can be simplified as

$$H_i(t) = \exp\left\{\frac{a^2\gamma_i^2}{2\beta_i^2}\left[\coth\frac{\hbar\omega_i}{2kT} + \operatorname{csch}\frac{\hbar\omega_i}{2kT}\cosh\left(it\omega_i + \frac{\hbar\omega_i}{2kT}\right)\right]\right\} \quad (4.21)$$

where $\beta_i^2 = \omega_i/\hbar$. Substituting Eq. (4.21) into Eq. (4.19), we obtain

$$K = \frac{C^2}{2\hbar\omega_s} e^{S_T} \int_{-\infty}^{\infty} dt \exp\left[-it\omega_s + \sum_i \frac{a^2\gamma_i^2}{2\beta_i^2} \operatorname{csch}\frac{\hbar\omega_i}{2kT}\cosh\left(it\omega_i + \frac{\hbar\omega_i}{2kT}\right)\right], \quad (4.22)$$

where $S_T = \frac{1}{2}\sum_i(a^2\gamma_i^2/\beta_i^2)\coth(\hbar\omega_i/2kT)$. The summation over i covers all the contributing normal modes in accepting the vibrational excitation energy. The integration in Eq. (4.22) can be carried out by using the saddle-point method or by expanding the exponential factor of Eq. (4.22) in power series of t (for the case $S_T > 1$).

As in the case of electronic relaxation, Eq. (4.22) for the vibrational relaxation rate constant can be used to study the temperature effect, energy gap dependence, and isotope effect (Lin, 1974; Nitzan *et al.*, 1975; Lin *et al.*, 1976; for a recent review, see Distler, 1979).

Another general model of the multiphonon mechanism of vibrational relaxation is based on an analogy with the Born–Oppenheimer adiabatic separation of electronic and nuclear motions (Lin, 1966; Fong, 1976). The basic idea is that since the vibrational motion of the relaxing mode is much faster than that of the accepting modes, the former adjusts instantaneously to the latter in exactly the same manner as the motion of electrons adjusts instantaneously to that of the nuclei. In the formal treatment, we partition the Hamiltonian of the total system consisting of vibrons (high frequency modes) and phonons (low frequency modes) as (Lin, 1976)

$$\hat{H} = \hat{T}_Q + \hat{T}_q + V(q, Q), \quad (4.23)$$

where the q's and Q's are normal coordinates of vibrons and phonons, respectively, and $\hat{T}_Q$ and $\hat{T}_q$ are the kinetic energy operators of phonons and vibrons, respectively.

To solve the Schrödinger equation of the vibron–phonon system

$$\hat{H}\psi = E\psi, \tag{4.24}$$

we first consider the solution of the Schrödinger equation of the vibron,

$$\hat{h}_q \phi_v(q, Q) = U_v(Q)\phi_v(q, Q), \tag{4.25}$$

where

$$\hat{h}_q = \hat{T}_q + V(q, Q). \tag{4.26}$$

Next we set

$$\psi = \sum_v \phi_v(q, Q)\theta_v(Q). \tag{4.27}$$

It follows that

$$[\hat{T}_Q + U_v(Q) + \langle\phi_v|\hat{H}'_{\mathrm{BO}}|\phi_v\rangle - E]\theta_v = -\sum_{v'}{}'\langle\phi_v|\hat{H}'_{\mathrm{B}}{}^{\mathrm{O}}|\phi_{v'}\rangle\theta_{v'}, \tag{4.28}$$

where $\hat{H}'_{\mathrm{BO}}$ represents the Born–Oppenheimer coupling:

$$\hat{H}'_{\mathrm{BO}}\phi_v\theta_v = -\hbar^2 \sum_\alpha \frac{\partial\phi_v}{\partial Q_\alpha}\frac{\partial\theta_v}{\partial Q_\alpha} - \frac{\hbar^2}{2}\sum_\alpha \frac{\partial^2\phi_v}{\partial Q_\alpha^2}\theta_v. \tag{4.29}$$

When $\langle\phi_v|\hat{H}'_{\mathrm{BO}}|\phi_{v'}\rangle$ for $v \neq v'$ is negligible, the solution becomes adiabatic:

$$\psi_{nv}(q, Q) = \phi_v(q, Q)\theta_{nv}(Q). \tag{4.30}$$

Thus, as in electronic relaxation, when the adiabatic wavefunctions are used as a basis set, the BO coupling can be used as the perturbation for vibrational relaxation.

Now suppose that $V(q, Q)$ takes the form (Burke and Small, 1974)

$$V(q, Q) = \frac{1}{2}\sum_i B_{ii}q_i^2 + \frac{1}{2}\sum_\alpha b_{\alpha\alpha}Q_\alpha^2 + \frac{1}{2}\sum_{i\alpha} B_{i\alpha}q_iQ_\alpha$$
$$+ \frac{1}{3!}\sum_{ij\alpha} D_{ij\alpha}q_iq_jq_\alpha + \frac{1}{3!}\sum_{i\alpha\beta} D_{i\alpha\beta}q_iQ_\alpha Q_\beta + \cdots. \tag{4.31}$$

In this case, in the lowest-order approximation we obtain

$$U_v(Q) = \sum_i (v_i + \tfrac{1}{2})\hbar\omega_i + \frac{1}{2}\sum_\alpha b_{\alpha\alpha}Q_\alpha^2 + \frac{1}{3!}\sum_{i\alpha}\frac{D_{ii\alpha}}{\omega_i}(v_i + \tfrac{1}{2})\hbar Q_\alpha, \tag{4.32}$$

which can be rewritten as

$$U_v(Q) = \sum_i (v_i + \tfrac{1}{2})\hbar\omega_i + \frac{1}{2}\sum_\alpha b_{\alpha\alpha}Q_{\alpha\alpha}^2, \tag{4.33}$$

where

$$Q_{v\alpha} = Q_\alpha + \frac{1}{6b_{\alpha\alpha}}\sum_i \frac{D_{ii\alpha}\hbar}{\omega_i}(v_i + \tfrac{1}{2}) = Q_\alpha + \Delta Q_{v\alpha}. \tag{4.34}$$

Equation (4.33) indicates that in the lowest-order approximation both vibron and phonon oscillators are harmonic; the origins of phonons oscillators are shifted by $\Delta Q_{v\alpha}$ depending on the vibron states.

Using the adiabatic approximation as the basis set, the rate constant for the vibron transition $v' \to v$ is given by

$$K_{v'\to v} = K_{v'v}(\beta) = \frac{2\pi}{\hbar} \sum_n \sum_m \rho_n |\langle mv | \hat{H}'_{\mathrm{BO}} | nv' \rangle|^2 \, \delta(E_{mv} - E_{nv'}). \tag{4.35}$$

We shall assume that the phonon heat bath is in thermal equilibrium, i.e., the vibrational relaxation of phonons is much faster than that of vibrons. Notice that

$$\begin{aligned} \langle mv | \hat{H}'_{\mathrm{BO}} | nv' \rangle = & -\hbar^2 \sum_\alpha \left\langle \phi_v \theta_{mv} \middle| \frac{\partial \phi_{v'}}{\partial Q_\alpha} \frac{\partial \theta_{nv'}}{\partial Q_\alpha} \right\rangle \\ & - \frac{\hbar^2}{2} \sum_\alpha \left\langle \phi_v \theta_{mv} \middle| \frac{\partial^2 \phi_{v'}}{\partial Q_\alpha^2} \theta_{nv'} \right\rangle. \end{aligned} \tag{4.36}$$

The second term in the BO coupling is usually smaller than the first term. Since the process required to simplify Eq. (4.35) is exactly the same as that involved in the electronic relaxation, we can simply write down the result (Lin, 1966)

$$\begin{aligned} K_{v'v}(\beta) = \sum_\alpha \frac{|R_\alpha(v'v)|^2}{\hbar^2} \int_{-\infty}^{\infty} dt \left[\frac{\omega_\alpha}{4\hbar} \left(\coth \frac{\hbar\omega_\alpha}{2kT} + 1 \right) e^{it\omega_\alpha} \right. \\ \left. + \frac{\omega_\alpha}{4\hbar} \left(\coth \frac{\hbar\omega_\alpha}{2kT} - 1 \right) e^{-it\omega_\alpha} \right] \exp\left(\frac{it}{\hbar} E_{vv'} \right) \prod_\sigma G_\sigma(t), \end{aligned} \tag{4.37}$$

where

$$R_\alpha(v'v) = -\hbar^2 \langle \phi_v | (\partial \phi_{v'}) / \partial Q_\alpha \rangle, \tag{4.38}$$

and

$$\begin{aligned} G_\sigma(t) = \exp\{ -\tfrac{1}{2} \beta_\sigma^2 \Delta Q_{vv',\sigma}^2 [\coth(\hbar\omega_\sigma/2kT) \\ - \operatorname{csch}(\hbar\omega_\sigma/2kT) \cosh(i\omega_\sigma t + (\hbar\omega_\sigma/2kT))]\}, \end{aligned} \tag{4.39}$$

with

$$\Delta Q_{vv',\sigma} = \frac{1}{6 b_{\sigma\sigma}} \sum_i \frac{D_{ii\sigma}\hbar}{\omega_i} (v_i - v'_i). \tag{4.40}$$

Equation (4.40) shows that the normal coordinate displacements of phonon modes depend linearly on the quantum numbers of the vibron (or vibrons) participating in the vibrational relaxation.

The similarity between Eqs. (4.22) and (4.37) should be noted. A detailed comparison between these two models of vibrational relaxation in dense media has been carried out (Lin, 1976; Knittel and Lin, 1978).

Recently the theoretical treatment of the vibrational–vibrational energy transfer in dense media has been carried out (Lin *et al.*, 1976; Gerber and Berkowitz, 1979; Blumen *et al.*, 1978; Zumofen, 1978). For the long range transfer, the treatment is similar to the Förster–Dexter theory for the transfer of electronic excitation (also see Section V).

In the following, we shall briefly review some recent experimental results of vibrational relaxation and dephasing of polyatomic molecules in condensed phases.

In binary liquids (CH_3CCl_3:CCl_4 and CH_3CCl_3:CD_3OD) a normal vibrational mode of component A is strongly excited and the relaxation via energy transfer to a normal mode of component B is studied on a time scale of several psec (Laubereau *et al.*, 1973). Comparison between different mixed systems indicates the importance of near resonance for efficient energy transfer. First the dependence of the vibrational lifetime τ' on concentration in binary liquid mixtures is investigated and, second, direct energy transfer to a different vibrational mode of the second molecular component is observed.

For the system CH_3CCl_3:CCl_4, the vibrational lifetime $\tau'(x)$ for the normal mode $\nu_H = 2939\ \text{cm}^{-1}$ as a function of the mole fraction x of CH_3CCl_3 obtained by Laubereau *et al.* (1973) is given in the following (also see Table 1):

$$\tau'\ (\text{psec}) = 5.2 \pm 0.8,\quad 8.0 \pm 1,\quad 15 \pm 2,\quad 29 \pm 2;$$
$$x = 1.0,\quad 0.8,\quad 0.6,\quad 0.4.$$

Strikingly different results were observed in the system CH_3CCl_3:CD_3OD:

$$\tau'\ (\text{psec}) = 10 \pm 2,\quad 6.5 \pm 1;$$
$$x = 0.4,\quad 0.6.$$

To show that the intermolecular energy transfer does take place in liquids, they have investigated the excess population of the ν_D (2200 cm^{-1}) vibration of CD_3OD molecules in CH_3CCl_3:CD_3OD($x_{CH_3CCl_3} = 0.6$). This excess population results from the decay of the laser excited ν_H mode of the CH_3CCl_3 molecules of the mixture. They obtain

$$\tau_1' = 25 \pm 10 \quad \text{psec},$$
$$\tau_0' = 6.5 \pm 1 \quad \text{psec},$$

where τ_0' and τ_1' are defined in the following equation:

$$dn_1/dt = -n_1/\tau_1' + \eta(N_0/N_1)(n_0/\tau_0').$$

That is, the initially excited mode with population $n_0(t)$ and relaxation time τ_0' decays to a lower frequency mode with the corresponding values $n_1(t)$ and τ_1'. η represents the efficiency of the energy transfer, and N_0 and N_1 are the number densities of the two molecular components.

The investigations of the system $CH_3CCl_3:CCl_4$ by Laubereau *et al.* (1973) suggest that the ν_H vibration decays via triple interactions with CH_3CCl_3 molecules. The δ_H bending mode at 1450 cm^{-1} appears to be a favorable decay channel. They have investigated the scattered signals from the δ_H bending mode in pure CH_3CCl_3 and obtained

$$\tau_1' = 4 \quad \text{psec},$$
$$\tau_0' = 5.2 \quad \text{psec}.$$

They have also determined the energy transfer efficiency η; they find $\eta = 2$ for the system $CH_3CCl_3:CCl_4$ and $\eta = 0.53$ for the system $CH_3CCl_3:CD_3OD$.

Spanner *et al.* (1976) have recently studied the vibrational relaxation of C_2H_5OH and CH_3I in CCl_4. Individual CH_3 stretching modes were excited by simple tunable IR pulses. The generated excess population of the molecules was monitored by subsequent probe pulses using spontaneous anti-Stokes Raman scattering. They believed that very rapid energy transfer between neighboring vibrational states is directly observed.

Experimentally, Spanner *et al.* (1976) made use of the recently developed parametric system (Laubereau *et al.*, 1974a) which allows the generation of tunable IR pulses and direct excitation of different vibrational states of the same polyatomic molecule.

We first discuss their experimental results of ethanol. Previous work with excitation via stimulated Raman scattering has given information on energy transfer and on possible decay routes of the high lying vibrational states of this molecule (Laubereau *et al.*, 1974b). In the experiments by Spanner *et al.*, ethanol was investigated in a 4×10^{-2} molar solution of CCl_4. The ethanol molecules were excited by an IR pulse at $\tilde{\nu}_{IR} = 2930$ cm^{-1} which corresponds to the frequency of the optically active CH_3 stretching vibration. The measured spontaneous anti-Stokes Raman signal curve shows three distinct features: A rapid rise, a fast decay with a time constant of approximately 2 psec and a slower decay with $T_1 = 40$ psec. According to Spanner *et al.*, the rise of the signal curve results from an IR excitation process and the first rapid decay indicates energy transfer to high lying neighboring vibrational energy states. Previous investigations on ethanol suggested rapid energy redistribution between neighboring CH stretching modes and overtones of CH bending modes. The energy exchange appears to occur very fast with a redistribution time of $T_r = 2$ psec. The final slow decay of the signal curve corresponds to the loss of excited state energy to lower lying energy levels. It has been shown that the δ_H bending modes are temporarily occupied during the decay process (Laubereau *et al.*, 1974b). Notice that in the neat liquid an effective energy relaxation time $T_1 = 22$ psec has been reported. The larger value of $T_1 = 40$ psec appears to be due to the low molar concentration of 4×10^{-2} in CCl_4.

Next we consider the experimental data on CH_3I reported by Spanner *et al.* The experimental data were obtained in a solution of 5×10^{-2} mole of CH_3I

in CCl_4. In a first experiment, the ν_1 mode (2950 cm^{-1}) of CH_3I was excited and probed, and in a second experiment, the ν_4 mode (3050 cm^{-1}) was excited but the ν_1 mode was probed. In these two experiments, the signal curves rise quickly during the excitation process and decay rapidly with an approximately equal time constant of 1.0 psec. In another experiment, they have measured the population lifetime for various concentrations of CH_3I (1%, 2%, and 5% of CH_3I in CCl_4); the experimental data indicate that the population lifetime does not depend strongly upon the concentration of CH_3I. Obviously, intramolecular processes are important for the fast decay of the excess population of the ν_1 mode.

Spanner *et al.*, have also measured the spontaneous Raman spectrum of the CH_3 stretching modes ν_1 and ν_4 of CH_3I dissolved in CCl_4 in the concentration range of 5–100 mole %. The spectra show a substantial difference of the linewidths of the two stretching modes. The observed spectral line broadening sets lower limits to the time constants. Using the equation $T_2 = (2\pi c\tilde{\nu}_{1/2})^{-1}$, they estimate time values of 0.2 and 1.0 psec from the full linewidths of the ν_4 and ν_1 modes, respectively. *A priori*, it is not possible to say which physical processes determine the observed linewidths.

More recently, Laubereau *et al.* (1978) have carried out a further investigation on the vibrational relaxation of CH stretching modes in liquids. CH stretching modes were first excited by psec IR pulses and the generated excess population was monitored by anti-Stokes scattering of subsequent ultrashort probe pulses. Experimental data are reported on five molecules: $CHCl_3$, CH_2Cl_2, CH_3CCl_3, CH_3CH_2OH, and CH_3I in the neat liquid and/or in solutions of CCl_4. The observed time constants vary between 1 and 100 psec depending upon the individual molecule and surrounding (see Table 1). Theoretical considerations show that rotational coupling, Fermi resonance, Coriolis coupling, and resonance energy transfer can strongly affect the vibrational population lifetime. The relevance of these processes is quite different for the various molecules investigated.

The vibrational modes of $CHCl_3$, CH_2Cl_2, CH_3CCl_3, CH_3I and CH_3CH_2OH are well established by IR and Raman spectroscopy. In Table 1, the experimental results obtained so far are listed. It is clearly seen in the last two rows that the two molecules $CHCl_3$ and CH_3I show just one very short time constant of 2.5 and 1 psec, respectively, while two time constants are observed for the other molecules. The short one is denoted as T'_1. It involves energy redistribution among strongly coupled states. The longer one is identified with T_1. It includes intermolecular transfer processes and migration of the energy. The longest lifetime observed so far in a polyatomic molecular at 300 K was found in the solution of 0.2 mole fraction of $CH_3CCl_3:CCl_4$. The population lifetime of the CH_3 stretching mode at 2939 cm^{-1} of CH_3CCl_3 is found to be concentration dependent in the solvent of CCl_4; its concentration dependence is shown in Table 1.

TABLE 1
Experimental Results

Substance	Concentration (mole fraction)	Vibration	Frequency (cm^{-1})	Excitation	T_1' (psec)	T_1 (psec)
$CHCl_3$	0.1	ν_1	3019	IR		2.5
CH_2Cl_2	0.1	ν_1	2985	IR	4 ± 2	40 ± 10
	1.0	ν_1	2985	IR	4	40
CH_3CCl_3	0.2	ν_1	2939	IR	4	100 ± 30
	0.4	ν_1	2939	R		29
	0.6	ν_1	2939	R		15
	0.8	ν_1	2939	R		8
	1.0	ν_1	2939	R		5.2
CH_3I	0.01	ν_1	2950	IR		1.5 ± 0.5
	0.05	ν_1	2950	IR		1.0
	0.05	ν_4	3050	IR		0.5
CH_3CH_2OH	0.04	—	2928	IR	2 ± 1	40
	1.0	—	2928	R	0.5	22

In Table 2, the four mechanisms (rotational coupling, Fermi resonance, Coriolis coupling, and resonance transfer) are listed which are thought to be relevant for the understanding of the population decay. The significance of the individual mechanism is indicated by the marks v (very important), m (minor importance), and u (unimportant).

Picosecond relaxation times of C—H stretching vibrations in a series of liquid hydrocarbons have been measured by Monson *et al.* (1974) using the Raman scattering technique. The results indicate that vibrational energy loss takes place primarily through the methyl groups in these molecules. In the Raman experiment, 1.06 μpsec pulses from a mode-locked N_d^{+3} glass laser were used as the excitation source. The liquid was then probed with 0.53 μsec

TABLE 2
Mechanisms

	Rotational coupling	Fermi resonance	Coriolis coupling	Resonance transfer
$CHCl_3$ (3019 cm^{-1})	m	v	m	u
CH_2Cl_2 (2985 cm^{-1})	m	m	u	m
CH_3CCl_3 (2939 cm^{-1})	u	v	u	v
CH_3CH_2OH (2928 cm^{-1})	u	m	u	m
CH_3I (2950 cm^{-1})	v	v	m	u

pulses, and the intensity of the anti-Stokes scattering, which is proportional to the population of the excited vibrational state, was measured as a function of the time delay between the exciting and probe pulses.

The dephasing times of four *n*-alkanes (C_7, C_{10}, C_{13}, C_{15}) were also obtained by Monson *et al.* from a stimulated (coherent) Raman experiment. The results show that the dephasing time increases with increasing *n*-alkane chain length. Vibrational relaxation times were obtained by collecting the (incoherent) Raman light at 90° to the incident light path. The results are given in Table 3. As shown in Table 3, the vibrational relaxation times for C—H vibrations in the normal alkanes increase with the increasing numbers of C atoms in the chain. In addition, two other effects have been observed. Isodecane shows about two-thirds the relaxation time of *n*-decane, while 1-heptane and 1-decane relax about half as fast as the corresponding *n*-alkanes. These results lead them to believe that the C—H stretching relaxation time depends in the main on the proportion of the methyl groups in the molecule. When the relaxation times are normalized to the ratio of the number of CH_3 groups to the total number of C atoms in the chain, a reasonable correlation is obtained for the compounds studied (see Table 3). Also listed in Table 3 is the relaxation time for 1.6-heptadiene, which has no CH_3 groups; this relaxation time is considerably longer, further confirming the above notion.

Because the Raman spectra of these types of molecules have three distinct maxima in the 3000 cm^{-1} region, a check was made by Monson *et al.* to see if the measured relaxation time depended upon detecting frequency. The relaxation data were found to be insensitive to the position of the 1.8 m spectrometer setting within a range of wavelengths corresponding to Raman shifts of 2900–3100 cm^{-1}.

TABLE 3

Experimental Results

Compound	Temperature (°C)	(a) Lifetime ±1.5 psec	(b) Ratio methyl carbons to total carbons	Normalized lifetime: (a) × (b)
n-heptane	20	11.0	2:7	3.14
	−25	14.0	—	—
	−70	16.0	—	—
isodecane	20	10.8	3:10	3.24
n-decane	20	16.0	2:10	3.20
	−25	21.0	—	—
n-tridecane	20	21.0	2:13	3.23
n-pentadecane	20	24.5	2:15	3.27
1-heptene	20	21.0	1:7	3.00
1-decene	20	33.5	1:10	3.35
1,6-heptadiene	20	60.0	—	—

Monson *et al.* also made some preliminary measurements in the C—H stretching region on the molecule CD_3—$(CH_2)_2$—CD_3 (at −42°C) and obtained a relaxation time of ~72 psec. If precise enough data were available, it might be possible to extract out the —CH_2— relaxation times from the *n*-alkane results by taking the total rate to be the sum of —CH_3 and —CH_2— rates normalized to the proportion of each such group in the molecule. The published data are too crude for this purpose. However, the probability of losing C—H stretch energy from a —CH_2— group in a normal alkane would seem from the other data to be less than implied by the 60 psec lifetime in 1.6-heptadiene.

The temperature dependence has been studied by Monson *et al.*, but it does not cover a wide enough range to suggest its exact origin. However, they are of the opinion that the participation in the relaxation mechanism of intermolecular collisions or internal rotations of CH_3 groups would give rise to a temperature effect similar to the one observed. Although the pathway for decay of the C—H stretch was not investigated, they felt that the vibrational relaxation most likely decays by way of two C—H bands at 1450 cm^{-1}, as observed by Laubereau *et al.* (1973) and Alfano and Shapiro (1972) in earlier work. It is interesting to note that proton spin–lattice relaxation in long chain hydrocarbons seems also to take place most effectively from —CH groups (Feigenson and Chan, 1974).

A new experimental technique to measure ultrashort vibrational relaxation times in liquids has recently been developed by Laubereau *et al.* (1975). A well defined vibrational state of polyatomic molecules is populated by resonant absorption of an IR psec pulse. The vibrational excitation is monitored with the help of a second pulse which generates transitions to a fluorescent state. Their technique is well suited for highly dilute systems. As an example, a vibrational energy relaxation time of coumarin 6 in CCl_4 is measured.

In their experimental system, an IR pulse at a frequency of $\tilde{\nu}_1 = 2970\ cm^{-1}$ has $\sim 10^{14}$ quanta and a pulse duration of ~3 psec and the probe light pulse has the $\tilde{\nu}_2 = 18{,}910\ cm^{-1}$. The 2970 cm^{-1} mode of coumarin 6 corresponds to the asymmetric CH_3 stretching mode of the ethyl groups.

Experimental results using the two-pulse technique for coumarin 6 in CCl_4 at a concentration of $3 \times 10^{-5} M$ are given below:

$$\tau' = 1.3 \pm 0.3 \quad \text{psec}, \qquad T = 205 \quad \text{K},$$

$$\tau' = 1.7 \pm 0.3 \quad \text{psec}, \qquad T = 253 \quad \text{K}.$$

Experimentally, Laubereau *et al.* (1975) find a time resolution of better than 0.5 psec. Measurements were made over a concentration range of $4 \times 10^{-6} M$ to $4 \times 10^{-4} M$. The observed relaxation time was found to be independent of concentration, suggesting intermolecular interactions between the coumarin molecules to be negligible. It should be noted that the linewidth $\Delta\tilde{\nu} \doteqdot 14\ cm^{-1}$ of the IR absorption band at 2970 cm^{-1} corresponds to a time constant of 0.4 psec. This value is determined by dephasing and energy relaxation processes (Fischer and Laubereau, 1975). Orientational motion of the coumarin 6 molecules is expected to be very slow ($\geq 10^{-10}$ sec) and does not affect the observed short relaxation times. In this connection, recent work on rhodamine 6G by Ricard and Ducing (1975) is of interest. These authors measured with a three-pulse technique the ground state recovery via vibrational relaxation. They found a time constant of several picoseconds.

Two types of time-resolved experiments have been performed on the intermediate sized polyatomic molecule diethylamine in the liquid phase by Weisman and Rice (1979) in order to elucidate the pathway for vibrational relaxation of the $v = 3$ level of the NH stretching mode, which has 9420 cm^{-1} of energy. With neither were transient populations in such modes observable. It is inferred that population relaxation in this highly excited room temperature system proceeds on the subpicosecond time scale to lower lying levels. The importance of the intramolecular channels for this decay is suggested.

Recently, using subpicosecond UV and visible pulses, Shank *et al.* (1977) have studied relaxations from highly excited molecular vibronic states with a resolution of 0.2 psec. The source of the subpicosecond optical pulses used in this experiment was a cavity-dumped, passively mode-locked dye laser (Shank and Ippen, 1974, 1975). Rhodamine B and rhodamine 6G have been investigated in various solvents and vibronic relaxation in coronene has been time resolved. The measurement technique utilizes two short optical pulses, one at frequency ω_1 and the other at ω_2. A short wavelength optical pulse ω_1 at 3076 Å is used to excite a molecule from its ground electronic state S_0 to a highly excited electronic and vibrational level. The molecule relaxes to the bottom of the S_1 state in a characteristic time τ. The S_1 level is probed by measuring optical gain for a second pulse at ω_3 (6150 Å). The delay between the rise of optical gain at ω_2 and the exciting pulse at ω_1 gives a measure of the relaxation time τ. The pumping and probing intensities are sufficiently low that the induced population of S_1 remains small and decays with its normal fluorescence tifetime in the nanosecond range.

Several experimenters have reported relaxation measurements of the vibration manifold in rhodamine dyes using picosecond techniques, but with a resolution about an order of magnitude more coarse than that reported by Shank *et al.*, (1977). Ricard (1975) observed delays of 4 psec or less in gain rise time for rhodamine 6G dependent upon solvent and wavelength. Ricard and Ducing (1975) measured the ground state vibrational relaxation to be 4 psec for rhodamine 6G. Malley and Mourou (1974) obtained time-resolved spectra of the fluorescence of rhodamine 6G in glycerol and observed a 20 psec time for the fluorescence to become homogeneously broadened. Mourou and Malley (1975 also reported the rise time of fluorescence in rhodamine 6G and rhodamine B in ethanol to be less than 1 psec.

The results of Shank *et al.* (1977), however, indicate that relaxation in the vibrational manifold occurs in a time scale of less than 0.2 psec with no measured difference between solvents. Relaxation on this rapid time scale suggests that the thermalization process is intramolecular and does not require an interaction with the solvent.

In an additional experiment, Shank *et al.* (1977) have also investigated vibronic relaxation in the molecule coronene. They excite into the S_3 state with a subpicosecond pulse at 3075 Å. Excited state absorption is then monitored as a function of time with the 6150 Å probe. The probe light experiences an absorption which is continually modified as the molecule relaxes to the S_1 state. A decay of about 2 psec is observed. A previous study by Anderson *et al.* (1975) indicated that the excited state S_1 absorption appeared with effectively zero delay following excitation to the higher state S_2.

C. Vibrational Relaxation in Isolated Molecules

The vibrational relaxation and dephasing in an isolated molecule has been reviewed recently (Rettschnick, this volume; Mukamel, 1978). Here we shall briefly discuss how to find the master equation and the rate constant in this case.

From Eq. (2.19), we have

$$\frac{d\rho_{n_s}}{dt} = \sum_{m_s} (k_{m_s n_s}\rho_{m_s} - k_{n_s m_s}\rho_{n_s}), \tag{4.41}$$

where, for example,

$$k_{n_s m_s} = \frac{2\pi}{\hbar} \sum_{n_b} \sum_{m_b} \rho_{n_b}^{(b)}(0) |H'_{n_s n_b : m_s m_b}|^2 \, \delta(E_{n_s n_b} - E_{m_s m_b}). \tag{4.42}$$

and $\rho_{n_b}^{(b)}(0)$ represents the initial probability distribution of the heat bath. For the case $\hat{H}' = QF_b$, Eq. (4.41) becomes

$$d\rho_{n_s}/dt = (n_s + 1)(k_{10}\rho_{n_s+1} - k_{01}\rho_{n_s}) + n_s(k_{01}\rho_{n_s-1} - k_{10}\rho_{n_s}), \tag{4.43}$$

where the rate constants k_{01} and k_{10} are defined by

$$k_{01} = \frac{\pi}{\omega_s} \sum_{n_b} \sum_{m_b} \rho_{n_b}^{(b)}(0) |\langle n_b | F_b | m_b \rangle|^2 \delta(E_{n_b} - E_{m_b} - \hbar\omega_s) \tag{4.44}$$

and

$$k_{10} = \frac{\pi}{\omega_s} \sum_{n_b} \sum_{m_b} \rho_{n_b}^{(b)}(0) |\langle n_b | F_b | m_b \rangle|^2 \delta(E_{n_b} - E_{m_b} + \hbar\omega_s), \tag{4.45}$$

respectively. Notice, for example, that k_{10} is for the $1 \to 0$ transition. If the heat bath is initially cold (i.e., the heat bath occupies the lowest state initially), then $k_{01} = 0$ and Eq. (4.43) reduces to

$$d\rho_{n_s}/dt = k_{10}[(n_s + 1)\rho_{n_s+1} - n_s\rho_{n_s}], \tag{4.46}$$

which is identical with Eq. (4.6) for the isothermal system at $T = 0$. The master equations of vibrational relaxation in an isolated molecule for other situations can be obtained similarly by choosing appropriate $\hat{H}'$.

Next we discuss the rate constant. As can be seen from Eqs. (4.44) and (4.45), to calculate k_{01} or k_{10} we have to know $\rho_{n_b}^{(b)}(0)$ first. For example, for the cold heat bath, we have $k_{01} = 0$ and

$$k_{01} = \frac{\pi}{\omega_s} \sum_{m_b} |\langle 0|F_b|m_b\rangle|^2 \delta(\hbar\omega_s - E_{m_b}). \tag{4.47}$$

This corresponds to the isothermal rate constant at $T = 0$ discussed in Section IVB. Another case of interest is that the heat bath is in equilibrium and has energy E. In this case, $\rho_{n_b}^{(b)}(0)$ is described by the so-called microcanonical distribution (also see Section VI)

$$\rho_{n_b}^{(b)}(0) = \delta(E - E_{n_b})/\textstyle\sum_{m_b} \delta(E - E_{m_b}). \tag{4.48}$$

A similar situation also takes place in electronic relaxation (Lin, 1972, 1973a)

Introducing the contour integral representation for the delta function $\delta(E - E_{n_b})$, Eq. (4.45) can be rewritten (see Section VI) as

$$k_{10} = \frac{1}{\rho^{(b)}(E)} \frac{1}{2\pi i} \int_c d\beta\, e^{\beta E} Q^{(b)}(\beta) K(\beta), \tag{4.49}$$

where $Q^{(b)}(\beta) = \sum_{n_b} \exp(-\beta E_{n_b})$, the canonical partition function, $\beta = 1/kT$, $K(\beta)$ represents the isothermal rate constant discussed in Section IVB, and $\rho^{(b)}(E)$ is the density of states of the heat bath with energy E, i.e.,

$$\rho^{(b)}(E) = \sum_{n_b} \delta(E - E_{n_b}) = \frac{1}{2\pi i} \int_c d\beta\, e^{\beta E} Q^{(b)}(\beta). \tag{4.50}$$

For the contour integral involved in Eqs. (4.49) and (4.50), the saddle-point method can be used (Lin, 1972, 1973a).

V. MIGRATION OF PARTICLES

In the previous sections, we derived the master equation in the Markoff approximation and showed the relation between the band shape function and the relaxation (and/or dephasing) processes. In this section, we shall derive the so-called generalized master equation (GME) for a system embedded in a heat bath; here the effect of the memory function is taken into consideration and the

Markoff approximation will be examined. This may be very important for an ultrafast process. It should be noted that the GME can be applied to both large and small systems and to canonical and microcanonical systems depending on the size and initial distribution function of the heat bath.

Defining the projection operator by $\hat{D}$ (Zwanzig, 1964) and applying $\hat{D}$ and $1 - \hat{D}$ to Eq. (AI.8), we find

$$P\hat{\rho}_1^{(S)}(P) - \hat{\rho}_1^{(S)}(0) = -\hat{D}\langle\hat{M}(P)\rangle\hat{\rho}_1^{(S)}(P) - \hat{D}\langle\hat{M}_c(P)\rangle\hat{\rho}_2^{(S)}(P) \tag{5.1}$$

and

$$P\hat{\rho}_2^{(S)}(P) - \hat{\rho}_2^{(S)}(0) = -i(1 - \hat{D})\hat{L}_s\hat{\rho}_2^{(S)}(P) - (1 - \hat{D})\langle\hat{M}_c(P)\rangle\hat{\rho}_1^{(S)}(P) \\ -(1 - \hat{D})\langle\hat{M}_c(P)\rangle\hat{\rho}_2^{(S)}(P), \tag{5.2}$$

where $\hat{\rho}_1^{(S)}(P) = \hat{D}\hat{\rho}^{(S)}(P)$ and $\hat{\rho}_2^{(S)}(P) = (1 - \hat{D})\hat{\rho}^{(S)}(P)$. Here the relations $\hat{D}\hat{L}_s = 0$ and $\hat{L}_s\hat{D} = 0$ have been used. Eliminating $\hat{\rho}_2^{(S)}(P)$ from Eqs. (5.1) and (5.2) yields

$$P\hat{\rho}_1^{(S)}(P) - \hat{\rho}_1^{(S)}(0) = -\hat{U}(P)\hat{\rho}_2^{(S)}(0) + \hat{W}(P)\hat{\rho}_1^{(S)}(P), \tag{5.3}$$

where

$$\hat{U}(P) = \hat{D}\langle\hat{M}_c(P)\rangle\hat{g}(P), \tag{5.4}$$

$$g(P) = 1/[P + i(1 - \hat{D})\hat{L}_s + (1 - \hat{D})\langle\hat{M}_c(P)\rangle], \tag{5.5}$$

and

$$\hat{W}(P) = -\hat{D}\langle\hat{M}_c(P)\rangle[1 - \hat{g}(P)(1 - \hat{D})\langle\hat{M}_c(P)\rangle]. \tag{5.6}$$

Carrying out the inverse Laplace transformation of Eq. (5.3) we obtain

$$\frac{\partial\hat{\rho}_1^{(S)}}{\partial t} = -\hat{U}(t)\hat{\rho}_2^{(S)}(0) + \int_0^t dt_1\, \hat{W}(t_1)\hat{\rho}_1^{(S)}(t - t_1), \tag{5.7}$$

where $\hat{U}(t)$ and $\hat{W}(t)$ represent the inverse Laplace transforms of $\hat{U}(p)$ and $\hat{W}(p)$, respectively. Because $\hat{\rho}_1^{(S)}$ is diagonal, from Eq. (5.7) we obtain the master equation,

$$\frac{\partial\rho_{n_sn_s}^{(S)}}{\partial t} = -[\hat{U}(t)\hat{\rho}_2^{(S)}(0)]_{n_sn_s} + \sum_{m_s}\int_0^t dt_1\, W_{n_sn_s:m_sm_s}(t_1)\rho_{m_sm_s}^{(S)}(t - t_1), \tag{5.8}$$

or

$$\frac{\partial\rho_{n_sn_s}^{(S)}}{\partial t} = -[\hat{U}(t)\hat{\rho}_2^{(S)}(0)]_{n_sn_s} + \sum_{m_s}\int_0^t dt_1\, [W_{n_sn_s:m_sm_s}(t_1)\hat{\rho}_{m_sm_s}^{(S)}(t - t_1) \\ - W_{m_sm_s:n_sn_s}(t_1)\rho_{n_sn_s}^{(S)}(t - t_1]. \tag{5.9}$$

Here the relation $\sum_{m_s} W_{m_s m_s : n_s n_s}(t) = 0$ has been used. Equation (5.9) is completely general; the only assumption that has been introduced is $\hat{\rho}(0) = \hat{\rho}^{(b)}(0)\hat{\rho}^{(S)}(0)$ and hence in the momory function $W_{n_s n_s : m_s m_s}(t)$ and $\hat{U}(P)$, only the average over the initial distribution of the heat bath is involved [see Eqs. (5.4)–(5.6)]. In other words, we can apply Eq. (5.9) to the system embedded in an isothermal heat bath (i.e., canonical systems) and to the system embedded in a heat bath of the finite size (e.g., the vibrational relaxation in an isolated molecule).

Under the initial diagonality condition (i.e., $\hat{\rho}_2^{(S)}(0) = 0$), we obtain

$$\frac{\partial P_n}{\partial t} = \int_0^t dt_1 \sum_m [W_{nm}(t - t_1)P_m(t_1) - W_{mn}(t - t_1)P_n(t_1)]$$

$$= \int_0^t dt_1 \sum_m [W_{nm}(t_1)P_m(t - t_1) - W_{mn}(t_1)P_n(t - t_1)], \qquad (5.10)$$

where $P_n = \rho_{n_s n_s}^{(S)}$ and $W_{nm}(t) = W_{n_s n_s : m_s m_s}(t)$. Equation (5.10) can often be put in the following form:

$$\frac{\partial P_n}{\partial t} = \sum_m [k_{nm} P_m(t) - k_{mn} P_n(t)], \qquad (5.11)$$

where k_{nm} and k_{mn} represent the rate constants. The passage from Eq. (5.10) to Eq. (5.11) makes use of the Markoffian approximation (Kenkre and Knox, 1974),

$$W_{nm}(t) = k_{nm}\,\delta(t). \qquad (5.12)$$

It must be emphasized that this passage from Eq. (5.10) to Eq. (5.11) does not hold for all interactions and for all times and its validity is claimed only for certain interaction Hamiltonians and only on a certain time scale (Kenkre and Knox, 1974; Oppenheim *et al.*, 1977).

In general the Markoffian approximation consists of replacing an equation of the kind (Kenkre and Knox, 1974)

$$A(t) = \int_0^t dt_1\, B(t - t_1)C(t_1) \qquad (5.13)$$

by

$$A(t) = \left[\int_0^\infty dt_1\, B(t_1)\right]C(t), \qquad (5.14)$$

which involves replacing $B(t)$ by

$$B(t) = \left[\int_0^\infty dt_1\, B(t_1)\right]\delta(t). \qquad (5.15)$$

While the theory of rate processes should really start from Eq. (5.10) in its full complexity, a simplification may often be possible whereby the time dependence of W_{nm} is assumed independent of the states n and m,

$$W_{nm}(t) = k_{nm} F(t). \tag{5.16}$$

This reduces Eq. (5.10) to the simpler form (Kenkre and Knox, 1974),

$$\frac{\partial P_n}{\partial t} = \int_0^t dt_1\, F(t - t_1) \sum_m [k_{nm} P_m(t_1) - k_{mn} P_n(t_1)]. \tag{5.17}$$

The nature of a rate process is then determined by the normalized memory function $F(t)$ as well as by the rate constants k_{nm}.

Let us now show the above master equation can provide a unified formulation of wavelike and diffusive transport. For this purpose, we use the site representations for n and m. On making the nearest neighbor approximation on k_{nm}, we find

$$\frac{\partial P_n}{\partial t} = \int_0^t dt_1\, F(t - t_1)[k_{n,n-1} P_{n-1}(t_1) + k_{n,n+1} P_{n+1}(t_1) - k_{n-1,n} P_n(t_1) - k_{n+1,n} P_n(t_1)]. \tag{5.18}$$

Taking the continuum limit, Eq. (5.18) reduces to

$$\frac{\partial P(x, t)}{\partial t} = \int_0^t dt_1\, M(t - t_1) \frac{\partial^2 P(x, t_1)}{\partial x^2}, \tag{5.19}$$

where

$$M(t) = k_{n,n+1} l^2 F(t), \tag{5.20}$$

assuming that $k_{n,n-1} = k_{n-1,n} = k_{n,n+1} = k_{n+1,n}$. If the normalized memory function $F(t)$ takes the form $F(t) = \delta(t)$, then Eq. (5.19) becomes

$$\partial P(x, t)/\partial t = D\, \partial^2 P(x, t)/\partial x^2, \tag{5.21}$$

where the diffusion constant $D = k_{n,n+1} l^2$. Equation (5.21) represents the diffusion limit. On the other hand, if $M(t) = c^2 \Theta(t)$, where $\Theta(t)$ is the Heaviside function [i.e., $\Theta(t) = 0$ for $t < 0$ and $\Theta(t) = 1$ for $t \geq 0$], then we obtain the wavelike limit

$$\partial^2 P(x, t)/\partial t^2 = c^2\, \partial^2 P(x, t)/\partial x^2, \tag{5.22}$$

where c is the wave velocity. It has been shown (Morse and Feshbach, 1953) that the characters of the motion predicted by Eqs. (5.21) and (5.22) can be combined into the equation

$$\partial^2 P(x, t)/\partial x^2 + (c^2/D)\, \partial P(x, t)/\partial t = c^2\, \partial^2 P(x, t)/\partial x^2. \tag{5.23}$$

This equation can be obtained from Eq. (5.19) by choosing the memory function as follows (Kemkre and Knox, 1974):

$$M(t) = c^2 \exp[-(c^2/D)t] \tag{5.24}$$

and assuming that $\partial P/\partial t = 0$ at $t = 0$.

From the above discussion, we can see that many rate processes like diffusion, dielectric relaxation, electron transfer reactions, isomerization reactions, nonradiative decay, energy transfer and many other thermally activated processes in dense media are closely related to each other theoretically and can be treated from a unified quantum mechanical viewpoint (Lin and Eyring, 1972; Fong, 1976).

From Eqs. (5.10)–(5.18) and Eqs. (3.6)–(3.14), we can see that in the Markoff approximation and assuming that $\hat{\rho}^{(b)}(0)$ is of the Boltzmann distribution, the rate constant involved in the transition from the n state (or position) to the m state (or position) for a thermally activated process can be expressed as

$$k_{nm}(\beta) = \frac{2}{\hbar} \sum_{v'} \sum_{v''} \rho_{nv'}^{(b)} |\langle nv'|\hat{H}'|mv''\rangle|^2 \frac{\Gamma_{mv'',nv'}}{\Gamma_{mv'',nv'}^2 + (E_{mv''} - E_{nv'})^2}, \tag{5.25}$$

where (v', v'') represent the quantum states of nuclear motion of the heat bath. The choice of $\hat{H}'$ depends on the rate process under consideration. As the damping constant $\Gamma_{mv'',nv'}$ approaches zero, Eq. (5.25) reduces to

$$k_{nm}(\beta) = \frac{2\pi}{\hbar} \sum_{v'} \sum_{v''} \rho_{nv'}^{(b)} |\langle nv'|\hat{H}'|mv''\rangle|^2 \delta(E_{mv''} - E_{nv'}). \tag{5.26}$$

Using the adiabatic approximation,

$$|nv'\rangle = \Phi_n(q, Q)\Theta_{nv'}(Q), \qquad |mv''\rangle = \Phi_m(q, Q)\Theta_{mv''}(Q), \tag{5.27}$$

we have

$$k_{nm}(\beta) = \frac{2\pi}{\hbar} \sum_{v'} \sum_{v''} \rho_{nv'}^{(b)} |\langle \Theta_{nv'}|H'_{nm}|\Theta_{mv''}\rangle|^2 \delta(E_{mv''} - E_{nv'}), \tag{5.28}$$

where $H'_{nm} = \langle \Phi_n|\hat{H}'|\Phi_m\rangle$. The matrix element H'_{nm} in general will depend on the nuclear coordinates. For the case in which H'_{nm} is independent of nuclear coordinates, the rate constant $k_{nm}(\beta)$ given by Eq. (5.28) can be simplified (Lin and Eyring, 1972):

$$k_{nm}(\beta) = \frac{1}{\hbar^2} |H'_{nm}|^2 \exp(-S) \int_{-\infty}^{\infty} dt \exp\left[it\omega_{mn} + \frac{1}{2} \sum_j \Delta_j^2 \right.$$
$$\left. \times \exp it\omega_j + \sum_j \bar{n}_j \Delta_j^2 (\cos \omega_j t - 1)\right], \tag{5.29}$$

where Δ_j represents the dimensionless normal coordinate displacement, $S = \frac{1}{2}\sum_j \Delta_j^2$, and $\bar{n}_j = [\exp(\hbar\omega_j/kT) - 1]^{-1}$. Here it has been assumed that the intermolecular motion in the condensed phase can approximately be treated as harmonic vibration and that the normal frequency modifications are negligible. Equation (5.29) corresponds to the so-called linear coupling case in the areas of exciton and polaron transport (Munn and Silbey, 1978; Blumen and Silbey, 1978).

The temperature dependence of the rate constant $k_{nm}(\beta)$ given by Eq. (5.29) has been discussed (Ma *et al.*, 1978; Fong, 1976; Lin and Eyring, 1972; Efrima and Metiu, 1978). For example, the Arrhenius form of $k_{nm}(\beta)$ can be obtained by expanding the exponential term in the integrand of Eq. (5.29) in power series of t and integrating the resulting equation:

$$k_{nm}(\beta) = (|H'_{nm}|^2/\hbar^2)[2\pi/\textstyle\sum_j (\bar{n}_j + \frac{1}{2})\Delta_j^2\omega_j^2]^{1/2} \times \exp[-(\omega_{mn} + \frac{1}{2}\sum_j \Delta_j^2\omega_j)^2/2\sum_j (\bar{n}_j + \frac{1}{2})\Delta_j^2\omega_j^2]. \tag{5.30}$$

When T is high, we have $\bar{n}_j = kT/\hbar\omega_j$ and Eq. (5.30) reduces to

$$k_{nm}(\beta) = (|H'_{nm}|^2/\hbar^2)(2\pi\hbar/kT\textstyle\sum_j \Delta_j^2\omega_j)^{1/2} \exp(-\Delta E/kT), \tag{5.31}$$

where

$$\Delta E = (\hbar/2\textstyle\sum_j \Delta_j^2\omega_j)\left(\omega_{mn} + \frac{1}{2}\sum_j \Delta_j^2\omega_j\right)^2. \tag{5.32}$$

For the motion between the two equivalent positions, $\omega_{mn} = 0$ and $\Delta E = \frac{1}{8}\hbar\sum_j \Delta_j^2\omega_j$, the minimum crossing in the multidimensional potential surface.

It should be noted that in order to evaluate $k_{nm}(\beta)$ given by Eq. (5.29), it is necessary to know the normal mode distribution of the local modes coupled to the motion of the system under consideration [i.e., the summation over j in Eq. (5.29)]. The commonly used models for the frequency distribution of local modes are the Einstein, Gaussian, exponential, and Lorentzian models (Fletcher *et al.*, 1978; Munn and Silbey, 1978). For example, in the Einstein model Eq. (5.29) becomes

$$k_{nm}(\beta) = \frac{2\pi|H'_{nm}|^2}{\hbar^2\omega}\exp[-S(1 + 2\bar{n})]\sum_{m=0}^{\infty}\frac{[S^2\bar{n}(\bar{n}+1)]^m}{(m!)^2}. \tag{5.33}$$

Next, we consider the damping effect on the rate constant (Ma *et al.*, 1978; Efrima and Metiu, 1978). In the linear coupling case, we have

$$k_{nm}(\beta) = \frac{|H'_{nm}|^2}{\hbar^2}\exp(-S)\int_{-\infty}^{\infty} dt\,\exp\left[it\omega_{mn} + \frac{1}{2}\sum_j \Delta_j^2\exp(it\omega_j) + \sum_j \bar{n}_j\Delta_j^2(\cos\omega_j t - 1) - \frac{1}{\hbar}\Gamma_{nm}|t|\right], \tag{5.34}$$

where Γ_{nm} represents the damping constant. Similarly, using the Einstein model we can obtain the rate constant with the damping effect being included as

$$k_{nm}(\beta) = \frac{|H'_{nm}|^2}{\hbar^2} \exp[-S(2\bar{n} + 1)] \sum_{m=0}^{\infty} \sum_{m'=0}^{\infty} \frac{S^{m+m'}\bar{n}^m(1 + \bar{n})^{m'}}{m!m'!}$$
$$\times \frac{2\Gamma_{nm}/\hbar}{(\Gamma_{nm}/\hbar)^2 + (m - m')^2\omega^2}. \tag{5.35}$$

The limitation and validity of the Arrhenius form of $k_{nm}(\beta)$ and the effect of the damping constants (or the linewidths) and anharmonicity on $k_{nm}(\beta)$ have recently been investigated by Efrima and Metiu (1978) and Ma *et al.*, (1978).

Recently in polaron migration experiments (Schein, 1977; Burshtein and Williams, 1978) it has been found that the mobility is very weakly dependent on temperature and cannot be represented by the Arrhenius form, and in measurements of exciton migration in anthracene (Ern *et al.*, 1972; Haarer and Wolf, 1970) it has been found that the diffusion coefficient decreases with temperature. These experimental results are consistent with the theoretical calculation discussed above (Ma *et al.*, 1978).

As mentioned above, a great number of rate processes can be treated from a unified viewpoint. In most cases, we are only interested in the rate of evolution of a few special degrees of freedom which are coupled to a dense manifold of other degrees of freedom that can be regarded as a heat bath. In all cases, we start with the Hamiltonian describing the motion of the special degrees of freedom when the heat bath degrees of freedom are held fixed. The Schrödinger equation corresponding to this Hamiltonian is then solved and a set of states and levels are provided which describe the motion of the special degrees of freedom. In most cases, not all those states are important in describing the process of interest. The next step is to allow the heat bath degrees of freedom to undergo motions (say vibrations for example). As a result, the main degrees of freedom are coupled to those of the heat bath. Since in most cases the motion of the heat bath degrees of freedom is much faster, the main degrees of freedom perceive it as a random thermal motion which can cause energy transfer into and/or energy removal from the special degrees of freedom or randomize the transition from one site to another (Shugard *et al.*, 1978a, b). The coupling strength between the special and bath degrees of freedom is generally to be determined by the molecular dynamics calculation. Thus in most cases this prevents detailed numerical calculations of the rate but allows us to predict its temperature and/or pressure dependence. From the above description, we can see that the treatment of a thermally activated process is similar to that of the radiationless transition (electronic relaxation) and hence the theoretical methods developed for the latter process can usually be used (Lin, 1976).

Recently coherent and incoherent energy transfer in solids (Silbey, 1976; Harris and Zwemer, 1978), electron tunneling (or transfer) in chemistry and biology (Libby, 1977) and chemical reactions at very low temperatures (Goldanskii, 1976, 1977) have been reviewed. Here we briefly review a related phenomenon, the proton transfer that can also be treated by using the theoretical methods described in this section.

Large Stokes shifts have been observed in the emission spectrum of a large number of aromatic ketones and alcohols (Weller, 1961; Vander Donckt, 1970). Molecules which possess an intramolecular hydrogen bond in the ground state often demonstrate an even larger red shift than is seen in aromatic molecules with just one functional group. For example, in methyl salicylate the fluorescence maximum is red shifted by about 10,000 cm^{-1} from the absorption maximum, while in phenol it is red shifted by about 4000 cm^{-1}. The large Stokes shift seen in the emission from methyl salicylate is due to intramolecular proton transfer in the excited state. Proton transfer is driven by a pK change of about -6 in the phenolic oxygen, and a pK change which may be as large as $+8$ for the carbonyl oxygen of the carboxyl group. This results in the formation of a zwitterion which fluoresces in the blue at about 450 nm. In addition to the blue fluorescence, there is a much weaker component at about 340 nm. Since methyl 0-methoxy benzoate emits at 320 nm, the near UV fluorescence was believed to originate from excited molecules in which the proton remains bound to the phenolic oxygen.

The 340 and 450 nm emissions of the methyl salicylate molecule were equally quenched by carbon disulfide. Thus it was speculated that the two forms of the excited molecule were in equilibrium (Weller, 1971). If they were in equilibrium, then the rate of proton translocation would have to be much faster than the deactivation of the singlet state. This sets a lower limit of 10^8 sec^{-1} on the intramolecular transfer rate. The relative contribution of the two components to the total fluorescence was strongly dependent on temperature and could also be altered by the solvent especially sensitive to the hydrogen bonding ability of the solvent (Sandros, 1976). Since the proton transfer appeared to proceed even at 4 K, it was felt that it occurred via a tunneling mechanism (Beens *et al.*, 1965). Sandros (1976) also found that the emission spectrum was dependent on the excitation wavelength.

Recently the time-resolved fluorescence from methyl salicylate and salicylic acid has been measured by Smith and Kaufmann (1978) to obtain additional information on the mechanism for proton transfer in the excited state. Methyl salicylate in methylcyclohexanone was excited with a 264 nm picopulse (8 psec in duration). Since the formation of the zwitterion is accompanied by fluorescence at 350 nm, measurements of the rise of the fluorescence at this wavelength should give a lower limit on the rate of intramolecular proton transfer. Deconvolution of the data indicates that the transfer rate must be greater than 10^{11} sec^{-1}. The data did not change even as the temperature was lowered to 4 K. Replacing the proton with a deuteron and cooling to 4 K also did not alter the data.

The fluorescence lifetime of methyl salicylate in methylcyclohexane was found to be about 280 psec at room temperature. The lifetime of the 450 nm fluorescence from methyl salicylate was measured over the range 40–35 K. The lifetime was nearly temperature independent between 40 and 160 K. Above 160 K the lifetime became shorter as the temperature was increased (see Table 4). Over a temperature range 253–333 K the relative quantum yield was found to decrease with increasing temperature. The change in the relative quantum yields of the 450 nm emission closely paralleled that observed in the lifetime data. This led Smith and Kaufmann to believe that the radiative rate remained constant while the nonradiative rate was a function of temperature. Using the 40 K fluorescence lifetime as an estimate for a temperature-independent radiative lifetime, they were able to calculate the temperature-dependent nonradiative rate using the formula

$$1/\tau = k_r + k_{nr}(T),$$

where τ is the measured fluorescence lifetime, k_r was chosen to be 1.2×10^8 sec^{-1} from low temperature work, and $k_{nr}(T)$ was the temperature-dependent nonradiative decay rate. A plot of the ln $k_{nr}(T)$ versus $1/\tau$ yields an activation energy of 3.7 kcal/mole. A single preliminary experiment for methyl salicylate in which the hydroxy proton has been replaced with a deuteron gives an identical energy of activation.

Methyl salicylate was also dissolved in heptylcyclohexane and octadecane. At 293 K, the former solvent has a viscosity about a factor of 4 larger than that

TABLE 4
Experimental Results

			Relative quantum yield	
Temperature (K)	τ (psec)	$k_{nr}(T)$, sec^{-1}	at 450 nm	at 340 nm
160	8300	—	—	—
184	5020	7.87×10^7	—	—
197	4650	9.46×10^7	—	—
212	2250	3.24×10^8	—	—
213	2092	3.58×10^8	—	—
233	1189	7.21×10^8	—	—
253	803	1.12×10^9	—	—
273	470	2.01×10^9	2.02	1.15
296	280	3.45×10^9	1.00	1.00
303	239	4.06×10^9	0.88	1.20
313	218	4.47×10^9	0.63	1.20
323	176	5.56×10^9	0.48	1.13
333	150	6.55×10^9	0.35	1.03
343	124	7.94×10^9	—	—
353	113	8.73×10^9	—	—

of methylcyclohexane, while the latter is a solid. The lifetime of the 450 nm fluorescence was measured to be about 350 and 450 psec, respectively, in these two solvents. Presumably these changes represent differences in the nonradiative rate.

The steady state fluorescence of the 340 nm component did not change measureably in intensity from 253 to 333 K, while in contrast (see Table 4) blue emission decreased by more than a factor of 6 as the temperature was increased. The excitation spectra to the 450 and 340 nm emission were slightly different as well. The amount of the 340 nm emission from methyl salicylate in methylcyclohexane was too small to accurately measure with the streak camera. However, in acetonitrile Smith and Kaufman found that the 340 nm light had a decay time of about 1 nsec, while the 450 nm luminescence lasted only 100 psec.

Salicylic acid is more complicated than methyl salicylate since it can exist in several ionic forms in the ground and excited states. In addition, it has a much greater tendency to form dimers than the methyl ester. The fluorescence spectrum and fluorescence lifetime of salicylic acid are strongly dependent on the solvent conditions. These data are summarized in Table 5 (Smith and Kaufman, 1978).

The measurement of intramolecular proton translocation sets a lower limit of 10^{11} sec^{-1} on the transfer rate. Since Smith and Kaufmann were unable to measure either a temperature effect or an isotope effect, they cannot confirm or deny the postulate that proton movement proceeds via a tunneling mechanism. The excited state species which emit at 340 and 450 nm are obviously not in equilibrium, since their fluorescence lifetimes differ by about a factor of 10 in

TABLE 5
Experimental Results

	λ_{max}(nm)		Lifetime
	Absorption	Fluorescence	(nsec)
Salicylic acid (MeOH $10^{-2}M$ + H^+)	304	350 + 438	0.38
Salicylic acid (MeOH $10^{-3}M$ + OH^-)	302	398	~3.4
Salicyclic acid (6 N KOH $10^{-2}M$)	304	400	~3.5
Salicylic acid (conc H_2SO_4)	305	410	0.33
Salicylic acid (MCH $10^{-3}M$)	313	415	0.62
Methyl salicylate (MCH $10^{-4}M$)	308	350 + 450	0.28

acetonitrile. In addition, while the fluorescence intensity of the 450 nm emission increases in parallel with an increased fluorescence lifetime as a consequence of lowering the temperature, the 340 nm emission remains relatively constant.

The primary process in visual excitation is initiated by a photochemical event, the absorption of a photon by the photoreceptor, rhodopsin, resulting in the formation of a new species, preluminorhodopsin. The characterization of the new species has been carried out by photostationary studies in low-temperature glasses and by picosecond kinetic studies near room temperatures. Preluminorhodopsin is formed within 6 psec following excitation of rhodopsin and has an absorption maximum at 543 nm which is bathochromically shifted compared to that of rhodopsin. This event has been classically described as the isomerization of the all-cis-retinal chromophore of rhodopsin to the all-trans-retinal form. That full isomerization of a bulky chromophore could occur within this time scale has been questioned (Warshel, 1976), and speculation still exists as to the nature of this photochemical event.

In order to characterize the origins of preluminorhodopsin and its nature, Peters *et al.* (1977) have recently carried out picosecond studies at low temperatures with optically clear glasses of rhodopsin/ethylene glycol. The risetime of preluminorhodopsin as monitored at 570 nm is still within 6 psec at 77 K, and moreover, no risetime for product appearance can be detected until the glass temperatures are below 30 K. At 20 K or below the formation of a relatively long-lived species, preluminorhodopsin is formed within 36 psec. It arises, however, not directly from ground-state rhodopsin but from an initial transient species (presumably the excited state of rhodopsin) which is formed too fast for them to monitor.

The remarkably fast formation of preluminorhodopsin at 4 K and the temperature dependence of formation (Table 6) has led them to consider mechanisms other than cis–trans isomerization as the photochemical event in the first step of visual transduction. A plausible alternative for the event would be proton translocation and if so, replacement of the proton involved with a deuterium should lead to a pronounced deuterium isotope effect. With the deuterated sample in which the proton of the protonated Schiff base has been shown to be exchanged for deuterium, a strong deuterium isotope effect on the rate of formation of preluminorhodopsin is observed (Table 6). For example, at 40 K the H species is formed in less than 6 psec, while the D species is formed within 17 psec, and similarly at 4 K, it is found to be 247 psec for the deuterated sample and 36 psec for the protonated sample. Ober the temperature range (4–40 K), the isotope effect on the rate k_H/k_D was found to be approximately 7.

The dependence of the rate of formation of preluminorhodopsin upon temperature is tabulated in Table 6. The data for both H- and D-rhodopsin show non-Arrhenius behavior (Peters *et al.*, 1977). As a working hypothesis, they assign the initial transient that decays to form preluminorhodopsin to the first singlet state of rhodopsin.

TABLE 6

Temperature Dependence of Formation of Preluminorhodopsin

$(1/T) \times 1000$	T (*K*)	ln k^a
	Rhodopsin	
250	4	22.1
50	20	22.5
33	30	23.7
25	40	24.8
	D-Rhodopsin	
250	4	24.1
100	10	24.3
40	25	25.5

[a] The value ln $k = 25.84$ corresponds to a lifetime of 6 psec.

From Table 6, we can see that the plot ln k versus $1/T$ is nonlinear and as $T \rightarrow 0$, k becomes temperature independent and asymptotically approaches a finite constant value. This nonclassical temperature dependence of the formation of preluminorhodopsin is characteristic of a tunneling phenomenon. Several lines of additional evidence have suggested that proton translocation is an attractive model to account for the appearance of preluminorhodopsin. The red-shifted absorption spectrum of preluminorhodopsin, compared to that of rhodopsin, suggests that preluminorhodopsin might be an even more highly protonated Schiff base form. Model studies of Waddell and Becker (1971) indicate that translocating the H^+ towards the Schiff base nitrogen could account for such a red shift. Resonance Raman data suggest that the preluminorhodopsin species is a protonated species like rhodopsin (Oseroff and Callender 1974). The phenonmenon of proton transfer in excited states has been well established in azole compounds (Williams and Heller, 1970) and azaindole dimers (Ingham and El-Bayoumi, 1974). Considering this body of evidence, an excellent candidate for proton translocation in rhodopsin would be the hydrogen of the protonated Schiff base that is formed between the lysine residue of opsin and the retinal chromophore (Salem, 1976; Mathies and Stryer, 1976; Kropf, 1976). Notice also that no experimental evidence exists that supports the isomerization of an olefin at 4 K; experimental studies with stilbene show that isomerization ceases to take place at 77 K.

Recently, El-Bayoumi *et al.* (1975) have measured the rate constant for intermolecular proton transfer in the excited state of the 7-azaindole (7-AI) hydrogen bonded dimer with nanosecond time resolution. Since double proton transfer in the excited dimer of 7-AI is too fast to be followed at room temperature, the study was performed at 77 K. A deuterated sample of 7-AI was used to further slow down the intermolecular proton transfer, whose rate constant is $1.9 \times 10^8\ \text{sec}^{-1}$ at 77 K. They suggest that the proton transfer at 77 K occurs via quantum mechanical tunneling.

Intramolecular proton transfer in the excited state of 2.4-bis(dimethylamino)-6-(2-hydroxy-5-methylphenyl)-*s*-triazine (TH)

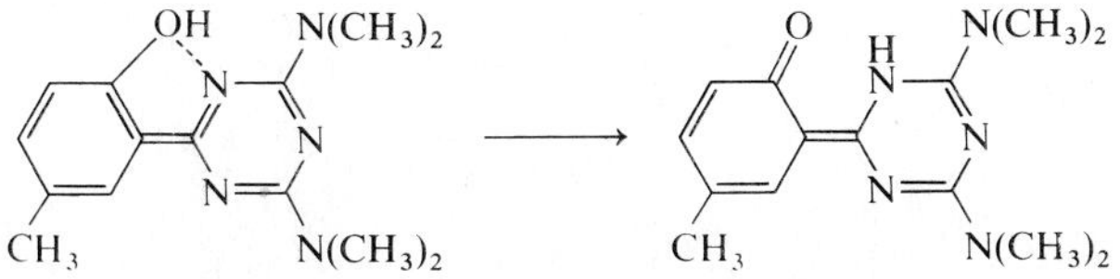

enol form keto form

has been measured by Shizuka *et al.* (1976) by means of picosecond spectroscopy (the pulse width is 15 psec). The lifetime of the excited singlet of TH (enol form) τ_{S_1} and the rate constant k_{PT} for the proton transfer in the S_1 state in cyclohexane ($4 \times 10^{-4} M$) at 298 K were obtained:

$$\tau_{S_1} = 6.3 \times 10^{11} \quad \text{sec} \qquad \text{and} \qquad k_{PT} = 1.1 \times 10^{10} \quad \text{sec}^{-1}.$$

The following data for TH are known: (1) the enthalpy change ΔH between S_1 (enol form) and S_1' (keto form) in TH is large enough to prevent the reverse proton transfer $S_1 \leftarrow S_1'$, and (2) the values of the spontaneous fluorescence quantum yield from S_1', the observed lifetime of S_1', and the radiative rate constant from S_1' are known to be

$$\phi_f' = 0.38, \qquad \tau_f' = 5 \quad \text{nsec}, \qquad \text{and} \qquad k_f' = 1.1 \times 10^8 \quad \text{sec}^{-1},$$

respectively, in degassed nonpolar solvents at 298 K.

According to Shizuka *et al.* (1976), the observation that the k_{PT} in the excited TH at 298 K is very fast can be explained as follows:

(1) The intramolecular hydrogen bond is formed in the ground state, which favors proton transfer in the excited state.

(2) The migration of the π electron from the oxygen atom to the benzene ring takes place in the S_1 state, resulting in an increase in acidity on the oxygen atom.

(3) The electron pair at the proper nitrogen atom in the triazine nucleus is localized as a proton acceptor (Kressge, 1975).

(4) The basicity of the nitrogen atom is increased in the S_1 state.

More recently Shizuka *et al.* (1977) have carried out the study of isotope and substituent effects on the intramolecular proton transfer in the excited singlet state of 6-(2-Hydroxy-5-methylphenyl)-*s*-triazines (see Table 7) by

TABLE 7

6-(2-Hydroxy-5-methylphenyl)-s-triazines

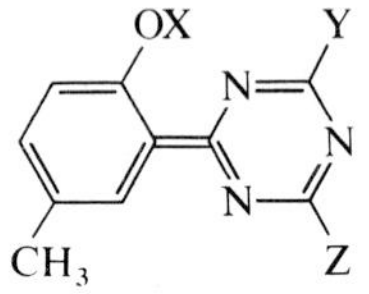

Notations	*X*	*Y*	*Z*
$(OO)_h$	H	OCH_3	OCH_3
$(ON)_h$	H	OCH_3	$N(CH_3)_2$
$(ON)_d$	D	OCH_3	$N(CH_3)_2$
$(NN)_h$	H	$N(CH_3)_2$	$N(CH_3)_2$
$(NN)_d$	D	$N(CH_3)_2$	$N(CH_3)_2$

means of picosecond and nanosecond time-resolved spectroscopy. The fluoroescence quantum yields Φ'_f and the lifetimes τ' of S'_1 at 298 and 77 K in MP have been measured. From the steady state approximation Φ'_f is given by

$$\Phi'_f = [k_{PT}/(k_{PT} + k_d)]k'_f\tau' = \gamma k'_f\tau',$$

where k_d is the rate constant in a rapid deactivation process of S_1 competing with k_{PT}, γ the proton transfer efficiency, and k'_f the radiative rate constant for $S'_1 \to S'_0$ (the ground state of keto form).

The experimental results of k_{S_1} ($k_{S_1} = k_{PT} + k_d$) at 298 K obtained by Shizuka *et al.* (1977) are

$(ON)_h$ and $(ON)_d$:

$$k_{S_1} = 2.1(\pm 0.4) \times 10^{10} \quad sec^{-1};$$

$(NN)_h$:

$$k_{S_1} = 2.6(\pm 0.5) \times 10^{10} \quad sec^{-1}.$$

Using the values of $\gamma_{ON} = 0.57$ and $\gamma_{NN} = 0.69$ (Shizuka *et al.*, 1975), they obtain at 298 K.

$(ON)_h$ and $(ON)_d$:

$$k_{PT} = 1.2(\pm 0.2) \times 10^{10} \quad sec^{-1}, \qquad k_d = 0.9(\pm 0.2) \times 10^{10} \quad sec^{-1};$$

$(NN)_h$:

$$k_{PT} = 1.8(\pm 0.4) \times 10^{10} \quad sec^{-1}, \qquad k_d = 0.8(\pm 0.2) \times 10^{10} \quad sec^{-1}.$$

From the experimental results, it can be said that

(1) there is no isotope effect for the intramolecular proton transfer in the excited singlet of 6-(2-hydroxy-5-methylphenyl)-*s*-triazines, and

(2) the substitution of electron-donating groups into the *s*-triazinyl moiety favors proton transfer in the excited state. The absence of an isotope effect on proton transfer leads Shizuka *et al.* to conclude that the process does not proceed via quantum mechanical tunneling but via a radiationless transition $S_1 \rightarrow S'_1$.

It has been observed by ESR (Iwasaki *et al.*, 1978) that thermal H atoms are trapped in CH_4 containing 0.5 mole % of C_2H_6 when X irradiated at 4.2 K and that upon warming to 10–20 K the trapped H atoms are freed and react selectively with C_2H_6 forming C_2H_5 radicals. The difference in the C—H bond dissociation energies between CH_4 and C_2H_6 must be responsible for this selectivity. The reaction of thermal H atoms of cryogenic temperatures is further confirmed by the 4.2 K photolysis of HI (0.15 mole %) in xenon matrices containing C_2H_6 (0.75 mole %). The H atoms trapped immediately after the photolysis are freed at 20–30 K and react with C_2H_6 forming C_2H_5 radicals. The tunneling effect must play an important role in the reaction at cryogenic temperatures (Iwasaki *et al.*, 1978; Toriyama and Iwasaki, 1978). Recently the tunnel effect in chemical reactions at cryogenic temperatures has received increasing attention (Goldanskii, 1976).

VI. MEMORY FUNCTION

In the above discussion [see Eqs. (5.1)–(5.10)] we have emphasized the importance of the memory function $W_{nm}(t)$ in determining the nature and mechanism of the rate processes and in evaluating the rate constant when the Markoff approximation holds. In this section, we shall show how to evaluate the memory function.

Notice that the memory function $\hat{W}(t)$ or $\hat{W}(P)$ is defined in Eq. (5.6), where it contains $\langle \hat{M}_c(P) \rangle$ or $Tr_b\,[\hat{M}_c(P)\hat{\rho}^{(b)}(0)]$. In other words, only the initial distribution of the heat bath is involved and hence the memory function defined by Eq. (5.6) can be applied to either the canonical system (isothermal system) or the microcanonical system (isolated molecule system). To second-order with respect to $\hat{L}'$ (or $\hat{H}'$), we find

$$\hat{W}(P) = -\hat{D}\langle i\hat{L}' \rangle + \hat{D}\langle (i\hat{L}')[1/(P + i\hat{L})](i\hat{L}') \rangle - \hat{D}\langle i\hat{L}' \rangle [1/(P + i\hat{L}_s)]\hat{D}\langle i\hat{L}' \rangle, \tag{6.1}$$

$$W_{n_s n_s : m_s m_s}(P) = -\langle \hat{L}'[1/(P + i\hat{L})]\hat{L}' \rangle_{n_s n_s : m_s m_s}, \tag{6.2}$$

and

$$W_{n_s n_s : n_s n_s}(P) = -\langle \hat{L}'[1/(P + i\hat{L})]\hat{L}' \rangle_{n_s n_s : n_s n_s}. \tag{6.3}$$

Inverting the Laplace transformation of Eqs. (6.2) and (6.3) yields

$$W_{n_s n_s : m_s m_s}(t) = -\langle \hat{L}' e^{-it\hat{L}} \hat{L}' \rangle_{n_s n_s : m_s m_s} \tag{6.4}$$

and

$$W_{n_s n_s : n_s n_s}(t) = -\langle \hat{L}' e^{-it\hat{L}} \hat{L}' \rangle_{n_s n_s : n_s n_s}, \tag{6.5}$$

respectively.

Equations (6.4) and (6.5) indicate that for the isothermal system [i.e., for $\hat{\rho}^{(b)}(0)$ of Boltzmann distribution], the memory functions $W_{n_s n_s : m_s m_s}(t)$ and $W_{n_s n_s : n_s n_s}(t)$ are closely related to the correlation functions (Gordon, 1968; Levine, 1966). Explicit expressions for $W_{n_s n_s : m_s m_s}(t)$ and $W_{n_s n_s : n_s n_s}(t)$ are given by

$$W_{n_s n_s : m_s m_s}(t) = \frac{2}{\hbar^2} \sum_{n_b} \sum_{m_b} \rho^{(b)}_{m_b m_b}(0) |H'_{n_s n_b, m_s m_b}|^2 \cos \omega_{m_s m_b, n_s n_b} t \tag{6.6}$$

and

$$\begin{aligned} W_{n_s n_s : n_s n_s}(t) &= -\frac{2}{\hbar^2} \sum_{n_b} \sum_{m_b} \sum_{m_s}^{n_s \neq m_s} \rho^{(b)}_{n_b n_b}(0) |H'_{n_s n_b, m_b m_b}|^2 \cos \omega_{m_s m_b, n_b n_b} t \\ &= -\sum_{m_s}' W_{m_s m_s : n_s n_s}(t), \end{aligned} \tag{6.7}$$

respectively. Here we have replaced $\hat{L}$ by $\hat{L}_s + \hat{L}_b$ in Eqs. (6.4) and (6.5). A somewhat better approximation will be to replace $1/(P + i\hat{L})$ in Eqs. (6.2) and (6.3) by $1/[P + i(\hat{L}_s + \hat{L}_b) + \hat{\Gamma}]$, i.e., by introducing the damping effect $\hat{\Gamma}$ (see Section II).

Notice that from Eqs. (5.12) and (6.6)

$$k_{nm} = \int_0^\infty W_{nm}(t)\, dt = \lim_{t \to \infty} \frac{2}{\hbar^2} \sum_{n_b} \sum_{m_b} \rho^{(b)}_{m_b m_b}(0) |H'_{n_s n_b, m_s m_b}|^2 \frac{\sin \omega_{m_s m_b, n_s n_b} t}{\omega_{m_s m_b, n_s n_b}}, \tag{6.8}$$

and $\lim_{t \to \infty} \sin \omega_{m_s m_b, n_s n_b} t / \omega_{m_s m_b, n_s n_b} = \pi \delta(\omega_{m_s m_b, n_s n_b})$. In other words, we obtain the Fermi golden rule expression for the rate constant k_{nm} in this case.

Next we shall show how to calculate a memory function $W_{m_s n_s}(t)$ of general interest. For this purpose, we shall choose to evaluate the memory function of the thermally activated processes discussed in Section V. Using the adiabatic approximation Eq. (5.27), Eq. (6.6) becomes

$$W_{mn}(t) = W_{m_s m_s : n_s n_s}(t) = \frac{2}{\hbar^2} \sum_{v'} \sum_{v''} \rho^{(b)}_{nv'} |\langle \Theta_{nv'} | H'_{nm} | \Theta_{mv''} \rangle|^2 \cos \omega_{mv'', nv'} t. \tag{6.9}$$

For simplicity we shall introduce the Condon approximation to rewrite Eq. (6.9) as

$$W_{mn}(t) = \frac{2}{\hbar^2} |H'_{nm}|^2 \operatorname{Re} \sum_{v'} \sum_{v''} \rho^{(b)}_{nv'} |\langle \Theta_{bv'} | \Theta_{mv''} \rangle|^2 \exp(it\omega_{mv'', nv'}). \tag{6.10}$$

If the wavefunctions $\Theta_{nv'}$ and $\Theta_{mv''}$ can be written as a product of harmonic oscillator wavefunctions,

$$\Theta_{nv'} = \prod_i X_{nv_i'}(Q_i'); \qquad \Theta_{mv''} = \prod_i X_{mv_i''}(Q_i''). \tag{6.11}$$

Substituting Eq. (6.11) into Eq. (6.10), we obtain

$$W_{mn}(t) = \frac{2}{\hbar^2} |H'_{nm}|^2 \operatorname{Re}\left[e^{it\omega_{mn}} \prod_i G_i(t)\right], \tag{6.12}$$

where $G_i(t)$ is defined by

$$G_i(t) = \sum_{v_i'} \sum_{v_i''} \rho_{nv_i'}^{(b)} |\langle X_{nv_i} | X_{mv_i''} \rangle|^2 \exp\{it[(v_i'' + \tfrac{1}{2})\omega_i'' - (v_i' + \tfrac{1}{2})\omega_i']\}. \tag{6.13}$$

If $\rho_{nv_i'}^{(b)}$ is of the Boltzmann distribution, the analytical expression for $G_i(t)$ has been obtained (Lin, 1966). In other words it is possible to obtain the analytical expression for the memory function in the quadratic coupling (Munn and Silbey, 1978) without the mode mixing.

For the case in which $\omega_i'' = \omega_i' = \omega_i$ (the linear coupling case), Eq. (6.13) can be simplified to (Lin, 1966)

$$G_i(t) = \exp\left[-\frac{\Delta_i^2}{2} \coth \frac{\hbar\omega_i}{2kT} + \frac{\Delta_i^2}{2} \coth \frac{\hbar\omega_i}{2kT} \cos\left(\omega_i t - \frac{i\hbar\omega_i}{2kT}\right)\right], \tag{6.14}$$

where Δ_i represents the dimensionless normal coordinate displacement of the ith mode between n and m states. Substituting Eq. (6.14) into Eq. (6.12) yields

$$W_{mn}(t) = W_{mn}(0) \operatorname{Re}\left\{\exp\left[it\omega_{mn} - \sum_i \frac{\Delta_i^2}{2} \coth \frac{\hbar\omega_i}{2kT} + \sum_i \frac{\Delta_i^2}{2} \operatorname{csch} \frac{\hbar\omega_i}{2kT} \cos\left(\omega_i t - \frac{i\hbar\omega_i}{2kT}\right)\right]\right\} \tag{6.15}$$

where $W_{mn}(0) = 2/\hbar^2 |H'_{nm}|^2$, the value of $W_{mn}(t)$ at $t = 0$. Equation (6.15) can be rewritten as

$$W_{mn}(t) = W_{mn}(0) \operatorname{Re}\left\{\exp\left[-S + it\omega_{mn} + \frac{1}{2} \sum_i \Delta_i^2 \exp(it\omega_i) + \sum_i \bar{n}_i \Delta_i^2 (\cos \omega_i t - 1)\right]\right\}, \tag{6.16}$$

or

$$W_{mn}(t) = W_{mn}(0) \exp\left[-\sum_i \frac{\Delta_i^2}{2} \coth \frac{\hbar\omega_i}{2kT} (1 - \cos \omega_i t)\right] \times \cos\left(\omega_{mn} t + \sum_i \frac{\Delta_i^2}{2} \sin \omega_i t\right), \tag{6.17}$$

where $S = \frac{1}{2} \sum_i \Delta_i^2$, which was defined in Section V.

The memory function given above has a wide variety of applications; it can be applied to the rate processes discussed in Section V and also to the vibrational relaxation in dense media (Lin, 1976). If the molecular rotation is important as the heat bath variables, then the above results for the memory function should be modified (Gerber and Berkowitz, 1979).

To evaluate Eq. (6.17) for $W_{mn}(t)$ in a dense medium, it is necessary to know the frequency distribution of local modes, i.e.,

$$S = \int f(\omega)\, d\omega. \tag{6.18}$$

For example, using the Einstein model $f(\omega)$ is given by (Fletcher *et al.*, 1978; Bondybey and Nitzan, 1977)

$$f(\omega) = S\,\delta(\omega - \omega_{\mathrm{E}}), \tag{6.19}$$

where ω_{E} is the Einstein frequency, and substituting Eqs. (6.18) and (6.19) into Eq. (6.17) yields

$$W_{mn}(t) = W_{mn}(0) \exp[-S \coth(\hbar\omega_{\mathrm{E}}/2kT)(1 - \cos \omega_{\mathrm{E}} t)] \cos(\omega_{mn} t + S \sin \omega_{\mathrm{E}} t). \tag{6.20}$$

Using the exponential model $f(\omega)$ is given by (Bondybey and Nitzan, 1977; Fletcher *et al.*, 1978)

$$f(\omega) = (S/n!\omega_{\mathrm{E}}^{n+1})\omega^n \exp(-\omega/\omega_{\mathrm{E}}), \tag{6.21}$$

where n and ω_{E} are constant. In this case, we find

$$\sum_i \frac{\Delta_i^2}{2} \coth \frac{\hbar\omega_i}{2kT} = \int_0^\infty f(\omega) \coth \frac{\hbar\omega}{2kT}\, d\omega = S + \sum_{m=1}^{\infty} \frac{2S}{[1 + (m\hbar\omega_{\mathrm{E}}/kT)]^{n+1}}, \tag{6.22}$$

$$\begin{aligned}
\sum_i \frac{\Delta_i^2}{2} \coth \frac{\hbar\omega_i}{2kT} \cos \omega_i t &= \int_0^\infty f(\omega) \coth \frac{\hbar\omega}{2kT} \cos \omega t\, d\omega \\
&= \frac{S}{2}\left[\frac{1}{(1 - it\omega_{\mathrm{E}})^{n+1}} + \frac{1}{(1 + it\omega_{\mathrm{E}})^{n+1}}\right] \\
&\quad + S \sum_{m=1}^{\infty} \left[\frac{1}{[1 - it\omega_{\mathrm{E}} + (m\hbar\omega_{\mathrm{E}}/kT)]^{n+1}} + \frac{1}{[1 + it\omega_{\mathrm{E}} + (m\hbar\omega_{\mathrm{E}}/kT)]^{n+1}}\right]
\end{aligned} \tag{6.23}$$

and

$$\sum_i \frac{\Delta_i^2}{2} \sin \omega_i t = \int_0^\infty f(\omega) \sin \omega t\, d\omega = \frac{S}{2i}\left[\frac{1}{(1 - it\omega_{\mathrm{E}})^{n+1}} - \frac{1}{(1 + it\omega_{\mathrm{E}})^{n+1}}\right]. \tag{6.24}$$

Similarly, for the Gaussian model (Bondybey and Nitzan, 1977; Fletcher, *et al.*, 1978)

$$f(\omega) = (S/\omega_E\sqrt{\pi}) \exp\{-[(\omega - \omega_M)/\omega_E]^2\}, \tag{6.25}$$

we have

$$\begin{aligned}\sum_i \frac{\Delta_i^2}{2} \coth \frac{\hbar\omega_i}{2kT} \cos \omega_i t &= \int_{-\infty}^{\infty} d\omega\, f(\omega) \coth \frac{\hbar\omega}{2kT} \cos \omega t \\ &= S \exp\left(-\frac{\omega_E^2 t^2}{4}\right) \cos \omega_M t + 2S \exp\left(-\frac{\omega_E^2 t^2}{4}\right) \\ &\quad \times \sum_{m=1} \exp\left(-\frac{m\hbar\omega_M}{kT} + \frac{m^2\hbar^2\omega_E^2}{4k^2T^2}\right) \cos \omega_E t\left(\frac{\omega_M}{\omega_E} - \frac{m\hbar\omega_E}{2kT}\right),\end{aligned} \tag{6.26}$$

$$\begin{aligned}\sum_i \frac{\Delta_i^2}{2} \coth \frac{\hbar\omega_i}{2kT} &= \int_{-\infty}^{\infty} d\omega\, f(\omega) \coth \frac{\hbar\omega}{2kT} \\ &= S + 2S \sum_{m=1} \exp\left(-\frac{m\hbar\omega_M}{kT} + \frac{m^2\hbar^2\omega_E^2}{4k^2T^2}\right),\end{aligned} \tag{6.27}$$

and

$$\sum_i \frac{\Delta_i^2}{2} \sin \omega_i t = \int_{-\infty}^{\infty} d\omega\, f(\omega) \sin \omega t = S \exp\left(-\frac{\omega_E^2 t^2}{4}\right) \sin \omega_M t. \tag{6.28}$$

For the Lorentzian model (Munn and Silbey, 1978)

$$f(\omega) = (S/\pi)\{\gamma/[\gamma^2 + (\omega - \omega_E)^2]\} \tag{6.29}$$

we have

$$\begin{aligned}\sum_i \frac{\Delta_i^2}{2} \coth \frac{\hbar\omega_i}{2kT} \cos \omega_i t &= \int_{-\infty}^{\infty} d\omega\, f(\omega) \coth \frac{\hbar\omega}{2kT} \cos \omega t \\ = S \exp(-\gamma|t|) \cos \omega_E t &+ 2S \cos \omega_E t \sum_{m=1}^{\infty} \exp\left(-\frac{m\hbar\omega_E}{kT} - \gamma\left|t + \frac{im\hbar}{kT}\right|\right),\end{aligned} \tag{6.30}$$

$$\sum_i \frac{\Delta_i^2}{2} \coth \frac{\hbar\omega_i}{2kT} = S \coth \frac{\hbar(\omega_E + \gamma)}{2kT}, \tag{6.31}$$

and

$$\sum_i \frac{\Delta_i^2}{2} \sin \omega_i t = \int_{-\infty}^{\infty} d\omega\, f(\omega) \sin \omega t = S \exp(-\gamma|t|) \sin \omega_E t. \tag{6.32}$$

As can be seen from Eqs. (6.18)–(6.32), the time-dependent behavior of the memory function $W_{mn}(t)$ is sensitive to the coupling mechanism [through $f(\omega)$] between the system and the heat bath, although the rate constant is not

(Fletcher *et al.*, 1978). A general time dependence of $W_{mn}(t)$ in the short time region (or for the large S value case) can be obtained from Eq. (6.17) by expanding $\cos \omega_i t$ and $\sin \omega_i t$ in power series of t. To the approximation of t^2, we find

$$W_{mn}(t) = W_{mn}(0) \exp\left(-t^2 \sum \frac{\Delta_i^2 \omega_i^2}{4} \coth \frac{\hbar\omega_i}{2kT}\right) \cos t\left(\omega_{mn} + \sum_i \frac{\Delta_i^2 \omega_i}{2}\right). \tag{6.33}$$

Next we consider the damping effect on the memory function $W_{n_s n_s : m_s m_s}(t)$. From Eqs. (6.2) and the results given in Section II, we find

$$W_{n_s n_s : m_s m_s}(P) = \frac{1}{\hbar^2} \sum_{n_b} \sum_{m_b} \rho^{(b)}_{m_b m_b}(0) |H'_{n_s n_b, m_s m_b}|^2$$
$$\times [1/(P + i\omega_{m_s m_b, n_s n_b} + \Gamma_{n_s m_s}) + 1/(P + i\omega_{n_s n_b, m_s m_b} + \Gamma_{n_s m_s})], \tag{6.34}$$

and

$$W_{n_s n_s : m_s m_s}(t) = \frac{2}{\hbar^2} \sum_{n_b} \sum_{m_b} \rho^{(b)}_{m_b m_b}(0) |H'_{n_s n\mu, m_s m_b}|^2 \exp(-\Gamma_{n_s m_s} t) \cos \omega_{m_s m_b, n_s n_b} t. \tag{6.35}$$

In this case, corresponding to Eq. (6.17) we have

$$W_{mn}(t) = W_{mn}(0) \exp\left[-\sum_i \frac{\Delta_i^2}{2} \coth \frac{\hbar\omega_i}{2kT}(1 - \cos \omega_i t) - \Gamma_{nm}|t|\right]$$
$$\times \cos\left(\omega_{mn} t + \sum_i \frac{\Delta_i^2}{2} \sin \omega_i t\right), \tag{6.36}$$

where the damping constant Γ_{nm} is defined in Section II.

As mentioned above, in calculating the memory function from Eqs. (6.1)–(6.3), only the initial distribution of the heat bath is involved and so far we have only considered the calculation of the memory function for an isothermal system (i.e., $\hat{\rho}^{(b)}(0)$ is the Boltzmann distribution). Let us now consider the case of isolated systems. If $\hat{\rho}^{(b)}(0)$ is of the equilibrium distribution, then for an isolated system with an energy E, the so-called microcanonical distribution $\rho^{(b)}_{m_b m_b}(0)$ is given by (Lin, 1972, 1973a)

$$\rho^{(b)}_{m_b m_b}(0) = \delta(E - E_{m_b})/\rho^{(b)}(E), \tag{6.37}$$

where $\rho^{(b)}(E)$ represents the density of states of the heat bath

$$\rho^{(b)}(E) = \sum_{m_b} \delta(E - E_{m_b}). \tag{6.38}$$

Introducing the contour integral representation for the delta function $\delta(E - E_{m_b})$,

$$\delta(E - E_{m_b}) = \frac{1}{2\pi i} \int_c d\beta\, e^{\beta(E - E_{m_b})}, \tag{6.39}$$

Eq. (6.38) can be expressed as (Lin, 1972, 1973a)

$$\rho^{(b)}(E) = \frac{1}{2\pi i}\int_c d\beta\, e^{\beta E} Q_b(\beta) = L^{-1}[Q_b(\beta)], \tag{6.40}$$

where $Q_b(\beta) = \sum_{m_b} e^{-\beta E m_b}$, the canonical partition function (with β being replaced by $1/kT$) and L^{-1} represents the inverse Laplace transformation. For practical calculation of $\rho^{(b)}(E)$, the saddle-point method is commonly used (Lin, 1972, 1973a).

Substituting Eqs. (6.39) and (6.37) into Eq. (6.6), we obtain the memory function $W_{n_s n_s : m_s m_s}(t)_E$ for the microcanonical system (or the isolated system) as

$$\begin{aligned} W_{n_s n_s : m_s m_s}(t)_E &= \frac{1}{\rho^{(b)}(E)}\frac{1}{2\pi_i}\int_c d\beta\, e^{\beta E} Q^{(b)}(\beta) W_{n_s n_s : m_s m_s}(t)_\beta \\ &= \frac{1}{\rho^{(b)}(E)} L^{-1}[Q^{(b)}(\beta) W_{n_s n_s : m_s m_s}(t)_\beta], \end{aligned} \tag{6.41}$$

where $W_{n_s n_s : m_s m_s}(t)_\beta$ denotes the memory function for an isothermal system (with $\beta = 1/kT$) that has been discussed in Eqs. (6.9)–(6.36).

Notice that if $W_{n_s n_s : m_s m_s}(t)_\beta$ does not vary rapidly with respect to β, then we may apply the saddle-point method to Eq. (6.4) to obtain

$$W_{n_s n_s : m_s m_s}(t)_E = W_{n_s n_s : m_s m_s}(t)_{\beta *} \tag{6.42}$$

where β^* represents the saddle-point value of β and is determined by (Lin, 1973a)

$$E = -\left[\frac{\partial}{\partial \beta}\log Q_b(\beta)\right]_{\beta = \beta^*}. \tag{6.43}$$

Another case of interest for the correlation function will be that associated with the single vibronic (or single rovibronic) level rate process. This case can be treated similarly as that given above and will not be discussed.

Another aspect of the memory function is that it can often be obtained from the spectral line function (Gordon, 1968; Kenkre and Knox, 1974; Lin, 1976). To demonstrate this point, let us consider the case of the energy transfer of electronic excitation as an example. Notice that in this case, the time dependent portion of Eq. (6.12) can be written as a product of the donor contribution (designated D) and the acceptor contribution (designated A)

$$\begin{aligned} W_{mn}(t) &= W_{mn}(0)\,\mathrm{Re}\left\{\left[\exp(it\omega_{mn}^{(A)})\prod_i G_i^{(A)}(t)\right]\left[\exp(it\omega_{mn}^{(D)})\prod_i G_j^{(D)}(t)\right]\right\} \\ &= W_{mn}(0)\,\mathrm{Re}\{C_A(t)C_D(t)\}. \end{aligned} \tag{6.44}$$

Equation (6.44) can be applied to the singlet–singlet and triplet–triplet transfers. Since in obtaining Eq. (6.44) the Condon approximation has been introduced, if the singlet–singlet transfer (or its equivalents like triplet–singlet transfer, singlet–triplet transfer, etc.) involves the symmetry-forbidden transition, then the normal coordinate dependence of the electric dipole transition moment should be taken into consideration (Lin, 1971, 1973b).

We will first express $C_A(t)$ in terms of the absorption coefficient of the acceptor. For this purpose, we use Eq. (3.15). As $\Gamma_{nm} \to 0$, Eq. (3.15) becomes

$$\alpha_{nm}^{(A)} = \frac{4\pi^2\omega_r}{3a\hbar c} |\mathbf{R}_{mn}^{(A)}|^2 \sum_{v'v''} \rho_{nv'}^{(A)} |\langle \Theta_{nv'}^{(A)} | \Theta_{mv''}^{(A)} \rangle|^2 \delta(\omega_r - \omega_{mv'', nv'}^{(A)}), \quad (6.45)$$

where $|\mathbf{R}_{mn}^{(A)}|^2$ represents the electronic transition moment of the acceptor and a is a function of refractive index introduced to correct the medium effect (Lin, 1973b). Introducing the integral expression for the delta function in Eq. (6.45), we obtain

$$\alpha_{nm}^{(A)}(\omega_r) = \frac{2\pi\omega_r}{3ach} |\mathbf{R}_{mn}^{(A)}|^2 \int_{-\infty}^{\infty} dt \exp(-it\omega_r) C_A(t). \quad (6.46)$$

Inversion of the Fourier transformation yields (Lin, 1973b)

$$C_A(t) = \int_{-\infty}^{\infty} \exp(it\omega_r) \alpha_{nm}^{(A)}(\omega_r) \frac{d\omega_r}{\omega_r} \bigg/ \int_{-\infty}^{\infty} \alpha_{nm}^{(A)}(\omega_r) \frac{d\omega_r}{\omega_r}, \quad (6.47)$$

or

$$C_A(t) = \int_{-\infty}^{\infty} \exp(it\omega_r) \alpha_{nm}^{(A)}(\omega_r)_N \frac{d\omega_r}{\omega_r}, \quad (6.48)$$

where

$$\alpha_{nm}^{(A)}(\omega_r)_N = \alpha_{nm}^{(A)}(\omega_r) \bigg/ \int_{-\infty}^{\infty} \alpha_{nm}^{(A)}(\omega_r) \frac{d\omega_r}{\omega_r}, \quad (6.49)$$

the normalized absorption coefficient. Equation (6.48) indicates that $C_A(t)$ can be obtained from the Fourier transformation of the normalized absorption coefficient of the acceptor.

Similarly, to determine $C_D(t)$ in Eq. (6.44), we start with the intensity distribution of the donor emission spectrum given by Eq. (3.21),

$$I_{nm}^{(D)}(\omega_r) = \frac{4\alpha\omega_r^3}{3\hbar c^3} |\mathbf{R}_{mn}^{(D)}|^2 \sum_{v'v''} \rho_{nv'}^{(D)} |\langle \Theta_{nv'}^{(D)} | \Theta_{mv''}^{(D)} \rangle|^2 \delta(\omega_{nv', mv''} - \omega_r), \quad (6.50)$$

where α is the correction factor for the medium effect. Repeating the derivation given by Eqs. (6.45)–(6.49), we obtain (Lin, 1973b)

$$C_D(t) = \int_{-\infty}^{\infty} I_{nm}^{(D)}(\omega_r)_N \exp(-it\omega_r) \frac{d\omega_r}{\omega_r^3}, \quad (6.51)$$

where

$$I_{nm}^{(D)}(\omega_r)_N = 1_{nm}^{(D)}(\omega_r) \bigg/ \int_{-\infty}^{\infty} I_{nm}^{(D)}(\omega_r) \frac{d\omega_r}{\omega_r^3}. \quad (6.52)$$

Recently, Eq. (6.44) has been applied by Kenkre and Knox (1975) to determine the memory function $W_{nm}(t)/W_{nm}(0)$ of the exciton migration in organic crystals like anthracene, bacteriochlorophyll, and adenosine phosphate at 300 K. They found that the main feature of the memories is their falloff in 0.01–0.03 psec, showing that all coherence is lost well before the picosecond spectroscopy region. They concluded that for these systems, any effects due to exciton motion may safely be described in terms of diffusion.

The relation between the memory function (or correlation function) and the spectral line function for other types of motion have been discussed by Gordon (1968) and Lin (1976).

Finally, let us discuss the Markoff approximation by using the memory function. Applying the Taylor expansion to $P_m(t - t_1)$ and $P_n(t - t_1)$ in Eq. (5.10) yields

$$\frac{dP_n}{dt} = \sum_m \sum_{l=0}^{\infty} \left[W_{nm}^{(l)}(t) \frac{d_l P_m}{dt^l} - W_{mn}^{(l)}(t) \frac{d_l P_n}{dt^l} \right], \tag{6.53}$$

where, for example,

$$W_{nm}^{(l)}(t) = \frac{1}{l!} \int_0^t dt_1 \, (-t_1)^l W_{nm}(t_1). \tag{6.54}$$

Notice that when the Markoff approximation is valid, the rate constant is given by

$$k_{nm} = \int_0^\infty dt \; W_{nm}(t). \tag{6.55}$$

Thus, the Markoff approximation holds provided that $k_{nm} = W_{nm}^{(0)}(t) = W_{nm}^{(0)}(\infty)$ and that the magnitude of $W_{nm}^{(l)}(t)\, d^l P_m/dt^l$ and $W_{mn}^{(l)}(t)\, d^l P_n/dt^l$ for $l \geq 1$ is negligible compared with that of dP_n/dt. These two conditions are satisfied if the memory function $W_{nm}(t)$ can be expressed as

$$W_{nm}(t) = k_{nm}\, \delta(t). \tag{5.12}$$

However, the condition given by Eq. (5.12) is perhaps too strict. For example, if in the long time region, $W_{nm}(t)$ is given by

$$W_{nm}(t) = W_{nm}(0) \exp(-t/T_{nm}) \tag{6.56}$$

where T_{nm} represents the correlation time, then the above conditions are satisfied provided $T_{mn}(k_{nm} + k_{mn}) \ll 1$. In other words, $T_{mn} \ll [1/W_{mn}(0)]^{1/2}$, where $W_{mn}(0) = (2/\hbar^2)|H'_{nm}|^2$.

In concluding the discussion of this section, we would like to point out the importance of the memory function. For the case in which the Markoff approximation holds, it is usually possible to obtain the analytical expression of the memory function for a rate process and to calculate the rate constant [see Eq. (6.8)], and one may employ the numerical method for the integration with respect to t whenever the analytical method for this purpose is too difficult; to compare different models, it is more informative to examine the memory functions than the corresponding rate constants (i.e., the rate constant can only provide the gross features of a rate process). For the case in which the Markoff approximation does not hold, the rate constants are not defined and one has to deal with the memory function directly in order to determine the time-dependent behavior of the system.

APPENDIX I

In this Appendix, we shall show how to eliminate the heat bath variables and to obtain the equation of motion for the density matrix of the system of interest. For this purpose, we apply the Laplace transformation to Eq. (2.1),

$$\hat{\rho}(P) = [1/(P + i\hat{L})]\hat{\rho}(0), \tag{AI.1}$$

where $\hat{\rho}(P)$ denotes the Laplace transform of $\hat{\rho}(t)$,

$$\hat{\rho}(P) = \int_0^\infty dt\, e^{-pt}\hat{\rho}(t)dt. \tag{AI.2}$$

Following Fano (1964) we introduce the transition operator $\hat{M}(P)$, defined by

$$\frac{1}{P + i\hat{L}} = \frac{1}{P + i\hat{L}_0}\left[1 + \hat{M}(P)\frac{1}{P + i\hat{L}_0}\right], \tag{AI.3}$$

where $\hat{L}$ and $\hat{L}_0$ are defined by Eq. (2.3). $\hat{M}(P)$ can be put into the following form:

$$\hat{M}(P) = (-i\hat{L}') + (-i\hat{L}')[1/(P + i\hat{L})](-i\hat{L}'). \tag{AI.4}$$

Substituting Eq. (AI.3) into Eq. (AI.1) and carrying out a trace over the quantum states of the heat bath, we obtain (Lin and Eyring, 1977b)

$$\hat{\rho}^{(s)}(P) = [1/(P + i\hat{L}_s)][1 + \langle\hat{M}(P)\rangle 1/(P + i\hat{L}_s)]\hat{\rho}^{(s)}(0), \tag{AI.5}$$

in which $\langle\hat{M}(P)\rangle = \mathrm{Tr}_b\,[\hat{M}(P)\hat{\rho}^{(b)}(0)]$ and the relation Eq. (AI.5) can be conveniently expressed as (Fano, 1964; Lin and Eyring, 1977b)

$$\hat{\rho}^{(s)}(P) = [1/(P + i\hat{L}_s + \langle\hat{M}_c(P)\rangle)]\hat{\rho}^{(s)}(0), \tag{AI.6}$$

where

$$\langle\hat{M}_c(P)\rangle = \{1/[1 + \langle\hat{M}(P)\rangle(P + i\hat{L}_s)^{-1}]\}(-\langle\hat{M}(P)\rangle). \tag{AI.7}$$

In Eqs. (AI.5) and (AI.6), the heat bath variables have been eliminated (Fano, 1964; Zwanzig, 1964; Lin and Eyring, 1977b).

Notice that Eq. (AI.6) can be rewritten as

$$P\hat{\rho}^{(s)}(P) - \hat{\rho}^{(b)}(0) = -i\hat{L}_s\hat{\rho}^{(s)}(P) - \langle \hat{M}_c(P)\rangle \hat{\rho}^{(s)}(P). \tag{AI.8}$$

Inverting the Laplace transformation of Eq. (AI.8), we find

$$\frac{\partial \hat{\rho}^{(s)}(t)}{\partial t} = -i\hat{L}_s\hat{\rho}^{(s)}(t) - \int_0^t dt_1 \, \langle \hat{M}_c(t_1)\rangle \hat{\rho}^{(b)}(t - t_1), \tag{AI.9}$$

in which

$$\langle \hat{M}_c(P)\rangle = \int_0^\infty dt \, e^{-Pt}\langle \hat{M}_c(t)\rangle. \tag{AI.10}$$

Equations (AI.5), (AI.6), and (AI.9) are equivalent and are useful for finding the density matrix of the system of interest at $t > 0$. Using the Taylor expansion

$$\hat{\rho}^{(s)}(t - t_1) = \sum_{n=1}^{\infty} \frac{1}{n!}(-t_1)^n \frac{\partial^n \hat{\rho}^{(s)}(t)}{\partial t^n}, \tag{AI.11}$$

Eq. (AI.9) becomes

$$\begin{aligned} \frac{\partial \hat{\rho}^{(s)}(t)}{\partial t} &= -i\hat{L}_s\hat{\rho}^{(s)}(t) - \sum_{n=0}^{\infty} \frac{1}{n!} \int_0^t dt_1 \, (-t_1)^n \langle \hat{M}_1(t_1)\rangle \frac{\partial^n \hat{\rho}^{(s)}(t)}{\partial t^n} \\ &= -i\hat{L}_s\hat{\rho}^{(s)} - \sum_{n=0}^{\infty} \hat{K}_n(t) \frac{\partial^n \hat{\rho}^{(s)}}{\partial t^n}. \end{aligned} \tag{AI.12}$$

Using the ordinary perturbation method, one can only find the proper expression for the equation of motion of the diagonal matrix element of $\hat{\rho}^{(s)}(t)$. To obtain the proper equations of motion for both diagonal and off-diagonal matrix elements of $\hat{\rho}^{(s)}(t)$ in the long time region, we may employ the so-called singular perturbation method (or the multiple time scale method) (O'Malley, 1974; Richardson *et al.*, 1973; Cukier and Deuch, 1969; Lin and Eyring, 1977b). According to the singular perturbation method, we replace the original single time variable t by $t_0, t_1, t_2 \ldots$, in which $t_n = \lambda^n t$ and λ represents the perturbation parameter associated with $\hat{L}'$. All t_n values are treated as independent variables. $\hat{\rho}^{(s)}$ and $\hat{K}_n(t)$ in Eq. (AI.12) are expanded as follows:

$$\hat{\rho}^{(s)} = \hat{\rho}_0^{(s)} + \lambda\hat{\rho}_1^{(s)} + \lambda^2\hat{\rho}_2^{(s)} + \cdots, \tag{AI.13}$$

and

$$\hat{K}_n(t) = \lambda\hat{K}_n^{(1)}(t) + \lambda^2\hat{K}_n^{(2)}(t) \cdots. \tag{AI.14}$$

Substituting Eqs. (AI.13) and (AI.14) into Eq. (AI.12) and using the Markoff approximation and the relation

$$\frac{\partial}{\partial t} = \frac{\partial}{\partial t_0} + \lambda\frac{\partial}{\partial t_1} + \lambda^2\frac{\partial}{\partial t_2} + \cdots. \tag{AI.15}$$

Eqs. (2.6)–(2.7) can be obtained (Lin and Eyring, 1977b).

APPENDIX II

In this Appendix we shall be concerned with the three-dimensional rotational effect on vibrational relaxation in dense media. The two-dimensional case has been extensively studied (Freed and Metiu, 1977; Freed *et al.*, 1977; Knittel and Lin, 1978; Diestler, 1979).

For simplicity we consider the system to consist of diatomics BC dissolved in monatomic crystals; the interaction between the host atom and guest atom is assumed to be of the Morse type

$$V = D_{\mathrm{B}}[1 - e^{-a_{\mathrm{B}}(X_{\mathrm{B}} - r_{\mathrm{B}0})}]^2 + D_{\mathrm{C}}[1 - e^{-a_{\mathrm{B}}(X_{\mathrm{B}} - r_{\mathrm{B}0})}]^2, \qquad \text{(AII.1)}$$

where

$$X_{\mathrm{B}}^2 = \alpha^2 r^2 + \bar{X}^2 - 2\bar{X}\alpha r \cos\delta \qquad \text{(AII.2)}$$

and

$$X_{\mathrm{C}}^2 = r^2(1-\alpha)^2 + \bar{X}^2 + 2\bar{X}r(1-\alpha)\cos\delta, \qquad \text{(AII.3)}$$

with $\alpha = m_{\mathrm{C}}/(m_{\mathrm{B}} + m_{\mathrm{C}})$. r is the instantaneous bond distance of BC. $\bar{X}$ represents the distance between the host atom and the center of mass of BC and δ is the angle between $\bar{X}$ and r with the two lines intersected at the center of mass of BC. When $\bar{X} > r$, Eq. (AII.2) and (AII.3) can be approximately expressed as

$$X_{\mathrm{B}} = \bar{X}[1 - \alpha(r/\bar{X})\cos\delta + \cdots] \qquad \text{(AII.4)}$$

and

$$X_{\mathrm{C}} = \bar{X}[1 + (1-\alpha)(r/\bar{X})\cos\delta + \cdots]. \qquad \text{(AII.5)}$$

Using Eqs. (AII.4) and (AII.5), Eq. (AII.1) becomes

$$V = D_{\mathrm{B}}[1 - e^{-a_{\mathrm{B}}(\bar{X} - r_{\mathrm{B}0})}e^{a_{\mathrm{B}}\alpha r\cos\delta}]^2 + D_{\mathrm{C}}[1 - e^{-a_{\mathrm{C}}(\bar{X} - r_{\mathrm{C}0})}e^{-a_{\mathrm{C}}(1-\alpha)r\cos\delta}]^2. \qquad \text{(AII.6)}$$

If we let $r = r_0 + \Delta r$, where Δr represents the displacement of the diatomic molecule, expand V in power series of Δr, and collect the linear terms, we find the perturbation for vibrational relaxation to be

$$\hat{H}' = \hat{H}'(R) + \hat{H}'(A), \qquad \text{(AII.7)}$$

where $\hat{H}'(R)$ denotes the contribution from the repulsive portion of V, and $\hat{H}'(A)$ represents the attractive contribution from V. They are given by (Lin, 1977; Lin *et al.*, 1979; Gerber and Berkowitz, 1979)

$$\hat{H}'(R) = 2\alpha\Delta r \sum_n \cos\delta_n \left[D_{\mathrm{B}} e^{-2a_{\mathrm{B}}(\bar{X}_n - r_{\mathrm{B}0})} e^{2a_{\mathrm{B}}\alpha r_0 \cos\delta_n} - D_{\mathrm{C}} \frac{(1-\alpha)}{\alpha} \times e^{-2a_{\mathrm{C}}(\bar{X}_n - r_{\mathrm{C}0})} e^{-2a_{\mathrm{C}}(1-\alpha) r_0 \cos\delta_n} \right], \qquad \text{(AII.8)}$$

and

$$\hat{H}'(A) = 2\alpha\Delta r \sum_n \cos\delta_n\left[-D_B e^{-a_B(\bar{X}_n - r_{B0})} e^{a_B \alpha r_0 \cos\delta_n} + D_C \frac{(1-\alpha)}{\alpha}\right.$$
$$\left.\times\, e^{-a_C(\bar{X}_n - r_{C0})} e^{-a_C(1-\alpha) r_0 \cos\delta_n}\right]. \tag{AII.9}$$

Here the contribution from all over the host atoms has been taken into account by the summation over n.

Using $\hat{H}'$ given by Eq. (AII.7), the rate constant of vibrational relaxation can be calculated,

$$K(\beta) = \frac{2\pi}{\hbar} \sum_{\{v\}} \sum_{\{v'\}} \rho_{\{v\}} |\langle 1\{v\}|\hat{H}'|0\{v'\}\rangle|^2 \delta(\hbar\omega_s + E_{\{v\}} - E_{\{v'\}}), \tag{AII.10}$$

for the transition $1 \to 0$ of the diatomics, where $\{v\}$ and $\{v'\}$ are the quantum numbers of the accepting degrees of freedom, and ω_s is the frequency of the molecular vibration. It is commonly believed that the repulsive portion $\hat{H}'(R)$ is more important in vibrational relaxation than $\hat{H}'(A)$.

Using the relation

$$e^{2a_B\alpha r_0 \cos\delta_n} = \sum_{J=0}^{\infty} (2J+1) j_J(2\alpha a_B r_0) P_J(\cos\delta_n) \tag{AII.11}$$

we find

$$\hat{H}' = 2a\Delta r \sum_n \sum_{J=0}^{\infty} F_J(\bar{X}_n)_n (2_J + 1) \cos\delta_n P_J(\cos\delta_n), \tag{AII.12}$$

where $j_J(y)$ is the Bessel function and

$$F_J(\bar{X}_n)_n = D_B e^{-2a_B(\bar{X}_n - r_{B0})} j_J(2\alpha a_B r_0) - \frac{(1-\alpha)}{\alpha} D_C e^{-2a_C(\bar{X}_n - r_{C0})}$$
$$\times\, j_J(2a_C(\alpha - 1)r_0) - D_B e^{-a_B(\bar{X}_n - r_{B0})} j_J(\alpha\, a_B r_0)$$
$$+ \frac{(1-\alpha)}{\alpha} D_C e^{-a_C(\bar{X}_n - r_{C0})} j_J(a_C(1-\alpha)r_0). \tag{AII.13}$$

Expressing $P_J(\cos\delta_n)$ in terms of spherical harmonics $Y_{JM}(\Theta_n\phi_n)$ and $Y_{JM}(\Theta\phi)$, Eq. (AII.12) becomes

$$\hat{H}' = 8\pi\alpha\Delta r \sum_n \sum_{J=0}^{\infty} F_J(\bar{X}_n)_n \left[\frac{J+1}{2J+3} \sum_{M=-J-1}^{J+1} Y_{J+1,M}(\Theta_n\phi_n) Y^*_{J+1,M}(\Theta\phi)\right.$$
$$\left. + \frac{J}{2J-1} \sum_{M=-J+1}^{J-1} Y_{J-1,M}(\Theta_n\phi_n) Y^*_{J-1,M}(\Theta\phi)\right]. \tag{AII.14}$$

This expression is convenient for studying the participation of molecular rotation in vibrational relaxation. It should be noted that $F_J(\overline{X}_n)_n$ will provide the coupling between the molecular vibration and the translation of the guest molecule and the vibration of host atoms; in other words, $\hat{H}'$ can provide us the information about the separate amount of the vibrational excitation going into molecular rotation and other degrees of freedom.

To be able to proceed with the derivation of $K(\beta)$, it is necessary to know the potential function for the molecular rotation. For simplicity, we shall assume that the rotation is free; for low barrier cases, this at least will provide us the zeroth order approximation. In this case, Eq. (AII.10) can be written as

$$K(\beta) = \frac{2\pi}{\hbar} \sum_{JM\{v\}} \sum_{J'M'\{v'\}} \rho_{J\{v\}} |\langle 1JM\{v\}|\hat{H}'|0J'M'\{v'\}\rangle|^2 \delta(\hbar\omega_s + E_{J\{v\}} - E_{J'\{v'\}}), \tag{AII.15}$$

where $\hat{H}'$ is given by Eq. (AII.14).

Let us consider the case $T = 0$. In this case, we are concerned with the calculation of

$$\langle 100\{v\}|\hat{H}'|0J'M'\{v'\}\rangle = 2\alpha\sqrt{4\pi}\langle 1|\Delta r|0\rangle \times \sum_n \left\{\frac{J'}{2J'+1}\langle\{v\}|F_{J'-1}(\overline{X}_n)_n|\{v'\}\rangle + [(J'+1)/(2J'+1)]\langle\{v\}|F_{J'+1}(\overline{X}_n)_n|\{v'\}\rangle\right\} Y_{J'M'}(\Theta_n\phi_n). \tag{AII.16}$$

In Eq. (AII.16), the terms involving $F_{J'-1}$ and $F_{J'+1}$ will describe the energy of vibrational excitation relaxed into the molecular translation and host phonons. For the case in which the $\overline{X}_n$'s are not displaced Eq. (AII.16) reduces to

$$\langle 100\{v\}|\hat{H}'|0J'M'\{v'\}\rangle = 2\alpha\sqrt{4\pi}\langle 1|\Delta r|0\rangle f_{J'\{v\}\{v'\}} I_{J'M'}, \tag{AII.17}$$

where

$$F_{J'\{v\}\{v'\}} = \delta_{\{v\}\{v'\}}\left[\frac{J'}{2J'+1}F_{J'-1}(\overline{X}) + \frac{J'+1}{2J'+1}F_{J'+1}(\overline{X})\right] \tag{AII.18}$$

and

$$I_{J'M'} = \sum_n Y_{J'M'}(\Theta_n\phi_n). \tag{AII.19}$$

Substituting Eq. (AII.19) into Eq. (AII.15) yields

$$K(\infty) = (32\pi^2\alpha^2/\hbar)|\langle 1|\Delta r|0\rangle|^2 f^2_{J'\{0\}\{0\}} \sum_{M'} |J_{J'M'}|^2 \rho(E_{J'}), \tag{AII.20}$$

where $\hbar\omega_s = E_{J'}$ and $\rho(E_{J'})$ is the density of states. To find the analytical expression for $K(\infty)$ given by Eq. (AII.20), we make use of the following approximation:

$$j_J(z) = \frac{(2z)^J J!}{(2J+1)!}\left[1 + \frac{z^2/2}{2J+3} + \frac{(z^2/2)^2}{2!(2J+3)(2J+5)} + \cdots\right]. \quad \text{(AII.21)}$$

It follows (Lin, 1977; Lin *et al.*, 1979) that

$$K(\infty) = \frac{32\pi\alpha^2}{\hbar^2\omega_r}|\langle 1|\Delta r|0\rangle|^2 G_{J'}^2 \frac{(1+\bar{a}^2\alpha^2 r_0^2)^2}{(4a\alpha r_0)^2}\left[\frac{J'!(4\alpha\bar{a}r_0)^{J'}}{(2J'-1)!}\right]^2 \quad \text{(AII.22)}$$

where $\omega_r = \hbar/2I$,

$$G_{J'} = D_B e^{-2a_B(\bar{X}-r_{B0})} + [(1-\alpha)/\alpha]D_C(-1)^{J'} e^{-2a_C(\bar{X}-r_{C0})}, \quad \text{(AII.23)}$$

and μ is the reduced mass of the diatomics. Equation (AII.22) can be put in the modified energy gap law form (Legay, 1978)

$$K(\infty) = \frac{2\pi\alpha^2}{\mu\hbar\omega_s\omega_r} G_{J'}^2 \frac{(1+\bar{a}^2\alpha^2 r_0^2)^2}{(a\alpha r_0)^2} J'(J'+1)\exp\{-2J'[\log J' - \log(\bar{a}e\alpha r_0)]\}. \quad \text{(AII.24)}$$

In Eqs. (AII.22) and (AII.24), an average quantity $\bar{a}$ has been introduced for a_B and a_C.

The rotational effect on the vibrational dephasing can be treated similarly. In this case, in addition to the inelastic process discussed above, we have to treat the elastic process

$$K_n(\beta) = \frac{2\pi}{\hbar}\sum_{\{v\}}\sum_{\{v'\}} \rho_{\{v\}}|\langle n\{v\}JM|H'|n\{v'\}J'M'\rangle|^2\delta(E_{J\{v\}} - E_{J'\{v'\}}). \quad \text{(AII.25)}$$

To simplify Eq. (AII.25) we may use Eq. (AII.6) directly without expanding V in power series of Δr and repeat the process given by Eqs. (AII.8)–(AII.21).

REFERENCES

Abragam, A. (1961). "The Principles of Nuclear Magnetism," Oxford Univ. Press., London and New York.

Albrecht, A. C. (1978). *In* "Advances in Laser Chemistry" (A. H. Zewail, ed.). Springer-Verlag, Berlin and New York.

Alfano, R. R., and Shapiro, S. L. (1972). *Phys. Rev. Lett.* **29**, 1655.

Anderson, R. W., Hochstrasser, R. M., Lutz, H., and Scott, G. W. (1975). *Chem. Phys. Lett.* **32**, 204.

Basu, S. (1964). *Adv. Quantum Chem.* **1**, 145.

Beck, S. M., Monts, D. M., Liverman, M. G., and Smalley, R. E. (1979). *J. Chem. Phys.* **70**, 1062.

Beens, H., Gellmann, K. H., Gurr, M., and Weller, A. (1965). *Disc. Faraday Soc.* **39**, 183.

Ben-Reuven, A. (1975). *Adv. Chem. Phys.* **33**, 235.

Berkowitz, M., and Gerber, R. B. (1977). *Phys. Rev. Lett.* **39**, 1000.

Blumen, A., and Silbey, R. (1978). *J. Chem. Phys.* **69**, 3589.

Blumen, A., Manz, J., and Yakhot, V. (1977). *Chem. Phys.* **26**, 287.

Blumen, A., Lin, S. H., and Manz, J. (1978). *J. Chem. Phys.* **69**, 881.
Bondybey, V. E., and Nitzan, A. (1977). *Phys. Rev. Lett.* **38**, 889.
Bray, M., and Berry, M. J. (1978). *J. Chem. Phys.* **68**, 4419.
Burke, E. P., and Small, G. J. (1974). *J. Chem. Phys.* **61**, 4588.
Burshtein, Z., and Williams, D. F. (1978). *J. Chem. Phys.* **68**, 983.
Cuckier, R. I., and Deutch, J. M. (1969). *J. Chem. Phys.* **50**, 36.
de Vries, H., de Bree, P., and Wiersma, D. A. (1977). *Chem. Phys. Lett.* **52**, 399.
Diestler, D. J., (1976). *Topics Appl. Phys.* **15**, 109.
Diestler, D. J., (1979). *Adv. Chem. Phys.* (in press).
Efrima, S., and Metiu, H. (1978). *J. Chem. Phys.* **69**, 5113.
Eisenthal, K. B. (1977). *Ann. Rev. Phys. Chem.* **28**, 207.
El-Bayoumi, M. A., Avonris, P., and Ware, W. R. (1975). *J. Chem. Phys.* **62**, 2499.
Ern, V., Suna, A., Tomkiewicz, Y., Avokian, P., and Groff, R. P. (1972). *Phys. Rev. B* **5**, 3222.
Fano, U. (1964). *In* "Lecture on the Many-Body Problem" (E. R. Caianiello, ed.). Academic Press, New York.
Feigenson, G. W., and Chan, S. I. (1974). *J. Am. Chem. Soc.* **96**, 1312.
Fischer, S. F., and Laubereau, A. (1975). *Chem. Phys. Lett.* **35**, 6.
Fleming, G. R., Gijzeman, O. L. J., and Lin, S. H. (1973). *Chem. Phys. Lett.* **21**, 527.
Fleming, G. R., Gijzeman, O. L. J., and Lin, S. H. (1974). *J. C. S. Faraday Trans. II* **70**, 37.
Fletcher, D., Fujimura, Y., and Lin, S. H. (1978). *Chem. Phys. Lett.* **57**, 400.
Fong, F. K. (ed.) (1976). *Topics Appl. Phys.* **15**.
Freed, K. F., and Metiu, H. (1977). *Chem. Phys. Lett.* **48**, 262.
Freed, K. F., Metiu, H., and Yeager, D. L. (1977). *Chem. Phys. Lett.* **49**, 19.
Fujimura, Y., and Lin, S. H. (1979a). *J. Chem. Phys.* **70**, 247.
Fujimura, Y., and Lin, S. H. (1979b). *J. Chem. Phys.* **71**, 3140.
Fujimura, Y., Lin, S. H., Shröderg, H., Neusser, H. J., and Schlag, E. W. (1979). *Chem. Phys.* (in press).
Genack, A. Z. Macfarland, R. M., and Brewer, R. G. (1976). *Phys. Rev. Lett.* **37**, 1078.
Gerber, R. P., and Berkowitz, M. (1979). *Chem. Phys.* **37**, 369.
Goldanskii, V. I. (1976). *Ann. Rev. Phys. Chem.* **27**, 85.
Goldanskii, V. I. (1977). *Accounts Chem. Res.* **10**, 153.
Gordon, R. G. (1968). *Adv. Magn. Reson.* **3**, 1.
Grovindjee (ed.) (1978). *Symp. Primary Photoprocesses Photosynthesis: Ultrafast Reactions, In Photochm. Photobiol.* **28**, No. 6.
Haarer, D., and Wolf, H. C. (1970). *Mol. Crystall. Liq. Cryst.* **10**, 359.
Harris, C. B., and Zwemer, D. A. (1978). *Ann. Rev. Phys. Chem.* **29**, 473.
Heitler, W. (1954). "Quantum Theory of Radiation." Oxford Univ. Press, London and New York.
Henry, B. R. (1977). *Accounts Chem. Res.* **10**, 207.
Hochstrasser, R. M., and Novak, F. A. (1978). *Chem. Phys. Lett.* **53**, 3.
Hochstrasser, R. M., and Nyi, C. A. (1979). *J. Chem. Phys.* **70**, 1112.
Ingham, K. C., and El-Bayoumi, M. A. (1974). *J. Am. Chem. Soc.* **96**, 1674.
Iwasaki, M., Toriyama, K., Muto, H., and Nunome, K. (1978). *Chem. Phys. Lett.* **56**, 494.
Kenkre, V. M., and Knox, R. S. (1974). *Phys. Rev. B* **9**, 5279.
Kenkre, V. M., and Knox, R. S. (1975). *J. Lumin.* **12**, 187.
Knittel, D. R., and Lin, S. H. (1978). *Mol. Phys.* **36**, 893.
Kressge, A. J. (1975). *Accounts Chem. Res.* **8**, 354.
Kropf, A. (1976). *Nature (London)* **264**, 92.
Laubereau, A., and Kaiser, W. (1975). *Ann. Rev. Phys. Chem.* **26**, 83.
Laubereau, A., and Kaiser, W. (1978). *In* "Chemical and Biological Applications of Lasers" (C. B. Moore, ed.), Vol. 2.
Laubereau, A., Kirschner, L., and Kaiser, W. (1973). *Opt. Commun.* **9**, 182.

Laubereau, A., Greiter, L., and Kaiser, W. (1974a). *Appl. Phys. Lett.* **25**, 87.
Laubereau, A., Kehl, G., and Kaiser, W. (1974b). *Opt. Commun.* **11**, 74.
Laubereau, A., Fischer, S. F., Spanner, K., and Kaiser, W. (1978). *Chem. Phys.* **31**, 335.
Lee, Y. T., Schulz, P. A., and Shen, Y. R. (1979). *Ann. Rev. Phys. Chem.* (in press).
Legay, F. (1978). *In* "Chemical and Biological Applications of Lasers" (C. B. Moore, ed.), Vol. 2.
Levine, R. D. (1966). "Quantum Mechanics of Molecular Rates Processes." Oxford Univ. Press, London and New York.
Libby, W. (1977). *Ann. Rev. Phys. Chem.* **28**, 105.
Lin, S. H. (1966). *J. Chem. Phys.* **44**, 3759.
Lin, S. H. (1971). *Mol. Phys.* **21**, 853.
Lin, S. H. (1972). *J. Chem. Phys.* **56**, 4155.
Lin, S. H. (1973a). *J. Chem. Phys.* **58**, 5760.
Lin, S. H. (1973b). *Proc. R. Soc. London Ser. A* **335**, 51.
Lin, S. H. (1974). *J. Chem. Phys.* **61**, 3810.
Lin, S. H. (1976). *J. Chem. Phys.* **65**, 1053.
Lin, S. H. (1977). *In* "Radiationless Processes." General Discussion Meeting, Amsterdam.
Lin, S. H., and Eyring, H. (1972). *Proc. Nat. Acad. Sci. U.S.* **69**, 3192.
Lin, S. H., and Eyring, H. (1974). *Ann. Rev. Phys. Chem.* **25**, 39.
Lin, S. H., and Eyring, H. (1977a). *Proc. Nat. Acad. Sci. U.S.* **74**, 3105.
Lin, S. H., and Eyring, H. (1977b). *Proc. Nat. Acad. Sci. U.S.* **74**, 3623.
Lin, S. H., Lin, H. P., and Knittel, D. R. (1976). *J. Chem. Phys.* **64**, 441.
Lin, S. H., Hays, T. H., and Eyring, H. (1979). *Proc. Nat. Acad. Sci. U.S.* **76**, 3571.
Louisell, W. H. (1972). "Quantum Statistical Properties of Radiation." Wiley (Interscience), New York.
Ma, S. M., Eyring, H., Lin, S. H., Wutz, D., and Fujimura, Y. (1978). *Chem. Phys. Lett.* **58**, 159.
Malley, M., and Mourou, G. (1974). *Opt. Commun.* **10**, 323.
Markham, J. J. (1959). *Rev. Mod. Phys.* **31**, 956.
Mathies, R., and Sryer, L. (1976). *Proc. Nat. Acad. Sci. U.S.* **73**, 2169.
Monson, P. R., Patumterapibal, S., Kaufmann, K. J., and Robinson, G. W. (1974). *Chem. Phys. Lett.* **28**, 315.
Montroll, E. W., and Shuler, K. E. (1957). *J. Chem. Phys.* **26**, 454.
Morse, P. M., and Feshbach, H. (1953). "Methods of Theoretical Physics," pp. 865–868. McGraw-Hill, New York.
Mourou, G., and Malley, M. (1975). *Chem. Phys. Lett.* **32**, 476.
Mukamel, S. (1978). *Chem. Phys.* **31**, 327.
Munn, R. W., and Silbey, R. (1978). *J. Chem. Phys.* **68**, 2439.
Nitzan, A., Mukamel, S., and Jortner, J. (1975). *J. Chem. Phys.* **63**, 200.
O'Malley, R. E., Jr. (1967). "Introduction to Singular Perturbations." Academic Press, New York.
Oppenheim, I., Shuler, K. E., and Weiss, G. H. (1977). "Stochastic Processes in Chemical Physics." MIT Press, Cambridge, Massachusetts.
Orlowski, T. E., and Zewail, A. H. (1979). *J. Chem. Phys.* **70**, 1390.
Oseroff, A., and Callender, R. (1974). *Biochemistry* **13**, 4243.
Oxtoby, D. W. (1979). *Adv. Chem. Phys.* (in press).
Perry, J. W., and Zewail, A. H. (1979). *J. Chem. Phys.* **70**, 582.
Peters, K., Applebury, M. L., and Rentzepis, P. M. (1977). *Proc. Nat. Acad. Sci. U.S.* **74**, 3119.
Reddy, K. V., Bray, R. G., and Berry, M. J. (1978). *In* "Advances in Laser Chemistry" (A. H. Zewail, ed.). Springer-Verlag, Berlin and New York.
Ricard, D. (1975). *J. Chem. Phys.* **63**, 3841.
Ricard, D., and Ducing, J. (1975). *J. Chem. Phys.* **62**, 3616.
Richardson, W. B., Volk, L., Lau, K. H., Lin, S. H., and Eyring, H. (1973). *Proc. Nat. Acad. Sci. U.S.* **70**, 1588.

Salem, L. (1976). *Science* **191**, 822.
Sandros, K. (1976). *Acta Chem. Scand.* **A30**, 761.
Schein, L. B. (1977). *Phys. Rev. B* **15**, 1024.
Schröder, H., Neusser, H. J., and Schlag, E. W. (1978). *Chem. Phys. Lett.* **54**, 4.
Shank, C. V., and Ippen, E. P. (1974). *Appl. Phys. Lett.* **24**, 373.
Shank, C. V., and Ippen, E. P. (1975). *Appl. Phys. Lett.* **27**, 488.
Shank, C. V., Ippen, E. P., and Teschke, O. (1977). *Chem. Phys. Lett.* **45**, 291.
Shizuka,H., Matsui, K., Okamura, T., and Tanaka, I. (1975). *J. Phys. Chem.* **79**, 2731.
Shizuka, H., Matsui, K., Hirota, Y., and Tanaka, I. (1976). *J. Phys. Chem.* **80**, 2070.
Shizuka, H., Matsui, K., Hirata, Y., and Tanaka, I. (1977). *J. Phys. Chem.* **81**, 2243.
Shugard, M., Tully, T. C., and Nitzan, A. (1978a). *J. Chem. Phys.* **69**, 336.
Shugard, M., Tully, J. C., and Nitzan, A. (1978b). *J. Chem. Phys.* **69**, 2424.
Silbey, R. (1976). *Ann. Rev. Phys. Chem.* **27**, 203.
Smalley, R. E., Levy, D. H., and Wharton, L. (1976). *J. Chem. Phys.* **64**, 3266.
Smith, K. K., and Kaufman, K. J. (1978). *J. Phys. Chem.* **82**, 2286.
Spanner, K., Laubereau, A., and Kaiser, W. (1976). *Chem. Phys. Lett.* **44**, 88.
Sun, H. Y., and Rice, S. A. (1965). *J. Chem. Phys.* **42**, 3826.
Swofford, R. L., Long, M. E., and Albrecht, A. C. (1977). *J. Chem. Phys.* **65**, 179.
Toriyama, K., and Iwasaki, M. (1978). *J. Phys. Chem.* **82**, 2056.
Vander Dorickt, E. (1970). *Prog. React. Kinet.* **5**, 273.
Waddell, W., and Becker, R. S. (1971). *J. Am. Chem. Soc.* **93**, 3788.
Warshel, A. (1976). *Nature (London)* **260**, 679.
Weisman, R. B., and Rice, S. A. (1979). *Chem. Phys. Lett.* **61**, 15.
Weller, A. (1961). *Prog. React. Kinet.* **1**, 188.
Williams, D. L., and Heller, H. (1970). *J. Phys. Chem.* **74**, 4473.
Zewail, A. H. (1977). *Chem. Phys. Lett.* **45**, 399.
Zumofen, G. (1978). *J. Chem. Phys.* **69**, 4264.
Zwanzig, R. (1964). *Physica* **30**, 1109.

Index